Brock/Springer Series in Contemporary Bioscience

Microbial Ecology of Leaves

Brock/Springer Series in Contemporary Bioscience

Series Editor: *Thomas D. Brock*
University of Wisconsin-Madison

Tom Fenchel
ECOLOGY OF PROTOZOA:
The Biology of Free-living Phagotrophic Protists

Johanna Döbereiner and Fábio O. Pedrosa
NITROGEN-FIXING BACTERIA IN NONLEGUMINOUS CROP PLANTS

Tsutomu Hattori
THE VIABLE COUNT: Quantitative and Environmental Aspects

Roman Saliwanchik
PROTECTING BIOTECHNOLOGY INVENTIONS:
A Guide for Scientists

Hans G. Schlegel and Botho Bowien (Editors)
AUTOTROPHIC BACTERIA

Barbara Javor
HYPERSALINE ENVIRONMENTS: Microbiology and Biogeochemistry

Ulrich Sommer (Editor)
PLANKTON ECOLOGY: Succession in Plankton Communities

Stephen R. Rayburn
THE FOUNDATIONS OF LABORATORY SAFETY:
A Guide for the Biomedical Laboratory

Gordon A. McFeters (Editor)
DRINKING WATER MICROBIOLOGY:
Progress and Recent Developments

Mary Helen Briscoe
A RESEARCHER'S GUIDE TO SCIENTIFIC AND MEDICAL ILLUSTRATIONS

Max M. Tilzer and Colette Serruya (Editors)
LARGE LAKES: Ecological Structure and Function

Jürgen Overbeck and Ryszard J. Chróst (Editors)
AQUATIC MICROBIAL ECOLOGY:
Biochemical and Molecular Approaches

(Continued on page 501)

John H. Andrews   Susan S. Hirano
Editors

Microbial Ecology of Leaves

With 77 Figures and 33 Tables

Springer-Verlag
New York Berlin Heidelberg London Paris
Tokyo Hong Kong Barcelona Budapest

1992

John H. Andrews
Department of Plant Pathology
Russell Laboratories
University of Wisconsin-Madison
Madison, WI 53706, USA

Susan S. Hirano
Department of Plant Pathology
Russell Laboratories
University of Wisconsin-Madison
Madison, WI 53706, USA

Cover illustration: A microbe's "eye view" of the leaf surface. Bacteria and fungi colonizing the trichomes on an apple leaf (magnification approximately 1,000 ×; courtesy of John H. Andrews).

Library of Congress Cataloging-in-Publication Data
Microbial ecology of leaves/John H. Andrews, Susan S. Hirano, editors.
p. cm.—(Brock/Springer series in contemporary bioscience)
Papers presented at the 5th International Symposium on the Microbiology of the Phyllosphere, held in Madison, Wis., July 31–Aug. 3, 1990.
Includes bibliographical references and index.
ISBN 0-387-97579-9.—ISBN 3-540-97579-9
1. Leaves—Microbiology—Congresses. 2. Microbial ecology—Congresses. 3. Plant-microbe relationships—Congresses.
I. Andrews, John H. II. Hirano, Susan S. III. International Symposium on the Microbiology of the Phyllosphere (5th:1990: Madison, Wis.) IV. Series.
QR351.M43 1992
576′.15—dc20 91-17252

Printed on acid-free paper.

Production managed by Linda H. Hwang.
Typeset by Asco Trade Typesetting Ltd., Hong Kong.
Printed and bound by Edwards Brothers, Inc., Ann Arbor, MI.
Printed in the United States of America.

9 8 7 6 5 4 3 2 1

ISBN 0-387-97579-9 Springer-Verlag New York Berlin Heidelberg
ISBN 3-540-97579-9 Springer-Verlag Berlin Heidelberg New York

To three of the early pioneers—Frederick Last, Curt Leben, and Jacoba Ruinen

Preface

This book is based on symposium addresses given at the 5th International Symposium on the Microbiology of the Phyllosphere, held in Madison, Wisconsin, from 31 July to 3 August 1990. The conference brought together about 100 scientists with diverse interests pertinent to the study of leaves and microbes, including bacteriology, mycology, medical microbiology, ecology, plant pathology, physiology, anatomy, molecular biology, statistics, aerobiology, and meteorology. What has been learned since the first conference of the series was convened in 1979 at Newcastle-upon-Tyne, England, by T.F. Preece and C.H. Dickinson?

Introductory remarks by N.J. Fokkema (Chapter 1) provide the perspective of a plant pathologist about how the discipline has evolved and where we stand now more than 30 years after the pioneering Dutch scientist J. Ruinen described the phyllosphere as an "ecologically neglected milieu."

The first major section of the book, Part 2, "The Habitat," is comprised of five chapters and considers leaf-microbe relationships in an aerial and an aquatic setting. B.E. Juniper (Chapter 2) sets the stage by reviewing the physical and chemical features of leaves that may influence microbial growth. Aerial movement of microbes to and from leaves is considered by D.E. Pedgley (Chapter 3). The aquatic analog to air as a medium is assessed by R. Goulder and J.H. Baker (Chapter 4). Leaf surface features that influence fungal infection are discussed by E.A. Allen, H.C. Hoch, J.R. Steadman, and R.J. Stavely (Chapter 5). M.K. Clayton and B.D. Hudelson (Chapter 6) discuss methods to analyze spatial patterns of microbial activity in the phyllosphere.

Part 3, "Microbial Community Components," consists of six chapters, beginning with a discussion by D.J. Drahos (Chapter 7) of the advances in molecular biology, which have revolutionized our ability to detect, quantify, and identify microbes. Then W.T. Starmer, J.C. Fogleman, and M.-A. Lachance (Chapter 8) review the yeast community of cacti. The stems of cacti, though strictly speaking not leaves, have much in common with the leaf surface; the yeast-*Drosophila*-cactus story is an excellent, instructive example of how an ecological study can span the range from the genetic and biochemical level through the ecosystem level. O. Petrini (Chapter 9) and M.R. Siegel and C.L. Schardl (Chapter 10) discuss the fungal endophytes of tree leaves and of grasses, respectively, another area where major advances have been made. In the course of two decades, endophytes have moved from being viewed suspiciously as artifacts, to the center of a major subdiscipline. Finally, some of the fungi of medical importance that are found on leaves are considered. The 1990 conference was the first one to include medical researchers. They report on leaves as a source of fungal allergens (Chapter 11; H.A. Burge, M.M. Muilenberg, and J.A. Chapman), and on the association between sphagnum moss and a fungus that causes skin lesions in humans (Chapter 12; D.M. Dixon and I.F. Salkin).

The six chapters in Part 4, "Microbial Community Processes and Evolution," concern the dynamic aspects of phyllosphere populations. L. Kinkel (Chapter 13) and S.S. Hirano and C.D. Upper (Chapter 14) review fungal and bacterial community dynamics, respectively. S.E. Lindow (Chapter 15) questions whether many of the bacterial traits presumed to be adaptive actually enhance survival in nature. P.K. Wolber and G.J. Warren (Chapter 16) discuss the evolutionary origin of the genes encoding ice nucleation and whether the ice phenotype appears to be selected for directly, or is a pleiotropic consequence of selection for a different but associated feature. The selection pressures promoting endophytism, and the costs and benefits to the plant of maintaining endophytes, are debated by K. Clay and G.C. Carroll (Chapters 17 and 18).

The leaf habitat can be changed incidentally (e.g., as a result of acid precipitation), or deliberately to achieve some goal such as enhanced plant growth. The five chapters of Part 5, "Modification of the Phyllosphere," address these issues. N. Magan and A.R. McLeod (Chapter 19) review recent progress towards a better understanding of the effects of air pollutants on phyllosphere communities. The complex interactions among nutrients on the leaf surface (e.g., as supplied by pollen or honeydew), necrotrophic pathogens (which require these exogenous nutrients in order to infect), yeasts (which antagonize necrotrophs by competing for nutrients), and fungicides (which reduce populations of both yeasts and pathogen) are analyzed by A.J. Dik (Chapter 21). J.W Fahey, M.B. Dimock, S.F. Tomasino, J.M. Taylor, and P.S. Carlson (Chapter 20) discuss the prospects for achieving biocontrol by use of genetically engineered endophytes, while H.C. Weltzien (Chapter 22) advocates use of compost extracts. In Chapter

23, C.D. Upper argues that the chemical control paradigm may be a poor model for efforts to achieve biological control.

The book concludes with a perspective by J.H. Andrews (Chapter 24) on the importance of the study of the microbial ecology of leaves and on directions for future research.

We take this opportunity to thank all the conference attendees for their participation, and the authors for their contributions to this text. Financial support from the following sponsors is acknowledged gratefully: Abbott Laboratories; Academic Press, Inc.; Agracetus; Agrigenetics; Calgene; DNA Plant Technology; Dow Elanco; E.I. DuPont de Nemours & Co.; Foundation for Microbiology; Herbert H. Kohl Charities, Inc.; Lipha Tech; Monsanto; Sandoz Crop Protection; Springer-Verlag New York, Inc.; United States Department of Agriculture (Agricultural Research Service and Competitive Research Grants Office, Plant Stress Panel); and the University of Wisconsin-Madison (College of Agricultural and Life Sciences, Department of Plant Pathology, Graduate School, Hewlett Fund, and University Lecturers' Committee). We also thank the many students, staff, and faculty who worked hard to help make the symposium a success, in particular, D. Arny, K. Canada, R. Durbin, B. Hudelson, D. Rouse, K. Smith, R. Spear, C. Upper, and L. Yourman. We acknowledge with appreciation the assistance of the Organizing Committee: C. Upper, N. Fokkema, G. Carroll, S. Lindow, and P. Surico.

John H. Andrews
Susan S. Hirano

Contents

Contributors

Eric A. Allen Biology Department, University of Victoria, Victoria, British Columbia, Canada

John H. Andrews Department of Plant Pathology, University of Wisconsin, Madison, WI 53706, USA

John H. Baker Natural Environment Research Council, Polaris House, North Star Avenue, Swindon, SN2 1EU, United Kingdom

Harriet A. Burge Department of Internal Medicine, University of Michigan, Ann Arbor, MI 48109, USA

Peter S. Carlson Crop Genetics International, 7249 National Drive, Hanover, MD 21076, USA

George C. Carroll Department of Biology, University of Oregon, Eugene, OR 97403, USA

Jean A. Chapman 35 Doctors Park, Cape Girardeau, MO 63701, USA

Keith Clay Department of Biology, Indiana University, Bloomington, IN 47405, USA

Murray K. Clayton Departments of Plant Pathology and Statistics, University of Wisconsin, Madison, WI 53706, USA

Aleid J. Dik Department of Plant Ecology and Evolutionary Biology, University of Utrecht, 3742 CP Baarn, The Netherlands

Michael B. Dimock Crop Genetics International, 7249 National Drive, Hanover, MD 21076, USA

Dennis M. Dixon Laboratories for Mycology, Wadsworth Center for Laboratories and Research, New York State Department of Health, Albany, NY 12201-0509, USA

David J. Drahos Monsanto Company, 700 Chesterfield Village Parkway, St. Louis, MO 63198, USA

Jed W. Fahey Crop Genetics International, 7249 National Drive, Hanover, MD 21076, USA

James C. Fogleman Department of Biological Sciences, University of Denver, Denver, CO 80208, USA

Nyckle J. Fokkema Research Institute for Plant Protection, 6700 GW Wageningen, The Netherlands

Ray Goulder Department of Applied Biology, University of Hull, Hull, HU6 7RX, United Kingdom

Susan S. Hirano Department of Plant Pathology, University of Wisconsin, Madison, WI 53706, USA

Harvey C. Hoch Department of Plant Pathology, Cornell University, New York State Agricultural Experiment Station, Geneva, NY 14456, USA

Brian D. Hudelson Department of Plant Pathology, University of Wisconsin, Madison, WI 53706, USA

Barrie E. Juniper Department of Plant Sciences, Oxford University, South Parks Road, Oxford, OX1 3RA, United Kingdom

Linda Kinkel Department of Plant Pathology, University of Minnesota, St. Paul, MN 55108, USA

Marc-Andre Lachance Department of Plant Sciences, University of Western Ontario, London, Ontario, N6A 5B7, Canada

Steven E. Lindow Department of Plant Pathology, University of California, Berkeley, CA 94720, USA

Naresh Magan Biotechnology Centre, Cranfield Institute of Technology, Bedford, MK43 0AL, United Kingdom

Andy R. McLeod National Power Technical and Environmental Centre, Kelvin Avenue, Leatherhead, Surrey KT22 7SE, United Kingdom

Michael M. Muilenberg Department of Internal Medicine, University of Michigan, Ann Arbor, MI 48109, USA

David E. Pedgley Natural Resources Institute, Chatham Maritime, Kent, ME4 4TB, United Kingdom

Orlando Petrini Mikrobiologisches Institut, ETH-Zentrum, CH-8092 Zurich, Switzerland

Ira F. Salkin Laboratories for Mycology, Wadsworth Center for Laboratories and Research, New York State Department of Health, Albany, NY 12201-0509, USA

Christopher L. Schardl Department of Plant Pathology, University of Kentucky, Lexington, KY 40546, USA

Malcolm R. Siegel Department of Plant Pathology, University of Kentucky, Lexington, KY 40546, USA

William T. Starmer Department of Biology, Syracuse University, Syracuse, NY 13244, USA

Rennie J. Stavely USDA-ARS, Departments of Microbiology and Plant Pathology, Beltsville, MD 20705, USA

James R. Steadman Department of Plant Pathology, University of Nebraska, Lincoln, NB 68583, USA

Jean M. Taylor Crop Genetics International, 7249 National Drive, Hanover, MD 21076, USA

Steven F. Tomasino Crop Genetics International, 7249 National Drive, Hanover, MD 21076, USA

Christen D. Upper USDA-ARS, Department of Plant Pathology, University of Wisconsin, Madison, WI 53706, USA

Gareth J. Warren DNA Plant Technology Corporation, 6701 San Pablo Avenue, Oakland, CA 94608, USA

Heinrich C. Weltzien Institut für Pflanzenkrankheiten, University of Bonn, Nusallee 9, D-5300 Bonn 1, Germany

Paul K. Wolber DNA Plant Technology Corporation, 6701 San Pablo Avenue, Oakland, CA 94608, USA

Part 1
Introduction

1

The Phyllosphere as an Ecologically Neglected Milieu: A Plant Pathologist's Point of View

Nyckle J. Fokkema

1.1 Introduction

Since the mid-1950s, phyllosphere research has been recognized as a special field of microbial ecology. It is appropriate to take the year 1955 as the starting date for phyllosphere research because that is when the term phyllosphere was introduced almost simultaneously by Last (1955) and by Ruinen (1956), respectively, as follows: "The presence of a highly developed non-parasitic flora near the living roots, the rhizosphere, has for long been recognized, but the recognition of a comparable flora of leaf surfaces is new. *Sporobolomyces, Tilletiopsis* and *Bullera* inhabit this environment, 'the phyllosphere'" (Last, 1955); and, "These observations suggested the existence of a characteristic milieu which is conditioned by the leaf, and may be called, in analogy with the rhizosphere, the 'phyllosphere'" (Ruinen, 1956).

Leben (1965, 1971) was one of the pioneers in the ecology of phyllosphere bacteria. He was, for a long time, the only American scientist active in this field, which was initially dominated by scientists in the United Kingdom, India, New Zealand, and The Netherlands. This situation has now changed dramatically, and the USA has become the focus of activity in phyllosphere research.

That phyllosphere research in fact existed became apparent following the initiatives of T.F. Preece and C.H. Dickinson who organized the first two symposia in this area held at Newcastle-upon-Tyne in 1970 and at Leeds in 1975. From these a tradition was born and meetings have been

Table 1.1 Previous phyllosphere symposia

1970	Newcastle-upon-Tyne, England Ecology of leaf-surface microorganisms Preece and Dickinson (1971)
1975	Leeds, England Microbiology of aerial plant surfaces Dickinson and Preece (1976)
1980	Aberdeen, Scotland Microbial ecology of the phyllosphere Blakeman (1981)
1985	Wageningen, The Netherlands Microbiology of the phyllosphere Fokkema and van den Heuvel (1986)
1990	Madison, Wisconsin, USA Microbial ecology of leaves Andrews and Hirano (1991, this volume)

held every five years. These symposia generated proceedings that provide the basic texts on phyllosphere ecology (Table 1.1). They form a unique document of the views of leading scientists during the past decades. Although the initial impetus in phyllosphere research came from plant pathologists, it gradually became characterized by the integration of disciplines being a unifying and inspiring goal for microbial ecologists, plant pathologists, bacteriologists, mycologists, and botanists. This goal has given us the opportunity to look beyond the narrow borders of our own specialities and to adopt methods and concepts from others.

Phyllosphere research can be consolidated around the following three issues: 1) the nature of the microorganisms themselves; 2) the factors influencing them; and 3) the effects of the microorganisms, i.e., what are they doing? The present state of the science will be presented in this book. My goal in this introductory chapter is to present a personal view as a plant pathologist on some selected topics that are not dealt with at length elsewhere in this volume. My comments will concern: 1) phyllosphere bacteria, 2) phyllosphere fungi, and 3) biological control of plant pathogens.

1.2 Phyllosphere Bacteria

The development of bacterial ecology has been stimulated by successful epiphytic colonization of the phyllosphere of the host plant by virulent strains of plant pathogenic bacteria, as well as by nonpathogenic isolates of the same or related species (Crosse, 1971; Hirano and Upper, 1983). The need for discrimination, especially between nonpathogenic and pathogenic strains, resulted in sophisticated molecular detection techniques as well as in many fundamental studies on population dynamics. In this section some remarks will be made on the role of classification in ecology, the contribution of scanning electron microscopy, and the potential significance of biosurfactant production by phyllosphere bacteria.

Taxonomic Classification and Ecology "No ecology without taxonomy" seems a logical statement, but nowhere else in phyllosphere biology is there a greater difference between 'lumpers' and 'splitters.' Some researchers are only interested in total numbers of aerobic bacteria. Frequently, separation of the pseudomonads from the rest of the bacterial flora is the highest level of taxonomic discrimination used (Fokkema and Schippers, 1986). Other researchers have used fatty acid profiles (e.g., Stead, 1989), protein fingerprints (e.g., Van Outryve et al., 1989), and DNA homology and serological properties (e.g., Van Vuurde, 1987) to classify phyllosphere bacteria at the species and subspecies levels. Bacterial taxonomists were well represented only at the 1975 symposium (Table 1.1) where review papers were given by Billing (1976) and Goodfellow et al. (1976). The latter introduced numerical taxonomy for phyllosphere bacteria. Detailed taxonomic studies are rarely conducted in phyllosphere research because they are too time consuming if numerous field samples need to be handled.

Modern characterization techniques allow us to answer intriguing ecological questions such as: "Is there a difference between phyllosphere and rhizosphere pseudomonads?" Van Outryve et al. (1989) made SDS-PAGE protein profiles of 590 bacterial isolates from witloof chicory leaves resulting in 149 different protein fingerprint types. Ninety percent of these types was only found once. Among 278 fluorescent *Pseudomonas* isolates, 20 different fingerprints could be distinguished, 175 isolates had the same fingerprint type, and this type was also frequently detected in isolates from roots. This indicates that rhizosphere and phyllosphere fluorescent pseudomonads may not be intrinsically different.

A similar diversity was detected by Morris and Rouse (1985) on snap bean leaves based on the differential ability of bacteria isolated from individual leaflets to utilize different single carbon and nitrogen sources. The aim of this study was to modify the antagonistic microflora by spraying selective nutrients that would promote their development but not that of the pathogen, *Pseudomonas syringae* pv. *syringae*. It would be interesting to see to what extent characterization on a nutritional basis matches the protein fingerprint types. This great diversity of strains of the same bacterial species was also observed in the population dynamics of *P. syringae* by O'Brien and Lindow (1989).

These examples of diversity among isolates of a species suggest that it will be unlikely that a single isolate may completely occupy the substratum after mass introduction in the field, an approach often used in biocontrol experiments. Such introductions frequently fail, perhaps due to the complex nature of the phyllosphere community, rather than to the absence of appropriate traits for colonization. One solution would be the introduction of mixtures of isolates or use of biocontrol agents that are as closely related as possible to the target organism, e.g., the deletion mutants of ice nucleation-active bacteria (Lindemann and Suslow, 1987).

Scanning Electron Microscopy At the phyllosphere symposium in 1970, the first results obtained with the scanning electron microscope (SEM) were presented. Samples could be prepared with relatively little disturbance. Microbes could be located on certain parts of the leaf and a wealth of different surface wax formations was evident (Martin and Juniper, 1970; Holloway, 1971). SE micrographs showed that wax crystals are not degraded by yeasts (Bashi and Fokkema, 1976) contrary to earlier conclusions by Van der Burg (1974) based on EM micrographs of carbon replicas. There also appeared to be an enormous amount of mucilage produced by yeasts, and ballistospore formation by the pink yeast *Sporobolomyces roseus* was occurring rather haphazardly (Bashi and Fokkema, 1976). This emphasized that the spore fall method for determining yeast populations is unsuitable.

The possibilities of the SEM, however, were best exploited by mycologists in studies of hyphal interactions and by bacteriologists in studies of colonization patterns of virulent and avirulent strains, e.g., *Xanthomonas campestris* pathovars on hosts and nonhosts (Mew et al., 1984; Mew and Cruz, 1986; De Cleene, 1989). Specific receptive sites for colonization could be distinguished. On various rice cultivars, only compatible strains of *X. campestris* pv. *oryzae* multiplied on the waterpores and not on the stomata, whereas avirulent strains triggered exudate production, which immobilized bacteria at the water pore. In contrast, virulent strains of *X. campestris* pv. *oryzicola* colonized the stomata but not the water pores. This phenomenon of site specificity of certain pathovars should be taken into account when biological control is envisaged. Restricted areas for virulent strains may also exclude antagonists from these areas, which may reduce the possibilities for antagonistic interaction if close contact is required.

Biosurfactants We know that the physicochemical properties of the leaf surface govern leaf wettability (Martin and Juniper, 1970) and that leaf diffusates contain both nutritional (Tukey, 1971) as well as toxic substances (Blakeman and Atkinson, 1981) for microbes. Epiphytic bacteria may change the nutrient situation and also release toxic substances. Furthermore, recent observations demonstrate that bacteria can increase leaf wettability by the production of biosurfactants (Bunster et al., 1989). This ability, which occurred in 50% of the *Pseudomonas* strains tested, can easily be determined by the shape of droplets of bacterial suspensions on polystyrene. Within 24 hr, droplets with surface-active strains will spread, while those with no surface-active strains will not. Surface-active strains applied to wheat leaves increased leaf wettabilty.

The significance of surface activity to bacterial diseases has clearly been demonstrated by Hildebrand (1989) in his studies on the etiology of broccoli head rot caused by species of *Pseudomonas* and *Erwinia*. Only some strains caused the water-soaked effect, and these strains also showed surfactant activity on polystyrene. Decay of nonwounded broccoli heads occurred only after inoculation with pectolytic *Pseudomonas* strains, which are also surfactant-active. Lack of one of these traits could be compensated by mixing of pectolytic, non-surface-active strains with nonpectolytic, surface-active strains. The surfactant involved has been identified as viscosin, which reduced surface tension of water from 73 to 27 mN/m, and which was able to cause substantial solute leakage from broccoli tissue (Hildebrand et al., 1991).

The ecological implications of this easily detectable characteristic for surface colonization need further exploration. Leben (1988) demonstrated the similarity between bacterial distribution patterns on cucumber leaves and areas with enhanced wettability. These areas, located mainly above the veins, are likely to be determined by physicochemical properties of the host but may to some extent be created by the bacteria themselves.

1.3 Phyllosphere Fungi

The ecology of hyphal fungi and yeasts has focused on six aspects: 1) monitoring techniques, 2) effects of naturally occurring chemicals on pathogens and saprophytes, 3) the role of fungi in biological control, 4) the effect of saprophytes on senescence, 5) the symptomless occurrence of endophytes and its meaning, and 6) ecological theory as it may pertain to fungal colonization. The following discussion concentrates on monitoring techniques and on the effect of leaf leachates on the mycoflora.

Cultural Methods: Leaf Washings vs. Washed Leaves It is interesting to note that in quantitative studies on phyllosphere fungi, two different cultural methods are used: the commonly used "plating of leaf washings" and the less frequently encountered "plating of washed leaves." The first method, based on culturing colony-forming units (CFUs) washed from leaves on agar plates, may underestimate the occurrence of nonsporeforming hyphal fungi and overestimate the sporeforming ones. However, it is excellent for quantifying yeasts, which are the dominant fungal leaf colonizers, because microscopic observations by Breeze and Dix (1981) also revealed that, in the summer, the biomass of yeasts on leaves of *Acer platanoides* was 50 times greater than that of hyphal fungi.

The plating on agar media of tiny leaf pieces after they have been washed will certainly exclude yeasts. The majority will be washed off from the leaves, a fact that has not always been realized (e.g., Cabral, 1985). However, the plating of leaf pieces after washing, with or without surface sterilization, gives a good impression of the balance between fungal epiphytes and endophytes. The hyphal epiphytes of *Eucalyptus* leaves (Cabral, 1985) were dominated by the omnipresent species of *Alternaria alternata, Cladosporium cladosporioides*, and *Epicoccum nigrum*. The endophytic population of *Eucalyptus* leaves—as in most perennials—seems rather host specific, whereas in a short-lived annual crop like wheat the endophytes (Sieber et al., 1988) are unspecialized, perhaps due to the short lifetime of the individual leaves. They consist of common phyllosphere fungi like *Alternaria, Cladosporium*, and *Epicoccum* spp., as well as the wheat pathogens *Septoria nodorum* and *Fusarium* spp. Even on symptomless leaves, *S. nodorum* was the most frequent endophyte. This is especially noteworthy in relation to the yield increases occasionally observed after fungicide application in the absence of visible disease (Fokkema, 1981). Plating of leaf pieces without washing was used by Thomas and Shattock (1986) to determine associations of hyphal fungi on dead leaf parts of *Lolium perenne. Cladosporium, Drechslera, Phoma, Epicoccum, Alternaria*, and *Leptosphaeria* were, in decreasing order, the dominant fungi on dead leaf tips. This study together with the studies by Andrews and Kinkel (1986) and Kinkel et al. (1989) of fungal immigration and community development are the most theoretical treatments available on fungal phyllosphere ecology.

Leaf Surface Chemicals Naturally occurring endogenous chemicals in the phyllosphere have received attention from the onset of phyllosphere research, and an entire section was devoted to this subject at the Aberdeen symposium organized by Blakeman (1981). Phyllosphere researchers are interested in the combined activities of nutritional and inhibitory substances in water droplets and films on leaf surfaces. The biological significance of toxic compounds only extractable from the leaf with organic solvents is, in my opinion, difficult to estimate because water is the only solvent naturally available in the phyllosphere. The early studies by Tukey (1971) on the substances that can be leached with water from the leaves are still relevant. Attempts to relate cultivar resistance and susceptibility to differences in sugar and amino acid content of leachates are seldom convincing because the differences are mostly small (Bal Kishan and Mehrotra, 1988). In addition to nutrients, chelating agents, which received much attention in studies on latent infections (Swinburne, 1981), terpenoids, phenolic compounds, and substances associated with wax (Blakeman and Atkinson, 1981) may all play major roles in the ecology of the phyllosphere. The presence of duvatrienediols on tobacco leaf surfaces was associated with induced as well as cultivar resistance against *Peronospora tabacina* (Tuzun et al., 1989).

Gallic acid, present in the phyllosphere of Norway maple, may restrict colonization by *Cladosporium, Cryptococcus*, and *Sporobolomyces* spp. (Irvine et al., 1978; Dix, 1979). *Penicillium* spp. and *Aureobasidium pullulans* were not restricted and this may account for their dominance in leaf litter. In fact, the role of nonnutritional chemicals on phyllosphere microorganisms is still poorly understood. This also applies to the presence of wax. The presence of wax, and particularly the shape of the wax crystals, together with the trichomes, will determine leaf wettability (Martin and Juniper, 1970; see also Chapter 2, this volume). Wax crystals may even promote deposition of airborne propagules (Forster, 1977). However, attempts to relate cultivar resistance of rape seed against *Alternaria brassicae* with the amount of wax revealed only minor differences (Conn and Tewari, 1989).

1.4 Biological Control of Fungal Pathogens

We now arrive at the practical question, why bother with these phyllosphere microorganisms? A full answer would include scientific curiosity. At least this is how phyllosphere research started in Europe, and perhaps it could have started only in Europe, where there was more academic freedom at that time than in the USA, where substantial dependence on external financing required practical results-oriented research.

Currently, all scientists have to justify their research with well-defined goals. For example, several chapters in this book deal with effects of air pollution on phyllosphere microorganisms. Changes in the microflora will reduce their beneficial effects as well as serve as bioindicators for air pollution (Dowding, 1986). On the other hand, the microflora themselves may, as aero-allergens, contribute to a form of air pollution. Endophytes may protect plants from insect damage but also may be dangerous for grazing animals.

Biological control of plant diseases by naturally occurring and introduced microorganisms in the phyllosphere has recently been reviewed (Mukerji and Garg, 1988); the most promising approach seems to be the biocontrol of postharvest fruit diseases (Wilson and Wisniewski, 1989). In addition, the complex role of naturally occurring yeasts and the fascinating effects of watery compost extracts on plant diseases will be reviewed by Dik (Chapter 21, this volume) and by Weltzien (Chapter 22, this volume), respectively. The following discussion, however, reviews the prospects of biocontrol of fungal diseases in the phyllosphere.

National governments are becoming increasingly aware of the environmental problems associated with many chemical control agents. Research toward the replacement of chemicals where possible by environmentally safer methods is currently emphasized. What can microbial ecologists offer with respect to diseases of aerial plant parts? There are three major prerequisites we should consider for biological control:

1. We should carefully study the life cycle of the pathogen and attack the pathogen in its most vulnerable stage.
2. The duration of the interaction between the biocontrol agent and the pathogen should be as long as possible. This may explain the relative success of biocontrol of soilborne diseases.

3. We should discriminate between control of necrotrophic and biotrophic pathogens, because a major mechanism like nutrient competition will not affect biotrophs.

In addition to these three factors, when biological control is discussed, it is appropriate to differentiate among the microhabitats where the agent is expected to operate. The most important of these are the healthy leaf surface, wounded plant tissue, and established infections.

Protection of the Healthy Leaf With respect to the biological control of aerial pathogens, in my opinion, it is not always useful to imitate the effect of fungicides by spraying healthy leaves with candidate biocontrol agents. This might be a useful approach for the control of saprophytically growing pathogenic bacteria (Lindow, 1985), but it is not an attractive approach for the control of fungal pathogens that can penetrate the leaf within 24 hr under suitable conditions, leaving a very short interference period. This implies that the successful biocontrol agent should be sufficiently established in the phyllosphere before the pathogen arrives, or that products responsible for inhibition should remain present at a concentration high enough for inhibition of the pathogen, or that the phyllosphere should remain deprived of nutrients able to stimulate infection. Are these requirements met in the biocontrol of infection by biotrophs, which depend on nutrients from living host cells?

There are several recent studies demonstrating that leaf rusts can be controlled by bacterial species such as *Bacillus subtilis* (Baker et al., 1985, Rytter et al., 1989), *Pseudomonas* spp. (Levy et al., 1989), and *Erwinia herbicola* (Kempf and Wolf, 1989). All of these studies have in common that antibiotic production was partly or completely responsible for the suppression. Field experiments by Baker et al. (1985), however, demonstrated that the bacteria as well as the culture filtrates had to be applied three times per week to control bean rust to a similar degree as a weekly spray with mancozeb, indicating that the bacterial population as well as the antibiotics were active for only a short period. This is in contrast to the control of coffee leaf rust by a commercial preparation of *Bacillus thuringiensis*. In this latter case the mechanism may be based on induced resistance. The protection lasted for five weeks when Thuricide HD was applied at a (very high) concentration of 20 mg per ml (Roveratti et al., 1989). Undiluted culture broth of *Bacillus subtilis* proved effective against a number of necrotrophic pathogens and barley powdery mildew; Rodgers (1989) considered it to be an interesting microbial fungicide.

What is the perspective for the biocontrol of necrotrophs, which benefit from the host after killing its tissue? Sensitivity of necrotrophs to nutrient competition by saprophytes is most easily tested by a positive response of these pathogens to added nutrients, such as 1% sucrose and 0.5% yeast extract. During my work with yeasts and *Cladosporium* spp. as antagonists, almost all phyllosphere saprophytes were effective in controlling infection by necrotrophic pathogens belonging to the genera *Cochliobolus, Septoria, Alternaria, Botrytis, Phoma,* and *Colletotrichum*. The only exception was *Rhynchosporium secalis,* which is a hemibiotroph similar to the apple scab fungus *Venturia inaequalis. R. secalis* seems to escape from interactions on the leaf surface by immediate development under the cuticle (Fokkema, unpublished).

Phyllosphere saprophytes are capable of removing infection-stimulating nutrients. There is generally no reason to introduce this type of antagonist deliberately because they are already naturally present in sufficient densities. Additional application will not provide further help (Fokkema et al., 1979). The natural behavior of saprophytes and what we might expect from them will be discussed by Dik (Chapter 21, this volume). Application of antagonists might be rewarding, however, when large amounts of naturally occurring exogenous nutrients such as remains of flowers or pollen grains are needed for infection by pathogens like *Sclerotinia sclerotiorum* and *Botrytis* spp. (e.g., Dubos, 1987; Nelson and Powelson, 1988; Boland and Inglis, 1989; Zhou and Reedeler, 1989). Antagonists may act as scavengers well before the pathogen arrives. Nevertheless, field applications seldom result in more than 50% control.

Bioprotection of Wounds Biological control of fresh wounds has the advantages of the absence of resident mycoflora competing with the antagonists and of the presence of favorable nutritional as well as microclimatic conditions. Timing of bioprotection of naturally occurring wounds such as leaf scars, wind and hail damage, and insect damage (Fermaud and Le Menn, 1989), however, is difficult, and these areas are similar to undamaged plant surfaces with high concentrations of exogenous nutrients.

The best likelihood for success is found where man-made wounds allow simultaneous application of the biocontrol agents. Such wounds may be made during harvesting and packing procedures of fruit and vegetables that lead to postharvest decay (Tronsmo, 1986; Wilson, 1989). The combined research efforts of the Appalachian Fruit Research Station (USA) and the Volcani Center (Israel) convincingly showed that *Bacillus* and *Candida* spp. can control soft rot of apples, stone fruits, and citrus fruits (Janisiewicz, 1988; Wilson and Wisniewski, 1989). Interestingly, application of the yeast in 2% calcium chloride increased efficacy about fivefold (McLaughlin et al., 1990). *Botrytis aclada,* which causes neck rot of onions during storage, invades in the field through wounds caused by removal of the leaves just before harvest. In a field experiment, wound treatment with *Trichoderma* reduced the percentage of neck rot from 35 to 24% (Köhl et al. 1990).

Although it is clear that biological control can operate under field conditions, the degree of protection is usually not sufficient. Perhaps we have concentrated too much on interactions with a short interference period, neglecting the exploitation of interactions with a longer duration.

Interaction with Established Infections Biocontrol with hyperparasites is only feasible when some degree of disease can be tolerated. This is the case in crops where the leaves are attacked but the fruits are not, e.g., mildew in cucumber, tomato, and fruit trees. The interference period is long and primarily the sporulation and thus the dissemination of the pathogen is affected. The hyperparasite *Ampelomyces quisqualis* has frequently been studied as a biocontrol agent against powdery mildew (Sundheim, 1986; Jayapal Gowdu and Balasubramanian, 1988). Unfortunately the pathogen was only parasitized at greater than 95% relative humidity (RH), which hampers its further exploitation.

Formulation, as well as selection for drought-resistant mutants, may overcome the need for high humidity. Philipp et al. (1990) demonstrated that 2% Hora Oleo 11 E (Ciba-Geigy), a formulated paraffin oil, allowed successful interaction at 70% RH. Mitchell et al. (1986) selected low-humidity, ethylmethanesulfonate mutants of *Dicyma pulvinata*, a hyperparasite of *Cercosporidium personatum* causing leaf spot of peanuts. Sztejnberg et al. (1989), however, controlled mildew of carrots, cucumber, and mango by *A. quisqualis* under field conditions without special adaptations because of favorable local climatic conditions. With necrotrophs and perhaps also with biotrophs, the interference period can be further extended to interactions on dead plant parts and plant debris. Here, hyperparasites may interfere with sporulation as well as with the formation of resting structures. Trutmann et al. (1982) and Gerlagh and Vos (1991) demonstrated that foliar applications of the hyperparasite *Coniothyrium minitans* to bean plant, although not reducing infection, did reduce the number of sclerotia as well as their viability. *Sclerotinia sclerotiorum* is an excellent model pathogen for biocontrol because reduction of the number of sclerotia eventually will reduce disease. This pathogen does not sporulate on the leaves, which excludes the possibility of secondary infections. We should realize, however, that we do not know the minimum number of sclerotia that can cause disease. Therefore, our current approach is to follow disease reduction by *C. minitans* and other antagonists over a number of years. Sclerotial diseases in arable crops are gradually becoming a problem. Why not be realistic and aim at an even, gradual disappearance?

Many necrotrophs sporulate abundantly on lesions as well as on dead leaves, and conidia contribute to epidemics in the same season. Biological control directed at sporulation is normal with biotrophs but has hardly been explored for necrotrophs. Biles and Hill (1988) proceeded along this line. They found that the sporulation of *Cochliobolus sativus* was reduced after treatment of lesions with *Trichoderma harzianum*. Similarly, preliminary experiments showed that various saprophytes could suppress sporulation of *B. aclada* on dead onion leaves (Köhl et al., 1990). Pfender (1988) observed suppression of ascocarp formation of *Pyrenophora tritici-repentis* in straw by an ascomycete. Whether these observations are applicable to the field situation remains to be investigated.

Encouraging in this respect is the earlier study by Cook (1970), who demonstrated the suppression of *Fusarium culmorum* on wheat straw by naturally occurring saprophytes. At the first phyllosphere symposium, Burchill and Cook (1971) reported on the possible control of apple scab by application of urea before leaf fall to suppress perithecia development. Microorganisms are likely to be involved. This approach was continued by Heye and Andrews (1983); application of *Chaetomium globosum* and *Athelia bombacina* to leaf litter reduced ascospore formation in spring (Young and Andrews, 1990). The promising effect of *A. bombacina* has recently been confirmed (Miedtke and Kennel, 1990). An interaction in leaf litter typically has a long interference period. Work with apple scab has convincingly demonstrated the general prospects for such an approach. The apple scab pathogen, however, disseminates very efficiently secondarily, which means that ascospore reduction alone cannot accomplish adequate control in practice. Nevertheless, integration with other control measures remains an attractive option.

1.5 Conclusions

This introductory chapter gives a plant pathologist's view of the developments in some fascinating areas of phyllosphere research during the past decades. Phyllosphere bacteriologists now are using an array of new serological and molecular techniques for the identification of species, pathovars, and isolates. These techniques have revealed a great diversity among saprophytic as well as pathogenic isolates of the same species, which, consequently, have differences in ecological behavior. Biosurfactant production may prove an important trait for surface colonization as well as pathogenesis. So far, fungal ecology has gained comparatively less from these new techniques. Various cultural methods can be used for the identification and quantification of populations of phyllosphere. The differences between epiphytic and endophytic species may increase with increasing lifetime of individual leaves. Saprophytic phyllosphere fungi together with pathogens and endophytes are the first to colonize moribund tissue. More knowledge on the antagonistic interactions in this niche may reveal an important place for biological control.

In summary, the prospects for biological control are primarily in:

1. protection of leaf infection by naturally occurring antagonists and induced resistance;
2. biological protection of man-made wounds against infection;
3. retardation of epidemics rather than protection of individual plants by reduction of inoculum production in living as well as dead tissue.

Finally, we may hope that a better understanding of the complex nature of the phyllosphere and its inhabitants may lead to additional novel and effective methods of disease control.

References

Andrews, J.H. and Kinkel, L.L. 1986. Colonization dynamics: the island theory. pp. 63–76 in Fokkema, N.J. and Van den Heuvel, J. (editors), *Microbiology of the Phyllosphere*. Cambridge University Press, Cambridge, UK.

Baker, C.J., Stavely, J.R., and Mock, N. 1985. Biocontrol of bean rust by *Bacillus subtilis* under field conditions. *Plant Disease* 69:770–772.

Bal Kishnan, N. and Mehrotra, R.S. 1988. Phyllosphere mycoflora of wheat in relation to leaf leachates and resistance to Helminthosporium blight. *Indian Phytopathology* 41:398–405.

Bashi, E. and Fokkema, N.J. 1976. Scanning electron microscopy of *Sporobolomyces roseus* on wheat leaves. *Transactions of the British Mycological Society* 67:500–505.

Biles, C.L. and Hill, J.P. 1988. Effect of *Trichoderma harzianum* on sporulation of *Cochliobolus sativus* on excised wheat seedling leaves. *Phytopathology* 78:656–659.

Billing, E. 1976. The taxonomy of bacteria on the aerial parts of plants. pp. 223–273 in Dickinson, C.H. and Preece, T.F. (editors), *Microbiology of Aerial Plant Surfaces*. Academic Press, London.

Blakeman, J.P. (editor). 1981. *Microbial Ecology of the Phylloplane*. Academic Press, London. 502 pp.

Blakeman, J.P. and Atkinson, P. 1981. Antimicrobial substances associated with the aerial surfaces of plants. pp. 245–263 in Blakeman, J.P. (editor), *Microbial Ecology of the Phylloplane*. Academic Press, London.

Boland, G.J. and Inglis, G.D. 1989. Antagonism of white mold (*Sclerotinia sclerotiorum*) of bean by fungi from bean and rapeseed flowers. *Canadian Journal of Botany* 67:1775–1781.

Breeze, E.M. and Dix, N.J. 1981. Seasonal analysis of the fungal community on *Acer platanoides* leaves. *Transactions of the British Mycological Society* 77:321–328.

Bunster, L., Fokkema, N.J., and Schippers, B. 1989. Effect of surface-active *Pseudomonas* spp. on leaf wettability. *Applied and Environmental Microbiology* 55:1340–1345.

Burchill, R.T. and Cook, R.T.A. 1971. The interaction of urea and microorganisms in suppressing the development of perithecia of *Venturia inaequalis* (Cke.) Wint. pp. 471–483 in Preece, T.F. and Dickinson, C.H. (editors), *Ecology of Leaf Surface Micro-organisms*. Academic Press, London.

Cabral, D. 1985. Phyllosphere of *Eucalyptus viminalis*: dynamics of fungal populations. *Transactions of the British Mycological Society* 85:501–511.

Cook, R.J. 1970. Factors affecting saprophytic colonization of wheat straw by *Fusarium roseum* f. sp. *cerealis* "Culmorum." *Phytopathology* 60:1672–1676.

Conn, K.L. and Tewari, J.P. 1989. Interactions of *Alternaria brassicae* conidia with leaf epicuticular wax of canola. *Mycological Research* 93:240–242.

Crosse, J.E. 1971. Interactions between saprophytic and pathogenic bacteria in plant disease. pp. 283–290 in Preece, T.F. and Dickinson, C.H. (editors), *Ecology of Leaf Surface Micro-organisms*. Academic Press, London.

De Cleene, M. 1989. Scanning electron microscopy of the establishment of compatible and incompatible *Xanthomonas campestris* pathovars on the leaf surface of Italian ryegrass and maize. *EPPO Bulletin* 19:81–88.

Dickinson, C.H. and Preece, T.F. 1976. *Microbiology of Aerial Plant Surfaces*. Academic Press, London. 669 pp.

Dix, N.J. 1979. Inhibition of fungi by gallic acid in relation to growth on leaves and litter. *Transactions of the British Mycological Society* 73:329–336.

Dowding, P. 1986. Leaf yeasts as indicators of air pollution. pp. 121–136 in Fokkema, N.J. and Van den Heuvel, J. (editors), *Microbiology of the Phyllosphere*. Cambridge University Press, Cambridge, UK.

Dubos, B. 1987. Fungal antagonism in aerial agrobiocenoses. pp. 107–135 in Chet, I. (editor), *Innovative Approaches to Plant Disease Control*. John Wiley & Sons, New York.

Fermaud, M. and Le Menn, R. 1989. Association of *Botrytis cinerea* with grape berry moth larvae. *Phytopathology* 79:651–656.

Fokkema, N.J. 1981. Fungal leaf saprophytes, beneficial or detrimental? pp. 433–454 in Blakeman, J.P. (editor), *Microbial Ecology of the Phylloplane*. Academic Press, London.

Fokkema, N.J. and Van den Heuvel, J. 1986. *Microbiology of the Phyllosphere*. Cambridge University Press, Cambridge, UK. 392 pp.

Fokkema, N.J. and Schippers, B. 1986. Phyllosphere versus rhizosphere as environments for saprophytic colonization. pp. 137–159 in Fokkema, N.J. and Van den Heuvel, J. (editors), *Microbiology of the Phyllosphere*. Cambridge University Press, Cambridge, UK.

Fokkema, N.J., Den Houter, J.G., Kosterman, Y.J.C., and Nelis, A.L. 1979. Manipulation of yeasts on field-grown wheat leaves and their antagonistic effect on *Cochliobolus sativus* and *Septoria nodorum*. *Transactions of the British Mycological Society* 72:19–29.

Forster, G.F. 1977. Effect of leaf surface wax on the deposition of airborne propagules. *Transactions of the British Mycological Society* 68:245–250.

Gerlagh, M. and Vos, I. 1991. Enrichment of soil with sclerotia to isolate antagonists of *Sclerotinia sclerotiorum*. pp. 165–171 in Beemster, A.B.R. (editor), *Biotic Interactions and Soil-Borne Diseases. Proceedings of the First Conference of the European Foundation for Plant Pathology*. Elsevier, Amsterdam.

Goodfellow, M., Austin, B., and Dawson, D. 1976. Classification and identification of phylloplane bacteria using numerical taxonomy. pp. 275–292 in Dickinson, C.H. and Preece, T.F. (editors), *Microbiology of Aerial Plant Surfaces*. Academic Press, London.

Heye, C.C. and Andrews, J.H. 1983. Antagonism of *Athelia bombacina* and *Chaetomium globosum* to the apple scab pathogen, *Venturia inaequalis*. *Phytopathology* 73:650–654.

Hildebrand, P.D. 1989. Surfactant-like characteristics and identity of bacteria associated with broccoli head rot in Atlantic Canada. *Canadian Journal of Plant Pathology* 11:205–214.

Hildebrand, P.D., Laycock, M.U., and Thibault, P. 1990. Biosurfactant production by pectolytic fluorescent pseudomonads and its role in broccoli head rot. *Canadian Journal of Plant Pathology* 12:334 (Abstr.).

Hirano, S.S. and Upper, C.D. 1983. Ecology and epidemiology of foliar bacterial plant pathogens. *Annual Review of Plant Pathology* 21:243–269.

Holloway, P.J. 1971. The chemical and physical characteristics of leaf surfaces. pp. 39–53 in Preece, T.F. and Dickinson, C.H. (editors), *Ecology of Leaf Surface Micro-organisms*. Academic Press, London.

Irvine, J.A., Dix, N.J., and Warren, R.C. 1978. Inhibitory substances in *Acer platanoides* leaves: seasonal activity and effects on growth of phylloplane fungi. *Transactions of the British Mycological Society* 70:363–371.

Janisiewicz, W. 1988. Biological control of diseases of fruits. pp. 153–165 in Mukerji, K.G. and Garg, K.L. (editors), *Biocontrol of Plant Diseases*, vol. II. CRC Press, Boca Raton, FL.

Jayapal Gowdu, B. and Balasubramanian, R. 1988. Role of phylloplane microorganisms in the biological control of foliar plant diseases. *Zeitschrift fur Pflanzenkrankheiten und Pflanzenschutz* 95:310–331.

Kempf, H.-J. and Wolf, G. 1989. *Erwinia herbicola* as a biocontrol agent of *Fusarium culmorum* and *Puccinea recondita* f. sp. *tritici* on wheat. *Phytopathology* 79:990–994.

Kinkel, L.L., Andrews, J.H., and Nordheim, E.V. 1989. Fungal immigration dynamics and community development on apple leaves. *Microbial Ecology* 18:45–58.

Köhl, J., Molhoek, W.M.L., and Fokkema, N.J. 1990. Biological control of *Botrytis aclada* in onions. *Phytopathology* 80:1049 (Abstr.).

Last, F.T. 1955. Seasonal incidence of *Sporobolomyces* on cereal leaves. *Transactions of the British Mycological Society* 38:221–239.

Leben, C. 1965. Epiphytic microorganisms in relation to plant disease. *Annual Review of Phytopathology* 3:209–230.

Leben, C. 1971. The bud in relation to the epiphytic microflora. pp. 117–127 in Preece, T.F. and Dickinson, C.H. (editors), *Ecology of Leaf Surface Microorganisms*. Academic Press, London.

Leben, C. 1988. Relative humidity and the survival of epiphytic bacteria with buds and leaves of cucumber plants. *Phytopathology* 78:179–185.

Levy, E., Eyal, Z., Carmely, S., Kashman, Y., and Chet, I. 1989. Suppression of *Septoria tritici* and *Puccinia recondita* of wheat by an antibiotic-producing fluorescent pseudomonad. *Plant Pathology* 38:564–570.

Lindemann, J. and Suslow, T.S. 1987. Competition between ice nucleation-active wild type and ice nucleation-deficient deletion mutant strains of *Pseudomonas syringae* and *P. fluorescens* biovar I and biological control of frost injury on strawberry blossoms. *Phytopathology* 77:882–886.

Lindow, S.E. 1985. Integrated control and role of antibiosis in biological control of fireblight and frost injury. pp. 83–115 in Windels, C.E. and Lindow, S.E. (editors), *Biological Control on the Phylloplane*. The American Phytopathological Society, St. Paul, MN.

Martin, J.T. and Juniper, B.E. 1970. *The Cuticles of Plants*. Edward Arnold, London. 347 pp.

McLaughlin, R.J., Wisniewski, M.E., Wilson, C.L., and Chalutz. E. 1990. Effect of inoculum concentration and salt solutions on biological control of postharvest diseases of apple with *Candida* spp. *Phytopathology* 80:456–461.

Mew, T.W. and Cruz, C.M.V. 1986. Epiphytic colonization of host and non-host plants by phytopathogenic bacteria. pp. 269–282 in Fokkema, N.J. and Van den Heuvel, J. (editors), *Microbiology of the Phyllosphere*. Cambridge University Press, Cambridge, UK.

Mew, T.W., Mew, I.C., and Huang, J.S. 1984. Scanning electron microscopy of virulent and avirulent strains of *Xanthomonas campestris* pv. *oryzae* on rice leaves. *Phytopathology* 74:635–641.

Miedtke, U. and Kennel, W. 1990. *Athelia bombacina* and *Chaetomium globosum* as antagonists of the perfect stage of the apple scab pathogen (*Venturia inaequalis*) under field conditions. *Zeitschrift fur Pflanzenkrankheiten und Pflanzenschutz* 97:24–32.

Mitchell, J.K., Taber, R.A., and Pettit, R.E. 1986. Establishment of *Dicyma pulvinata* in *Cercosporidium personatum* leaf spot of peanuts: effect of spray formulation, inoculation time and hours of leaf wetness. *Phytopathology* 76:1168–1171.

Morris, C.E. and Rouse, D.I. 1985. Role of nutrients in regulating epiphytic bacterial populations. pp. 63–82 in Windels, C.E. and Lindow, S.E. (editors), *Biological Control on the Phylloplane*. The American Phytopathological Society, St. Paul, MN.

Mukerji, K.G. and Garg K.L. 1988. *Biocontrol of Plant Diseases*, vol. I and II. CRC. Press, Boca Raton, FL. 211 and 198 pp.

Nelson, M.E. and Powelson, M.L. 1988. Biological control of grey mold of snap beans by *Trichoderma hamatum*. *Plant Disease* 72:727–729.

O'Brien, R.D. and Lindow, S.E. 1989. Effect of plant species and environmental conditions on epiphytic population sizes of *Pseudomonas syringae* and other bacteria. *Phytopathology* 79:619–627.

Pfender, W.F. 1988. Suppression of ascocarp formation in *Pyrenophora tritici-repentis* by *Limonomyces roseipellis*, a basidiomycete from reduced-tillage wheat straw. *Phytopathology* 78:1254–1258.

Philipp, W.-D., Beuther. D.H., Hermann, D., Klinkert, F., Oberwalder, C., Schmidtke, M., and Straub, B. 1990. Zur Formulierung des Mehltauhyperparasiten *Ampelomyces quisqualis* Ces. *Zeitschrift fur Pflanzenkrankheiten und Pflanzenschutz* 97:120–132.

Preece, T.F. and Dickinson, C.H. 1971. *Ecology of Leaf Surface Micro-organisms*. Academic Press, London. 640 pp.

Rodgers, P.B. 1989. Potential of biological control organisms as a source of antifungal compounds for agrochemical and pharmaceutical product development. *Pesticide Science* 27:155–164.

Roveratti, D.S., Teixeira, A.R., and Moraes, W.B.C. 1989. *Bacillus thuringiensis*—a new perspective for induced protection to coffee leaf rust. *Journal of Phytopathology* 126:149–159.

Ruinen, J. 1956. Occurrence of *Beijerinckia* species in the phyllosphere. *Nature* 177:220–221.

Rytter, J.L., Lukezic, F.L., Craig, R., and Moorman, G.W. 1989. Biological control of geranium rust by *Bacillus subtilis*. *Phytopathology* 79:367–370.

Sieber, T., Riesen, T.K., Muller, E., and Fried, P.M. 1988. Endophytic fungi in four winter wheat cultivars (*Triticum aestivum* L.) differing in resistance against *Stagonospora nodorum* (Berk.) Cast. & Germ. = *Septoria nodorum* (Berk.) Berk. *Journal of Phytopathology* 122:289–306.

Stead, D.E. 1989. Grouping of *Xanthomonas campestris* pathovars of cereals and grasses by fatty acid profiling. *EPPO Bulletin* 19:57–68.

Sundheim, L. 1986. Use of hyperparasites in biological control of biotrophic plant pathogens. pp. 333–347 in Fokkema, N.J. and Van den Heuvel, J. (editors), *Microbiology of the Phyllosphere*. Cambridge University Press, Cambridge, UK.

Swinburne, T.R. 1981. Iron and iron chelating agents as factors in germination, infection and aggression of fungal pathogens. pp. 227–243 in Blakeman, J.P. (editor), *Microbial Ecology of the Phylloplane*. Academic Press, London.

Sztejnberg, A., Galper, S., Mazar, S., and Lisker, N. 1989. *Ampelomyces quisqualis* for biological and integrated control of powdery mildews in Israel. *Journal of Phytopathology* 124:285–295.

Thomas, M.R. and Shattock, R.C. 1986. Filamentous fungal associations in the phylloplane of *Lolium perenne*. *Transactions of the British Mycological Society* 87:255–268.

Tronsmo. A. 1986. Use of *Trichoderma* spp. in biological control of necrotrophic pathogens. pp. 348–362 in Fokkema, N.J. and Van den Heuvel, J. (editors), *Microbiology of the Phyllosphere*. Cambridge University Press, Cambridge, UK.

Trutmann. P., Keane, P.J., and Merriman, P.R. 1982. Biological control of *Sclerotinia sclerotiorum* on aerial parts of plants by the hyperparasite *Coniothyrium minitans*. *Transactions of the British Mycological Society* 78:521–529.

Tukey, H.B., Jr. 1971. Leaching of substances from plants. pp. 67–80 in Preece, T.F. and Dickinson, C.H. (editors), *Ecology of Leaf Surface Micro-organisms*. Academic Press, London.

Tuzun, S., Reuveni, M., Siegel, M.R., and Kuc, J. 1989. The effect of removing leaf surface components with acetone from immunized and nonimmunized resistant tobacco plants on their susceptibility to blue mold. *Phytopathology* 79:1024–1027.

Van der Burg, A.C. 1974. The occurrence of *Sporobolomyces roseus*, a red yeast, on leaves of *Phragmites australis*. Dissertation, Free University, Amsterdam. 74 pp.

Van Outryve. M.F., Gossele, F., and Swings, J. 1989. The bacterial microflora of witloof chicory (*Cichorum intybus* L. var. *foliosum* Hegi) leaves. *Microbial Ecology* 18:175–186.

Van Vuurde, J.W.L. 1987. New approach in detecting phytopathogenic bacteria by combined immunoisolation and immunoidentification assays. *EPPO Bulletin* 17:139–148.

Wilson, C.L. 1989. Managing the microflora of harvested fruits and vegetables to enhance resistance. *Phytopathology* 79:1387–1390.

Wilson, C.L. and Wisniewski, M.E. 1989. Biological control of postharvest diseases of fruits and vegetables: an emerging technology. *Annual Review of Phytopathology* 27:425–441.

Young, C.S. and Andrews. J.H. 1990. Inhibition of pseudothecial development of *Venturia inaequalis* by the basidiomycete *Athelia bombacina* in apple leaf litter. *Phytopathology* 80:536–542.

Zhou, T. and Reedeler. R.D. 1989. Application of *Epicoccum purpurescens* spores to control white mold of snap bean. *Plant Disease* 73:639–642.

Part 2
The Habitat

2

The Leaf from the Inside and the Outside: A Microbe's Perspective

Barrie E. Juniper

2.1 Introduction

A microbial spore that arrives by air, rain splash, vector, or brush contact on any plant surface, like a pollen grain but unlike most seeds, has probably not been selected in the form in which it is spread for sustained viability. Therefore, on the exposed, desiccating, radiation-prone, potentially crowded, suppurating, vibrating surface of a leaf, a spore's effective life may be very short. The spore must germinate and, in many cases, then penetrate the leaf. The requirements of this development are rigid. It may grow epiphytically, but may be required to penetrate or be forced, as in bacteria or viruses, through layers of waxes, cutin, and sometimes polyisoprene, and the pectocellulosic wall, which may occasionally be lignified. Only then may it be able to reach the softer walls, i.e., those lacking the encrusting substances listed above, and the more sheltered, more nitrogen-rich interior. (Only the smallest traces of proteins have been found in the waxes isolated from plant surfaces [Hallam and Juniper, unpublished; see Section 2.4].) All permeable plant surfaces, whether external as in glands or nonwaxy surfaces, or internal as in uncutinized walls, are leaky and nutrient-rich (Tukey, 1970) and will more readily be colonized (Lippincott et al., 1977; Juniper and Jeffree, 1983). Within the leaf or other tissue, although physically safer, the growing hyphal tip, virus population, or bacterial colony may be impeded by induced physical or chemical barriers. Fungitoxic substances, phytoalexins, bacteriotoxins, and even callose barriers (Templeton et al., 1990; Goodman et al., 1986) may come into play. Every aspect of the architecture of the outside and inside of a plant's surface is crucial to this successful progression.

2.2 Arrival on the Plant Surface

The pathogen floats in the airstream over the surface of the leaf. Its rate of fall will be slow by virtue of its small size and low density, but it will eventually enter the still or less turbulent area of the boundary layer, the skin of quiet air that lies over the surface of a leaf (Monteith, 1973; Burrage, 1970, 1976). The thickness of this layer has proved exceedingly difficult to calculate (Jones, 1983) because of edge effects and the local turbulence produced by trichomes and epicuticular waxes, for example. None, however, dispute its presence.

Other microbes may arrive in a water drop, either harvested by a rain drop falling through an aerosol of particles, or gathered by rain splash from a neighboring leaf or another part of the canopy (Thompson, 1969); they may be deposited onto the leaf surface by any of a range of vectors, pollinating insects, phytophagous lepidoptera, or ovipositing females of a vast array of arthropods and, for viruses of certain field crops, by agricultural machinery (Parry, 1990), and even by rabbits distributing the particles on their fur!

Adhesion Whether the microbe lodges on the leaf surface will depend principally upon the characteristics of that leaf surface. There is a general impression that most plant surfaces are unwettable, but this is not necessarily so. Figure 2.1 shows the surface of the gametophyte of the moss *Rhacocarpus purpurascens*, which has no xylem, but does have an ectohydric water-conducting system (Barthlott, 1990). Similar coarsely fractal surfaces, i.e., ones possessing a very complex topography, may be found on the surfaces of some pollen grains and spores, and some seeds and other plant parts such as petals and stigmas. Such wettable surfaces are frequently contaminated with both inorganic and organic dusts and microbes (Figure 2.2). Bacteria persistent on such surfaces may even play a role in frost damage to the plant by the formation of ice nuclei (Lindow, 1983), and such damage might allow pathogen penetration (see Section 2.5).

Repulsion and Loss Highly hydrophobic surfaces (Figures 2.3, 2.4, and 2.5), i.e., those with large contact angles (Figure 2.6C) usually bear crystalline waxes with individual crystal measurements at or just below the resolution of the light microscope (Jeffree and Sandford, 1982). These surfaces are often surprisingly clean, even outside the greenhouse (Figure 2.4). However, sticky contaminants or particles, such as the hydrocarbon-rich soot in the stomatal cavity (figure 5.1 in Juniper and Jeffree, 1983), can lodge on the most hydrophobic surface. Dry spores that are wettable themselves and have no firm adhesion on such surfaces may possibly be gathered by a rain droplet with the kinetic energy generated from a strike on a hard surface and be washed clear (Thompson, 1969). High contact

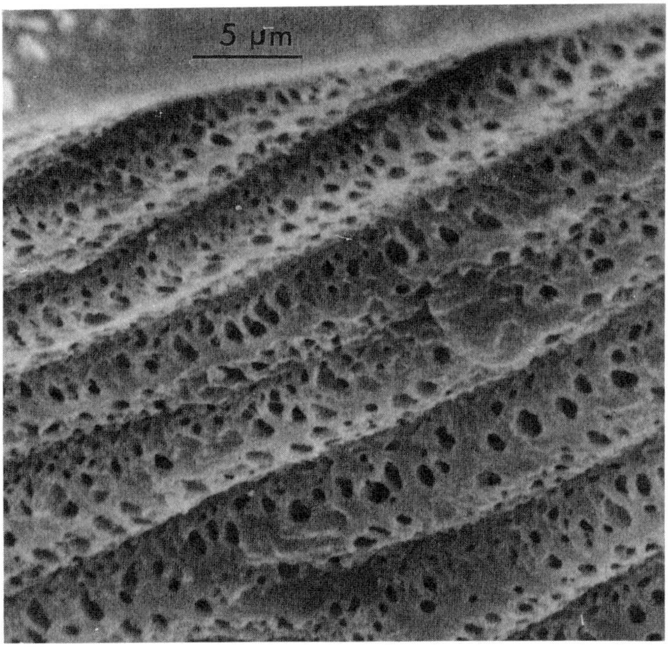

Figure 2.1 SEM of the surface of the bryophyte *Rhacocarpus purpurascens*. Courtesy of W. Barthlott.

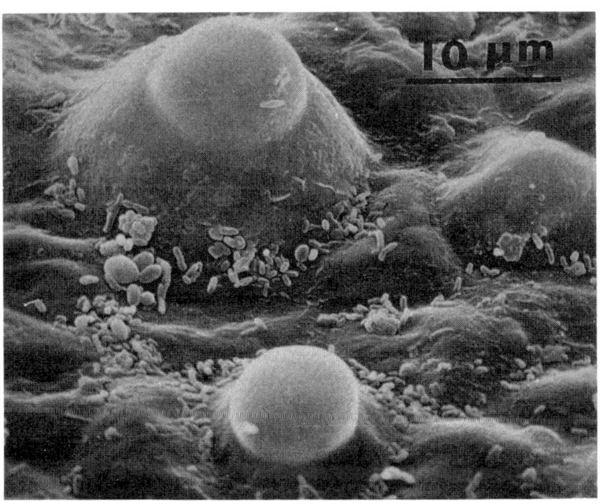

Figure 2.2 SEM of ocelli on the adaxial leaf surface of *Fittonia verschaffeltii*. These are cutin accumulations thought to serve as light-concentrating devices. Courtesy of J. Whatley.

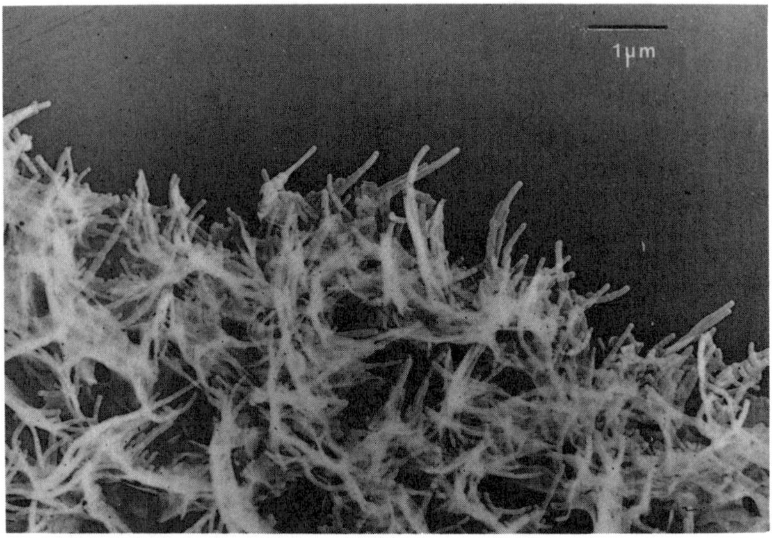

Figure 2.3 TEM of a transverse view of the epicuticular wax crystals on a *Lupinus* leaf. Each wax crystal is about 2 μm long.

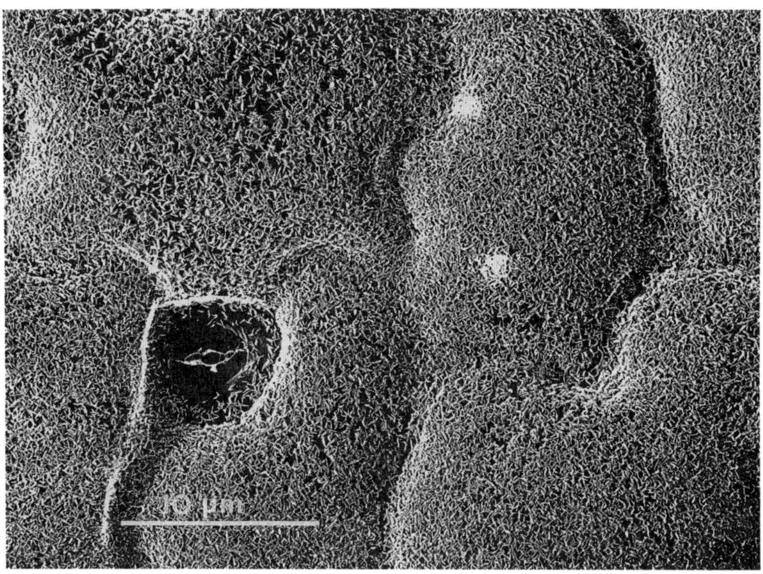

Figure 2.4 TEM of a surface view of the adaxial surface of a pea (*Pisum sativum*) leaf.

Figure 2.5 Low-magnification SEM of the adaxial leaf surface of wheat (*Triticum*). Courtesy of P. Harris.

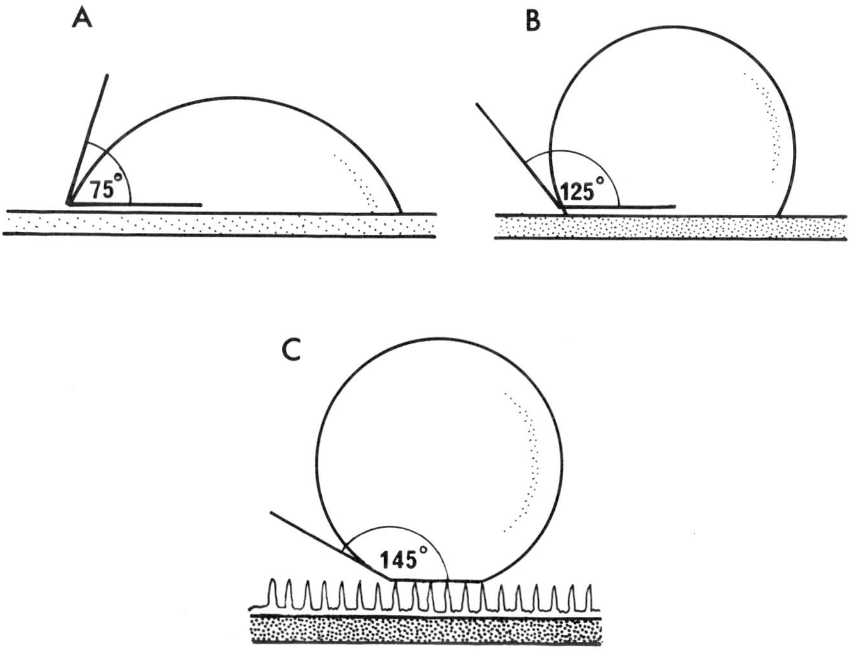

Figure 2.6 Effect of contact angle size on a water drop on a leaf surface. (A) Wax-free plant cuticle: small contact angle; water drop covers a large surface angle. (B) Smooth wax surface: larger contact angle; droplet covers a smaller area. (C) Crystalline wax surface or ornamented cuticle: largest contact angle; droplet is in contact with the tips of the projections only.

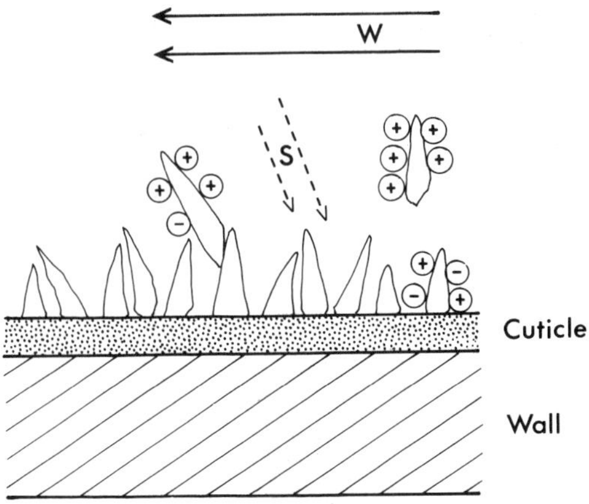

Figure 2.7 A proposed mechanism to explain the loss of wax and other particles from leaf surfaces either by the energy gathered from wind (W) movements or photons (S) falling on the leaf surface. Each wax particle might act as a dipole bringing about a localized repulsion either from its neighboring crystallites or the more homogeneous surface below.

angles (Figure 2.6C) having the ability to cause water droplets to "pearl off" may be a significant way of keeping plant surfaces clean.

Another feature causing displacement of surface particles, and undoubtedly microbes, is the continuous loss of particulate material, principally wax particles but possibly also polyester and polyisoprene particles, from the plant surface (Moorby and Squire, 1963; Beauford et al., 1975). These authors noticed that radioactive tracers and heavy metals were discharged from leaf surfaces, even in still air, and these pollutants were associated with wax particles from the epicuticular wax layer. We know of no mechanism for this active discharge of extracellular material. However, as Figure 2.7 suggests, each wax particle might act as a dipole; wind (W) moving over the surface or even the energy from a photon (S) falling upon a wax crystal might generate sufficient charge to create small dipole moments and drive material from the surface. The very small amounts of protein present (see Section 2.4) might act as local conducting skins to generate such charges. Such discharges might contribute to the vast quantities of particulate matter in the atmosphere, which form the blue haze seen over many mountain regions. Went (1974) has speculated eloquently on the origin and possible roles of these natural aerosols.

Retention Three factors, principally, may retain the microbe on the leaf surface: 1) the cell may itself be hydrophilic or "sticky," e.g., soot particles or similar hydrocarbons, or pollenkitt-coated, insect-transmitted pollen (Figure 2.8; the lipo-protein layer found on many pollen grains); 2) the originally hydrophobic leaf surface may have been altered by weathering, mechanical or chemical degradation, or even asymmetric patterns of growth; 3) broken epidermal cells or trichomes may, by exudation of some of their contents and subsequent ingestion of cytoplasm, draw microbes back into living cell systems.

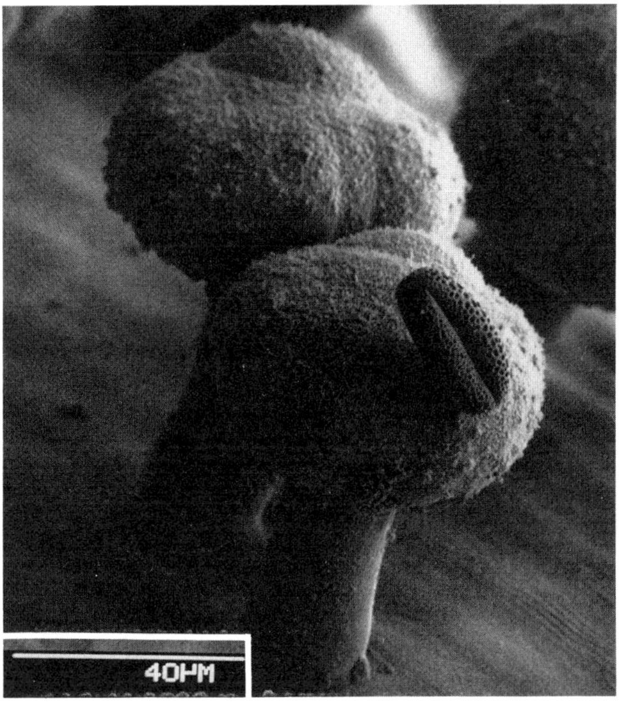

Figure 2.8 SEM of an unidentified pollen grain on a stalked gland of *Pinguicula vulgaris*.

As Figure 2.6 shows, the nature of the surface may affect the adhesion of a particle. A leaf surface with a relatively small contact angle (Figure 2.6A) not only will enhance the speed of drying on a surface, but will encourage spread, in some cases by swimming zoospores of a solution of pathogens over a larger area (Royle and Kremheller, 1981). The properties of leaf surfaces in this regard have been extensively reviewed (Silva-Fernandes, 1965; Hallam and Juniper, 1970; Holloway, 1970, 1982; Martin and Juniper, 1970; Baker and Holloway, 1971; Juniper and Jeffree, 1983.)

In addition to all the physical phenomena described earlier, there may also be an active role for certain spores *prior* to germination. It seems that, for example, the urediniospores of *Uromyces viciae-fabae* are surrounded by a thin coating of colloidal, iron-positive, extracellular matrix when they are released at the surface of the uredinium. On contact with a substratum, whether artificial or natural, this matrix may provide sufficient adhesion to bind the spore to that surface (Beckett and Porter 1988; Beckett et al., 1990; see also Hamer et al., 1988).

2.3 Diversity of the Surface

Damage aside (see Section 2.5), very few plant surfaces are uniform in chemistry or micromorphology. Those that are, e.g., the gametophytic structure seen in Figure 2.1 or a grape (*Vitis*), have proved immensely valuable in experimentation (Grncarevic and Radler, 1967). Even if the microbe achieves lodgement, however, the diversity of the surface, both horizontally and vertically, will present problems to the next stage of its development.

Diversity of the Waxes In Section 2.2 and Figure 2.6, we assumed in considering contact and repulsion, particularly in relation to the wax layer, the surface to be uniform in its characteristics. This assumption is unfounded. Wax detail may vary at the primary, secondary, or tertiary sculptural level (Barthlott, 1990). The cotyledon and the juvenile and mature leaves may differ markedly in their waxiness and/or in the fine structure of their waxes (Martin and Juniper, 1970). Apart from the fact that the plant surface may vary immensely in topography, i.e., in the convexity of the tabular epidermal cells (Figures 2.2, 2.4 and 2.5), or in the presence or absence of trichomes or emergences (Figure 2.5), some surfaces (e.g., *Rosa*) are glabrous above, but markedly glaucous below. The pattern and diversity of the epicuticular wax, even its dimorphism on a single surface as in certain species of *Eucalyptus* (Carr and Carr, 1987; Carr et al., 1985), may have an influence on the success or failure of the microbe. Even the crude amount of wax developed on leaves of the same plant in different light regimes is known to vary markedly. The quantity is generally a function of light (Martin and Juniper, 1970; Steinmuller and Tevini, 1985).

Diversity of the Cutins Apart from the diversity of the wax, the cutin itself, whether it is of polyester, the most common constituent, or of polyisoprene-rubber (Carr and Carr, 1987; Cutler et al., 1982), may be uniform over a leaf surface. More commonly, however, it is thicker over the outer radial walls of the epidermal cells, and thicker or thinner, but rarely the same, over and around the various trichomes or emergences from the plant surface (Martin and Juniper, 1970; Fahn, 1979; Holloway, 1982). Sometimes, as over certain glands (Figure 2.8) or stigmas, the cuticle may be so thin as to be virtually undetectable (Fahn, 1979; Juniper et al., 1989). There seems however to be little correlation between cuticular thickness and the resistance of the leaf to desiccation, damage, or the assault of pathogens (Jeffree, 1986).

Stomata Stomata are most frequent on the abaxial surface (Meidner and Mansfield, 1968). Their role as ports of entry for bacteria will be discussed in Section 2.5 and are dealt with in the context of fungal infections by Allen et al. (see Chapter 5). Stomata may be sunk in crypts, as in *Rhipsalis dissimilis*;

Figure 2.9 SEM of two stomata on the adaxial leaf surface of *Drosera capensis*. Courtesy of C. Merriman.

approximately flush with the level of the epidermal cells, as in Figures 2.5, 2.9, and 2.10; or raised above the surface, as in the striking *Euphorbia tirucalli* (Figure 2.11) (Juniper and Jeffree, 1983). Even more extraordinary are the phylloclades as typified by *Colletia cruciata* (Rhamnaceae). Here each stoma is surrounded by a chimney-like hollow wax tube often so tall that it can be easily seen with the light microscope (Barthlott, 1990).

Extrafloral Nectaries, Hydathodes, and Lenticels Extrafloral nectaries are widespread in higher plants and are by no means confined to the angiosperms. They are found, for example, in the fern (*Pteridium*). They, hydathodes, and lenticels, are areas on a plant surface where the delicate interior is undefended by a cuticle (Fahn, 1979). As a result, they may provide sites for the entry of pathogens (Hirano and Upper, 1983), and are known to be the preferred sites of entry for the infection of certain fruit trees, e.g., *Pyrus* and *Malus*, by the bacterium *Erwinia amylovora* (fireblight), which is vectored by bees.

Trichomes and Emergences Trichomes and emergences are, in enormous variety, widespread in the plant kingdom (Wilkinson, 1979; Jeffree, 1986) (Figures 2.2 and 2.11). Many trichomes are dead at maturity; those that remain alive, with the exception of the secretory trichomes (Figure 2.8), are often heavily thickened with cutin, or even silica. Their many roles in

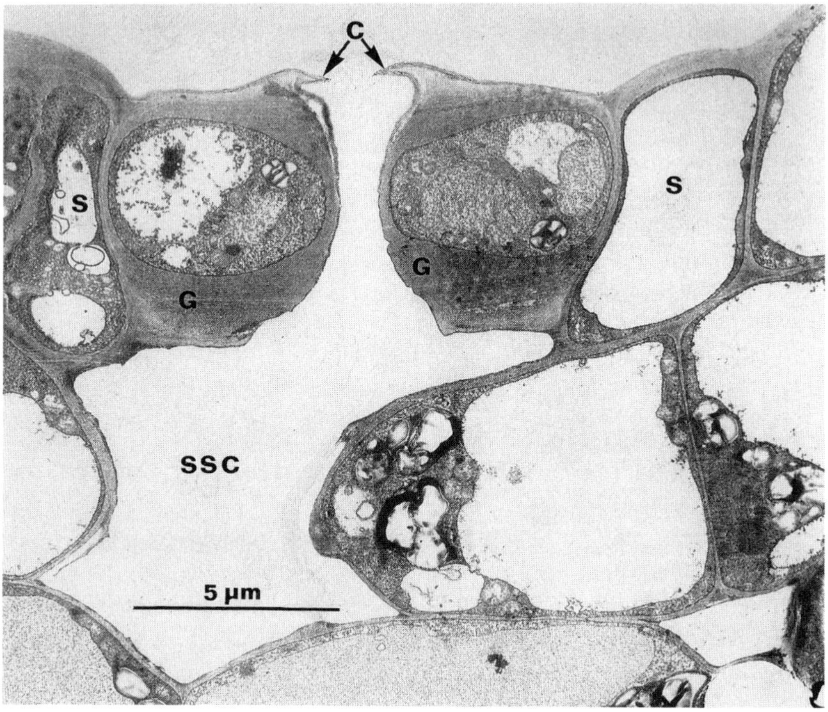

Figure 2.10 TEM section through a stoma of *Citrus*. C = cuticular lips; G = guard cells, S = subsidiary cells, SSC = substomatal cavity. Courtesy of E. Pacini.

the interaction of the plant surface with insects have been extensively studied (Juniper and Southwood, 1986). Broken trichomes are, however, well known as ports of entry for bacteria and viruses (Hirano and Upper, 1983), and as described in Section 2.2 above, whether intact or broken, they will disturb the airflow over a plant surface and thus the deposition of spores.

2.4 Microenvironment Factors

Moisture In general, spores will neither germinate in air nor will pathogens develop under conditions of low relative humidity. Slow desiccation until dead is probably the fate of the majority of spores, although some, like certain seeds, may survive longer as a result of a secreted extracellular matrix (Beckett, et al., 1990). Some will only germinate in a pool of water (see below). If the leaves of a plant are dry, as all gardeners know, fungal infections are markedly reduced. But even dew may be sufficient. Generally speaking, dew fall is usually at least an order of magnitude less than potential rates of evaporation (Jones, 1983) and so rarely contributes

Figure 2.11 SEM of raised stoma on the adaxial leaf surface of *Euphorbia tirucalli*. Courtesy of W. Barthlott.

significantly to the water balance of a plant. However, there are areas of the world such as the dew deserts of Chile, the Sonoran Desert of the southwestern United States, and possibly also the warm coasts adjacent to cold seas, e.g., parts of California and the Iberian peninsula, where dew or mist condensation may equal or exceed rainfall (Kerr and Beardsell, 1975; Hull et al., 1979; Juniper and Jeffree, 1983; Juniper et al., 1989). Whatever its significance in overall leaf water potential, the importance of mist or dew lies in creating local water pools in which many spores are able successfully to germinate. The duration of leaf wetness from dew may be significant. During the summer in England, dew may persist on a plant surface for 6 to 12 hours (Jones, 1983). Here again the detail of the leaf surface enabling it to shed or pearl off water droplets or, conversely, retain them as a film (Figure 2.6) may be crucial. Thus a pea (*Pisum*) leaf (Figure 2.4) will not retain dew. Urediniospores inoculated onto such a surface cannot imbibe water from misting and will only swell and germinate on such a surface if inoculated in a humidity chamber (Beckett et al., 1990).

Feeding damage caused by the typhlocybine leaf hopper (*Ossiannilssonola*) on sycamore (*Acer pseudoplatanus*) has been shown to cause a water loss of 18% in the day time and 75% or greater at night (Whittaker, 1984). It seems possible that such puncturing by many types of phytophagous insects might not only cause wounds leading to penetration by bacteria and viruses, but also cause local increases in leaf surface humidity sufficient to cause a fungal spore to germinate. However, for some parasitic fungi (e.g., certain of the downy mildews such as downy mildew of hop *Pseudoperonospora humuli*), dew is not sufficient and rainfall (prolonged free water) is a necessity (Royle and Kremheller, 1981). In this case, the swimming zoospores settle over stomatal openings on the lower (abaxial) surfaces of the leaf because the shape of the abaxial leaf concentrates the water droplets over these areas. Perhaps, were the abaxial surface of the leaf glaucous and unwettable, as in some *Rosa* species, such infections might be reduced.

Temperature Temperature on the leaf surface may be critical. For *Phytophthora infestans* on potato leaves, uninucleate zoospores are produced below 15°C, but above 20°C, multinucleate germ tubes emerge (Webster, 1986). If there is free water on the plant surface, the zoospores may be able to swim around for some time. Depending on the temperature, they may then settle, encyst, and form germ tubes, which may in turn form an appressorium leading to the penetration of the cuticle. Penetration can occur within two hours (Webster, 1986). Thus the degree of solar reflectance from a leaf surface, which is in major part due to the light-scattering from epicuticular wax (Barber and Jackson, 1957), might be critical in this matter.

Stimulants All leaf surfaces are leaky. The leachates, which are particularly rich after long periods without rain, may consist of sugars, amino acids, proteins, salts, and phenolics. Among them are fungitoxic and viral-suppressing compounds (Purnell, 1971). The quantities of leachate can be considerable. Apple (*Malus*) leaves may lose 25–30 kg of potassium, 9 kg of sodium, and 10 kg of calcium per hectare per year, as well as sugars, amino acids, and phenolics (Tukey, 1970). Much of this leachate is recycled, but all of it must profoundly affect the phylloplane flora (Lovett and Sagar, 1978). The secretion of sucrose, glucose, and fructose from petals appears to stimulate the germination of spores of *Botrytis cinerea*. This phenomenon may explain the ease with which these fungi attack flower heads in wet summers (Deverall, 1981). Even if these leachates (e.g., sucrose) are washed off, they are rapidly replaced (Purnell and Preece, 1971). We know very little about the origins, role(s), and functions of proteins on the leaf surface, but small quantities of at least three similar, low-molecular-weight proteins (molecular weight of 20, 30, and 45,000) regularly appear in extracted leaf waxes of such different genera as *Eucalyptus* and *Acacia* (Hallam and Juniper, unpublished).

Apart from the possible stimulants provided unwittingly by the host itself, it is now thought that other inhabitants of the leaf surface (e.g., pollen grains) (Figure 2.8) and bacteria, including nitrogen-fixing bacteria, may also provide stimulation (Ruinen, 1975; Silvester, 1976; Deverall, 1981; also see Chapter 1, this volume).

Inhibitors A range of inhibiting substances, both soluble on the surface and in the wax layer, has been detected (Peries, 1962; Hargreaves et al.,

1982). Fungitoxic substances have even been detected in the stomata (Koshino et al., 1987). Bacteriotoxic substances inhibiting *Agrobacterium tumefaciens*, for example, have been detected in the cell walls of *Phaseolus* (Lippincott et al., 1977). The literature is very extensive and has been reviewed by Martin (1964), Emmett and Parberry (1975), Bailey (1986), Cervone et al. (1989), and Goodman et al. (1986).

For some time there has been evidence that fungitoxic substances might be present in the wax layer (Blakeman, 1971; Blakeman and Atkinson, 1976). Fat-soluble fungitoxic substances in the wax layer, such as the unsaturated fatty acids detected in the stomata by Koshino et al. (1987), might inhibit or slow the growth of the germinating spore. Watkinson (1984) has isolated the fat-soluble fungal inhibitor, α-amino-isobutyric acid from *Iris tingitana*. Viruses too may be destroyed on leaf surfaces. Nuclear polyhedrosis virus is apparently inactivated on cotton (*Gossypium hirsutum*) leaves by an exudate of the leaf glands (Elleman and Entwistle, 1985). The exudate has a high pH, but the inhibitory effect is thought to be due principally to the unusually high concentrations of potassium, magnesium, and calcium ions, which in other circumstances may be stimulants, as noted above.

2.5 Journey into a Leaf

Penetration of the Wax Layer The epicuticular wax crystals (Figures 2.3, 2.4, and 2.5), which, as we have seen, generally have only a loose attachment to the surface (Figure 2.7), may be able to keep the viable spore away from the more hydrophilic cuticle proper and away from any stimulation, as suggested in Section 2.4. As noted earlier (see Section 2.3) the amount of wax on a leaf surface is known often to be a function of light. Thus, whether a spore lands on a shade leaf or a sun leaf, with their differing morphology, cuticular thickness, and pigmentation (Juniper and Jeffree, 1983), might be critical. Steinmuller and Tevini (1985) showed that enhanced levels of ultraviolet B light can markedly increase the total wax. In barley (*Hordeum*), bean (*Phaseolus*), and cucumber (*Cucumis*) there was an increase over controls of about 25%. Aldehydes in the barley and cucumber were doubled. There is plenty of evidence from the haloes of disturbed crystals that surround growing hyphal tips on leaf surfaces indicating that "waxases" of some form or another must exist in pathogenic fungi and perhaps also in saprophytic fungi (Martin and Juniper, 1970). However, although cutinases have been extensively studied, there seems to be very little work on wax-destroying organisms.

Finding a Crack A crack in the cuticle or a gap or scraping in the epicuticular wax, which is frequently seen under the microscope, may be detected by the deposited spore or developing hyphal tip in several ways. In the first instance, water vapor loss will be greater from such areas than on an undamaged part of the leaf (Whittaker, 1984). Such a leakage may enhance germination (see Section 2.4). Leakage of gases will also be greater at this point. Perhaps most important of all, leachates, particularly the sugars or other stimulating compounds (see Section 2.4), presumably will be greatest over these areas.

Finding a Stoma A stoma will present a target not only by its peculiar shape, but also by the enhanced flow of water vapor from its aperture. The guard cells are, in addition, almost the only superficial cells to possess full developed chloroplasts and thus might possibly be richer in leachates (see Section 2.4). The stoma as a target is discussed in Chapter 5 (this volume) and by Dickinson and Lucas (1982).

Penetration of the Cuticle Some microbes seem only able to penetrate through cracks or other defects in the ectoderm, e.g., species of the fungus *Mycosphaerella*, which is only able to infect paw-paw (*Carica papaya*) fruits if the cuticle is damaged. Others, such as *Fusarium solani pisi*, possess a cutinase. Dickman et al. (1989) have shown that it is possible to transform *Mycosphaerella* isolates to a cutinase-producing form by cloning the *Fusarium* cutinase gene and transferring it to the *Mycosphaerella* cell.

The cuticle and/or the epidermal wall may be subject to enzymic softening. Commonly, apparent enzymic softening of the cuticle can be detected by the "halo" that surrounds the actual point of penetration (Webster, 1986). Sargent (1981) has shown that in *Bremia lactucae*, which causes downy mildew of lettuce, the cuticle of the host is sensitive to the presence of the fungus. Directly beneath the forming appressorium, the host cell wall stains very readily, indicating the movement into the wall at this position, or the formation within it of compounds possessing abundant osmiophilic groups. The host cytoplasm, rich in crystalline material, also accumulates beneath the appressorium indicating a very generalized reponse to the pathogen's presence. In the case of successful penetration, the host cuticle is not only dissolved, but becomes swollen and thrown into folds around the periphery of the pore. There is no evidence of physical penetration since neither cuticle nor wall layers are turned inwards at this point. However, simple physical penetration, in some infections, does occur. *Botrytis cinerea*, for example, has been shown to push its way through thin gold films and paraffin wax membranes (Deverall, 1981).

Hard leaves with massive cuticles, the sclerophylls of Grubb (1986), probably evolved principally as a defense against leaf-chewing phytophagous insects. But these properties of hardness and thickness of the biopolyester also may be significant in resistance to fungal penetration. The cuticle is usually, however, very susceptible to damage and is easily penetrated (Wullschleger and Osterhuis, 1987). Insect attack also causes significant water loss (see Section 2.4). We must assume that wax, cuticular, pectin, and pectocellulosic wall repair mechanisms are active and effective in responding to this continuous assault.

2.6 Internal Environment

Having germinated successfully and achieved the first penetration, the hyphal tip of a fungus that has penetrated the surface can take shelter within the tissue of the host.

Within the Cuticle Many fungi, bacteria, and viruses are common inhabitants of plant tissue. Some of the fungi and bacteria, the so-called endophytes, apparently cause little harm, and may provide some indirect benefits, e.g., insect resistance; these groups are discussed in Chapter 17 (this volume). A very special group of fungi, again with very low pathogenicity, occupies the cuticle of some higher plants. Species of the ascomycete *Vizella* may complete their whole life cycle in the thick cuticles of genera such as *Banksia* and *Olearia* (Swart, 1971). The details of the penetration and life cycle are difficult to observe in this essentially two-dimensional habitat, but the thin, ca. 5-μm hyphae ramify in or very occasionally under the thick cuticles of the Australian native plant *Banksia*. Related species are also able to colonize the similarly robust cuticles of *Olearia argophylla*, which is a native of the Mediterranean. Another species, *Vizella interrupta*, is found in the cuticles of African Proteaceae. Thus, a range of pachycaulous species (in the sense of the term used by Grubb, 1986), having massive cuticles, is able to be penetrated by fungi. It does seem surprising that the biopolyester, with its limited nutritional content, is able to support hyphal growth. These invasions do not appear to progress to the richer nutritional reservoirs of the pectin layer, the pectocellulosic wall, or the cytoplasm below. The internal structures, although nutritionally more rewarding, might expose the exploratory hyphae to phytoalexins or polymer containment, whereas such responses of the living leaf are improbable in the inert external cuticle. The genus *Vizella* ramifies exclusively within the cuticle, forms ascocarps in dilations of the adaxial cuticle, and finally bursts through the cuticle to release the ascospores.

Beneath the Cuticle Many fungi do, however, take the next step of invading the tissues below. Typical of these pectinophyllic inhabitants are the fungal parasites *Diplocarpon rosae*, the causative agent of black spot of roses, and *Venturia inequalis*, the cause of apple scab (Deverall, 1981). Other pathogens seem to live a little deeper and are actually within the wall itself (Barckhaus and Weinert, 1974).

Through the Pectocellulosic Wall Once within the "phytotelm" (Gk., a plant pond) of the leaf, defined as an additional environment created by any part of a plant (Frank and Lounibos, 1983), the germinal tube or bacterium or virion enters a completely different habitat (Figures 2.10, 2.12, and 2.13; Pegg, 1985). Some fungi proceed no farther than the epidermal cells. Certain powdery mildews are apparently able to obtain all their nutrients from the colonization of epidermal cells. Wall-degrading enzymes capable

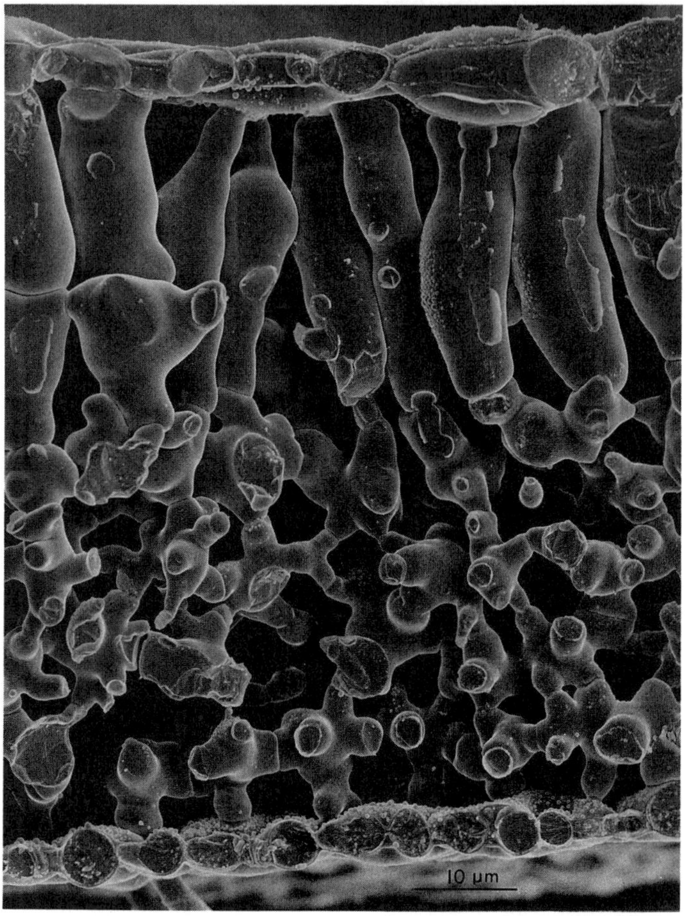

Figure 2.12 Cryo-SEM of a frozen and fractured whole leaf of *Phaseolus*. Courtesy of C. Jeffree.

of digesting external or sister cell pectocellulosic walls are virtually universal and the passage of a fungal hypha from one cell to another within a host must present no difficulty. Viruses readily pass through plasmodesmata, often modifying them as they go (Maule et al., 1988). The temperature within the leaf, as opposed to the outer surface will be the result of the solar gain on the exposed adaxial surface, which is a function of the albedo of the leaf, and the cooling effect of transpiration from the leaf's surfaces.

Inside The internal environment (Figure 2.12) will be: more humid, almost immune to ultraviolet radiation due to the impermeable nature of the polyester cuticle, and at a much lower light intensity, a damped temper-

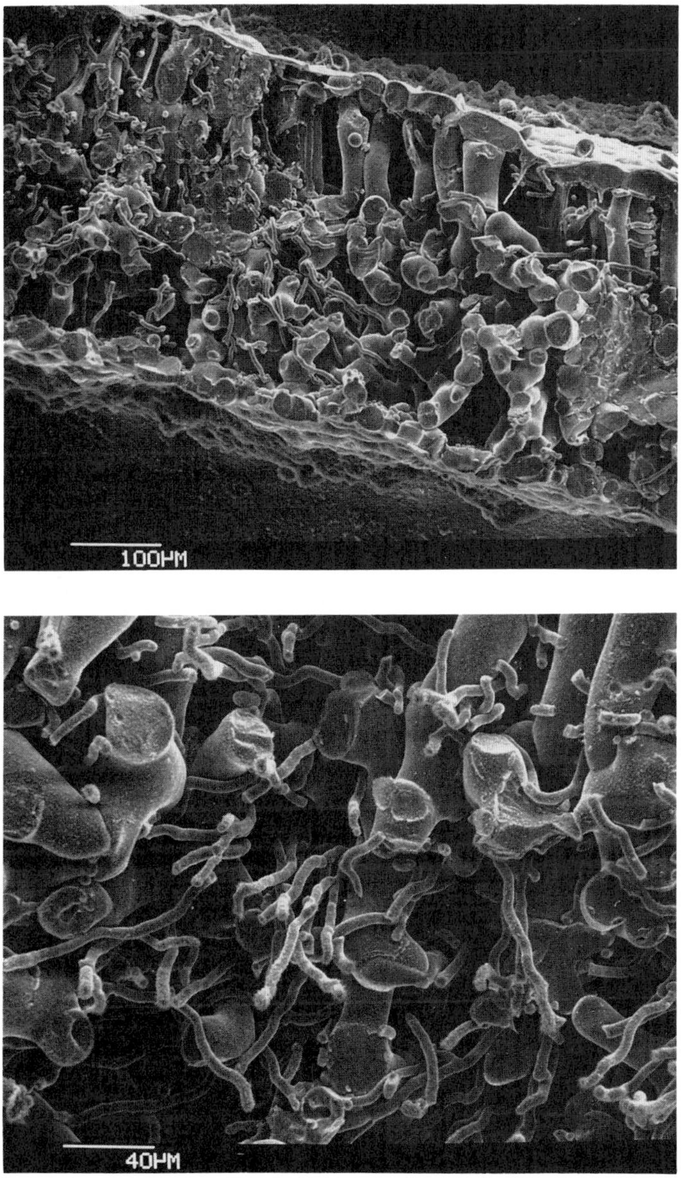

Figure 2.13 Cryo-SEM at two magnifications of a frozen and fractured whole leaf of *Vicia faba* showing a *Uromyces viciae-fabae* infection. Courtesy of C. Merriman.

ature range, and a much higher amplitude of CO_2 concentration than the external environment. The leaching of substances (see Section 2.4) has only been studied, for obvious reasons, from the adaxial and abaxial outer surfaces of leaves, and occasionally other plant parts. But we can safely assume that the internal wall surfaces, mostly without encrusting substances, will also contain significant quantities of "lost" sugars, amino acids, etc.

The spaces between the cells of the palisade and mesophyll (Figures 2.10, 2.12, and 2.13) may have arisen by cell lysis and/or cell separation in any given tissue (Jeffree et al., 1986). The distinction, although significant in ontogenetic terms, is probably academic to the passage of a fungal hypha between or into cells. In all cases, apparently unprotected, uncutinized, and unlignified pectocellulosic walls will be exposed. Chitinase and lysozymes may offer some defense, and other defense systems may also have a role (Templeton et al., 1990) Often the spaces between the cells are traversed by pectic strands (Carr and Carr, 1975) but these offer little or no barrier to hyphal progression.

Internal Surfaces The cutin of the cuticle, whether polyester or polyisoprene, will descend, while attenuating, into the substomatal chamber (Figure 2.10). It is noticeable that stomatal-penetrating fungi e.g., *Uromyces viciae-fabae* (figure 2 in Beckett and Woods, 1987) explore deep into the substomatal cavity, avoiding the cutinized layers, and only make initial contact on a chloroplast-containing mesophyll cell. Below the substomatal chamber, the exposed surfaces of the palisade and mesophyll cells (Figures 2.12 and 2.13) appear smooth, unstructured, and homogeneous in the scanning and transmission electron microscopes. All the crystalline deposits or apparent droplets on these surface seen in the scanning electron microscope are now thought to be artifacts of the preparative process (Jeffree et al., 1987). Such surfaces are likely to be a complex and somewhat variable copolymer of the mixed sugars of the hemicelluloses and pectins. Their only consistent feature will be a markedly hydrophilic surface. However, apparently, no detailed analyses of these surfaces have been attempted.

Within the Tissues of the Living Leaf The hyphal tube of a fungus may penetrate: 1) down a stoma (Beckett and Woods, 1987); 2) down the anticlinal walls between epidermal cells, taking advantage of the commonly thinner cuticle and thicker pectic layers at this junction; 3) directly into an epidermal cell; or 4) the product of an epidermal cell, e.g., a nectary or trichome (Dickinson and Lucas, 1982). In the first two instances, there will probably be no major reaction. In the latter two cases, the plant may respond in a number of ways, as the searching hyphal tip makes contact with an internal wall it would appear, immediately, to secrete at least two matrix materials (Beckett and Woods, 1987; Beckett et al., 1990). Some of this matrix material may be of host origin. These matrix materials may contain enzymes and, at this point of extension, appressorium-like structures may be formed leading to the penetration of the host cell and the destruction of its tissue (Beckett and Woods, 1987; Figure 2.13).

However attacked, and by whatever pathogen, four hundred million years of evolution have provided the leaf with enough physical and chemical defenses to make any general envious.

The author thanks Dr. Alan Beckett, Dr. Christopher Jeffree, and Dr. Sarah Watkinson for their help in writing this review, but all the views expressed are my responsibility.

References

Bailey, J.A. (editor). 1986. *Biology and Molecular Biology of Plant-Pathogen Interactions* NATO ASI Series, vol H1. Springer-Verlag, New York. 415 pp.

Baker, E.A. and Holloway, P.J. 1971. Scanning electron microscopy of waxes on plant surfaces. *Micron* 2:364–380.

Barber, H.N. and Jackson, W.D. 1957. Natural selection in action in *Encalyptus*. *Nature* 179:1267–1269.

Barckhaus, R. and Weinert, H. 1974. Die fleischfressende Pflanzen *Sarracenia purpurea*. Licht und elektronmikroskopische Untersuchungen. *Mikrokosmos* 63:38–47.

Barthlott, W. 1990. Scanning electron microscopy of the epidermal surface in plants. pp. 69–94 in Claugher, D. (editor), *Scanning Electron Microscopy in Taxonomy and Functional Morphology*, special vol. 41. The Systematics Association, Clarendon Press, Oxford.

Barthlott, W. and Schultze-Motel, W. 1981 Zur Feinstruktur der Blattoberflachen und Systematischen Stellung der Laubmoosgattung *Rhacocarpus* und andere Hedwigiaceae. *Willdenowia* 11:3–11.

Beauford, W., Barber, J. and Barringer, A.R. 1975. Heavy metal release from plants into the atmosphere. *Nature* 256:35–37.

Beckett, A. and Porter, R. 1988. The use of complementary fractures and low-temperature scanning electron microscopy to study hyphal-host cell surface adhesion between *Uromyces viciae-fabae* and *Vicia faba*. *Canadian Journal of Botany* 66:645–651.

Beckett, A. and Woods, A.M. 1987. The pattern of colony development and the formation of the uredinium of *Uromyces viciae-fabae* on *Vicia faba*. *Canadian Journal of Botany* 65:1998–2006.

Beckett, A., Tatnell, J.A. and Taylor, N. 1990. Adhesion and pre-invasion behaviour of urediniospores of *Uromyces viciae-fabae* during germination on host and synthetic surfaces. *Mycological Research*. 94:865–875.

Blakeman, J. 1971. The chemical environment of the leaf surface in relation to growth of pathogenic fungi. pp. 255–268 in Preece, T.F. and Dickinson, C.H. (editors), *Ecology of Leaf Surface Microorganisms*. Academic Press, London.

Blakeman, J.P. and Atkinson, P. 1976. Evidence for a spore germination inhibitor co-extracted with wax from leaves. pp. 441–449 in Dickinson, C.H. and Preece, T.F. (editors), *Microbiology of Aerial Plant Surfaces*. Academic Press, London.

Burrage, S.W. 1970. The micro-climate at the leaf surface. pp. 91–101 in Preece, T.F. and Dickinson, C.H. (editors), *Ecology of Leaf Surface Microorganisms*. Academic Press, London.

Burrage, S.W. 1976. Aerial microclimate around plant surfaces. pp. 173–184 in Dickinson, C.H. and Preece, T.F. (editors), *Microbiology of Aerial Plant Surfaces*. Academic Press, London.
Carr, D.J. and Carr, S.G.M. 1987. *Eucalyptus 11. The Rubber Cuticle, and other Studies of the Corymbosae*. Phytoglyph Press, Canberra, Australia. 372 pp.
Carr, D.J., Carr, S.G.M. and Lenz, J.R. 1985. Oriented arrays of epicuticular wax plates in *Eucalyptus* species. *Protoplasma* 124:205–212.
Carr, S.G.M. and Carr, D.J. 1975. Intercellular pectic strands in parenchyma: studies on plant cell walls by scanning electron microscopy. *Australian Journal of Botany* 23:95–105.
Cervone, F., Hahn, M.G. DeLorenzo, G., Darvill, A. and Albersheim, P. 1989. Host-pathogen interactions. Plant Physiology 90:542–548.
Cutler, D.L., Alvin, K.L. and Price, C.E. 1982. *The Plant Cuticle*. Linnean Society Symposium Series no. 10. Academic Press, London. pp. 461.
Deverall, B.J. 1981. *Fungal Parasitism*, 2nd ed. Edward Arnold, London. 66 pp.
Dickinson, C.H. and Lucas, J.A. 1982. *Plant Pathology and Plant Pathogens*, 2nd ed. Blackwell Scientific Publication, Oxford 229 pp.
Dickman, M.B., Podila, G.K. and Kolattukudy, P.E. 1989. Insertion of cutinase gene into a wound pathogen enables it to infect intact host. *Nature* 342:446–448.
Elleman, C.J. and Entwistle, P.F. 1985. Inactivation of a nuclear polyhedrosis virus on cotton by the substances produced by the cotton leaf surface glands. *Annals of Applied Biology* 106:83–92.
Emmett, R.W. and Parberry, D.G. 1975. Appressoria. *Annual Review of Phytopathology* 13:147–167.
Fahn, A. 1979. *Secretory Tissues in Plants*. Academic Press, London. 302 pp.
Frank, J.H. and Lounibos, J.P. (editors). 1983. *'Phytotelmata': Terrestrial Plants as Hosts for Aquatic Insect Communities*. Plexus, NJ. 322 pp.
Goodman, R.N., Kiraly, Z. and Wood, K.R. 1986. *The Biochemistry & Physiology of Plant Disease*. University of Missouri Press, Columbia, MO. 433 pp.
Grncarevic, M. and Radler, F. 1967. The effect of wax components on cuticular transpiration-model experiments. *Planta* 75:23–27.
Grubb, P.J. 1986. Sclerophylls, pachyphylls and pycnophylls: the nature and significance of hard leaf surfaces. pp. 137–150 in Juniper, B.E. and Southwood, T.R.S. (editors), *Insects and the Plant Surface*. Edward Arnold, London.
Hallam, N.D. and Juniper, B.E. 1970. The anatomy of the leaf surface. pp. 3–37 in Preece, T.F. and Dickinson, C.H. (editors), *Ecology of Leaf Surface Microorganisms*. Academic Press, London.
Hamer, N.D., Howard, R.J., Chumley, F.G. and Valent, B. 1988. A mechanism for surface attachment in spores of a plant pathogenic fungus. *Science* 239: 288–290.
Hargreaves, J.A., Brown, G.A. and Holloway, P.J. 1982. The structural and chemical characteristics of the leaf surface of *Lupinus albus* L. in relation to the distribution of antifungal compounds. pp. 331–340. in Cutler, D.F., Alvin, K.L. and Price, C.E. (editors), *The Plant Cuticle*. Linnean Society Symposium no. 10. Academic Press, London.
Hirano, S.S. and Upper, C.D. 1983. Ecology and epidemiology of foliar bacterial pathogens. *Annual Review of Phytopathology* 21:243–269.
Holloway, P.J. 1970. The chemical and physical characteristics of leaf surfaces. pp. 39–53 in Preece, T.F. and Dickinson, C.H. (editors), *Ecology of Leaf Surface Microorganisms*. Academic Press, London.
Holloway, P.J. 1982. Structure and histochemistry of plant cuticular membranes: an overview. pp. 1–32 in Cutler, D.F., Alvin, K.L. and Price, C.E. (editors), *The Plant Cuticle*. Linnean Society Symposium Series no. 10. Academic Press, London.

Hull, H.M., Went, F.M. and Bleckmann, C.A. 1979. Environmental modification of epicuticular wax structure of *Prosopis* leaves. *Journal of Arizona-Nevada Academy of Science* 14:39–42.

Jeffree, C.E. 1986. The cuticle, epicuticular waxes and trichomes of plants, with reference to their structure, functions and evolution. pp. 23–64 in Juniper, B.E. and Southwood, T.R.S. (editors), *Insects and the Plant Surface*. Edward Arnold, London.

Jeffree, C.E. and Sandford, A.P. 1982. Crystalline structure of plant epicuticular waxes demonstrated by cryostage scanning electron microscopy. *New Phytologist* 91:549–559.

Jeffree, C.E., Dale, J.E. and Fry, S.C. 1986. The genesis of intercellular spaces in developing leaves of *Phaseolus vulgaris* L. *Protoplasma* 132:90–98.

Jeffree, C.E., Read N.D., Smith. J A.C. and Dale, J.E. 1987. Water droplets and ice deposits in leaf intercellular spaces: redistribution of water during cryofixation for scanning electron microscopy. *Planta* 172:20–37.

Jones, H.G. 1983. *Plants and Microclimate*. Cambridge University Press, Cambridge. 323 pp.

Juniper, B.E. and Jeffree, C.E. 1983. *Plant Surfaces*. Edward Arnold, London. 93 pp.

Juniper, B.E. and Southwood, T.R.E. (editors). 1986. *Insects and the Plant Surface*. Edward Arnold, London. 360 pp.

Juniper, B.E., Robins, R.J. and Joel, D.M. 1989. *The Carnivorous Plants*. Academic Press, London. 353 pp.

Kerr, J.P. and Beardsell, M.F. 1975. Effect of dew on leaf water potentials and crop resistances in *Paspalum* pasture. *Agronomy Journal* 67:596–599.

Koshino, H., Togiya, S., Yoshihara, T., Sakamura, S., Shimanuki, T., Sato, T. and Tajimi, A. 1987. Four fungitoxic C-18 hydroxy unsaturated fatty acids from stomata of *Epichloe typhina*. *Tetrahedron Letters* 28:73–76.

Lindow, S.E. 1983. The role of bacterial ice nucleation in frost injury to plants. *Annual Review of Phytopathology* 21:363–384.

Lippincott, B.B., Whatley, M.H. and Lippincott, J.A. 1977. Tumor induction by *Agrobacterium* involves attachment of the bacterium to a site on the host plant cell wall. *Plant Physiology* 59:388–390.

Lovett, J.V. and Sagar, G.R. 1978. Influence of bacteria in the phyllosphere of *Camelina sativa* (L.) *Crantz* on germination of *Linum usitatissimum* L. *New Phytologist* 81:617–625.

Martin, J.T. 1964. The role of the cuticle in the defense against plant disease. *Annual Review of Phytopathology* 2:81–102.

Martin, J.T. and Juniper, B.E. 1970. *The Cuticles of Plants*. Edward Arnold, London. 347 pp.

Maule, A.J., Wilson, I.G. Harker, C.L. and Linstead, P.J. 1988. Caulimoviruses: the multiplication and spread of cauliflower mosaic virus in infected plants. pp. 57–61 in *John Innes Institute Annual Report*, Norwich, East Anglia, England.

Meidner, H. and Mansfield, T.A. 1968. *Physiology of Stomata*. McGraw-Hill, London. pp. 179.

Monteith, J.L. 1973. *Principles of Environmental Physics*. Edward Arnold Contemporary Biology Series no. 18. Edward Arnold, London. 241 pp.

Moorby, J. and Squire, H.M. 1963. The loss of radioactive isotopes from the leaves of plants in dry condition. *Radiation Botany* 3:163–167.

Parry, D.W. 1990. *Plant Pathology in Agiculture*. Cambridge University Press, Cambridge. pp. 361.

Pegg, G.F. 1985. Life in a black hole—the microenvironment of the vascular pathogen. *Transactions of the British Mycological Society* 85:1–20.

Peries, O.S. 1962. Studies on strawberry mildew caused by *Sphaerotheca macularis*

11 Host parasite relationships on foliage of strawberry varieties. *Annals of Applied Biology* 50:225–233.
Purnell, T.J. 1971. Effects of pre-inoculation washing of leaves with water on subsequent infections by *Erysiphe cruciferarum*. pp. 269–275 in Preece, T.F. and Dickinson, C.H. (editors), *Ecology of Leaf Surface Microorganisms*. Academic Press, London.
Purnell, T.J. and Preece, T.F. 1971. Effects of foliar washing on infection of leaves of swede (*Brassica napus*) by *Erysiphe cruciferarum*. *Physiological Plant Pathology* 1:123–132.
Royle, D.J. and Kremheller, H.T. 1981. Downy mildew of the hop. pp. 395–419 in Spencer, D.M. (editor), *The Downy Mildews*. Academic Press, London.
Ruinen, J. 1975. Nitrogen fixation by free-living microorganisms. pp. 85–100 in Stewart, W.D.P. (editor), *International Biological Programme*, no. 6, Cambridge University Press, Cambridge.
Sargent, J.A. 1981. The fine structure of the downy mildews. pp. 183–236 in Spencer, D.M. (editor), *The Downy Mildews*. Academic Press, London.
Silva-Fernandes, A.M.S. 1965. Studies on plant cuticle. VIII Surface waxes in relation to water-repellency. *Annals of Applied Biology* 56:297–304.
Silvester, W.B. 1976. Symbiotic nitrogen fixation in plants. pp. 521–538 in Nutman, P.S. (editor), *International Biological Programme*, no. 7. Cambridge University Press, Cambridge.
Steinmuller, D. and Tevini, M. 1985. Action of ultraviolet radiation (UV-B) upon cuticular waxes in some crop plants. *Planta* 164:557–564.
Swart, H.J. 1971. Australian leaf-inhabiting fungi 1. Two species of *Vizella*. *Transactions of the British Mycological Society* 57:455–464.
Templeton, M.D., Dixon, R.A., Lamb, C.J. and Lawton, M.A. 1990. Hydroxyproline-rich glycoprotein transcripts exhibit different spatial patterns of accumulation in compatible and incompatible interactions between *Phaseolus vulgaris* and *Colletotrichum lindemuthianum*. *Plant Physiology* 94:1265–1269.
Thompson, F.B. 1969. The effect of rain on plants, pests and pesticides. pp. 1495–1498 in *Physicochemical and Biophysical Panel of the Society of Chemical Industry*. Symposium 18th, October 1969. Society of Chemical Industry, London.
Tukey, H.B. 1970. Leaching of substances from plants. pp. 67–80 in Preece, T.F. and Dickinson, C.H. (editors), *Ecology of Leaf Surface Microorganisms*. Academic Press, London.
Watkinson, S. 1984. Inhibition of growth and development of *Serpula lacrimans* by the non-metabolised amino-acid analogue α-aminoisobutyric acid. *FEMS Microbiology Letters* 24:247–250.
Webster, J. 1986. *Introduction to the Fungi*, 2nd ed. Cambridge University Press, Cambridge. 669 pp.
Went, F. 1974. Reflections and speculations. *Annual Review of Plant Physiology* 25:1–26.
Whittaker, J.B. 1984. Responses of Sycamore (*Acer pseudoplatanus*) leaves to damage by a typhlocybine leaf hopper, *Ossiannilssonola callosa*. *Journal of Ecology* 72:455–462.
Wikinson, H.P. 1979. The plant surface Part V. The cuticle. pp. 140–156 in Metcalfe, C.R. and Chalk, L. (editors), *The Anatomy of the Dicotyledons*, Clarendon Press, Oxford.
Wullschleger, S.D. and Oosterhuis, D.M. 1987. Electron microscope study of cuticular abrasion on cotton leaves in relation to water potential measurements. *Journal of Experimental Botany* 38:660–667.

3

Aerobiology: The Atmosphere as a Source and Sink for Microbes

David E. Pedgley

3.1 Introduction

It is well known that microbes can be carried on the wind from one leaf to another, whether those leaves are centimeters or hundreds of kilometers apart. Each step in the dispersal sequence, take-off—transport—deposition, is influenced by atmospheric behavior. For leaves where take-off is occurring, the atmosphere is a sink; conversely, where landing is occurring, it is a source. In this chapter we will review our understanding of the influences of atmospheric behavior on *passive* dispersal of *windborne* microbes. Neither dispersal involving vectors such as insects and humans, nor *active* dispersal such as the discharge of fungal spores into the air from fruiting bodies is considered. Most research has been concerned with microbes that cause disease in crop plants but many of the results should be applicable to other organisms, such as pollen and yeasts.

The aim of this chapter is to help answer the fundamental question: what is the probability of windborne microbes reaching point A from a source at B? This question is asked by many: by those wishing to grow disease-free crops isolated from sources of infection, or seed crops far from sources of unwanted pollen; by those providing warnings of pathogen spread to growers; by those wishing to apply pesticides at the most appropriate time; by those wishing to reduce within-crop spread; by those assessing the value of quarantine measures; by those researching sudden outbreaks of disease; and by those releasing new genetically engineered microbes.

Understanding the windborne spread of pathogenic microbes is particularly important if new strains are involved or if persistence at a given place is unlikely, due to cold, drought or lack of hosts. Two influences of atmospheric behavior on microbe dispersal will be considered, wind and rain.

3.2 Wind

We know from experience that wind speed and direction at a given place are continually changing and over a wide range of time scales. Wind changes are caused by the passage of an unceasing sequence of atmospheric eddies. Like a river, the atmosphere is full of eddies, small and large, weak and strong, brief and persistent, each evolving and interacting with its neighbors. Meteorologists classify these eddies by their sizes: micro-scale ones are smaller than a few kilometers across, and they include turbulence; macro-scale ones are larger than a few hundreds of kilometers, and they include cyclones and anticyclones familiar from weather maps; meso-scale ones are those inbetween, and include sea breezes, thundersqualls, and flow around mountains (Orlanski, 1975; Ray, 1986). The structure and behavior of individual eddies are studied with networks of measuring stations making simultaneous observations at set times, from which the evolution of the three-dimensional structure can be described with more or less precision. All countries have routine networks, but the spacing is usually adequate for monitoring only the macro-scale eddies. Those on a meso-scale are studied by means of specialized networks and by aircraft, with the assistance of radar and satellites. Micro-scale eddies are more difficult to study because they are small and fleeting; their combined effects are often treated stochastically, as fluctuations from a mean for a period much longer than the lifetime of an individual eddy. Some micro-scale eddies can be studied by radar or other remote-sensing techniques, or by rapid-response instruments in a dense network or on towers. Micro-scale eddies affect the take-off and landing of microbes, as well as their dispersion once they are airborne, whereas meso- and macro-scale eddies affect transport over distances from tens to thousands of kilometers.

Turbulence Wind turbulence is experienced as gusts and lulls, each lasting no more than a few tens of seconds. Gusts are a mechanism for releasing microbes from leaf surfaces through fluttering and rubbing (Aylor, 1978), or even directly by sweeping the surfaces (Aylor and Parlange, 1975). Wind speeds averaged over one second are typically twice the speeds averaged over one minute; hence gusts strong enough to release microbes may be recorded only by sufficiently sensitive instruments. Within a few millimeters of a surface—in the laminar boundary layer—turbulence is suppressed by the air's viscosity, or by internal friction due to molecular diffusion. Flow in this layer is weak and smooth and is unable to dislodge microbes, but it is occasionally affected by gusts (Aylor and Parlange, 1975).

Turbulence is due to both convection and vertical wind shear. It modifies the mean structure of the lower atmosphere—the planetary boundary layer. Over land, daytime atmospheric convection results from the sun heating the ground, and this is by far the major cause of turbulence. Daytime convection in dry weather is in the form of columns of buoyant, warm air rising to 1–2 km, increasing to a maximum in the afternoon (and even reaching 5 km in arid lands during the hot time of year). Mixing of these columns with their surroundings leads to a decrease of temperature with height in the convective boundary layer of about 10°C per kilometer. Cloudy convection can sometimes penetrate to much greater heights, even 10–15 km; microbes may then be taken high into the atmosphere as the tall clouds evaporate by mixing with their dry surroundings, or they may fall out in rain.

Nighttime turbulence is usually much weaker; it results from wind shear, when the wind speed and direction change with height due principally to surface friction as the wind blows past vegetation and other elements contributing to the roughness of the ground. Some of the wind's kinetic energy is then converted to heat, and the rest passes into smaller eddies that are shed downwind. The nighttime turbulent layer is much shallower—often less than a few hundred meters—and its structure is more complex than by day, being modified by gravity waves, air drainage, and barrier effects of topography. Local convection can still develop at night over warm water, cities, and power plants. Ray (1986) provides a comprehensive account of these meso-scale wind systems in the lower atmosphere, with sections on observing, forecasting, and modeling techniques.

Both take-off from and deposition on vegetation are enhanced by strong gusts, when vertical components of the wind are much greater than the settling speeds of microbes (Aylor et al., 1981). A plant canopy, whether field crop, orchard, or forest, creates numerous, small, turbulent eddies as wakes in the lee of individual elements such as leaves and stems. It can also wave in the wind, intermittently storing kinetic energy from the wind before releasing it to the eddies (Raupach and Thom, 1981). Turbulence inside a canopy is due to both shear and wake production; it therefore increases with canopy density and structural complexity and is characterized by occasional eddies producing strong gusts separated by longer quiet spells (Shaw et al., 1979). The horizontal wind component is therefore positively skewed (Legg, 1983; Baldocchi and Meyers, 1988), whereas the vertical wind component is negatively skewed, with the strongest gusts being downward from above the canopy top (Gao et al., 1989; Amiro and Davis, 1988; Baldocchi and Meyers, 1988), and there is an absence of strong upcurrents from the canopy floor.

Strong gusts can release microbes even when the average wind speed is too weak to do so: see, for example, Hammett and Manners (1974). Gusts are particularly intense in the squalls accompanying showers. Detailed study of the structure of gusts penetrating down into a forest has shown that they are usually preceded by weak upcurrents (Gao et al., 1989). Airborne microbes released by earlier gusts that have been slowed by canopy drag can presumably be carried above the canopy top in this way, and the proportion that escapes increases with the gust speed (Waggoner, 1983). It is the take-off of fungal spores that has been most studied, but bacterial cells can become airborne in the same way (Lindemann and Upper, 1985; Lindemann et al., 1982). Of course, plant canopies are not uniform; their structures vary both vertically (e.g., crown space and understory) and horizontally (e.g., forest clearings and crop rows), leading to complex vertical profiles of mean wind speed and turbulence, as well as to the formation of intermittent roll eddies lasting up to a few tens of seconds as gusts penetrate downwards into clearings and between rows (Arkin and Perrier, 1974; Bergen, 1975; Weiss and Allen, 1976). For a general introduction to wind behavior within plant canopies, see Shaw (1982); and for microbe movement within canopies, see Legg (1983).

It is well known that dry deposition is by a combination of sedimentation and impaction (Aylor, 1975; McCartney and Aylor, 1987). Turbulence assists deposition by bringing windborne microbes down from the planetary boundary layer into the plant canopy, where sedimentation dominates during quiet spells but impaction dominates during the strong gusts, when microbes deviate most from the streamlines around obstacles (Legg, 1983). Allowance must also be made for the structure and behavior of the plants themselves. Microbes can be expected to impact only on the finer plant parts such as hairs, spines, and upwind edges, judged by the behavior in the laboratory of particles identifiable by fluorescence (Hirst and Stedman, 1971) or radioactivity (Little, 1977), but retention depends on surface smoothness of both microbe and plant (Forster, 1977; Ferrandino and Aylor, 1985). Impaction efficiency is greater within the upper canopy due to vertical variations in structure of both canopy and turbulence (McCartney and Aylor, 1987). Moreover, alignment of the vegetation may change with wind speed, which therefore modifies impaction. For a general introduction to particle capture by vegetation, see Chamberlain and Little (1981).

Microbes carried upwards by turbulence may be taken downwind in a plume. Very little is known about the structure and behavior of microbe plumes because of the great difficulty of obtaining field observations. Microbe plumes are probably similar to those of smoke or of invisible atmospheric pollutants, which have been widely studied (Pasquill and Smith, 1983). However, microbe plumes are not buoyant like those of some pollutants from hot sources such as fires, smoke stacks, and power plant cooling towers. Two characteristics of a plume that are important when considering downwind effects on deposition are spreading and meandering.

A plume spreads vertically and horizontally by turbulence. Daytime convective mixing leads to upward bulges in a plume near its source, resembling meandering in the vertical plane, but eventually mixing will spread it vertically, although not necessarily uniformly, through the whole depth of the planetary boundary layer. Convective turbulence can be expected to bring some of the microbes back to near the ground, therefore enhancing deposition, but other microbes in the convective layer may need several hours to reach the ground by sedimentation once convection has died away, typically late in the afternoon. With a settling speed of about 1 cm/sec, microbes will fall 360 m in 10 hours. Because nighttime turbulence is shallower and weaker, microbes released at night presumably remain close to the ground.

Turbulence, generated during the day or night, also causes spreading of a plume horizontally. The combined effect of innumerable small eddies is to produce an average plume concentration that decreases downwind (Smith, 1983). Despite great dilution far downwind, deposition of a few microbes may be enough to start disease. As plume width and depth increase, larger eddies begin to influence spread. On meeting a meso-scale eddy, a plume can be deflected or severely disrupted. For example, winds among mountains at night tend to blow along the valleys; there, a night-time plume would resemble the course of a mountain river. Again, a plume meeting the leading edge of sea breeze or storm outflow would be taken aloft bodily and become involved in complex circulations in the vertical plane. Plumes are also patchy, due not only to the vagaries of the mixing process but also to fluctuations in source strength (Sawford, 1987).

Wind Trajectories A plume meanders laterally under the influence of eddies that are larger than the plume width. Meandering is made visible in smoke near its source, but hundreds of kilometers downwind it can be detected only with the aid of added tracers and extensive field measurements by aircraft. Often some of the meso-scale eddies present may be unknown, thus adding uncertainty to displacements calculated from wind trajectories.

There are several ways of calculating wind trajectories, based largely on observations from the routine macro-scale network (e.g., Heffter, 1980, 1983; Scott and Achtemeier, 1987). They have been used in studies of spread of atmospheric pollution and of migrant insects. Their accuracy is not easy to assess, but attempts have been made from comparisons with experiments using chemical tracers and constant-level balloons (Pack et al., 1978; Clarke et al., 1983; Reisinger and Mueller, 1983; Stocker et al., 1990; Chock and Kuo, 1990). When calculating trajectories of clouds of microbes, however, further difficulties arise from uncertainties in height and timing of displacements. Because of the vertical shear of wind speed and direction, especially at night in the planetary boundary layer, trajectories can vary markedly with vertical separations of less than one kilometer. Day-released microbes can perhaps be assumed to be confined to the planetary boundary layer, unless there is good evidence to the contrary, such as the presence of tall convection clouds. Two or more representative heights within that layer might then be chosen for the calculations.

Trajectories have been used to provide circumstantial evidence for long-distance spread of plant pathogens. The earliest example is the northward spread of wheat stem rust across North America in the 1920s and 1930s (Stakman and Christensen, 1946); and it has been deduced that a similar spread occurred across India from sources in the southern hills to deposition in rain over the northern plains (Nagarajan and Singh, 1976; Nagarajan et al., 1976; Nagarajan and Joshi, 1980, 1985). Another example is the northward spread of tobacco blue mold from the Caribbean to the eastern USA (Davis and Main, 1984, 1989; Davis et al., 1985; Main and Davis, 1989). There is circumstantial evidence that many serious fungal pathogens are spread by long-distance transport of spores. The evidence comes from biogeographical mapping of spread, from the random distribution of disease foci in a crop canopy, and from the trapping of microbes in air far from sources. For a convincing case to be made, all of the following information is needed:

1. timing of deposition at disease site derived either directly from observations of spore arrival (by rain sampling or trapping) or indirectly from positions of the uppermost of the oldest pustules, which will give an indication of the time elapsed since deposition if it is assumed that they were due to spores landing on the youngest growth;
2. existence of calculated backward trajectories from disease site to a known source of inoculum;
3. evidence that the observed spores came from that source, e.g., by using a natural marker such as a distinctive strain; and
4. evidence that sufficient numbers of spores could survive aerial transport to induce disease.

One example for which all such information was available is the spread of live spores of cereal mildews and rusts for some 500 km across the North Sea from England to Denmark (Hermansen et al., 1978) using: 1) pots of seedlings to trap spores on arrival; 2) backward trajectories (up to one and a half days) at 1000 mbar (near the earth's surface) and 850 mbar (about 1.5 km above sea level); 3) knowledge that England was the dominant source, because crop varieties with comparable resistance had been little cultivated in Europe outside Denmark and England; 4) knowledge that disease was in fact induced.

Another example, based on circumstantial but convincing evidence, is the southwestward spread of coffee rust in South America, from its first detection in 1970 in the state of Bahia, Brazil, to the states of São Paulo and Paraná in 1971, and then to Paraguay and northern Argentina by 1974 (Waller, 1979). This spread was consistent with dominant wind directions and with later trapping of spores by aircraft over Brazilian plantations. In contrast, usually little of this information is available, leading to much speculation on the possibility of transoceanic and intercontinental displacements. Such has happened with, for example, southern corn leaf blight (Moore, 1970), Sigatoka disease of bananas (Stover, 1962), sugarcane rust (Purdy, 1985), crown rust of oats (Dinoor, 1981), leaf rust of wheat (Subba Rao et al., 1989), and stem rust of wheat in the southern hemisphere (McEwan, 1966; Watson and de Sousa, 1983).

Anticyclonic weather, when winds are often weak and the planetary boundary layer may have a sharply defined top marked by a temperature inversion, can be expected to favor the accumulation of airborne microbes over several days, in much the same way as with air pollution. In such weather, the large diurnal variations of temperature and relative humidity will enhance spore development and release (Gottwald, 1983; Lacy and Pontius, 1983), but the associated prolonged sunshine may reduce survival. As the microbe-laden planetary boundary layer leaves the anticyclone, it will be lifted by gentle but persistent and widespread updrafts associated with neighboring cyclones, especially in poleward-flowing air ahead of cold fronts, where the resulting cloudiness can be expected to enhance survival. Such movements provide opportunities for the annual reintroduction of pathogens that are unable to survive temperate-latitude winters. Because of the accompanying variation of wind direction with height, microbe-free air can flow underneath, leading to the formation of layers of microbe-laden air aloft.

In tropical latitudes, some microbes can survive the dry season on alternative hosts or as drought-resistant spores, but others cannot. Those with windborne spores reinvade annually as the zone of seasonal rains and plant growth oscillates across the equator. This rainy zone coincides with convergence of the trade winds from opposite hemispheres at the Inter-Tropical Convergence Zone (ITCZ), which lies in the summer hemisphere. Windborne spores that are released in the rainy zone will tend to be confined there by the convergent winds, so that during the seasons when the ITCZ is changing latitude (in spring and fall), spores can be carried from places where rains (and plant growth) are ending to others where they are beginning (Waller, 1976). The position of the ITCZ is affected by the distribution of continents and oceans. In particular, if it shifts to more than about 10° latitude from the equator, the trade wind blowing from the winter hemisphere turns into a westerly monsoon after crossing the equator. Also, on any one day there will be numerous meso- and macro-scale systems modifying the wind field and distribution of rain.

3.3 Rain

Falling raindrops release microbes into the air by the well-known mechanisms of tap and puff (Hirst and Stedman, 1963) as well as splash. The last is the result not only of rain but also of overhead irrigation and fungicide spraying, and of drip from trickling rain, dew, and perhaps cloud and fog water filtered out by plants. In the past 10 years or so there has been a revival of interest in rain splash; most work has been done with fungal spores but bacteria are also involved (Quinn et al., 1980; Venette, 1982).

Splash The splash mechanism has been examined by Stow and Stainer (1977) and by MacDonald and McCartney (1988). The numbers and sizes of resulting droplets are affected by drop size and speed, and by target roughness, wetness, curvature, and inclination. A drop with radius less than about 1.0 mm gives no splash droplets, but a larger one spreads over a dry surface to form a circular wall that breaks into jets, leading to the characteristic crown shape. The jets break down into hundreds of small droplets, most with radii of less than 50 μm, the number increasing with drop kinetic energy and with greater surface roughness. When there is a water film, the number of droplets decreases with increasing film thickness up to about 0.5 mm. Spread by splash is greater if the microbes lie in the water film on plant surfaces rather than in the falling drop (Fitt and Lysandrou, 1984).

Some of the rain falling on a plant canopy reaches the soil directly, particularly the large drops falling most nearly vertically (and others when winds are light), and some is intercepted by the plants, where it splashes from leaf surfaces, trickles over them (picking up microbes), and drips from the margins or runs down the stems (Quinn and Laflen, 1983). The proportions vary with plant type and cover, and hence with age; canopy configuration may also change as the rain water accumulates. Hence the drop spectrum and kinetic energy at soil level are altered by the presence of the canopy. Splash droplets are small and have little kinetic energy but drip drops are large (even 6–7 mm in diameter) and, although they may not have fallen far enough to reach their settling speeds, their kinetic energy can be comparable with that of the smaller, faster drops from which they formed. Indeed, in a forest, the kinetic energy of drip onto the floor may be greater than that of the small- to medium-sized drops raining onto the forest canopy because the kinetic energy of a drop increases with the cube of its radius and with the square of its falling speed, which itself increases with radius.

Splash is particularly important for short-distance spread within a crop (Fitt et al., 1989), where it is greatest in the upper part of the canopy (Griffiths and Ao, 1976). The larger droplets (radius greater than about 0.1 mm) carry most of the microbes (Fitt and Bainbridge, 1983; Drinkall and Price, 1983; Brennan et al., 1985a, 1985b). Because they have greater settling speeds and are airborne only briefly, they are little affected by turbulence, and therefore they behave more or less ballistically (Walklate, 1989). These droplets are formed mostly by splashing of the larger raindrops, but no relationship has been found between splash intensity and rainfall intensity. This is not surprising because there is no simple relationship between drop size spectrum and rainfall intensity. An exponential size distribution, with the smallest drops most frequent and the largest (diameter about 5 mm) limited by aerodynamic break-up, which might be expected from simple growth of drops by accretion of droplets within rain clouds, is modified by various processes before the drops can reach the ground (Ulbrich, 1983). Gravity sorting, evaporation, and coalescence all selectively remove the smallest drops, whereas break-up (whether aerodynamically or following collision) removes the largest.

Much rainfall is derived from snow or hail melting as it falls from higher and colder levels in the atmosphere, leading to drop spectra different from those involving an all-water process. When snow flakes fall through the melting layer, there is enhanced coalescence because of the differences in settling speeds of flakes and drops. In a given rainfall, the largest drops present may not be the largest possible if they have had insufficient time to grow, because they have fallen from a cloud that is shallow or that has weak updrafts and a small water content. Hence splash effects will vary with rainfall origin, even with the same rainfall rate.

Droplets splashed from rigid or plant surfaces rise to 40–70 cm (Griffiths and Ao, 1976; Kuske and Benson, 1983; Fitt and Lysandrou, 1984; Brennan et al., 1985a; MacDonald and McCartney, 1988; Walklate et al., 1989), as is illustrated by the height of mud stain at the base of walls. Those rising furthest have diameters 0.2–0.6 mm and are best produced by large drops (Shaw, 1987). Few droplets splash horizontally more than 50 cm in still air. Because leaves can be deflected by raindrop impact, not all of a drop's kinetic energy need be imparted to the droplets, and the splash intensity is reduced. Smaller droplets, although carrying fewer microbes, can be carried by turbulence to the upper part of a tall canopy, as seems to happen, for example, in cocoa plantations (Gregory et al., 1984). They can also be taken tens or hundreds of meters horizontally on the wind, providing a significant dispersal mechanism within field and plantation crops, for fungal spores (Bertrand and English, 1976; Brennan et al., 1985b) and bacteria (Venette, 1982; Harrison and Brewer, 1982). Some droplets have been found to remain airborne for up to an hour (Graham and Harrison, 1975; Carnegie, 1980) with the obvious potential for downwind drift over many kilometers, but the droplets may then evaporate and any microbes in them may not survive, particularly in the sunshine between daytime showers. For a review of the role of rain in the dispersal of pathogens, see Fitt et al. (1989).

Wash-out and Rain-out Falling raindrops not only release microbes, they also remove them from the air, by wet deposition in spore showers. Two mechanisms are involved: wash-out and rain-out. The former is a scrubbing process where a microbe-laden wind blows under a rain cloud; the latter involves microbes in the condensation of water vapor as air rises and cools to form rain clouds. Both processes can occur together. Both water vapor and microbes enter the atmosphere from the earth's surface, and because the average residence time of water vapor in the atmosphere is about one week (based on global, average, annual rainfall rate and global, average water vapor content of the atmosphere), presumably the average residence time for those microbes that have been mixed into the planetary boundary layer is also about one week because dry deposition is slow. However, residence time will be reduced where wash-out hastens deposition. Of course, there will be large geographical variations in residence time: in regions with frequent rains it may be no more than one to two days, whereas in arid regions it may be as much as several weeks.

The efficiency of capture of microbes of given size increases with smaller raindrops and larger settling speeds. Because settling speed is less for smaller drops, capture efficiency has been estimated to reach a maximum, for microbes of radii 2–6 μm, with drops of radius about 0.5 mm, typical of frontal rains rather than showers and drizzle (Starr and Mason, 1966). Using these efficiencies, and typical raindrop spectra, it can be calculated (Starr, 1967) that half the microbes with radius 2 μm would be washed out in 4 hr with rain of intensity 1 mm hr^{-1}, and in 1 hr at 5 mm hr^{-1}. Corresponding times for radius 6 μm are 1 hr and 0.25 hr. Since microbe clouds can take 10 hr in passing through frontal rain, it is clear that the great majority will be washed out. The pattern of deposition from a given rainstorm, whether meso-scale or macro-scale, is likely to be complex, depending on the varying distributions of rainfall rate and drop spectra as well as microbe plume structure. There seems to have been little attempt to measure, let alone understand, such patterns of deposition.

3.4 Modeling Dispersal

To attempt an answer to our original question we need to quantify our description of the dispersal stages. Within a canopy, dispersal is dominated by the rates of sedimentation (especially of the larger splash droplets) and impaction, so that deposition of many microbes occurs within a few meters of their points of take-off, and within seconds or minutes. By contrast, long-distance dispersal of microbes within the planetary boundary layer, although it involves the same stages, takes place over periods of hours or even days.

Legg and Powell (1979) used a simple two-dimensional numerical model of spore movement within and above a crop, based on a starting concentration profile and measurements of windspeed, turbulent diffusion, and foliage distribution. They found good agreement with observations of *Lycopodium* spores released from a line of point source of spores, but the model failed with a narrow strip of barley infected with mildew, possibly because the spores were clumped and therefore fell faster than single spores. Aylor (1982) showed that agreement could be improved if the model is modified to take account of the crucial role of gusts. A study by Aylor (1989) of the relative contributions of filtration by, and escape from, a 1-m-tall canopy of wheat has shown that as many as half of the microbes escape within 1–2 m downwind of their points of take-off.

Once airborne, a cloud of microbes diffuses under the influence of turbulence so we might suppose that diffusion theory, widely used in pollution studies, would be applicable to a plume of microbes in the planetary boundary layer. But it is not, because it is invalid close to the source, it ignores the dominance of strongest gusts, and it ignores downstream diffusion (in contrast to cross-stream) (Legg, 1983). More promise is shown by random walk models, which simulate a trajectory by a sequence of short steps, each partly random and partly determined by one or more preceding steps (McCartney and Fitt, 1985). However, because of their inertia, airborne particles do not follow exactly the same trajectories as the moving air in which they are embedded. As a result, larger microbes and splash droplets tend to move between eddies, although the relative motion is reduced by viscous drag. Walklate (1987) has examined the consequences of this inertia on the random walk model of Legg (1983). We must also remember that plumes are patchy and do not have the simple Gaussian cross-plume concentration profiles assumed in some diffusion models. Moreover, plumes from various sources may overlap, and a small local source may produce aerial concentrations as great as those from a large distant source. Our current limited understanding of plume behavior does not allow the introduction of improved models, particularly for plumes in the nighttime boundary layer (Weil, 1985). For a review of long-range transport of spores, see Pedgley (1986); for the application of transport models to disease epidemiology, see Davis (1987).

Deposition models used in pollution studies may be applicable to microbes. However, these models require extensive and continuous monitoring of pollutant concentration. Comparable monitoring of airborne microbes may be very costly. Microbe viability data also need to be incorporated. Since wet deposition rate away from sources will often exceed the dry rate (because falling raindrops rapidly scavenge a cloud of microbe-sized particles) and will be episodic (because rainfall is erratic in space and time), some progress might be made by monitoring and giving advance warning of the coexistence of microbe cloud and rainfall. This is particularly so over mountains, where falls tend to be both heavier and more persistent than over plains. With major pathogens, for example, it might be feasible to install a network of aerial samplers, especially over the principal sources, whose records would be fed regularly into a dispersion model that calculated the continually changing geographical distribution, depth, and density of the pathogen cloud. Combined with forecasts of rainfall (drop size, intensity, and duration), the regions most at risk from wet deposition could be estimated daily or even hourly.

3.5 Conclusions

To summarize, we are aware of, and partially understand, the effects of atmospheric behavior on the various stages of microbe dispersal, but quantification is still far away (for reviews, see McCartney and Fitt, 1985; and Aylor, 1986). The various stages are difficult to measure and therefore to model. There are great uncertainties in all three stages, and the application of theory is still in its infancy: the validity of physical models of diffusion in interpreting the behavior of microbial clouds is limited by lack of verification in the field. In particular, there is still much to be learned about take-off, survival, and wet deposition.

Although theory is still inadequate for quantitative modeling, we can nevertheless end on a positive note by suggesting some answers to four practical questions based on the one posed in Section 3.1. Question: "How can I reduce the risks of crop disease due to windborne microbe X?" Answer: through analysis of past occurrences of the spread of X on the wind. If this has not already been done, then multidisciplinary research on the epidemiology of a number of case studies should be performed, if sufficient records of disease occurrence exist and if enough is known about atmospheric influences on dispersal. Question: "How can I be warned of the possible arrival of X so that I can undertake timely control measures?" Answer: through a warning and forecasting service based on: 1) routine and regular monitoring of sources and their development (perhaps using numerical models); 2) trajectories (using either observed or forecast winds) ranging in scale from field to continent; and 3) forecasts of the distribution of rain (or the grower's own sampling of rain water for the presence of X). Question: "How can I reduce the spread of X in my crop once it has arrived?" Answer: by having taken steps to reduce the spread of plumes from disease foci before they become established, e.g., by using wind breaks or mixed crops, and by timely application of pesticide. Question: "How can I reduce the risk that my crop will become a source for X spreading to another grower downwind?" Answer: by destroying crop residues, preferably on calm days so that spores released in mechanical operations do not drift far.

Until theory becomes adequate for modeling, we must rely on such empiricism. Of course, whether the effort would be worthwhile depends on the relative importance of windborne pathogens in comparison to other pressures on the grower. At least it should provide some preliminary rules of thumb to those waiting for answers to the original question.

The author thanks Dr. M.J. Jeger and Dr. J.M. Waller for their comments on an earlier draft.

References

Amiro, B.D. and Davis, P.A. 1988. Statistics of atmospheric turbulence within a natural black spruce canopy. *Boundary-Layer Meteorology* 44:267–283.

Arkin, G.F. and Perrier, E.R. 1974. Vorticular air flow within an open row crop canopy. *Agricultural Meteorology* 13:359–374.

Aylor, D.E. 1975. Deposition of particles in a plant canopy. *Journal of Applied Meteorology* 14:52–57.

Aylor, D.E. 1978. Dispersal in time and space: aerial pathogens. pp. 159–180 in Horsfall, J.G. and Cowling, E.B. (editors), *Plant Disease: an Advanced Treatise*, vol. II. Academic Press, New York.

Aylor, D.E. 1982. Modeling spore dispersal in a barley crop. *Agricultural Meteorology* 26:215–219.

Aylor, D.E. 1986. A framework for examining inter-regional aerial transport of fungal spores. *Agricultural and Forest Meteorology* 38:263–288.

Aylor, D.E. 1989. Aerial spore dispersal close to a focus of disease. *Agricultural and Forest Meteorology* 47:109–122.

Aylor, D.E. and Parlange, J.-Y. 1975. Ventilation required to entrain small particles from leaves. *Plant Physiology* 56:97–99.

Aylor, D.E., McCartney, H.A., and Bainbridge, A. 1981. Deposition of particles liberated in gusts of wind. *Journal of Applied Meteorology* 20:1212–1221.

Baldocchi, D.D. and Meyers, T.P. 1988. Turbulence structure in a deciduous forest. *Boundary-Layer Meteorology* 43:345–364.

Bergen, J.D. 1975. Air movement in a forest clearing as indicated by smoke drift. *Agricultural Meteorology* 15:165–179.

Bertrand, P.F. and English, H. 1976. Release and dispersal of conidia and ascospores of *Valsa leucostoma*. *Phytopathology* 66:987–991.

Brennan, R.M., Fitt, B.D.L., Taylor, G.S., and Colhoun, J. 1985a. Dispersal of *Septoria nodorum* pycnidiospores by simulated raindrops in still air. *Phytopathologische Zeitschrift* 112:281–290.

Brennan, R.M., Fitt, B.D.L., Taylor, G.S., and Colhoun, J. 1985b. Dispersal of *Septoria nodorum* pycnidiospores by simulated rain and wind. *Phytopathologische Zeitschrift* 112:291–297.

Carnegie, S.F. 1980. Aerial dispersal of the potato gangrene pathogen, *Phoma exigua* var. *foveata*. *Annals of Applied Biology* 94:165–173.

Chamberlain, A.C. and Little, P. 1981. Transport and capture of particles by vegetation. pp. 147–173 in Grace, J., Ford, E.D., and Jarvis, P.G. (editors), *Plants and Their Atmospheric Environment*. Blackwell, Oxford.

Chock, D.P. and Kuo, Y.H. 1990. Comparison of wind-field models, using the CAPTEX data. *Journal of Applied Meteorology* 29:76–91.

Clarke, J.F., Clark, T.L., Ching, J.K.S., Haagenson, P.L., Husar, R.B., and Patterson, D.E. 1983. Assessment of model simulation of long-distance transport. *Atmospheric Environment* 17:2449–2462.

Davis, J.M. 1987. Modeling the long-range transport of plant pathogens in the atmosphere. *Annual Review of Phytopathology* 25:169–188.

Davis; J.M. and Main, C.E. 1984. A regional analysis of the meteorological aspects of the spread and development of blue mold on tobacco. *Boundary-Layer Meteorology* 28:271–304.

Davis, J.M. and Main, C.E. 1989. The aerobiology of the sporangiospores of *Peronospora tabacina*. pp. 264–267 in *9th Conference of Biometeorology and Aerobiology*. American Meteorological Society, Boston.

Davis, J.M., Main, C.E., and Nesmith, W.C. 1985. The biometeorology of blue mold of tobacco: Part 2. The evidence for long-rang sporangiospore transport. pp. 473–498 in MacKenzie, D.R., Barfield, C.S., Kennedy, G.G., Berger, R.D., and Toranto, D.J. (editors), *The Movement and Dispersal of Agriculturally Important Biotic Agents*. Claitor's Publishing Division, Baton Rouge, LA.

Dinoor, A. 1981. Epidemics caused by fungal pathogens in wild and crop plants. pp. 143–158 in Thresh, J.M. *Plant Pathogens and Vegetation*. Pitman, Boston.

Drinkall, M.J. and Price, T.V. 1983. Dispersal of *Synchytrium psophocarpi* in Papua New Guinea. *Plant Pathology* 32:229–237.

Ferrandino, F.J. and Aylor, D.E. 1985. Deposition of particles to a wheat canopy: the importance of inertial impaction. *Agricultural and Forest Meteorology* 34:195–204.

Fitt, B.D.L. and Bainbridge, A. 1983. Dispersal of *Pseudocercosporella herpotrichoides* spores from infected wheat straw. *Phytopathologische Zeitschrift* 106:214–225.

Fitt, B.D.L. and Lysandrou, M. 1984. Studies on mechanisms of splash dispersal of spores, using *Pseudocercosporella herpotrichoides* spores. *Phytopathologische Zeitschrift* 111:323–331.

Fitt, B.D.L., McCartney, H.A., and Walklate, P.J. 1989. The role of rain in dispersal of pathogen inoculum. *Annual Review of Phytopathology* 27:241–270.

Foster, G.F. 1977. Effect of leaf surface wax on the deposition of airborne propagules. *Transactions of the British Mycological Society* 68:245–250.

Gao, W., Shaw, R.H., and Paw U, K.T. 1989. Observation of organised structure in turbulent flow within and above a forest canopy. *Boundary-Layer Meteorology* 47:349–377.

Gottwald, T.R. 1983. Factors affecting spore liberation by *Cladosporium carpophilum*. *Phytopathology* 73:1500–1505.

Graham, D.C. and Harrison, M.D. 1975. Potential spread of *Erwinia* spp. in aerosols. *Phytopathology* 65:739–741.

Gregory, P.H., Griffin, M.J., Maddison, A.C., and Ward, M.R. 1984. Cocoa black pod: a reinterpretation. *Cocoa Growers' Bulletin* 35:5–22.

Griffiths, E. and Ao, H.C. 1976. Dispersal of *Septoria nodorum* spores and spread of glume blotch of wheat in the field. *Transactions of the British Mycological Society* 67:413–418.

Hammett, K.R.W. and Manners, J.G. 1974. Conidium liberation in *Erysiphe graminis*: III. Wind tunnel studies. *Transactions of the British Mycological Society* 62:267–282.

Harrison, M.D. and Brewer, J.E. 1982. Field dispersal of soft rot bacteria. pp. 31–53 in Mount, M.S. and Lacy, G.H. (editors), *Phytopathogenic Prokaryotes*. Academic Press, New York.

Heffter, J.L. 1980. Air Resources Laboratories atmospheric transport and dispersion model. *NOAA Technical Memorandum* ERL-ARL-81.

Heffter, J.L. 1983. Branching atmospheric trajectory model. *NOAA Technical Memorandum*. ERL-ARL-121.

Hermansen, J.E., Torp, U., and Prahm, L.P. 1978. Studies of transport of live spores of cereal mildew and rust fungi across the North Sea. *Grana* 17:41–46.

Hirst, J.M. and Stedman, O.J. 1963. Dry liberation of fungal spores by raindrops. *Journal of General Microbiology* 33:335–344.

Hirst, J.M. and Stedman, O.J. 1971. Patterns of spore dispersal in crops. pp. 229–237 in Preece, T.F. and Dickinson, C.H. (editors), *Ecology of Leaf Surface Microorganisms*. Academic Press, New York.

Kuske, C.R. and Benson, D.M. 1983. Survival and splash dispersal of *Phytophthora parasitica*, causing dieback in Rhododendron. *Phytopathology* 73:1188–1191.

Lacy, M.L. and Pontius, G.A. 1983. Prediction of weather-related release of conidia of *Botrytis squamosa* from onion leaves in the field. *Phytopathology* 73:670–676.

Legg, B.J. 1983. Movement of plant pathogens in the crop canopy. *Philosophical Transactions of the Royal Society of London* B302:559–574.

Legg, B.J. and Powell, F.A. 1979. Spore dispersal in a barley crop: a mathematical model. *Agricultural Meteorology* 20:47–67.

Lindemann, J. and Upper, C.D. 1985. Aerial dispersal of epiphytic bacteria over bean plants. *Applied Environmental Microbiology* 50:1229–1232.

Lindemann, J., Constantinidou, H.A., Barchet, W.R., and Upper, C.D. 1982. Plants as sources of airborne bacteria, including ice nucleation-active bacteria. *Applied Environmental Microbiology* 44:1059–1063.

Little, P. 1977. Deposition of 2.75, 5.0 and 8.5 μm particles to plant and soil surfaces. *Environmental Pollution* 12:293–305.

MacDonald, O.C. and McCartney, H.A. 1988. A photographic technique for investigating the splashing of water drops on leaves. *Annals of Applied Biology* 113:627–638.

Main, C.E. and Davis, J.M. 1989. Epidemiology and biometeorology of tobacco blue mold. pp. 201–215 in McKeen, W.E. (editor), *Blue Mold of Tobacco*. Amer. Phytopathol. Soc.

McCartney, H.A. and Aylor, D.E. 1987. Relative contributions of sedimentation and impaction to deposition of particles in a crop canopy. *Agricultural and Forest Meteorology* 40:343–358.

McCartney, H.A. and Fitt, B.D.L. 1985. Construction of dispersal models. *Advances in Plant Pathology* 3:107–143.

McEwan, J.M. 1966. The source of stem-rust infecting New Zealand wheat crops. *New Zealand Journal of Agricultural Research* 9:536–541.

Moore, W.F. 1970. Origin and spread of southern corn leaf blight in 1970. *Plant Disease Reporter* 54:1104–1108.

Nagarajan, S. and Joshi, L.M. 1980. Further investigations on predicting wheat rusts appearance in central and peninsular India. *Phytopathologische Zeitschrift* 98:84–90.

Nagarajan, S. and Joshi, L.M. 1985. Epidemiology in the Indian subcontinent. pp. 371–402 in Roelfs, A.P. and Bushnell, W.R. (editors), *The Cereal Rusts*. Academic Press, New York.

Nagarajan, S. and Singh, H. 1976. Preliminary studies on forecasting wheat stem rust appearance. *Agricultural Meteorology* 17:281–289.

Nagarajan, S., Singh, H., Joshi, L.M., and Saari, E.E. 1976. Meteorological conditions associated with long-distance dissemination and deposition of *Puccinia graminis tritici* uredospores in India. *Phytopathology* 66:198–203.

Orlanski, I. 1975. A rational subdivision of scales for atmospheric processes. *Bulletin of the American Meteorological Society*. 56:527–530.

Pack, D.H., Ferber, G.J., Hefter, J.L., Telegadas, S.K., Angell, J.K., Hoecker, W.H., and Machta, L. 1978. Meteorology of long-range transport. *Atmospheric Environment* 12:426–444.

Pasquill, F. and Smith, F.B. 1983. *Atmospheric Diffusion*, 3rd edition. Ellis Horwood, Chichester, UK.

Pedgley, D.E. 1986. Long distance transport of spores. pp. 346–365 in Leonard, K.J. and Fry, W.E. (editors), *Plant Disease Epidemiology*, vol. I. McGraw Hill, New York.

Purdy, L.H. 1985. Sugarcane rusts. pp. 237–256 in Roelfs, A.P. and Bushnell, W.R. (editors), *The Cereal Rusts*. Academic Press, New York.

Quinn, C.E., Sells, I.A., and Graham, D.C. 1980. Soft rot *Erwinia* bacteria in the atmospheric bacterial aerosol. *Journal of Applied Bacteriology* 49:175–181.

Quinn, N.W. and Laflen, J.M. 1983. Characteristics of raindrop throughfall

under corn canopy. *Transactions of the ASAE* 26:1445–1450.
Raupach, M.R. and Thom, A.S. 1981. Turbulence in and above plant canopies. *Annual Review of Fluid Mechanics* 13:97–129.
Ray, P.S. (editor). 1986. *Mesoscale Meteorology and Forecasting*. American Meteorological Society, Boston. 793pp.
Reisinger, L.M. and Mueller, S.F. 1983. Comparison of tetroon and computed trajectories. *Journal of Climate and Applied Meteorology* 22:664–672.
Sawford, B.L. 1987. Conditional concentration statistics for surface plumes in the atmospheric boundary layer. *Boundary-Layer Meteorology* 38:209–223.
Scott, R.W. and Achtemeier, G.L. 1987. Estimating pathways of migrating insects carried in atmospheric winds. *Environmental Entomology* 16:1244–1254.
Shaw, M.W. 1987. Assessment of upward movement of rain splash using a fluorescent tracer method and its application to the epidemiology of cereal pathogens. *Plant Pathology* 36:201–213.
Shaw, R.H. 1982. Wind movement within canopies. pp. 17–41 in Hatfield, J.L. and Thomason, I.J. (editors), *Biometeorology in Integrated Pest Management*. Academic Press, New York.
Shaw, R.H., Ward, D.P., and Aylor, D.E. 1979. Frequency of occurrence of fast gusts of wind inside a corn canopy. *Journal of Applied Meteorology* 18:167–171.
Smith, F.B. 1983. Meteorological factors influencing the dispersion of airborne diseases. *Philosophical Transactions of the Royal Society of London* B302:439–450.
Stakman, E.C. and Christensen, C.M. 1946. Aerobiology in relation to plant disease. *Botanical Review* 12:205–253.
Starr, J.R. 1967. Deposition of particulate matter by hydrometeors. *Quarterly Journal of the Royal Meteorological Society* 93:516–521.
Starr, J.R. and Mason, B.J. 1966. The capture of airborne particles by water drops and simulated snow crystals. *Quarterly Journal of the Royal Meteorological Society* 92:490–499.
Stocker, R.A., Pielke, R.A., Verdon, A.J., and Snow, J.T. 1990. Characteristics of plume releases as depicted by balloon launchings and model simulations. *Journal of Applied Meteorology* 29:53–62.
Stover, R.H. 1962. Intercontinental spread of banana leaf spot (*Mycosphaerella musicola* Leach). *Tropical Agriculture, Trinidad* 39:327–338.
Stow, C.D. and Stainer, R.D. 1977. The physical products of a splashing water drop. *Journal of the Meteorological Society of Japan* 55:518–532.
Subba Rao, K.V., Snow, J.P., and Berggren, G.T. 1989. Detection, viability, and possible sources of uridiniospores of *Puccinia recondita* f. sp. *tritici* in Louisiana. *Plant Disease* 73:437–442.
Ulbrich, C.W. 1983. Natural variations in the analytical form of the raindrop size distribution. *Journal of Climate and Applied Meteorology* 22:1764–1775.
Venette, J.R. 1982. How bacteria find their hosts. pp. 3–30 in Mount, M.S. and Lacy, G.H. (editors), *Phytopathogenic Prokaryotes*. Academic Press, New York.
Waggoner, P.E. 1983. The aerial dispersal of the pathogens of plant disease. *Philosophical Transactions of the Royal Society of London* B302:451–462.
Walklate, P.J. 1987. A random-walk model for dispersion of heavy particles in turbulent air flow. *Boundary-Layer Meteorology* 39:175–190.
Walklate, P.J. 1989. Vertical dispersal of plant pathogens by splashing. Part I: the theoretical relationship between rainfall and upward rain splash. *Plant Pathology* 38:56–63.
Walklate, P.J., McCartney, H.A., and Fitt, B.D.L. 1989. Vertical dispersal of plant pathogens by splashing. Part II: experimental study of the relationship between raindrop size and maximum splash height. *Plant Pathology* 38:64–70.
Waller, J.M. 1976. The influence of climate on the incidence and severity of some diseases of tropical crops. *Review of Plant Pathology* 55:185–191.

Waller, J.M. 1979. The recent spread of coffee rust (*Hemileia vastatrix*) and attempts to control it. pp. 275–283 in Ebbels, D.L. and King, J.E. (editors), *Plant Health*. Blackwell, Oxford.

Watson, I.A. and de Sousa, C.N.A. 1983. Long distance transport of spores of *Puccinia graminis tritici* in the southern hemisphere. *Proceedings of the Linnean Society of New South Wales*. 106:311–321.

Weil, J.C. 1985. Updating applied diffusion models. *Journal of Climate and Applied Meteorology* 24:1111–1130.

Weiss, A. and Allen Jr, L.H. 1976. Air-flow patterns in vineyard rows. *Agricultural Meteorology* 16:329–342.

4

Submerged Leaf Surfaces as a Microbial Habitat

Raymond Goulder and John H. Baker

4.1 Introduction

The economic importance of plant pathogens in the phyllosphere of terrestrial crop plants has encouraged research on the microbial ecology of aerial leaf surfaces. The present volume and the four previous volumes (Preece and Dickinson, 1971; Dickinson and Preece, 1976; Blakeman, 1981; Fokkema and van den Heuvel, 1986) are quite narrowly restricted to the aerial phyllosphere. Information on the microbiology of submerged leaf surfaces is, in contrast, scattered in ecological, microbiological, and freshwater journals and is probably not well known to students of the aerial phyllosphere. The microbial community at submerged leaf surfaces is of ecological importance in many fresh waters. Submerged leaf surfaces are sites of primary and secondary production by microalgae and bacteria which can rival that of phytoplankton and bacterioplankton in the water column. The community serves as food for grazing invertebrates and protozoa, it contributes to biopurification of organically-polluted watercourses, and can be a substantial source of planktonic microorganisms. We, therefore, make no apology for devoting the bulk of this review to submerged, freshwater, leaf surfaces, except for our concluding comparison of the aerial and aquatic phylloplane.

4.2 The Habitat at Submerged Leaf Surfaces

Freshwater bodies possess well-documented (e.g., Wetzel, 1983) seasonal, diurnal, and spatial variation of, for example, dissolved inorganic and organic nutrients, oxygen, CO_2, temperature, and light. Superimposed on this variation level is a smaller-scale variation in the local habitat within stands of submerged plants, brought about by the physical presence and biological functioning of the submerged plants and their epiphytes. Thus the following may occur:

1. Reduction in velocity of flowing water. Madsen and Warncke (1983) observed a 58–92% reduction within a bed of *Callitriche stagnalis* Scop. in a small stream; Sand-Jensen et al. (1989a) found a 1.3 to 3.5-fold decrease in mean velocity in a stream with rich submerged vegetation dominated by *Potamogeton pectinatus* L. and *Sparganium emersum* Rehmann.
2. Attenuation and change in spectral composition of light. Westlake (1966) found 99% attenuation of surface irradiance at ca. 11-cm depth in a stand of *Ranunculus penicillatus* (Dumort.) Bab. and 90% at ca. 44 cm in *Potamogeton pectinatus* compared to only about 10% and 60% at the same depths in open water. Westlake (1964) also observed an increase in the proportion of nonphotosynthetically useful yellow-green light within weed beds.
3. Enhancement of vertical temperature gradients. Dale and Gillespie (1977) noted that the water temperature within the top of a bed of *Vallisneria americana* Michx. was higher than in surrounding water and temperature decrease with depth was more pronounced.
4. Change in pH, bicarbonate alkalinity, and thus availability of dissolved inorganic carbon. Goulder (1969) described a pH raised to ca. 10 and alkalinity depressed to 0.8 meq 1^{-1} within a stand of *Ceratophyllum demersum* L. during a summer afternoon, compared to pH 7.7 and alkalinity 2.7 meq 1^{-1} in open water.
5. Change in dissolved oxygen concentrations relative to those in open water. Buscemi (1958) showed increased midafternoon oxygen concentration towards the top of a bed of *Elodea canadensis* Michx. together with rapid depth decrease. Also, diurnal changes in oxygen concentration, with daytime increase and nighttime depletion, which occur in weed beds (e.g., Goulder, 1970), are much less pronounced in the surrounding water.

The importance of these macrophyte-mediated changes is emphasized by Morin and Kimball (1983). They studied epiphytes of *Myriophyllum heterophyllum* Michx. in a New Hampshire, USA, lake and found that in early summer the macrophyte occupied only the lower water column. Chemical conditions within the vegetated area resembled those of the surrounding water, and the microalgae of the epiphyton, that is the microbial community at submerged plant surfaces, were dominated by diatoms from the plankton. By midsummer, in contrast, vegetation occupied the whole water column and there were marked local vertical physicochemical gradients, with midday temperature, dissolved oxygen, pH, and light being maximal at the water surface. These conditions were accompanied by a shift in epiphyte species composition towards blue-green (Cyanophyceae) and filamentous green (Chlorophyceae) algae on apical and midstem foliage with diatoms on lower-stem foliage; also, algal populations became less dense on lower stems.

In addition to the macrophyte-mediated change in the bulk water within plant stands, there is a distinct microclimate at the surface of submerged foliage. A diffusive boundary layer, present even in well-stirred conditions, is described by Sand-Jensen et al. (1985), Sand-Jensen (1989), and Prins and Elzenga (1989). These authors regard this as an essentially unmixed layer within which transport of dissolved chemical species is by molecular diffusion. Its thickness may be substantially increased by the presence of an epiphyte community. Thus, Sand-Jensen (1989) suggests that a diffusive boundary layer of thickness about 100 μm on an uncolonized leaf under well-stirred conditions may be increased to 1100 μm by a 1000-μm-thick epiphyte community. The epiphyte community is thus envisaged as occupying the diffusive boundary layer where, because of slowness of diffusion, there may be depletion of essential elements and the accumulation of waste metabolites. For example, the diffusion rates of CO_2 and oxygen in water are about 10,000 times less than in air. Thus Prins and Elzenga (1989) consider that within diffusive boundary layers, ranging in thickness from 10 μm in rapidly-stirred solution to 500 μm in stagnant water, there is marked CO_2 depletion following photosynthetic demand. This also leads to a local increase in pH and consequent shift in remaining inorganic carbon from CO_2 to bicarbonate ions, forcing host plants and their epiphytic algae to utilize bicarbonate as photosynthetic substrate, and leading to further pH elevation and reduced availability of dissolved inorganic carbon because of carbonate generation.

The oxygen microclimate within epiphyte communities on several hydrophyte species is described by Sand-Jensen et al. (1985). They used vertically travelling microelectrodes of 2–10 μm in diameter within mixed epiphyte communities of bacteria and diatoms on plants in stirred-water conditions, under artificial light-dark cycles. They found, for example, on leaves of *Potamogeton crispus* L. collected in early June and with an epiphyte community ca. 1130 μm thick, that oxygen concentration in the dark decreased through the ca. 300-μm-thick diffusive boundary layer, present above the epiphyte layer, to ca. 87% of concentration in the bulk-water phase. There was then a further decrease through the epiphyte community, to ca. 33% at the leaf surface. In the light (467 μmol m^{-2} s^{-1}), in contrast, oxygen concentration at the top of the epiphyte layer was about 148% of that in the bulk water and increased to ca. 241% at the leaf surface. It was also observed that the shift between oxygen depletion and supersaturation was rapid, being completed within 15–20 min of change in light status.

Clearly the diffusive boundary layer ensures that the microhabitat of epiphyton is very different from conditions in the surrounding water.

In running waters, the thickness of the diffusive boundary layer depends on current velocity. Westlake (1967) showed that the photosynthetic rate of *Ranunculus penicillatus* increased with increase in current velocity, over a range of velocity (0–5 mm s^{-1}) likely to be found in weed beds. The increase in photosynthetic rate was related to a reduction in thickness of the diffusive boundary layer, which restricted supply of dissolved inorganic carbon from the surrounding water. A further important feature of boundary layers in flowing waters is that they afford protection from forces tending to dislodge epiphytic microorganisms. This is considered by Silvester and Sleigh (1985) who applied mathematical models from hydraulic and aeronautical engineering and, for mathematical purposes, considered the boundary-layer thickness as extending from the surface to the shell where current velocity equals 0.9 times that of bulk-water flow. Their models describe relationships between current velocity and predicted boundary-layer thickness (δ) at plane surfaces (leaves) and cylinders (trailing stems) parallel to direction of flow. They show how δ decreases with increase in current velocity, and increases with downstream distance from leading (upstream-facing) edges. For example, at a plane surface parallel to water flow (perhaps a trailing leaf of *Sparganium emersum*) in a current velocity of 0.1 m s^{-1}, $\delta =$ ca. 3.3 mm at 10 cm and 6.9 mm at 40 cm behind the leading edge, while at 0.01 m s^{-1}, $\delta =$ ca. 10.6 mm at 10 cm and 21.5 mm at 40 cm behind the leading edge. Epiphytic microorganisms occupying such boundary layers tradeoff the advantage of low viscous drag forces, and less risk of being washed away, against lower water-exchange rates and concomitant depletion of nutrients, oxygen, and CO_2. Nevertheless, forces on epiphytic microorganisms may still be sufficient to dislodge cells, especially during high-flow conditions or near to leading edges. Thus Silvester and Sleigh (1985) calculated predicted shear stress (τ_0) at a plane surface, at different distances from the leading edge, over a range of current velocity (0.05–1.0 m s^{-1}) which was realistic for streams. Predicted τ_0 values (ca. 6–1000 pN μm^{-2}) were compared with experimentally determined values, in the literature, of stress (force per unit area) needed to detach various microorganisms, including bacteria and diatoms. It was concluded that shear-stress values likely to be found in streams were indeed within the range required to dislodge microorganisms.

A further complexity of the microclimate at some submerged leaf surfaces is the phenomenon of convective gas flow through the gas films found between hydrophobic plant surfaces and the surrounding water. Raskin and Kende (1983, 1985) and Beckett et al. (1988) describe such gas flow through a continuous gas film that coats submerged leaf surfaces of deep-water rice (*Oryza sativa* L.) and connects with the atmosphere where the leaves emerge at the water surface. The gas flow arises because CO_2 is more soluble in water than oxygen. In the dark, CO_2 produced by respiration of underwater plant parts is lost from the gas film into the surrounding water, causing a decrease in gas pressure, which is compensated by convective inflow of air from the atmosphere. In the light during photosynthesis, the flow can be stopped or reversed. The mean thickness of the gas film on deep-water rice, calculated from data in Raskin and Kende (1983), is ca. 45 μm on adaxial leaf surfaces and 15 μm on abaxial surfaces. We have frequently observed such films as a silvery sheen on, for example, the trailing leaves of *Glyceria fluitans* (L.) R. Br. in calcareous streams. Thus, epiphytic microorganisms on some submerged leaves are at a hydrophobic surface in a well-aerated gaseous phase, which is continuous with the atmosphere. Evidently their microhabitat is very different from that at wetted, submerged, leaf surfaces.

4.3 The Microbial Community at Submerged Leaf Surfaces

Composition and Function The epiphytic microflora of freshwater macrophytes consists largely of heterotrophic bacteria and autotrophic microalgae. In situ observations on plants from calcareous streams, by bright-field microscopy following phenolic aniline blue (PAB) staining and by scanning electron microscopy (SEM), showed numerous, short, rod-shaped bacteria closely adhering to epidermal leaf surfaces of *Berula erecta* (Hudson) Coville, *Ranunculus trichophyllus* Chaix, and *Lemna minor* L. (Lorch and Ottow, 1985, 1986; Baker, 1988). Also observed were chains of rods, and coccoid, filamentous, and stalked bacterial cells. SEM also provided some evidence of an associated polymeric polysaccharide matrix (Rogers and Breen, 1981), which is a frequent feature of bacterial communities at aquatic surfaces (Costerton et al., 1987). Extensive colonization by microalgae may also be shown by SEM. Allanson (1973) described a community dominated by pennate diatoms on the underside of leaves of *Potamogeton natans* L. from a pond in Oxford, England. This consisted of diatom cells attached to the leaf over the whole lower-frustule surface (e.g., *Cocconeis* spp.) or borne on stalks (e.g., *Gomphonema* spp.) together with diatoms that formed a loosely woven upper story (e.g., *Synedra* spp.) not directly attached to the leaf surface but seemingly supported by filamentous connectives. It appears that epiphytic microflora, perhaps especially in standing waters, consist of both firmly attached (adnate) and loosely attached components (Burkholder and Wetzel, 1989) and that the loosely attached component can be lost by careless sampling. This problem was overcome by Haines et al. (1987) who devised a sampler to collect undisturbed epiphyton on submerged stems (culms) of *Phragmites australis* (Cav.) Trin. ex Steudel, together with surrounding water. Epiphyton that separated spontaneously from its host stem after collection was referred to as loose epiphyton while that which remained was regarded as firmly attached. They showed that bacterial abundance and secondary production, and algal biomass and primary production, of loose epiphyton mostly exceeded that of firmly attached epiphyton. The loosely attached epiphytes are, however, potentially more transient; they are liable to be dislodged by water movement and are perhaps more available to grazers.

Not much is known about the taxonomic composition of freshwater epiphytic bacterial populations. Hossell and Baker (1979a) tentatively identified 120 isolates associated with *Ranunculus penicillatus* from the River Frome, a chalk stream in S. England. These isolates were cultured on casein-peptone-starch (CPS) agar (Jones, 1970) and were mainly (94%) Gram-negative rods with a preponderance of only a few genera: 24%, *Pseudomonas;* 24%, *Flavobacterium;* 6%, *Xanthomonas;* 5%, *Acinetobacter;* 3%, *Aeromonas,* and 3%, *Moraxella.* This study, however, had the limitation that it included only the small proportion of the total bacteria that could be isolated on CPS agar.

More is known about numerical abundance of epiphytic bacteria. Abundance may be expressed as colony-forming units (CFU) from colony counts on agar plates. However, it may be better expressed as total bacteria from direct counts made, for example, by bright-field microscopy on in situ bacteria after PAB staining (Hossell and Baker, 1979b), or by epifluorescence microscopy on fluorochrome-stained bacteria in suspensions prepared by treatment of plant shoots in a stomacher (Fry et al., 1985). Information on the abundance of CFU and total bacteria at submerged surfaces of freshwater plants from diverse sites has been collated by Al-Hadithi and Goulder (1989a). Examples on leaves of plants from English chalk streams are: CFU: $2.2–11.8 \times 10^4$ cm^{-2} on *Ranunculus penicillatus* (Hossell and Baker, 1979a); total bacteria: $0.07–1.8 \times 10^7$ cm^{-2} on *R. penicillatus,* $0.03–1.5 \times 10^7$ cm^{-2} on *Veronica beccabunga* L. (Baker and Orr, 1986), $0.5–2.6 \times 10^7$ cm^{-2} on *Apium nodiflorum* (L.) Lag., $0.6–5.5 \times 10^7$ cm^{-2} on *Nasturtium officinale* R. Br., and $1.1– 4.8 \times 10^7$ cm^{-2} on *Glyceria fluitans* (Rimes and Goulder, 1986a).

The heterotrophic activity of freshwater epiphytic bacteria, expressed as the potential for glucose mineralization (V_{max}) was determined by measuring $^{14}CO_2$ released during incubation of epiphyte suspensions over a range of added ^{14}C-glucose concentration (Rimes and Goulder, 1986b). The procedure was based on the method of Fry and Humphrey (1978) and used a model for substrate utilization derived from enzyme kinetics (Wright and Hobbie, 1966). V_{max} for glucose mineralization by the epiphytic bacteria of a mixed community of submerged macrophytes (mainly *Nasturtium officinale, Apium nodiflorum* and *Veronica anagallis-aquatica* L.), in a calcareous stream in N.E. England, ranged from 1.7–25.2 $\mu g\ h^{-1}$ per gram dry weight of vegetation. This was in early summer when the standing crop of total epiphytic bacteria was $3.8–6.4 \times 10^{10}$ g^{-1} dry weight. The results, converted to glucose-mineralization potential per unit area of vegetated stream bed, represented 0.5–8.2 mg m^{-2} h^{-1} and, interestingly, were fairly similar to values calculated from Allen and Ocevski (1982) for glucose-assimilation capacity of submerged epiphytic bacteria per unit area occupied by littoral vegetation in a lake in Yugoslavia; i.e., 1.7 mg m^{-2} h^{-1} on *Phragmites australis*, 1.9 mg m^{-2} h^{-1} on *Potamogeton perfoliatus* L., and 0.07 mg m^{-2} h^{-1} on *Nuphar lutea* (L.) Sm. These values may be compared with glucose-mineralization potentials of bacterioplankton populations, e.g., 0.04–7.5 μg l^{-1} h^{-1} in the calcareous River Hull in N.E. England (Goulder, 1980). In well-vegetated streams the heterotrophic activity of epiphytic bacteria may well exceed that of bacterioplankton.

Secondary productivity of epiphytic bacteria is more immediately relevant to the understanding of carbon flux at submerged plant surfaces. Haines et al. (1987) used the ^{3}H-thymidine method (Fuhrman and Azam, 1982), in which incorporation of the ^{3}H-labelled nucleoside thymidine into DNA is an indicator of secondary carbon fixation, and they obtained mean bacterial production rates of 0.32–0.45 μg C cm^{-2} h^{-1} for loose epiphyton, and 0.03–0.15 μg C cm^{-2} h^{-1} for firmly attached epiphyton on submerged stems of *Phragmites australis* in a South African marsh.

The release of extracellular enzymes is another important function of epiphytic bacteria and probably also of microalgae. Organic polymers at submerged plant surfaces require degradation to oligomeric or monomeric molecules, by extracellular enzyme activity, before they can be transported into epiphytic bacterial cells. Goulder (1990) used fluorogenic analog substrates (compounds of 4-methylumbelliferone) to determine the potential for extracellular enzyme activity associated with the epiphytic microflora of *Phragmites australis* stems in gravel-pit ponds of N.E. England. Examples of activities obtained were: phosphatase, 4.5–23.7 nmol cm^{-2} h^{-1}; β-D-glucosidase, 1.2–12.6 nmol cm^{-2} h^{-1}; and β-D-galactosidase, 0.1–1.4 nmol cm^{-2} h^{-1}. Interestingly, the relative magnitude of the different extracellular enzyme activities was constant on *P. australis* stems from eight different ponds. The study of extracellular enzyme activity in natural waters is currently a growth area which has been reviewed by Hoppe (in press). Most studies to date have been on water samples. Further work on the extracellular enzyme activity of epiphytic communities will allow assessment of its importance in whole freshwater bodies.

Epiphytic microalgae, unlike bacteria, can be readily observed, and often identified, using conventional bright-field microscopy. Thus much is known about the taxonomic composition of algal epiphyton on submerged freshwater plants. A great diversity of algal taxa is usually found. Pennate diatoms are invariably important, sometimes overwhelmingly dominant, but other taxa, particularly members of the Chlorophyceae and Cyanophyceae, are frequently conspicuous (e.g., Bownik, 1970; Hickman, 1971; Allanson, 1973; Eminson and Moss, 1980; Sand-Jensen and Sondergaard, 1981; Morin, 1986; Burkholder and Wetzel, 1989). Algal units (cells, filaments, and colonies) can be counted in suspensions or concentrates of dislodged epiphyton; thus the relative frequency or numerical abundance of the different taxa has been determined. For example, Eminson and Moss (1980) examined epiphytic algae on *Potamogeton pectinatus* in summer in Hickling Broad, E. England. They recognized 39 algal taxa and found that the relative frequencies of the most common species were: *Diatoma elongatum* Agardh., 26.6%; *Achnanthes minutissima* Kütz., 9.2% (Bacillariophyceae); *Monoraphidium* sp., 13.0%; and *Scenedesmus quadricauda* (Turp.) Bréb., 9.2% (Chlorophyceae). Also, Hickman (1971) found numerical abundance of total algal cells on submerged stems of *Equisetum fluviatile* L., over two years in a small pond in S. England, of up to ca. 4.8×10^6 cm^{-2}. This comprised mainly diatoms and species of *Oedogonium* and *Bulbochaete* (Chlorophyceae).

The biomass of epiphytic algae is frequently expressed as chlorophyll *a* density, as determined from suspensions of dislodged cells. This, unlike, for example, use of organic dry weight, excludes heterotrophic epiphyton and nonliving organic matter. Chlorophyll *a* densities of ca. 1–27 μg cm^{-2} on *Equisetum fluviatile* (Hickman, 1971) and means of 1.9–4.3 μg cm^{-2} on *Scirpus validus* Vahl. and *Phragmites australis* in a eutrophic Canadian lake (Jenkerson and Hickman, 1986) have been reported.

Primary productivity of epiphytic algae is measured by assessment of oxygen production (e.g., Meulemans, 1988) or, more usually, by modification of the method devised for phytoplankton photosynthesis by Steemann Nielsen (1952). Here, suspensions of dislodged epiphyton (Hickman, 1971; Hickman and Klarer, 1973) or, preferably, submerged shoots with intact epiphyton (Cattaneo and Kalff, 1979; Haines et al., 1987) are incubated, in transparent chambers submerged in natural waters, with ^{14}C-labelled dissolved inorganic carbon as a known portion of total inorganic carbon. Photosynthetic carbon fixation is calculated from ^{14}C uptake into algal cells. Hickman (1971) obtained photosynthetic rates of ca. 1–20 μg C cm^{-2} h^{-1} on *Equisetum fluviatile*, while Cattaneo and Kalff (1979) found rates of up to ca. 11 μg C cm^{-2} day^{-1} for loose epiphyton on *Potamogeton richardsonii* (A. Benn) Rydb. in a lake in Quebec. Haines et al. (1987) measured rates up to 0.3 μg C cm^{-2} h^{-1} for loose and 0.4 μg C cm^{-2} h^{-1} for firmly attached epiphyton on *Phragmites australis*. Such measurements facilitate comparisons with primary productivity of the host macrophytes and phytoplankton, quantities which are extensively reviewed by, for example, Wetzel (1983).

Spatial and Temporal Variation Epiphytic bacteria are not uniformly distributed on the submerged surfaces of freshwater plants (Figure 4.1). Kudryavtsev (1984) found that mean densities of total bacteria on stems of water lily (*Nuphar* or *Nymphaea* sp.), pondweed (*Potamogeton* sp.) and bistort (*Polygonum* sp.) were 8.0–12.3 × 10^6 cm^{-2} but densities on submerged leaves were only 30–75% of these values. Similarly, Rimes and Goulder (1986a) noted that total bacteria were more abundant on stems plus petioles

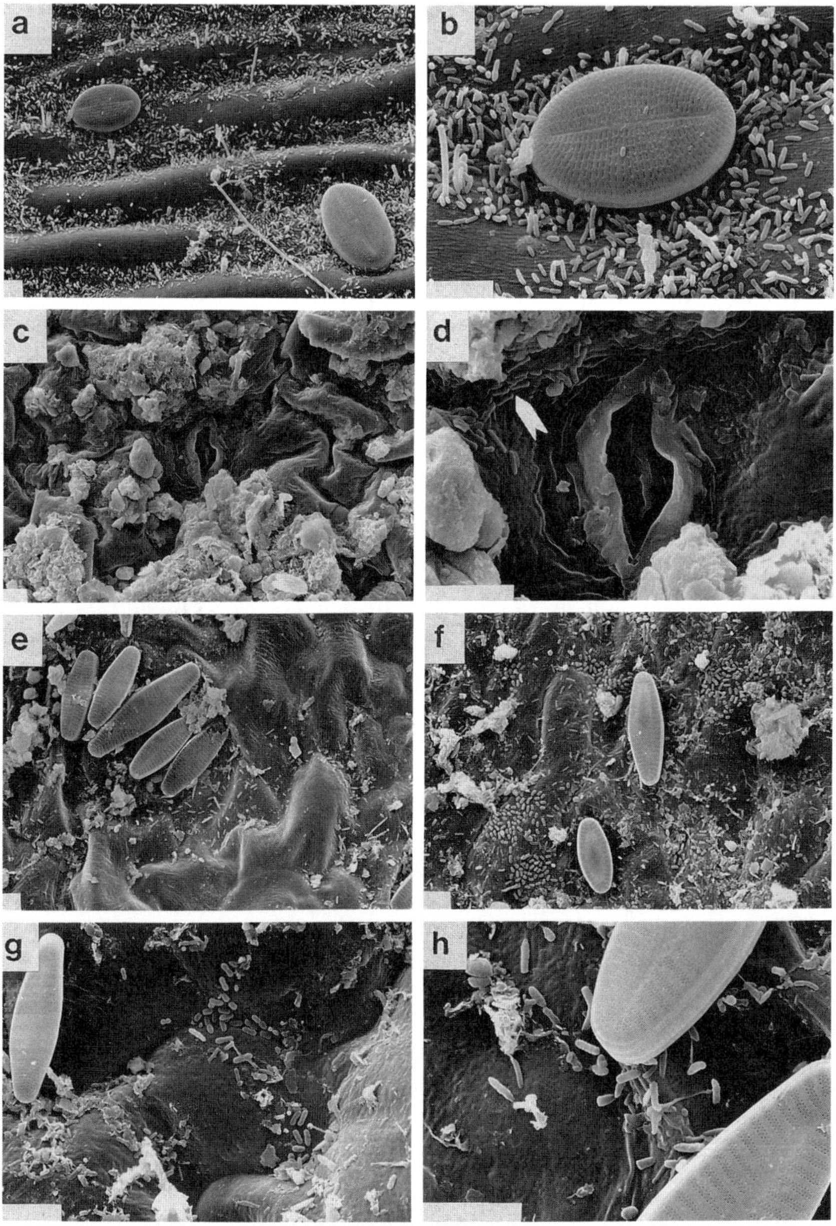
a
b
c
d
e
f
g
h

than on submerged leaves of *Apium nodiflorum, Nasturtium officinale* and *Glyceria fluitans*; i.e., mean densities of 3.1, 5.9, and 6.7×10^7 cm^{-2} respectively compared to 1.2, 2.5, and 2.6×10^7 cm^{-2}. The distribution on the surface of individual leaves is not uniform. Lorch and Ottow (1985) found concentrations of bacteria, several cells deep, along grooves between epidermal cells of *Berula erecta*. Baker and Orr (1986) observed that abaxial (lower) surfaces of submerged leaves of *Nasturtium officinale* and *Veronica beccabunga* supported more total bacteria than adaxial (upper) surfaces. Especially notable is the increase in bacterial density as leaves age. Ramsay (1974) found that CFU and total bacteria were more abundant on mature than young leaves of *Elodea canadensis*. Hossell and Baker (1979a) showed that total bacteria on a shoot of *Ranunculus penicillatus* increased from 3.9×10^4 cm^{-2} on the apical leaf (leaf 1) to 6.0×10^6 cm^{-2} on leaf 6. Figure 4.2 demonstrates change in bacterial abundance with leaf age. Here, one of each pair of opposite leaves was collected from 15 separate submerged plants of *Veronica beccabunga*, and bacteria were counted in situ after PAB staining. Mean density of bacteria increased with distance of leaf from the shoot apex, on both abaxial and adaxial surfaces. Note also that densities were greater on abaxial surfaces.

Various reasons may account for the uneven distribution of bacteria on submerged leaf surfaces. Lorch and Ottow (1985) suggested that, in a river environment, colonization of epidermal grooves might protect against currents and mechanical scraping. Baker and Orr (1986) suggested several reasons for nonuniform distribution:

1. Age of leaf. Older leaves have had a longer time for colonization, both through attachment of planktonic bacteria and through in situ cell division. Also, organic exudates are released from submerged foliage (e.g., Hough and Wetzel, 1975; Baker and Farr, 1987). Such release, which probably stimulates growth of epiphytic bacteria, might increase with leaf age.

Figure 4.1 Scanning electron micrographs of leaf surfaces. (a) and (b) Abaxial surface of submerged leaf of the aquatic moss *Fontinalis antipyretica* Hedw. Abundant rod-shaped bacteria are concentrated in the depressions between leaf cells; the diatoms (*Cocconeis*) are virtually free of bacteria. (c) and (d) Adaxial surface of leaf 2 or 3 at the apex of an aerial shoot of *Nasturtium officinale*. The microflora are sparse, being restricted to a few bacteria around the stoma (arrow). (e) Adaxial surface of leaf 2 or 3 of a submerged shoot; the microflora, composed of bacteria and diatoms, are more abundant. (f) Abaxial surface of leaf 11 of submerged shoot; there is dense bacterial colonization. (g) and (h) Adaxial and abaxial surfaces, respectively, of leaf 11 of submerged shoot; bacteria are concentrated in grooves between epidermal cells. Leaves were critical-point-dried and gold-sputter-coated (Baker, 1988) by S. Lythe and J. Halder. Shoots were collected in March from the River Hull (*Fontinalis*) at Hempholme and from Mill Beck (*Nasturtium*) at Goodmanham, Northeastern England. Bars equal 5 μm.

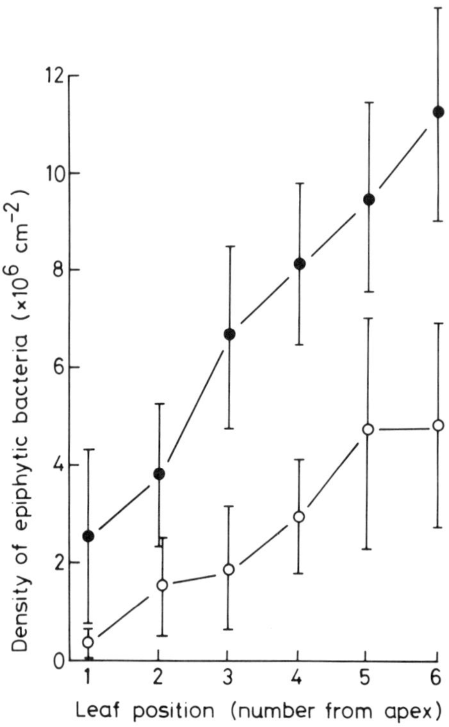

Figure 4.2 Mean population density of bacteria on submerged leaves of *Veronica beccabunga* collected in winter 1981 from Bere Stream, Southern England; ○, adaxial surfaces; ●, abaxial surfaces; vertical bars indicate 95% confidence intervals (n = 15 leaves). Bacteria were counted in situ using bright-field microscopy after phenolic aniline blue (PAB) staining. Modified from Baker and Orr (1986) with permission from Blackwell Scientific Publications.

2. Nonuniform grazing of bacteria by invertebrates and protozoa. Ciliated protozoa associated with *Ranunculus penicillatus*, for example, are concentrated at nodes (Baldock et al., 1983).
3. Adverse pH regime at adaxial leaf surfaces. Submerged leaves of some macrophytes show polarity in release of hydroxyl ions following utilization of bicarbonate as a photosynthetic substrate (e.g., *Elodea, Egeria, Hydrilla,* and *Potamogeton* spp.). Hydroxyl ion release from these plants occurs only at the adaxial surface, thus large pH differences develop across leaves. In extreme cases, pH at the adaxial surface may reach 10–11 compared to 4–7 at the abaxial surface (Prins and Elzenga, 1989). Such a locally high pH might inhibit bacteria. It is interesting to note that Baker and Orr (1986) found no difference between bacterial abundance on adaxial and abaxial surfaces of leaves of *Veronica baccabunga* in a softwater stream in which photosynthetic utilization of bicarbonate was unlikely to be important. More information is needed about which species of submerged macrophytes have leaves which exhibit polarity in use of bicarbonate.

Epiphytic microalgae are also unevenly distributed on submerged plants. Rogers and Breen (1981) used SEM to show an obvious relationship between algal abundance and leaf age on *Potamogeton crispus* in a shallow lake. Here, the youngest leaves were devoid of epiphytic algae, slightly older leaves supported a sparse population of the diatom *Cocconeis placentula* (Ehrenb.) Cleve, while mature leaves had a dominant population of *C. placentula* along with other diatoms (*Gomphonema* and *Navicula* spp.) and filamentous Cyanophyceae. The oldest leaves of all had markedly increased density of *C. placentula* but the diversity of diatoms did not appear to change. Shift in the species composition of microalgae on *Myriophyllum heterophyllum* with increase in water depth, in response to macrophyte-mediated change in physicochemical conditions in the water column (Morin and Kimball, 1983), has already been described. Baker and Orr (1986) found that diatoms, like bacteria, were more abundant on the abaxial leaf surfaces of *Nasturtium officinale* and *Veronica beccabunga*. Similarly, Cattaneo (1978) observed that the diatoms *Eunotia* spp. and *Coleochaete orbicularis* Pringsheim (Chlorophyceae) were more abundant on abaxial surfaces of *Potamogeton richardsonii* in a Canadian lake, but *Coleochaete nitellarum* Jost preferred adaxial surfaces. Possible reasons given for this distribution are, again, differential pH and the observation that precipitated calcium carbonate on adaxial surfaces might hinder growth of attached algae through shading or crowding. Cattaneo (1978) also observed that diatoms (*Cocconeis placentula* and *Eunotia* spp.) were concentrated at the margin of *P. richardsonii* leaves, while *Coleochaete orbicularis* and *C. nitellarum* were most abundant on the lamina. This distribution was also found on the leaves of plastic plants and was, therefore, ascribed to physical rather than biological variation across the leaf surface.

Temporal variation in abundance of epiphytic bacteria on submerged foliage is not much explored. Baker and Orr (1986) followed total bacteria on *Veronica beccabunga* over 15 weeks (November–February), in a calcareous stream in S. England, but found no significant change with time on either abaxial or adaxial leaf surfaces. However, they also investigated bacteria on the fourth leaf from the apex of *Ranunculus penicillatus* at intervals over three periods during 1980–1981 (April–May, September–October, May–September). There were no significant differences between the three periods, hence, there was no evidence of seasonal trends. Significant differences were found, however, within each period (Figure 4.3) but these were not related to discharge or temperature, which were relatively constant over the periods studied. Hence this short-term fluctuation was perhaps related to change in unmeasured biotic variables. Change in abundance of total bacteria on *Nasturtium officinale*, *Apium nodiflorum*, and *Glyceria fluitans*, over a spring growing season (January–June), was observed in a calcareous stream in N.E. England by Rimes and Goulder (1986a). Here, bacterial abundance, on all three species, decreased in early February and then increased, more or less, until June. The initial decrease coincided with increase in water flow, which may have dislodged bacteria, while the subsequent increase was accompanied by decreasing flow and increase in water temperature (Figure 4.4).

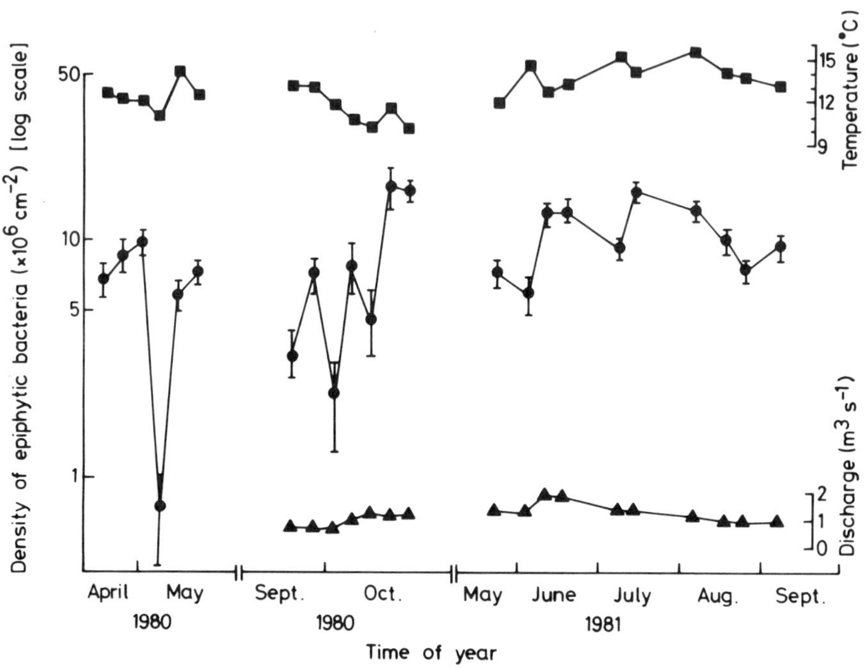

Figure 4.3 Variation in: ●, density of total bacteria with time on the fourth leaf from the apex of submerged shoots of *Ranunculus penicillatus*, with 95% confidence intervals; ■, water temperature; and ▲, discharge in the Bere Stream, S. England. Bacteria were counted in situ after PAB staining. From Baker and Orr (1986) with permission from Blackwell Scientific Publications.

Seasonal variation in heterotrophic activity, as V_{max} for glucose and acetate uptake, of bacteria at submerged surfaces of *Phragmites australis, Potamogeton perfoliatus,* and *Nuphar lutea* was investigated in a lake in Yugoslavia, by Allen and Ocevski (1982). Maximum V_{max} values tended to occur in May–June, but there were differences between macrophyte species in seasonal patterns, which were not closely linked to physicochemical variables, including temperature, and may have been controlled by availability of organic substrates at the plant surfaces.

Temporal variation in populations of epiphytic microalgae is more fully described than for bacteria. For example, marked shifts in species composition, especially among diatoms, have been found on submerged plants in English standing waters (Hickman, 1971; Moss, 1981). Changes over the year, in abundance and photosynthetic rate, per unit macrophyte surface area or dry weight, have also been reported. Hickman (1971) showed conspicuous but irregular peaks of total algal cells, chlorophyll *a*, and photosynthesis on *Equisetum fluviatile* in a small pond. These occurred, for example, in January, March, July, and September. Photosynthesis, which quite

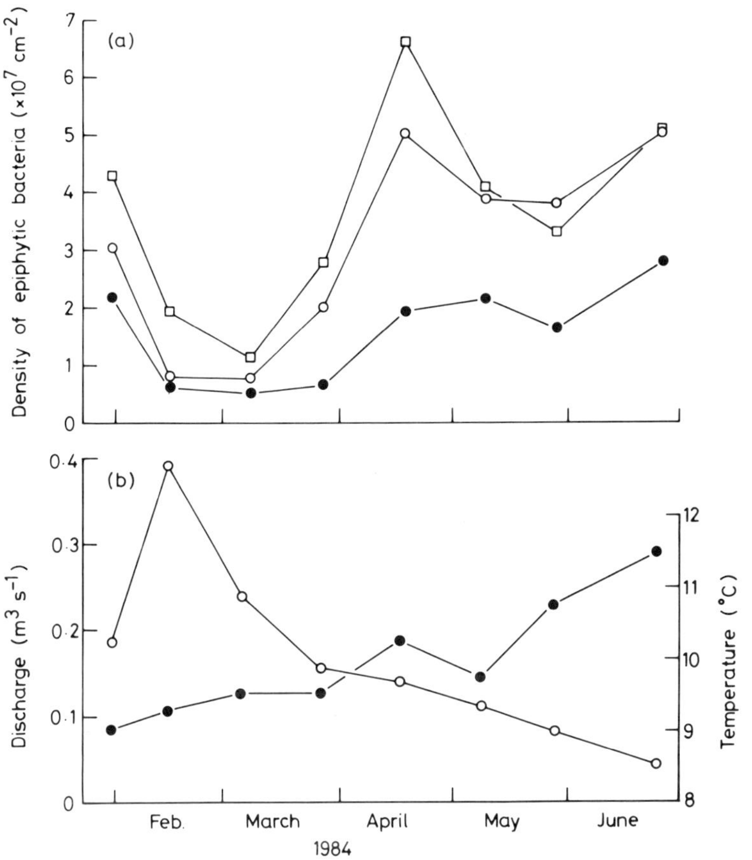

Figure 4.4 (a) Total bacteria on whole submerged shoots of: ○, *Nasturtium officinale;* ●, *Apium nodiflorum;* and □, *Glyceria fluitans*. Suspensions of epiphytic bacteria were prepared by stomaching six shoots; bacteria were stained with acridine orange and counted by use of epifluorescence microscopy; 95% confidence intervals were plus or minus <10% of the mean count. (b) ○, Discharge and ●, water temperature. Mill Beck, Goodmanham, N.E. England, 1984. Modified from Rimes and Goulder (1986a) with permission from Blackwell Scientific Publications.

closely followed cell numbers and chlorophyll *a*, was also influenced by light conditions on the sampling days. Sand-Jensen and Sondergaard (1981) found substantial seasonal changes in epiphyton chlorophyll *a* levels on *Littorella uniflora* (L.) Ascherson in several lakes in Jutland, Denmark, but there was no common pattern. In two lakes, there was rapid decrease from a spring peak to low summer levels, followed by an autumn increase; other lakes had a summer maximum or irregular variation. Between-lake differences were related to availability of inorganic nutrients. In a eutrophic river in Denmark, Sand-Jensen et al. (1989b) found that epiphyton biomass and photosynthesis on *Potamogeton pectinatus* decreased rapidly in spring to low summer and autumn values. The decline was ascribed to spring macrophyte growth producing uncolonized surface, and to scouring during conditions of high discharge; low populations persisted in summer, possibly because of shading by the host macrophyte.

Relations of the Epiphytic Community with Planktonic Populations and Macrophyte Hosts An important question is whether epiphytic bacteria and microalgae are related to, or are independent of, populations in the surrounding water. Suspended bacteria in calcareous headstreams seem to resemble epiphytic bacteria and may originate through dislodgement from submerged vegetation. Evidence for this is as follows: 1) Baker and Farr (1977) found that the taxonomic composition of planktonic bacteria isolated from the River Frome and Tadnoll Brook in S. England (87% Gram-negative rods, with *Pseudomonas, Flavobacterium, Acinetobacter,* and *Moraxella* numerically most important) was similar to that of epiphytic bacteria on *Ranunculus penicillatus* in the River Frome (Hossell and Baker, 1979a). 2) There is a pronounced downstream increase in concentration of suspended total bacteria in well-vegetated calcareous headstreams (Goulder, 1984; Rimes and Goulder, 1986c). Rimes and Goulder (1987) used image analysis (Fry and Davies, 1985) to show that, along with this increase, there is downstream change in cell dimensions, so that the dimensions of suspended bacteria come to resemble those of epiphytic bacteria. Mean cell volume (with 95% confidence intervals) of suspended bacteria, in March, at the spring-fed source of Mill Beck, Goodmanham, N.E. England, was 0.11 (0.09–0.13) μm^3 but this increased to 0.22 (0.19–0.25) μm^3 at 360 m downstream of the source, becoming similar to volumes of 0.25 (0.22–0.27) μm^3 for epiphytic bacteria on *Nasturtium officinale* and 0.19 (0.17–0.20) μm^3 on *Apium nodiflorum*. 3) Rimes and Goulder (1986b) measured drift loss of suspended total bacteria and of glucose-mineralization capacity, from the 48-m length of Mill Beck immediately downstream of its source. Here the drift loss of bacteria per day equalled the increase in concentration of suspended bacteria (cells m^{-3}) as water flowed along the 48 m length of stream times the flow rate (m^3 day^{-1}). The drift loss per day of glucose mineralization capacity equalled increase in mineralization capacity (mg m^{-3} h^{-1}) in water during its flow along the 48 m of stream times the flow rate (m^3 day^{-1}). Mean daily drift losses, during May–June 1983, were $4.3–6.7 \times 10^{14}$ bacteria day^{-1} and 24.0–39.2 mg h^{-1} of glucose-mineralization capacity per day. They also estimated total epiphytic bacteria on submerged vegetation ($3.6–6.2 \times 10^{15}$), and their total glucose-mineralization capacity (284–1060 mg h^{-1}), for the entire 306 m^2 of stream bed occupied by vegetation (mainly *Nasturtium officinale, Apium nodiflorum,* and *Veronica anagallis-aquatica*). The mean drift losses per day represented 11–12% of total epiphytic bacteria and 8–18% of epiphytic glucose-mineralization capacity. Hence, the epiphytic bacterial population should have been able to sustain the drift losses. Similarly, Rimes and Goulder (1987) worked further downstream on a 135-m richly vegetated length, over January–June 1984, and found that daily drift loss of bacteria averaged 53% (range 9–132%) of total epiphytic bacteria and again concluded that this loss could be accounted for by sloughing of epiphytic bacteria.

The interdependence of epiphytic and suspended bacteria in calcareous streams is reinforced by a degree of reattachment of planktonic cells to plant surfaces, a process that is probably especially important in the colonization of newly grown foliage. Thus, Figure 4.5a shows rapid colonization at the abaxial surfaces of *Apium nodiflorum* leaves submerged in Waterston Stream, a small calcareous watercourse in Dorset, S. England. The population density of total bacteria increased by about 9.0×10^6 cm^{-2} over 7 days. The source of largely uncolonized leaves for this experiment was aerial foliage; emergent leaves maintained as controls showed no increase in density of epiphytic bacteria over seven days. However, the increase in Waterston Stream is, in part, due to in situ cell division and there might also have been losses through, for example, protozoan grazing or sloughing. These problems were overcome by Rimes and Goulder (1985) who used *A. nodiflorum* shoots grown in the laboratory from surface-sterilized seeds and exposed them for only short durations (up to 10 h) in Church Beck, a calcareous headstream at North Newbald, N.E. England. Figure 4.5b gives an example of the results obtained; here the increase was ca. 1.1×10^6 bacteria cm^{-2} over 10 h. The attachment rate was, however, related to the concentration of suspended bacteria in the surrounding water. Thus, when *A. nodiflorum* shoots were submerged in Church Beck at different distances (75–400 m) downstream of the source, and hence at different concentrations of suspended bacteria—since there was marked downstream increase in concentration—there was a linear relationship between attachment rate and concentration of suspended bacteria (Figure 4.6). The attachment rate relative to bacterial concentration was 1.1×10^4 bacteria cm^{-2} h^{-1} for each 10^5 bacteria ml^{-1} in the surrounding water or, since bacteria were counted in suspensions prepared by stomaching, 1.7×10^4 bacteria cm^{-2} h^{-1} for each 10^5 bacteria ml^{-1}, when correction was made for efficiency of release of bacteria during stomaching. Colonization by reattachment is rapid. However, Rimes and Goulder (1987) calculated that attachment by suspended bacteria in a 135-m, richly vegetated length of Mill Beck, Goodmanham, was small compared to losses through sloughing of epiphytic bacteria. Attachment through February–June represented only 1.3–3.7% of total epiphytic bacteria per day, compared to a net daily drift loss of 9–132% of total epiphytic bacteria.

In larger rivers and standing fresh waters, there may be looser links between epiphytic and planktonic bacteria. Al-Hadithi and Goulder (1989a) compared the physiological state of bacteria on stems of *Phragmites australis*, in a gravel-pit pond in N.E. England, with that of suspended bacteria. The following criteria indicated that epiphytic bacteria were physiologically

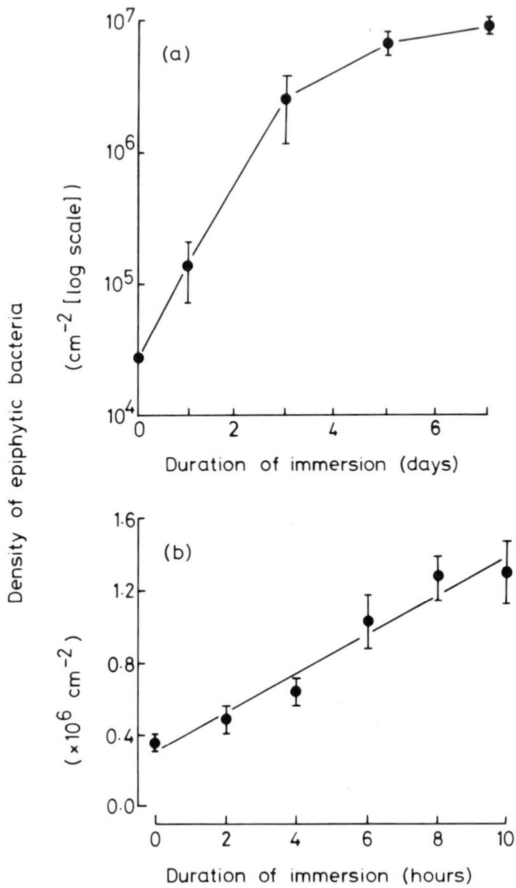

Figure 4.5 (a) Bacterial colonization of abaxial surfaces of *Apium nodiflorum* leaves submerged over seven days in Waterston Stream, S. England, June 1985. Bacteria were counted in situ after PAB staining. Based on unpublished data from J.H. Baker. (b) Bacterial colonization of *Apium nodiflorum* leaves submerged over 10 hours in Church Beck, North Newbald, N.E. England, at 400 m downstream of the source at a concentration of suspended bacteria 1.01×10^6 ml^{-1}, 3 October, 1984. Bacteria were assayed by acridine-orange direct counts on suspensions prepared by stomaching. Vertical bars indicate 95% confidence intervals. From Rimes and Goulder (1985) with permission from Blackwell Scientific Publications.

more active than planktonic bacteria: 1) Values of colony-forming units (CFU) on agar plates, as a percentage of total bacteria from direct counts, were much greater for epiphytic bacteria; mean (and range) = 9.2% (0.3–22.6%) compared to 0.10% (0.03–0.23%). 2) The percentage of cells capable of intracellular reduction of 2-(*p*-iodophenyl)-3-(*p*-nitrophenyl)-5-

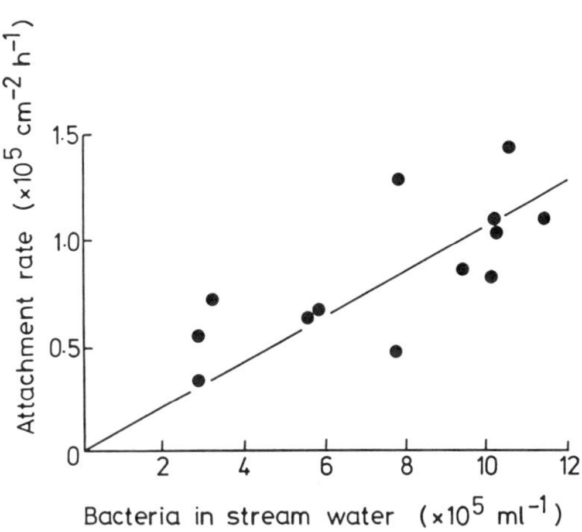

Figure 4.6 Relationship between attachment rate of bacteria to submerged *Apium nodiflorum* leaves and concentration of total bacteria in the surrounding stream water; Church Beck, N.E. England, September–October 1984. Total bacteria were assayed by acridine-orange direct counts on suspensions prepared by stomaching and on stream water. Modified from Rimes and Goulder (1985) with permission from Blackwell Scientific Publications.

phenyltetrazolium chloride (INT), with deposition of opaque granules of INT-formazan (Zimmermann et al., 1978; Tabor and Neihof, 1982), was greater for epiphytic bacteria; 60.2% (11.1–90.5%) compared to 19.0% (5.5–47.0%). 3) The percentage of cells capable of growth on agar films to form microcolonies (Fry and Zia, 1982) was greater for epiphytic bacteria; 25.1% (9.8–89%) compared to 14.8% (4.8–40%). Furthermore, epiphytic bacteria were bigger; overall, 20.8% of epiphytic bacteria were >2.5 μm long, compared with only 1.5% of planktonic bacteria, while 28.3% of epiphytic bacteria were <0.82 μm compared to 50.3% of planktonic bacteria. In six additional gravel-pit ponds, CFU as percentage of total bacteria, and percentage of cells capable of INT reduction, were again greater for epiphytic bacteria. Al-Hadithi and Goulder (1989a) suggested that the physiological differences between epiphytic and planktonic bacteria imply that there was not continuous exchange, by more or less identical cells, between the epiphytic and planktonic mode. Thus, either 1) epiphytic and planktonic populations were distinct and originated separately by cell division either on plant surfaces or in the water column, or 2) epiphytic and planktonic populations were interrelated and underwent exchange, but attached cells were larger, actively growing and dividing, whereas planktonic cells were smaller, nongrowing, and in transit between more-favorable surface habitats. The second hypothesis was developed in relation to marine bacteria by Kjelleberg et al. (1987). In further work on watercourses enriched by sewage-works effluent in the lowland Yorkshire River Derwent and Driffield Canal, N.E. England, Al-Hadithi and Goulder (1989b) compared epiphytic bacteria on submerged surfaces of *Callitriche* sp., *Glyceria maxima* (Hartman) Holmberg, and *Sparganium erectum* L. with planktonic bacteria. CFU as a percentage of total bacteria, and percentage of cells able to reduce INT, were again greater for epiphytic bacteria but the difference between epiphytic and planktonic bacteria was not so extreme as in the nonenriched gravel-pit ponds. This was ascribed to greater availability of dissolved organic nutrients in the enriched watercourses.

Interrelationships are also found between epiphytic and planktonic algae in freshwaters. Moss (1981) and Jenkerson and Hickman (1986), working in Hickling Broad, E. England, and Hastings Lake, Alberta, observed that, while some algae were largely epiphytic, others were conspicuous in both epiphyton and phytoplankton. Epiphytic populations of species also found in the plankton represented: 1) senescent cells caught up from the phytoplankton, 2) actively growing cells which acted as a source of cells to the plankton, and 3) representatives of species which actively grew and divided in both epiphytic and planktonic habitats.

There has been controversy over the extent to which epiphytic algae rely upon their macrophyte host for inorganic nutrients, and possibly even on organic exudates as a carbon source. Several studies have compared composition, abundance, or photosynthetic rate of epiphytic algae on real and plastic plants and have concluded that hosts are largely neutral substrata (e.g., Cattaneo and Kalff, 1979; Fontaine and Nigh, 1983; Morin, 1986). Thus the surrounding water seems to be the predominant nutrient source. Host plants do, however, to some degree supply nutrients to epiphytic algae (e.g., phosphorus; Carignan and Kalff, 1982) and it is, perhaps, most likely that nutrients from host plants are more important in oligotrophic than in richer waters (Eminson and Moss, 1980; Burkholder and Wetzel, 1989).

4.4 Comparison of Aerial and Aquatic Phylloplanes

Following the above review of habitat and microbiota at submerged leaf surfaces, comparison can be made with the aerial phylloplane. At first impression, the two habitats, aerial and potentially desiccating conditions, or permanent surrounding water, seem so very different that few features in common might be expected. The truth of this impression is now appraised.

Conditions in the atmosphere at large, that is weather, influence the habitat at aerial leaf surfaces just as conditions throughout whole water bodies influence the submerged phylloplane habitat. There is also in aerial plant stands, as in aquatic ones, a local climate mediated by the presence of the plants themselves. Monteith and Unsworth (1990), for example, illustrate vertical variation of net radiation, windspeed, air temperature, vapor pressure, and CO_2 concentration within cereal stands, and they describe diurnal change within this local climate. Aerial, like submerged, leaves also have a distinct microclimate at their immediate surfaces. There is a boundary layer (Monteith and Unsworth, 1990) within which air flow is reduced and the leaf itself influences levels of, for example, temperature, moisture, and windspeed (Burrage, 1971, 1976). These variables exhibit horizontal, temporal, and between-leaf variation which is related to the nature of the leaf (thickness, shape, roughness, hairiness, etc.) as well as to surrounding conditions.

Aerial, like submerged, leaves have a heterotrophic microflora of bacteria, but mycelial and yeast-like fungi are also important. There is little discussion of fungi on submerged leaves but this may reflect lack of information rather than their absence. Smith et al. (1989) describe experimental colonization of the submerged hydrophyte *Myriophyllum spicatum* L. by several fungal species. Especially notable is the absence from the aerial phyllosphere of an extensive autotrophic community of microalgae, presumably because conditions are usually too dry. The bacterial community on aerial leaves resembles that of submerged leaves in that isolates are mostly Gram-negative rods and are dominated by relatively few taxa. For example, Stott (1971) found that isolates from sugar-beet leaves (*Beta vulgaris* L.), were largely Gram-negative rods. Similar observations were made by Dickinson et al. (1975), who found that the predominant genera on leaves of *Lolium perenne* L. were *Chromobacterium*, *Corynebacterium*, *Pseudomonas*, and *Xanthomonas*, and by Corpe and Rheem (1989) who found *Erwinia*, *Pseudomonas*, *Xanthomonas*, *Flavobacterium*, and pink-pigmented facultative methylotrophs predominant on leaves of *Trifolium repens* L. The observation that aerial leaf-surface bacteria comprise relatively few important taxa is also supported by numerical taxonomy of isolates from *Lolium perenne* (Goodfellow et al., 1976; Austin et al., 1978) and by protein-fingerprint typing (using SDS- PAGE) of isolates from chicory, *Cichorium intybus* L. (Van Outryve et al., 1989).

The numerical abundance of bacteria on aerial leaves is often assessed as CFU per unit area but comparison between studies, and with data for submerged leaves, is problematic because different dislodgement techniques and culture media are often used. Nevertheless, CFU densities reported for aerial leaves tend to be less than for submerged leaves. Values as low as $<10^5$ cm^{-2}, e.g., $0.06–6.1 \times 10^4$ cm^{-2} have been reported on beech leaves, *Fagus* sp. (Jensen, 1971), and they are usually $<10^6$ cm^{-2}, e.g., $10^3–2.5 \times 10^5$ cm^{-2} on *Lolium perenne* (Dickinson et al., 1975); $1.2 \times 10^2–1.5 \times 10^5$ cm^{-2} on first leaves of wheat, *Triticum* sp. (Fokkema and Schippers, 1986); and $6.2 \times 10^1–6.8 \times 10^5$ on chicory (Van Outryve et al., 1989), although densities $>10^6$ cm^{-2} may occur, e.g., $1.0 \times 10^2–4 \times 10^6$ cm^{-2} on sugar beet (Stott, 1971). In contrast, CFU density on submerged leaves is usually $>10^5$ cm^{-2} and frequently $>10^6$ cm^{-2} (Al-Hadithi and Goulder, 1989a). These differences could in part be attributed to a difference in host plant species, but Baker and Orr (1986) compared leaves of the same age from emergent and submerged shoots of *Veronica beccabunga*. They found that the density of total bacteria on the submerged leaves was two orders of magnitude greater than on the corresponding emergent leaves. There is no known chemical or anatomical difference between emergent and submerged leaves of *V. beccabunga*. A similar large difference in bacterial population density was found between the upper aerial surface and the lower submerged surface of *Lemna minor* (Hossell and Baker, 1979b). Here the difference is even more remarkable insofar as the aerial surface is at the water surface and therefore at 100% relative humidity. These very marked differences could be due to changes in attachment, detachment, growth, or predation of the epiphytic bacteria. Of these, the most likely to cause the observed disparity are attachment and/or growth. Attachment rate might be greater in the water because of the higher number of bacteria per unit volume; often $>10^5$ ml^{-1} in fresh water, compared to perhaps 0 to 10^6 m^{-3} in air (Edmonds, 1979).

Microorganisms on aerial leaves, as on submerged leaves, have a nonuniform horizontal distribution. Leben (1965) found accumulation of bacteria in depressions at the junctures of epidermal cells on cucumber (*Cucumis sativus* L.) leaves and suggested these were regions of greater exudation from the leaves. Birkby and Preece (1987) observed extensive populations of bacteria and fungi on ligules and adjacent leaf blade in the grass *Dactylis glomerata* L., while the remainder of the leaf blade was sparsely populated. They attributed these patterns to accumulation of debris, water, and nutrients in the junction of ligule and leaf blade. Also, Corpe and Rheem (1989) used SEM to show concentration of bacteria at leaf margins of *Trifolium repens* where bacteria avoided wax-coated areas of the lamina. There is also between-leaf variation in bacterial abundance on aerial foliage; a 1000-fold variation in population size, from leaf to leaf, has been reported (Hirano and Upper, 1986). Such variation may to some extent be related to position of leaves in the canopy, e.g., on apple (*Malus pumila* Mill.) trees (Andrews et al., 1980) and, as with submerged leaves, to leaf age, e.g., on *Lolium perenne* (Dickinson et al., 1975) and chicory (Van Outryve et al., 1989). The temporal variation of bacteria on aerial leaf surfaces, unlike that on freshwater foliage, has been extensively explored. Here, increase in abundance with progression of season from spring to autumn is a frequent observation, e.g., on beech (Jensen, 1971), *Lolium perenne* (Dickinson et al., 1975), apple (Andrews et al., 1980), wheat (Fokkema and Schippers, 1986), and diverse cultivated plants (Hirano and Upper, 1986).

Exchange of freshwater bacteria between the planktonic and epiphytic mode may be important, especially in calcareous headstreams. It appears that this parallels the situation at surfaces of terrestrial foliage, where leaves are surrounded by moving air. Detachment of bacteria and fungal propagules from aerial leaves (Grace and Collins, 1976; Lindemann et al., 1982) contributes to the diverse microflora present in air (Gregory, 1973; Edmonds, 1979), while the trapping of airborne microorganisms at leaf surfaces (Gregory, 1971) may enhance phylloplane populations. Lindow et al. (1988) showed trapping of airborne *Pseudomonas syringae* van Hall by oat (*Avena sativa* L.) leaves, and Kinkel et al. (1989) measured immigration of fungal propagules to disinfected apple leaves at 300 to 2.9×10^4 per leaf during 12-hour periods. This represented perhaps 1–15% of individuals naturally present on the leaves.

Finally, we draw attention to the role of the surrounding fluid in the nutrition of phylloplane microorganisms. Epiphytic bacteria in freshwaters can utilize dissolved organic and inorganic nutrients in the surrounding water as well as exudates from their macrophyte host. This appears to be especially important in organically enriched waters (e.g., waters receiving sewage-works effluent). Epiphytic microalgae also obtain inorganic nutrients from the surrounding water, and there is controversy over whether hosts act merely as inert surfaces. In contrast, microbiota on aerial leaves must rely largely on their hosts as nutrient source, along with, probably to a lesser extent, input from dust, pollen, rainwater, seaspray, and atmospheric pollution. Hence organic and inorganic exudates of aerial leaves (e.g., Tukey, 1971; Godfrey, 1976) are of vital importance to aerial leaf microflora whereas similar exudates may be less critical to the freshwater phylloplane microflora.

References

Al-Hadithi, S.A. and Goulder, R. 1989a. Physiological state of epiphytic bacteria on submerged stems of the reed *Phragmites australis* compared with planktonic bacteria in gravel-pit ponds. *Journal of Applied Bacteriology* 66:107–117.

Al-Hadithi, S.A. and Goulder, R. 1989b. A note on the physiological state of epiphytic bacteria compared with planktonic bacteria in two organically-enriched watercourses. *Journal of Applied Bacteriology* 67:293–297.

Allanson, B.R. 1973. The fine structure of the periphyton of *Chara* sp. and *Potamogeton natans* from Wytham Pond, Oxford, and its significance to the macrophyte-periphyton metabolic model of R G Wetzel and H L Allen. *Freshwater Biology* 3:535–542.

Allen, H.L. and Ocevski, B.T. 1982. Comparative seasonal uptake of glucose and acetate by epiphytic bacteria on the surfaces of three species of macrophyte in the littoral zone of Lake Ohrid, Yugoslavia. *Archiv für Hydrobiologie* 93:423–445.

Andrews, J.H., Kenerley, C.M., and Nordheim, E.V. 1980. Positional variation in phylloplane microbial populations within an apple tree canopy. *Microbial Ecology* 6:71–84.

Austin, B., Goodfellow, M., and Dickinson, C.H. 1978. Numerical taxonomy of phylloplane bacteria isolated from *Lolium perenne*. *Journal of General Microbiology* 104:139–155.

Baker, J.H. 1988. Epiphytic bacteria. pp. 171–191 in Austin, B. (editor), *Methods in Aquatic Microbiology*. John Wiley & Sons, New York.

Baker, J.H. and Farr, I.S. 1977. Origins, characterization and dynamics of suspended bacteria in two chalk streams. *Archiv für Hydrobiologie* 80:308–326.

Baker, J.H. and Farr, I.S. 1987. Importance of dissolved organic matter produced by duckweed (*Lemna minor*) in a southern English river. *Freshwater Biology* 17:324–330.

Baker, J.H. and Orr, D.R. 1986. Distribution of epiphytic bacteria on freshwater plants. *Journal of Ecology* 74:155–165.

Baldock, B.M., Baker, J.H., and Sleigh, M.A. 1983. Abundance and productivity of protozoa in chalk streams. *Holarctic Ecology* 6:238–246.

Beckett, P.M., Armstrong, W., Justin, S.H.F.W., and Armstrong, J. 1988. On the relative importance of convective and diffusive gas flows in plant aeration. *New Phytologist* 110:463–468.

Birkby, K.M. and Preece, T.F. 1987. The ligules of green leaves of cock's-foot grass, *Dactylis glomerata* L., are a micro-habitat for bacteria and fungi. *Journal of Applied Bacteriology* 63:505–511.

Blakeman, J.P. (editor). 1981. *Microbial Ecology of the Phylloplane*. Academic Press, London.

Bownik, L.J. 1970. The periphyton of the submerged macrophytes of Mikolajskie Lake. *Ekologia Polska* 18:503–520.

Burkholder, J.M. and Wetzel, R.G. 1989. Microbial colonization on natural and artificial macrophytes in a phosphorus-limited, hardwater lake. *Journal of Phycology* 25:55–65.

Burrage, S.W. 1971. The micro-climate at the leaf surface. pp. 91–101 in Preece, T.F. and Dickinson, C.H. (editors), *Ecology of Leaf Surface Micro-organisms*. Academic Press, London.

Burrage, S.W. 1976. Aerial microclimate around plant surfaces. pp. 173–184 in Dickinson, C.H. and Preece, T.F. (editors), *Microbiology of Aerial Plant Surfaces*. Academic press, London.

Buscemi, P.A. 1958. Littoral oxygen depletion produced by a cover of *Elodea canadensis*. *Oikos* 9:239–245.

Carignan, R. and Kalff, J. 1982. Phosphorus release by submerged macrophytes: significance to epiphyton and phytoplankton. *Limnology and Oceanography* 27:419–427.

Cattaneo, A. 1978. The microdistribution of epiphytes on the leaves of natural and artificial macrophytes. *British Phycological Journal* 13:183–188.

Cattaneo, A. and Kalff, J. 1979. Primary production of algae growing on natural and artificial aquatic plants: a study of interactions between epiphytes and their substrate. *Limnology and Oceanography* 24:1031–1037.

Corpe, W.A. and Rheem, S. 1989. Ecology of methylotrophic bacteria on living leaf surfaces. *FEMS Microbiology Ecology* 62:243–250.

Costerton, J.W., Cheng, K.-J., Geesey, G.G., Ladd, T.I., Nickel, J.C., Dasgupta, M., and Marrie, T.J. 1987. Bacterial biofilms in nature and disease. *Annual Review of Microbiology* 41:435–464.

Dale, H.M. and Gillespie, T.J. 1977. The influence of submersed aquatic plants on temperature gradients in shallow water bodies. *Canadian Journal of Botany* 55:2216–2225.

Dickinson, C.H. and Preece, T.F. (editors). 1976. *Microbiology of Aerial Plant Surfaces*. Academic Press, London.

Dickinson, C.H., Austin, B., and Goodfellow, M. 1975. Quantitative and qualitative studies of phylloplane bacteria from *Lolium perenne*. *Journal of General Microbiology* 91:157–166.

Edmonds, R.L. (editor). 1979. *Aerobiology, The Ecological Systems Approach*. Dowden, Hutchinson & Ross, Stroudsburg, Pennsylvania.

Eminson, D. and Moss, B. 1980. The composition and ecology of periphyton communities in freshwaters. I. The influence of host type and external environment on community composition. *British Phycological Journal* 15:429–446.

Fokkema, N.J. and Schippers, B. 1986. Phyllosphere versus rhizosphere as environments for saprophytic colonization. pp. 137–159 in Fokkema, N.J. and van den Heuvel, J. (editors), *Microbiology of the Phyllosphere*. Cambridge University Press, Cambridge, UK.

Fokkema, N.J. and van den Heuvel, J. (editors). 1986. *Microbiology of the Phyllosphere*. Cambridge University Press, Cambridge, UK.

Fontaine, T.D. and Nigh, D.G. 1983. Characteristics of epiphyte communities

on natural and artificial submersed lotic plants: substrate effects. *Archiv für Hydrobiologie* 96:293–301.
Fry, J.C. and Davies, A.R. 1985. An assessment of methods for measuring volumes of planktonic bacteria, with particular reference to television image analysis. *Journal of Applied Bacteriology* 58:105–112.
Fry, J.C. and Humphrey, N.C.B. 1978. Techniques for the study of bacteria epiphytic on aquatic macrophytes. pp. 1–29 in Lovelock, D.W. and Davies, R. (editors), *Techniques for the Study of Mixed Populations*. Society for Applied Bacteriology, Technical Series No. 11. Academic Press, London.
Fry, J.C. and Zia, T. 1982. A method for estimating viability of aquatic bacteria by slide culture. *Journal of Applied Bacteriology* 53:189–198.
Fry, J.C., Goulder, R., and Rimes, C.A. 1985. A note on the efficiency of stomaching for the quantitative removal of epiphytic bacteria from submerged aquatic plants. *Journal of Applied Bacteriology* 58:113–115.
Fuhrman, J.A. and Azam, F. 1982. Thymidine incorporation as a measure of heterotrophic bacterioplankton production in marine surface waters: evaluation and field results. *Marine Biology* 66:109–120.
Godfrey, B.E.S. 1976. Leachates from aerial parts of plants and their relation to plant surface microbial populations. pp. 433–439 in Dickinson, C.H. and Preece, T.F. (editors), *Microbiology of Aerial Plant Surfaces*. Academic Press, London.
Goodfellow, M., Austin, B., and Dawson, D. 1976. Classification and identification of phylloplane bacteria using numerical taxonomy. pp. 275–292 in Dickinson, C.H. and Preece, T.F. (editors), *Microbiology of Aerial Plant Surfaces*. Academic Press, London.
Goulder, R. 1969. Interactions between the rates of production of a freshwater macrophyte and phytoplankton in a pond. *Oikos* 20:300–309.
Goulder, R. 1970. Day-time variations in the rates of production by two natural communities of submerged freshwater macrophytes. *Journal of Ecology* 58:521–528.
Goulder, R. 1980. Seasonal variation in heterotrophic activity and population density of planktonic bacteria in a clean river. *Journal of Ecology* 68:349–363.
Goulder, R. 1984. Downstream increase in the abundance and heterotrophic activity of suspended bacteria in an intermittent calcareous headstream. *Freshwater Biology* 14:611–619.
Goulder, R. 1990. Extracellular enzyme activities associated with epiphytic microbiota on submerged stems of the reed *Phragmites australis*. *FEMS Microbiology Ecology* 73:323–330.
Grace, J. and Collins, M.A. 1976. Spore liberation from leaves by wind. pp. 185–198 in Dickinson, C.H. and Preece, T.F. (editors), *Microbiology of Aerial Plant Surfaces*. Academic Press, London.
Gregory, P.H. 1971. The leaf as a spore trap. pp. 239–243 in Preece, T.F. and Dickinson, C.H. (editors), *Ecology of Leaf Surface Micro-organisms*. Academic Press, London.
Gregory, P.H. 1973. *The Microbiology of the Atmosphere*, 2nd edition. Leonard Hill Books, Aylesbury.
Haines, D.W., Rogers, K.H., and Rogers, F.E.J. 1987. Loose and firmly attached epiphyton: their relative contributions to algal and bacterial carbon productivity in a *Phragmites* marsh. *Aquatic Botany* 29:169–176.
Hickman, M. 1971. The standing crop and primary productivity of the epiphyton attached to *Equisetum fluviatile* L. in Priddy Pool, North Somerset. *British Phycological Journal* 6:51–59.
Hickman, M. and Klarer, D.M. 1973. Methods for measuring the primary pro-

ductivity and standing crops of an epiphytic algal community attached to *Scirpus validus* Vahl. *Internationale Revue der Gesamten Hydrobiologie* 58:893–901.

Hirano, S.S. and Upper, C.D. 1986. Temporal, spatial, and genetic variability of leaf-associated bacterial populations. pp. 235–251 in Fokkema, N.J. and van den Heuvel, J. (editors), *Microbiology of the Phyllosphere*. Cambridge University Press, Cambridge, UK.

Hoppe, H.-G. in press. Microbial extracellular enzyme activity: a new key parameter in aquatic ecology. In Chróst, R.J. (editor), *Microbial Enzymes in Aquatic Environments*. Springer-Verlag, New York.

Hossell, J.C. and Baker, J.H. 1979a. Epiphytic bacteria of the freshwater plant *Ranunculus penicillatus*: enumeration, distribution and identification. *Archiv für Hydrobiologie* 86:322–337.

Hossell, J.C. and Baker, J.H. 1979b. A note on the enumeration of epiphytic bacteria by microscopic methods with particular reference to two freshwater plants. *Journal of Applied Bacteriology* 46:87–92.

Hough, R.A. and Wetzel, R.G. 1975. The release of dissolved organic carbon from submersed aquatic macrophytes: diel, seasonal, and community relationships. *Verhandlungen der Internationalen Vereinigung für Theoretische und Angewandte Limnologie* 19:939–948.

Jenkerson, C.G. and Hickman, M. 1986. Interrelationships among the epipelon, epiphyton and phytoplankton in a eutrophic lake. *Internationale Revue der Gesamten Hydrobiologie* 71:557–579.

Jensen, V. 1971. The bacterial flora of beech leaves. pp. 463–469 in Preece, T.F., and Dickinson, C.H. (editors), *Ecology of Leaf Surface Micro-organisms*. Academic Press, London.

Jones, J.G. 1970. Studies on freshwater bacteria: effect of medium composition and method on estimates of bacterial population. *Journal of Applied Bacteriology* 33:679–686.

Kinkel, L.L., Andrews, J.H., and Nordheim, E.V. 1989. Fungal immigration dynamics and community development on apple leaves. *Microbial Ecology* 18:45–58.

Kjelleberg, S., Hermansson, M., and Marden, P. 1987. The transient phase between growth and non-growth of heterotrophic bacteria, with emphasis on the marine environment. *Annual Review of Microbiology* 41:25–49.

Kudryavtsev, V.M. 1984. Bacteria on vascular aquatic plants. *Hydrobiological Journal* 19:50–55.

Leben, C. 1965. Influence of humidity on the migration of bacteria on cucumber seedlings. *Canadian Journal of Microbiology* 11:671–676.

Lindemann, J., Constantinidou, H.A., Barchet, W.R., and Upper, C.D. 1982. Plants as sources of airborne bacteria, including ice nucleation-active bacteria. *Applied and Environmental Microbiology* 44:1059–1063.

Lindow, S.E., Knudsen, G.R., Seidler, R.J., Walter, M.V., Lambou, V.W., Amy, P.S., Schmedding, D., Prince, V., and Hern, S. 1988. Aerial dispersal and epiphytic survival of *Pseudomonas syringae* during a pretest for the release of genetically engineered strains into the environment. *Applied and Environmental Microbiology* 54:1557–1563.

Lorch, H.-J. and Ottow, J.C.G. 1985. Use of bacteria attached to submerged macrophytes and glass slides as indicators of an increasing water pollution. *Verhandlungen der Internationalen Vereinigung für Theoretische und Angewandte Limnologie* 22:2297–2303.

Lorch, H.-J. and Ottow, J.C.G. 1986. Scanning electron microscopy of bacteria and diatoms attached to a submerged macrophyte in an increasingly polluted stream. *Aquatic Botany* 26:377–384.

Madsen, T.V. and Warncke, E. 1983. Velocities of currents around and within submerged aquatic vegetation. *Archiv für Hydrobiologie* 97:389–394.
Meulemans, J.T. 1988. Seasonal changes in biomass and production of periphyton growing upon reed in Lake Maarsseveen. *Archiv für Hydrobiologie* 112: 21–42.
Monteith, J.L. and Unsworth, M.H. 1990. *Principles of Environmental Physics*, 2nd edition. Edward Arnold, London.
Morin, J.O. 1986. Initial colonization of periphyton on natural and artificial apices of *Myriophyllum heterophyllum* Michx. *Freshwater Biology* 16:685–694.
Morin, J.O. and Kimball, K.D. 1983. Relationship of macrophyte-mediated changes in the water column to periphyton composition and abundance. *Freshwater Biology* 13:403–414.
Moss, B. 1981. The composition and ecology of periphyton communities in freshwaters. II. Inter-relationships between water chemistry, phytoplankton populations and periphyton populations in a shallow lake and associated experimental reservoirs ('Lund tubes'). *British Phycological Journal* 16:59–76.
Preece, T.F. and Dickinson, C.H. (editors). 1971. *Ecology of Leaf Surface Microorganisms*. Academic Press, London.
Prins, H.B.A. and Elzenga, J.T.M. 1989. Bicarbonate utilization: function and mechanism. *Aquatic Botany* 34:59–83.
Ramsay, A.J. 1974. The use of autoradiography to determine the proportion of bacteria metabolizing in an aquatic habitat. *Journal of General Microbiology* 80:363–373.
Raskin, I. and Kende, H. 1983. How does deep water rice solve its aeration problem. *Plant Physiology* 72:447–454.
Raskin, I. and Kende, H. 1985. Mechanism of aeration in rice. *Science* 228:327–329.
Rimes, C.A. and Goulder, R. 1985. A note on the attachment rate of suspended bacteria to submerged aquatic plants in a calcareous stream. *Journal of Applied Bacteriology* 59:389–392.
Rimes, C.A. and Goulder, R. 1986a. Temporal variation in density of epiphytic bacteria on submerged vegetation in a calcareous stream. *Letters in Applied Microbiology* 3:17–21.
Rimes, C.A. and Goulder, R. 1986b. Quantitative observations on the ability of epiphytic bacteria to contribute to the populations of suspended bacteria in two dissimilar headstreams. *Freshwater Biology* 16:301–311.
Rimes, C.A. and Goulder, R. 1986c. Suspended bacteria in calcareous and acid headstreams; abundance, heterotrophic activity and downstream change. *Freshwater Biology* 16:633–651.
Rimes, C.A. and Goulder, R. 1987. Relations between suspended bacteria, epiphytic bacteria and submerged vegetation over the spring growing season in a calcareous headstream. *Freshwater Biology* 17:291–305.
Rogers, K.H. and Breen, C.M. 1981. Effects of epiphyton on *Potamogeton crispus* L. leaves. *Microbial Ecology* 7:351–363.
Sand-Jensen, K. 1989. Environmental variables and their effect on photosynthesis of aquatic plant communities. *Aquatic Botany* 34:5–25.
Sand-Jensen, K. and Sondergaard, M. 1981. Phytoplankton and epiphyte development and their shading effect on submerged macrophytes in lakes of different nutrient status. *Internationale Revue der Gesamten Hydrobiologie* 66:529–552.
Sand-Jensen, K., Revsbech, N.P., and Jorgensen, B.B. 1985. Microprofiles of oxygen in epiphyte communities on submerged macrophytes. *Marine Biology* 89:55–62.

Sand-Jensen, K., Jeppesen, E., Nielsen, K., Van der Bijl, L., Hjermind, L., Nielsen, L.W., and Iversen, T.M. 1989a. Growth of macrophytes and ecosystem consequences in a lowland Danish stream. *Freshwater Biology* 22:15–32.

Sand-Jensen, K., Borg, D., and Jeppesen, E. 1989b. Biomass and oxygen dynamics of the epiphyte community in a Danish lowland stream. *Freshwater Biology* 22:431–443.

Silvester, N.R. and Sleigh, M.A. 1985. The forces on microorganisms at surfaces in flowing water. *Freshwater Biology* 15:433–448.

Smith, C.S., Chand, T., Harris, R.F., and Andrews, J.H. 1989. Colonization of a submersed aquatic plant, Eurasian water milfoil (*Myriophyllum spicatum*), by fungi under controlled conditions. *Applied and Environmental Microbiology* 55:2326–2332.

Steemann Nielsen, E. 1952. The use of radio- active carbon (C^{14}) for measuring organic production in the sea. *Journal du Conseil Permanent International pour l'Exploration de la Mer*. 18:117–140.

Stott, M.A. 1971. Studies on the physiology of some leaf saprophytes. pp. 203–210 in Preece, T.F. and Dickinson, C.H. (editors), *Ecology of Leaf Surface Micro-organisms*. Academic Press, London.

Tabor, P.S. and Neihof, R.A. 1982. Improved method for determination of respiring individual microorganisms in natural waters. *Applied and Environmental Microbiology* 43:1249–1255.

Tukey, H.B. 1971. Leaching of substances from plants. pp. 67–80 in Preece, T.F. and Dickinson, C.H. (editors), *Ecology of Leaf Surface Micro-organisms*. Academic Press, London.

Van Outryve, M.F., Gosselé, F., and Swings, J. 1989. The bacterial microflora of witloof chicory (*Cichorium intybus* L. var. *foliosum* Hegi) leaves. *Microbial Ecology* 18:175–186.

Westlake, D.F. 1964. Light extinction, standing crop and photosynthesis within weed beds. *Verhandlungen der Internationalen Vereinigung für Theoretische und Angewandte Limnologie* 15:415–425.

Westlake, D.F. 1966. The light climate for plants in rivers. pp. 99–119 in Bainbridge, R., Evans, G.C., and Rackham, O. (editors), *Light as an Ecological Factor*. Blackwell Scientific Publications, Oxford.

Westlake, D.F. 1967. Some effects of low-velocity currents on the metabolism of aquatic macrophytes. *Journal of Experimental Botany* 18:187–205.

Wetzel, R.G. 1983 *Limnology*, 2nd edition. W.B. Saunders Company, Philadelphia.

Wright, R.T. and Hobbie, J.E. 1966. Use of glucose and acetate by bacteria and algae in aquatic ecosystems. *Ecology* 47:447–464.

Zimmermann, R., Iturriaga, R., and Becker-Birck, J. 1978. Simultaneous determination of the total number of aquatic bacteria and the number thereof involved in respiration. *Applied and Environmental Microbiology* 36:926–935.

5

Influence of Leaf Surface Features on Spore Deposition and the Epiphytic Growth of Phytopathogenic Fungi

Eric A. Allen, Harvey C. Hoch, James R. Steadman, and Rennie J. Stavely

5.1 Introduction

Leaf surfaces provide an important substrate for the growth of a wide variety of fungal organisms. Saprophytic fungi obtain water and nutrients from these surfaces and the leaf surface provides a physical environment suitable for growth and reproduction. For phytopathogenic fungi, the leaf surface represents a temporary, supportive environment, but a surface that must be breached before a successful pathogenic interaction can become established. Thus, the physical and chemical characteristics of the leaf surface play an important role in governing the success or failure of fungal growth on, and subsequently in, the leaf. Numerous reviews have been written about the growth of fungi in association with plant surfaces (e.g., Preece, 1976; Aist, 1981; Wynn, 1981; Staples and Macko, 1984; Hoch and Staples, 1987, 1991). This chapter focuses on features of the leaf surface as they influence the establishment and growth of various fungi. Attention will be given to the time and place of initial spore deposition on the leaf, to growth over the surface, and in the case of pathogenic fungi, to their penetration into the leaf. Since the authors' research experience lies in the growth and development of rust fungi, considerable emphasis will be placed on examples involving these organisms.

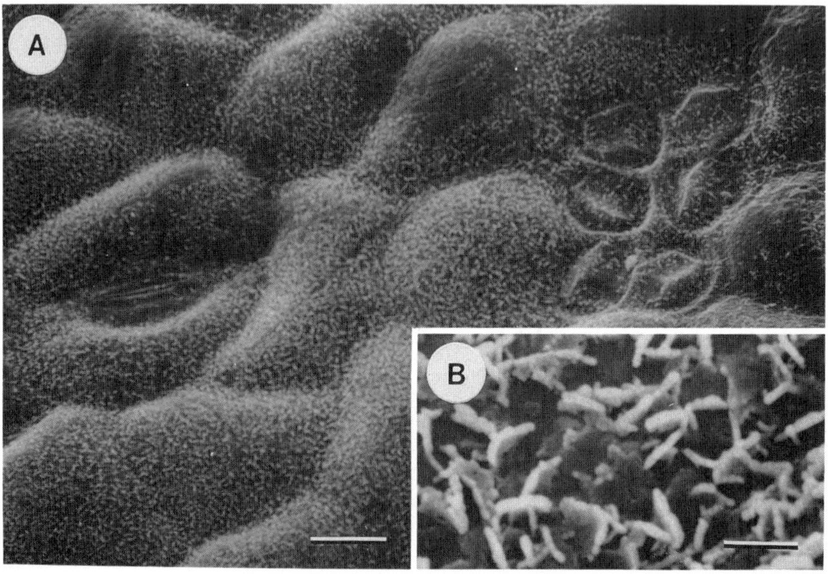

Figure 5.1 Epicuticular wax crystals on the adaxial surface of *Arachis*. SEM preparation; glutaraldehyde-fixed, critical-point-dried, gold-coated. (A) Bar equals 10 μm. (B) Bar equals 1 μm.

5.2 Leaf Surface Features

The vast amount of variation observed in the physical characteristics of leaves in nature makes it difficult to generalize about fungal responses to leaf surfaces. Leaf surface features vary among plant species, cultivars of the same species, leaves on the same plant, and commonly, between the upper and lower surfaces on the same leaf. Nevertheless, several common features are shared by leaves; themes on which nature has played many variations. Several excellent and detailed reviews devoted to leaf surface morphology are those by Esau (1977), Hallam and Juniper (1971), and Cutter (1976).

Epidermal cells form the outer layer of the leaf surface and are covered with a cuticular layer of variable thickness. The cuticle provides protection from water loss and, relevant to the current discussion, acts as a barrier to pathogenic organisms (Martin and Juniper, 1970; Köller, 1990). Wax crystals, which appear in many forms (Jeffree et al., 1976), cover the cuticle to varying degrees and contribute to the hydrophobic properties of the leaf surface (Figure 5.1). Epidermal cells occur in different shapes and are frequently modified to form a variety of cell types including trichomes or hairs, secretory glands, and stomata. Trichomes are projections from the leaf surfaces that exhibit extreme variability in size, morphology, and func-

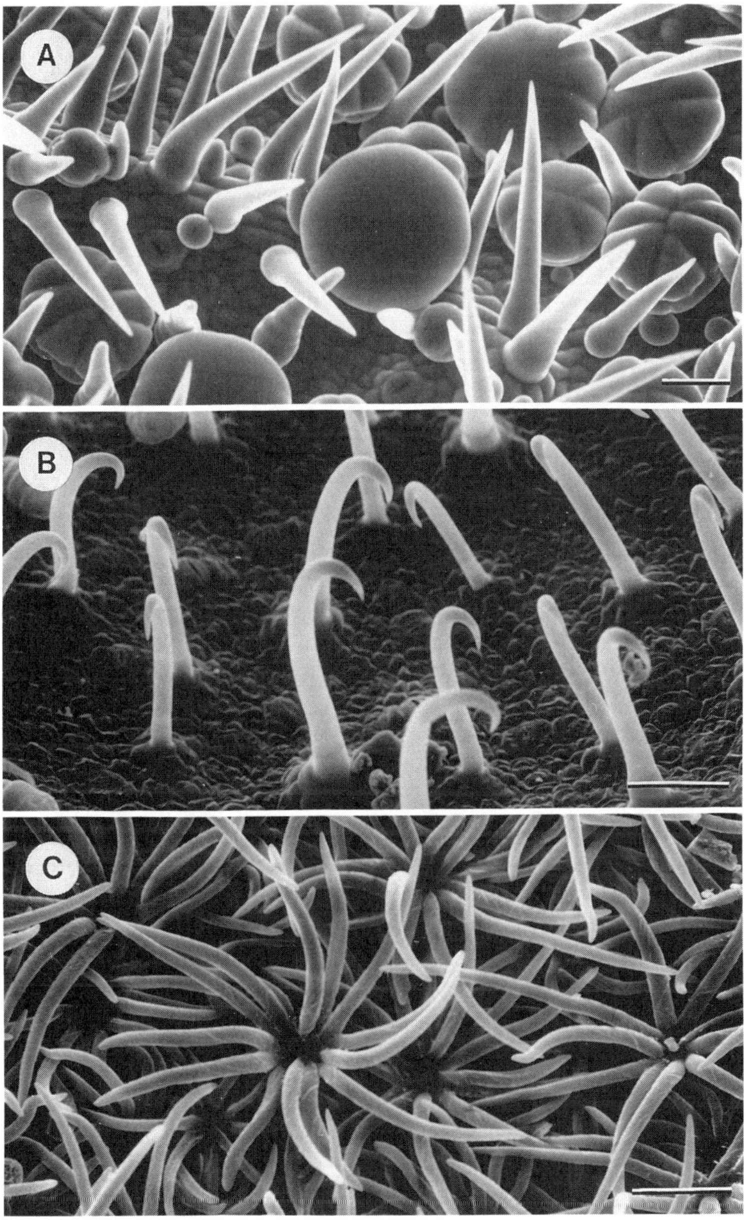

Figure 5.2 Various morphological forms of trichomes on leaf surfaces. (A) Straight hairs and glandular trichomes on the adaxial surface of *Phaseolus anisotrichus*. SEM preparation; fresh leaf, no further preparation. Bar equals 25 μm. (B) Hooked trichomes on the adaxial surface of a trifoliolate leaf of *Phaseolus vulgaris*. SEM preparation; glutaraldehyde-fixed, critical-point-dried, gold-coated. Bar equals 50 μm. (C) Stellate trichomes on the abaxial surface of *Quercus bicolor*. SEM preparation; fresh leaf, gold-coated. Bar equals 50 μm.

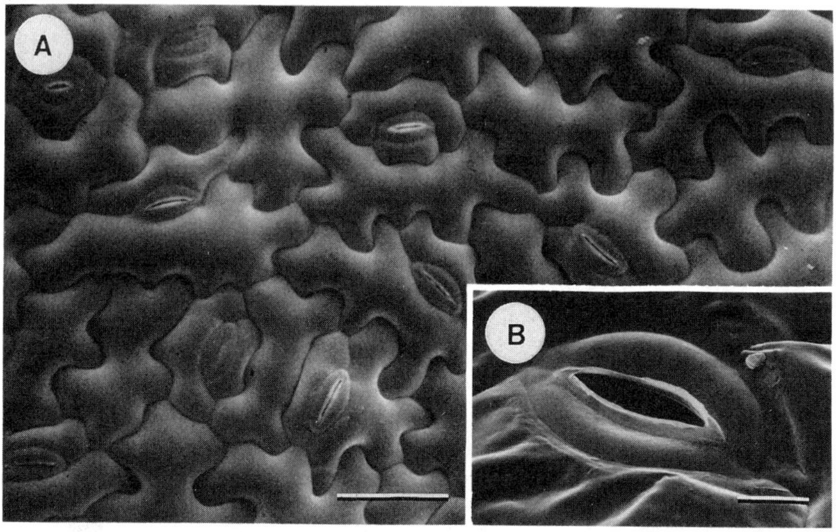

Figure 5.3 (A) Adaxial surface of *Phaseolus vulgaris*. Stomata are interspersed among irregularly shaped epidermal cells. SEM preparation; frozen-hydrated, gold-coated. Bar equals 50 μm. (B) Stoma of *Phaseolus vulgaris* showing the guard-cell "lips." SEM preparation; glutaraldehyde-fixed, freeze-dried, gold-palladium-coated. Bar equals 5 μm.

tion. These hair-like structures may be straight (Figure 5.2A), hooked (Figure 5.2B), or may form complex stellate structures (Figure 5.2C). Glands are often stalked on trichomes (Figure 5.2A) or are sessile and secrete a wide range of substances, some of which stimulate or inhibit the growth of microorganisms (Blakeman and Atkinson, 1981). Stomata are openings in the epidermis that function in gas exchange through the opening and closing of the stomatal aperture(s) (Figure 5.3). There is considerable variation in both the morphology and epidermal position of stomata, important factors in fungal interactions with these structures (Palevitz, 1981; Wilkinson, 1979).

5.3 Spore Deposition

The initial interaction between phylloplane fungi and the plants on which they grow begins when spores or other fungal propagules land on the leaf surface. Deposition of spores on leaves is very complex and is influenced by the interaction of factors such as spore size, wind speed, and the moisture conditions of both the spore and the leaf (Ingold, 1971). Leaves can be viewed as spore traps (Gregory, 1971) with trapping efficiency dependent largely upon their physical characteristics. For example, leaf shape varies dramatically, from cylindrical conifer leaves to laminate leaves of varying width and levels of dissection. Leaf shape can exert a significant influence on the aerodynamic properties of the leaf, and therefore on how and where spores are deposited. Carter (1965) studied ascospore deposition of *Euphyta armenicae* on petioles, stems, and leaf lamina of apricot. He demonstrated that spore impaction was more efficient on leaf petioles (1 mm in diameter) than on stems (5 mm in diameter) or the flat leaf laminae. His study on natural plant surfaces complemented the earlier work of Gregory (1951) who showed that the efficiency of spore impaction was highest on narrow cylindrical objects at high wind speeds. On laminate leaves, variation in air currents over the leaf surface due to edge effects or movement of the leaf blade strongly influence the dynamics of spore deposition (Gregory and Stedman, 1953).

Hairs and other related structures also act to prevent spores from reaching the leaf surface. This may occur in a number of ways; for example, in the simplest case, hairs form a passive screen on which spores deposited by sedimentation are caught. On some leaf surfaces with numerous long (1–2 mm) hairs such as those found on *Phaseolus vulgaris* cv. Pompadour (PC-50), this screen of hairs is thought to result in a significant reduction in the number of bean rust spores reaching the epidermis (Steadman and Mmbaga, unpublished). Hairs may also provide surfaces upon which spores are impacted. Gregory (1973) suggested that spore deposition on leaves was increased by impaction of small spores on hairs, owing to the high trapping efficiency of these structures. When spores entrapped on hairs germinate, the resulting germ tubes often become entangled in the canopy, and as a result they do not reach the leaf surface (Heath, 1974; Wynn, 1976; von Alten, 1983). Fungal germ tubes, exemplified by *Uromyces appendiculatus*, can frequently be observed suspended between leaf hairs of pubescent leaves (Figure 5.4).

The dynamics of the deposition and evaporation of water from rain or dew on pubescent leaf surfaces are important to the successful germination of spores, survival of the resulting germlings, and the probability of initiating the infection process. Burrage (1969) observed that condensation of water, in the form of dew on wheat leaves, began on leaf hairs. As more water condensed, droplets increased in size, eventually touching the cuticle where they remained, forming a high contact angle with the leaf surface. Hydrophobic spores, restricted to the droplet surface, migrate passively by surface tension and gravitational forces to the water/leaf interface. When droplets remain discrete, the spores usually germinate close to the leaf surface. Increasing hair density, however, produces more droplet foci, which eventually coalesce to form a continuous water film. Burrage suggested that hydrophobic spores would continue to be held on the surface of such water films, physically separated from the leaf surface (Figure 5.4). The structure, density, and hydrophobicity of trichomes of *Salvinia* is such that water droplets or films remain suspended within the hairs above the

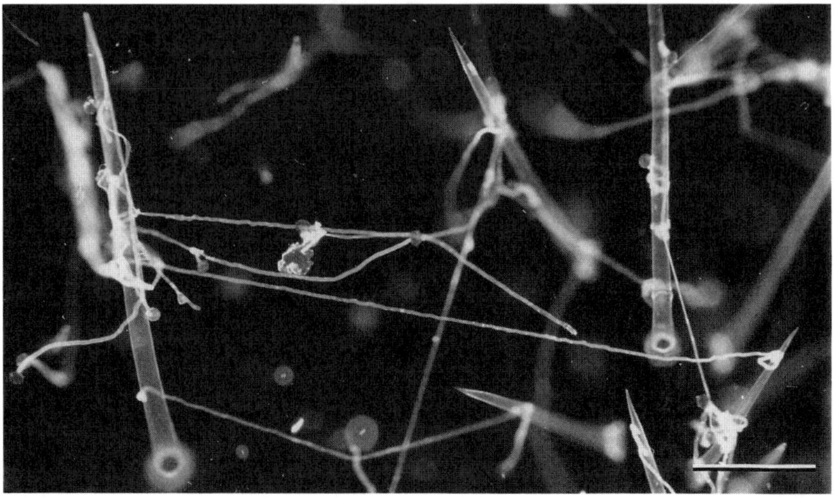

Figure 5.4 Germ tubes of *Uromyces appendiculatus* suspended among trichomes on the adaxial surface of *Phaseolus vulgaris*. Calcofluor-stained fresh leaf viewed with UV-fluorescence illumination. Bar equals 200 μm.

surface, separated from the leaf epidermis (Martin and Juniper, 1970). Germlings that remain in the suspended film likely perish when the water evaporates. Survival of fungal structures could also be enhanced by the presence of hairs. For example, when spores penetrate the pubescence "screen,"and are present on the leaf surface at the time of dew formation, moisture necessary for spore germination would be provided by the water film, while the hairs would prevent the spores from being dislodged by water runoff.

Leaf pubescence has been considered as a potential mechanism for nonspecific plant resistance to fungal pathogens (Steadman and Shaik, 1988; Zaiter et al., 1990). Shaik (1985) and Steadman and Shaik (1988), for example, reported that low-intensity development of rust pustules was correlated with bean leaf pubescence and that this correlation was not altered by rust races. Rust pustule intensity was lowest when inoculation was made on young, unexpanded leaves where hair density was highest. Rust intensity was also lower on the third or higher set of trifoliolate leaves (with high straight-hair density) than on primary or first trifoliolate leaves (no abaxial straight hairs).

Sticky exudates produced by glands and glandular trichomes are known to act as traps for pollen (Konar and Linskens, 1966) and insects (Southwood, 1986). It is quite likely therefore, that fungal spores as well as other forms of biota are also trapped on these structures.

The interaction between insects and leaf surface features may also affect spore deposition as many insects are potential vectors of fungal propagules (Ingold, 1971). Surface features such as trichomes or cuticular waxes have been demonstrated to repel insects (Webster, 1975; Southwood, 1986) and could reduce the deposition of spores carried by these insects.

5.4 Spore Retention

After spores have reached the leaf surface, their attachment and retention on the surface are once again affected by the physical characteristics of the leaf. Just as leaf hairs, glands, or other structures that protrude above the leaf surface affect spore deposition, spore retention is also influenced. The height, density, and distribution of these leaf features determine the thickness of the laminar boundary layer, and thus the deposition of spores whether by sedimentation or impaction on the leaf surface.

Other physical characteristics of the leaf surface are related to spore retention and water runoff. The hydrophobicity of a leaf surface is primarily determined by the nature of the cuticular waxes, their chemical composition, and the size, shape, and distribution of wax crystals (Martin and Juniper, 1970). Other factors such as the size and distribution of trichomes can also be involved in determining the wettability of the leaf surface (Burrage, 1969). On highly hydrophobic leaf surfaces, i.e., with low wettability, water droplets maintain a high contact angle and are readily removed if the leaf is shaken or inclined (Heather, 1967a, 1967b). Davies (1961) found that the wettability of both spores and leaf surfaces influenced the deposition and distribution of spores. He reported that nonwettable spores held on the surface of water droplets were deposited on the hydrophobic leaf surfaces over which they passed, whereas wettable spores were carried away with the droplets. In contrast, Cook (1980) observed a low retention of highly water-repellent peanut rust spores on the hydrophobic adaxial leaf surfaces of peanut (Figure 5.1). Often, spores are ultimately deposited in locations on leaves and stems where runoff water collects, such as along veins of laminate leaves or in axils of leaf, bud, or flower parts (Johnstone, 1931). Alternatively, a wettable surface tends to hold water droplets or films, and spores associated with such surfaces are more readily retained, possibly accounting for increased infection efficiencies (Burrage, 1969). This "infection-wettable surface" association was demonstrated by Fisher and Corke (1971) who attempted to explain differences between the infection of red and green apple fruit surfaces by *Venturia inequalis*, the apple scab fungus, on the basis of differences in the amount and form of cuticular waxes on the two surfaces, corresponding to the hydrophobicity of the surfaces. The more wettable red fruit surface exhibited higher levels of infection than the more hydrophobic green fruit surface.

The attachment of fungal propagules to leaf surfaces is often accomplished through interactions of the leaf surface with adhesive materials secreted by the fungus. In some cases, spores arrive at the plant in a "sticky" condition, and at times in groups, through the prior wetting of mucilaginous substances bound to the spore surface (Ingold, 1978). Other fungi release adhesive compounds after contact with the leaf surface. Paus and Raa (1973) and Hamer et al. (1988) described the secretion of adhesive materials by spores prior to germination. The production of adhesive substances by some spores, e.g., *Nectria haematococca* (Jones and Epstein, 1989) and *Colletotrichum lindemuthianum* (Young and Kauss, 1984), requires respiration and protein synthesis. In contrast, adhesive substances may be preformed, as in *Magnaporthe grisea*. Conidia of this fungus have been shown to attach in the presence of metabolic inhibitors (Hamer et al., 1988). Many other fungi produce adhesives only after the initiation of germ tube growth (McKeen, 1974).

Secretions of fungal spores and germlings function as more than just binding agents; they may also modify the leaf surface in preparation for spore germination and/or germ tube penetration of the leaf epidermis. A recent study by Kunoh et al. (1988) and Nicholson et al. (1988) showed that within 10 min of contact with a barley leaf surface, but prior to germ tube emergence, *Erysiphe graminis* conidia released a liquid film onto the leaf surface. Nicholson et al. (1988) demonstrated that the secretion contained a mixture of proteins, some of which had esterase activity and were involved in the degradation of leaf surface waxes (Nicholson and Epstein, 1991). Enzymes have also been identified in spore mucilage of *Colletotrichum graminicola* and are thought to prepare the leaf surface for fungal penetration (Nicholson and Morales, 1980; Kolattukudy and Soliday, 1985).

Electrostatic charges on spores (and other particulate matter) may be involved in spore deposition and retention on leaf surfaces (Gregory, 1971). Although there are reports regarding the influence of electrostatic charges on droplets as they apply to pesticide application (Law and Lane, 1981), and numerous researchers have suggested electrostatic discharge as a mechanism of spore release (Gregory, 1957; Leach, 1976; Meredith, 1965), there is little experimental documentation of electrostatic attraction of spores to leaf surfaces.

5.5 Hyphal Growth and Adhesion on the Leaf Surface

Many fungi, particularly plant pathogens, maintain close contact with the surfaces on which they grow. As a result, they are able to detect specific characteristics of the surface that are necessary for establishing a successful host-parasite interaction. Both chemical and physical (topographical) features of the host surface are used by the pathogen to direct fungal cell growth as well as to signal for infection events (Hoch and Staples, 1987, 1991). This recognition of chemical and topographic signals by fungal germ tubes or hyphae is termed chemotropism and thigmotropism, respectively (Staples et al., 1984). Plant pathogens, particularly many of the rust fungi, demonstrate significant thigmotropic responses, which are necessary for normal growth and development (Staples and Macko, 1984; Staples et al., 1985a, 1985b; Hoch and Staples, 1991).

For many fungi, hyphal growth on the leaf surface is oriented by topographical features. More than 60 years ago, it was noted that germ tube growth of *Puccinia graminis* urediniospores occurred at right angles to the longitudinal axis of wheat leaves (Johnson, 1934). Johnson also noted that growth of *Puccinia helianthi* germ tubes occurred perpendicularly to the veins of sunflower leaves. He attributed these directional growth patterns to a thigmotropic stimulus. Johnson further observed that stomata on wheat leaves are arranged in longitudinal rows and suggested that the transverse growth habit of germ tubes increased the probability of stomatal contact and subsequent host infection. The mechanisms responsible for thigmotropic responses of rust and powdery mildew fungi were studied in great detail by Dickinson (1949a, 1949b, 1949c, 1969, 1970, 1971, 1972, 1977, 1979). From these studies he showed that the directed growth of germ tubes was due to contact with a repetitive series of structural features on the substratum. Using a variety of artificial membranes chemically modified to obtain different surface features, he demonstrated that germ tube morphology, such as branching patterns and zigzag growth, was determined by the height and frequency of ridges on the membranes. Close similarity in the fungal responses on artificial membranes to those observed on host plants were noted (Dickinson, 1971). More recently, Staples et al. (1983) demonstrated that *Puccinia graminis* grew at right angles to undefined grooves on surfaces of polyethylene, polystyrene, and aluminum. Lewis and Day (1972) suggested that the regular lattice of wax crystals present on the cuticle of wheat leaves provided a thigmotropic signal by which germ tubes were oriented. However, in light of our current knowledge that oriented growth will occur on either natural or artificial surfaces lacking such wax crystals, it is likely that the larger topographies, provided by the ridges and grooves on these surfaces provide the signal for growth orientation.

The physical parameters of the signal that orients growth were recently further defined, using microfabricated surfaces (Hoch et al., 1987). Grooves of precise dimensions were etched into silicon wafers using electron beam lithography, and from these, thin polystyrene replicas were made. The resulting polystyrene membranes bore uniformly sized ridges, 0.5 μm high by 1.4 μm wide. Germlings of *Uromyces appendiculatus* normally exhibited curved and wandering growth on smooth surfaces, but on membranes bearing ridges spaced 0.5 μm apart, growth was straight and perpendicular to the ridges (Figure 5.5A). Ridge spacings of 15.0 μm and wider still induced the germlings to grow at right angles to the ridges, especially after ridge contact was made. However, as the germling tip grew into the space between the ridges, growth was randomly oriented until contact was made

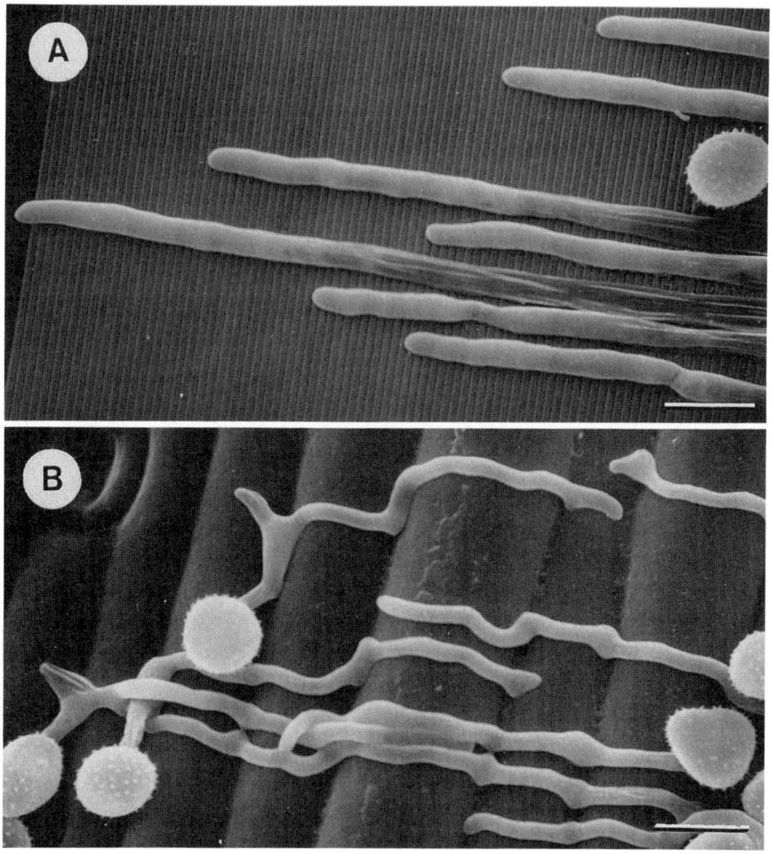

Figure 5.5 Oriented germ tube growth of *Uromyces appendiculatus*. (A) Germ tube growth is straight and oriented on polystyrene membranes bearing ridges 0.5 μm high, 2.0 μm wide, and spaced 0.5 μm apart. (B) Germ tube growth is similarly straight and oriented on polystyrene replicas of adaxial epidermal cells of graminaceous leaf surfaces, exemplified here by *Saccharum*. SEM preparation; OsO_4- fixed, critical-point-dried, gold-coated. Bars both equal 20 μm.

with the next ridge. Signaling for growth orientation was equally efficient on both ridges and grooves.

These experiments on germ tube growth on artificial surfaces closely reflect growth on natural leaf surfaces. The ridges and grooves providing the signal for directed growth on artificial membranes appear to be equivalent to the depressions, or "valleys," located at the anticlinal wall junctions formed between the epidermal cells on leaves (Figure 5.5). Germ tube tips coming in contact with such junctions perceive signals directing growth orientation. On graminaceous plant leaves, therefore, germ tube growth tends to be perpendicular to the parallel cell junctions between epidermal

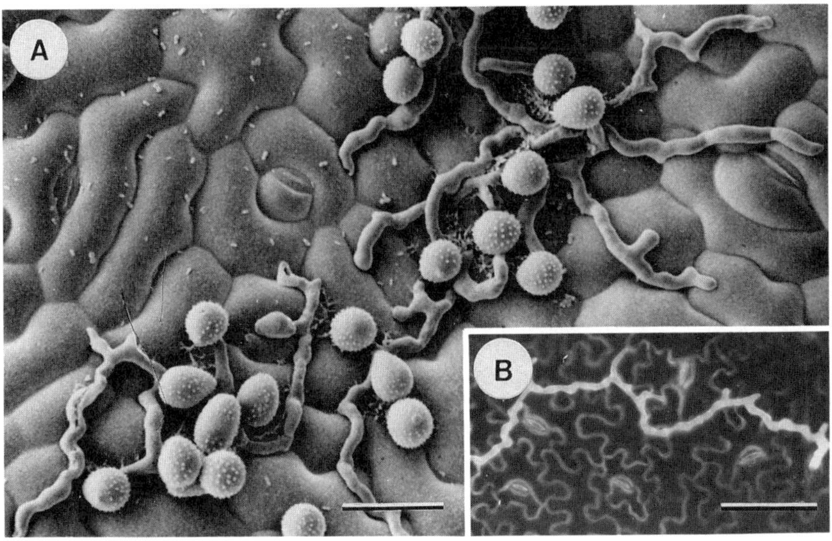

Figure 5.6 Germ tube growth of *Uromyces appendiculatus* on a nonfully expanded adaxial leaf surface of *Phaseolus vulgaris* illustrating oriented growth perpendicular to and along the anticlinal "valleys" formed at epidermal cell junctions. (A) SEM preparation; frozen-hydrated, gold-coated. Bar equals 50 μm. (B) Fresh specimen stained with 0.01% Auromine O and 0.1% Calcofluor and viewed with fluorescence illumination. Bar equals 100 μm.

cells (Figure 5.5B). The epidermal cells of dicotyledonous plants, however, are more irregular in outline resulting in more complex oriented growth patterns (Figure 5.6). On such leaves, germ tubes appear to take very convoluted paths, but in fact, they generally follow the "perpendicular rule." A convincing example of directed growth is seen on the cells at the base of trichomes where germ tubes circle the trichome as they grow over the cells that radiate out from the hair.

Fungal germ tubes are also commonly observed growing in the valleys between epidermal cells (Figure 5.6) (Mercer et al., 1971; Pring, 1980; Armentrout and Downer, 1987; Hopkin et al., 1988; Christiansen and Smedgaard, 1990). This entrapped type of growth generally continues until another signal is received, i.e., a cell junction is reached or the depth of the valley changes, allowing the germ tube to "escape." The depth of epidermal cell junctions varies with age and among plant species. On *Phaseolus vulgaris* leaves, for example, the depth of the anticlinal valley can be as great as 8.0 μm with an acute base. Stereo-pair views of epidermal cells can be advantageously used to reveal the dramatic relief of such surfaces (Figure 5.7). Germ tube or hyphal branching is also stimulated by contact with epidermal cell junctions (Figure 5.6).

It is clear, therefore, that many fungi possess thigmotropic sensing

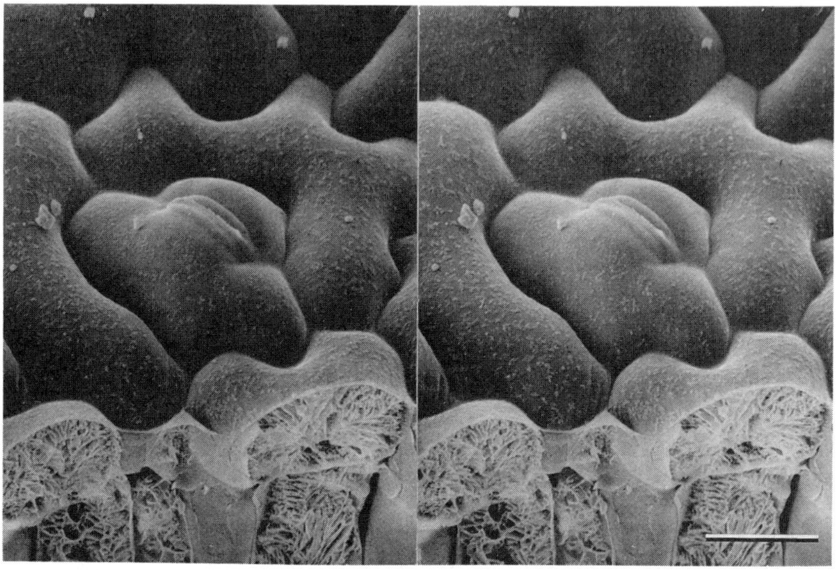

Figure 5.7 Stereo-pair view of stoma and epidermal cells of *Phaseolus vulgaris* illustrating the depth of leaf surface topography. SEM preparation; frozen-hydrated, gold-coated. Bar equals 25 μm.

mechanisms, and that leaf surface topography can provide the necessary signals to orient growth. This interaction could play a role in defining host specificity. For example, Heath (1974) reported that the efficiency of germ tube growth toward stomata differed on host and non-host surfaces suggesting directional growth as a factor in non-host resistance. Thus the development of resistant plants can be envisaged through the modification of host plant surface characteristics such as epidermal cell size and shape, to resemble that of non-hosts.

Most epiphytic fungi will grow on many different plant surfaces as well as on a variety of artificial substrates. Host specificity appears to be less discriminating during prepenetration growth of most pathogens, yet differences in growth have been observed. Wynn (1976) reported that *Puccinia graminis* germ tubes grew tightly appressed to, and at right angles to, wheat leaf epidermal cells, whereas *Uromyces appendiculatus* was less tightly attached and exhibited less controlled directional growth on the same leaves. In comparison, the bean rust fungus grew tightly appressed to polystyrene replicas of wheat leaves. Directional growth of these germ tubes was also maintained perpendicular to the longitudinal axes of the cell outlines. Wynn suggested that the cuticular waxes on the wheat leaves, not present on the wheat leaf replicas or bean leaves, prevented complete germ tube adherence.

The adhesion of germ tubes or hyphae to leaf surfaces is often achieved through the production of an extracellular matrix or mucilage (Jacobi et al., 1982; Gold and Mendgen, 1984; Mims and Richardson, 1989). The qualities of this matrix and its interaction with the cuticle contribute to the success of adhesion, growth, and surface recognition. Nicholson and Epstein (1991) described the composition of the adhesive compounds and noted that they generally occur in two classes: polysaccharides and proteins, usually as glycoproteins. Lapp and Skoropad (1978) reported that a hemicellulose was involved in the adhesion of *Colletotrichum graminicola* appressoria and a galactosaminoglycan has been reported to be present in the mucilage produced by both *Neurospora crassa* (Reissig et al., 1975) and *Bipolaris sorokiniana* (Pringle, 1981). Glycoproteins are involved in the adherence of bean rust germlings (Epstein et al., 1985), and *Buergenerula spartinae* hyphopodia (Onyile et al., 1982) to their substrates. Other components of the matrix appear to be involved in modification of the cuticle in preparation for growth and penetration by the fungus. Enzymatic activity in association with fungal mucilage has been demonstrated (Staub et al., 1974; Kolattukudy, 1985; Ramadoss et al., 1985; Nicholson et al., 1988).

Characteristics of the cuticle also have an effect on germ tube adherence to the leaf surface and are of critical importance to surface growth and the recognition of tropic signals. The chemical nature of the surface is important in the cuticle—fungal interaction and may evoke the release of adhesive compounds by fungi or, more commonly, function in site recognition processes (Nicholson and Epstein, 1991). Another important characteristic of adhesion relates to the hydrophobicity of the cuticle. The involvement of hydrophobic bonds in adhesion has been demonstrated on artificial substrates. The conidia of *Magnaporthe grisea,* the rice blast fungus, adhere more tightly to teflon than to glass, a much less hydrophobic surface (Hamer et al., 1988). Similarly, *U. appendiculatus* grows tightly appressed to such hydrophobic surfaces as polyethylene, polycarbonate (Epstein et al., 1987), collodion, and polystyrene (Wynn, 1976). On clean glass surfaces, germ tubes of *U. appendiculatus* exhibit poor adhesion. However, when the glass is treated with silane compounds, e.g., dimethyldicholorosilane, which adds a hydrophobic molecular surface to the glass, germ tube adhesion is improved (Hoch, unpublished). Natural leaf surfaces vary in hydrophobicity (Silva Fernandes, 1965) and have similar effects on germ tube adherence and growth to that observed on artificial surfaces. Germ tubes of *P. sorghi* grew away from the surface of mutant, waxless, corn leaves, but grew tightly appressed to, and showed directional growth on waxy leaves (Wynn and Staples, 1981). Not all fungi, however, show a preference for hydrophobic cuticles. Macroconidia of *Nectria haematocca,* for example, adhere to both hydrophobic and hydrophilic surfaces (Hickman and Epstein, 1988).

Although studies regarding these interactions have focused on pathogenic organisms, it is likely that non-pathogens would be affected similarly. The growth of fungi on plant surfaces, then, is affected in part by the characteristics of extracellular materials of fungal origin, and their interaction with the chemical characteristics of the plant cuticle, particularly those which influence hydrophobicity. Strategies designed to control the growth of fungal organisms could potentially focus on modifying these parameters through genetic manipulation of the plants or through the application of materials that altered the characteristics of the plant surface.

5.6 Development of Infection Structures

Plant pathogenic fungi must enter and infect their hosts in order to obtain nutrients necessary for their survival. To accomplish this, these fungi have developed various means of crossing the barrier presented by the host epidermis. Many use enzymes or mechanical force to penetrate directly through the cuticle. Other fungi gain access through natural openings, such as lenticels and hydathodes, or more commonly through stomata (Royle, 1976; Dodman, 1979). Some fungal pathogens have very specific entry portals. *Gibberidea heliopsis*, for example, enters its host directly through the tip cell of leaf trichomes (Mason, 1973). Such site-specific penetration requires that the fungus be able to locate and recognize leaf surface features. Various mechanisms have been implicated in this process, including chemical signals (chemotropism), growth along water or vapor pressure gradients (hydrotropism), or through the recognition of physical topography (thigmotropism). For example, Edwards and Bowling (1986) suggested that germ tubes of *Uromyces viciae-fabae* were directed to stomata by following pH gradients, and germ tubes of *Cercospora* are thought to be attracted to stomata by a hydrotropic mechanism (Rathaiah, 1977). Upon reaching a preferred penetration site, changes are initiated in germ tube morphology resulting in the formation of appressoria (Emmett and Parbery, 1975), often the first in a series of infection structures (Wynn and Staples, 1981; Hoch and Staples, 1987). The location of the preferred penetration sites varies among fungi. Direct-penetrating fungi such as *Perenospora parasitica*, *Erysiphe polygoni* (Preece et al., 1967), *Colletotrichum* spp. (Anderson and Walker, 1962; Lapp and Skoropad, 1978), *Puccinia psidii* (Hunt, 1968), *Cochliobolus* spp. (Hau and Rush, 1979), *Botrytis* spp. (Clark and Lorbeer, 1976), and *Endocronartium harknessii* (Hopkin et al., 1988) preferentially form appressoria over the anticlinal wall junctions between epidermal cells. Likewise, *Physopella zeae* and *Phakopsora pachyrhizi* develop appressoria at this location, although both also exhibit a low percentage of penetrations at other sites, e.g., directly through epidermal cells (Bonde et al., 1976, 1982). *P. pachyrhizi* will also form appressoria upon contact with hard surfaces such as glass (Kitani and Inoue, 1960; Koch and Hoppe, 1988). The nature of the signal provided by the leaf that induces appressorium formation at epidermal cell junctions is not clear. Various chemical stimuli have been shown to induce appressorium formation, including plant exudates in the form of guttation fluids (Endo and Amacher, 1964; Froussard, 1981) or other exogenous nutrients (Clark and Lorbeer, 1976; Blakeman and Atkinson, 1981). It has been suggested that due to breaks (Wheeler, 1977) or thinning of the cuticle at these sites (Schönherr and Bukovac, 1970) such compounds may be secreted or accumulate above anticlinal wall junctions (Wynn, 1981). Studies with several fungi using plastic replicas of leaf surfaces, however, indicate that the topography of the inert surface alone can be effective at signaling appressorium initiation (Clark and Lorbeer, 1976; Lapp and Skoropad, 1978; Hau and Rush, 1979; Wynn and Staples, 1981). Other evidence suggests that compounds secreted by the germ tube tip accumulate in the epidermal cell junctions, and that these autotropic (self-induced) signals are responsible for stimulating appressorium formation. Although chemical signals appear to play a part in some systems, thigmotropism is recognized as the most prominent component in the differentiation process for initiating appressorium formation in many fungi (Wynn, 1981; Armentrout and Downer, 1987). In many instances, thigmotropic recognition of surfaces is quite nonspecific. Evidence for this is found in the similarity of growth and differentiation responses that many thigmotropic fungi exhibit on host plants, on nonhosts, and on artificial surfaces (Maheshwari et al., 1967; Heath, 1974; Wynn, 1976; Niks, 1981; Patton and Spear, 1989). These observations further strengthen the concept that many interactions between fungi and plant surfaces rely predominantly on physical stimuli.

Perhaps no fungi exhibit thigmotropic responses better than the rust fungi. Spores of these fungi, particularly urediniospores and aeciospores, must find and recognize very specific sites, usually stomata, to form appressoria. Not surprisingly, characteristics specific to stomata are thought to provide the signal(s) that triggers the fungus to cease further polarized apical growth and to initiate infection structure formation. Wynn (1976) showed that appressorium formation in *U. appendiculatus* was induced through contact with stomata on natural bean leaf surfaces (Figure 5.8A), but perhaps more importantly, on plastic replicas of leaves. The rust germlings were very accurate, because on both replicas and real leaves over 92% of the germlings formed appressoria over stomata on either abaxial or adaxial surfaces. Using serially produced leaf replicas to eliminate the possibility of the carryover of chemicals from the original leaves, Wynn clearly demonstrated that appressoria of *U. appendiculatus* were induced by a purely thigmotropic signal associated with the stomatal complex. Other rust fungi grown on leaf replicas of their respective hosts have given mixed results. Wynn reported as unpublished results (in Wynn and Staples, 1981), that *P. sorghi* formed appressoria on the stomata of corn leaf replicas, whereas *P. graminis* grown on wheat leaf replicas failed to recognize the stomata. Similarly, when grown on plastic membranes with defined artificial topographies (0.5-μm-high microfabricated ridges), *U. appendiculatus*

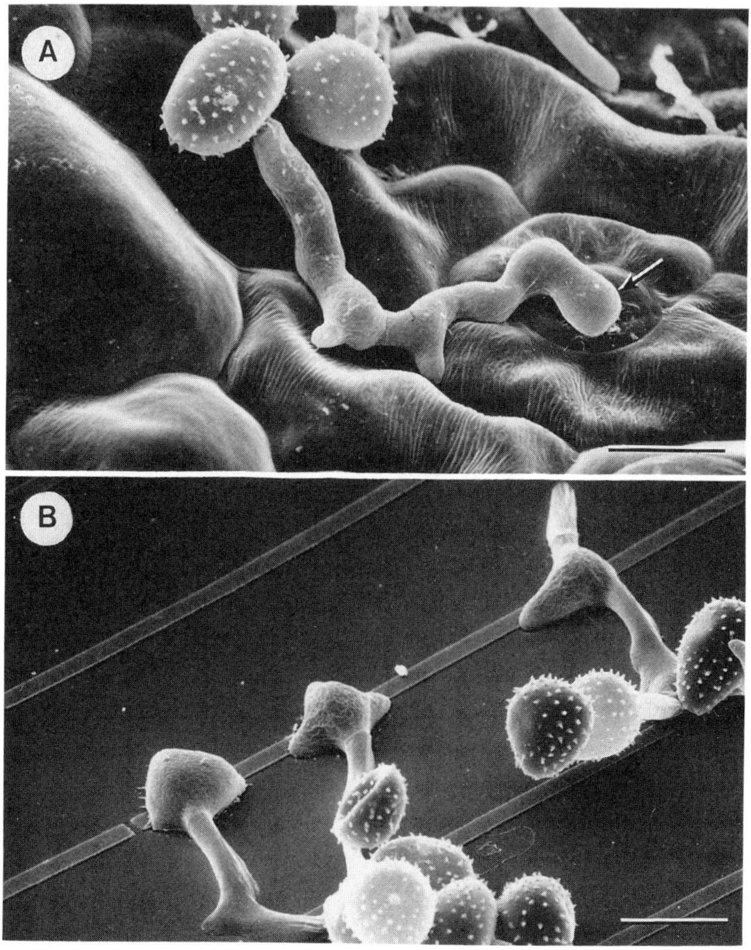

Figure 5.8 Appressoria of *Uromyces appendiculatus* formed in response to topographic signals. (A) Early developmental stage of an appressorium (arrow on the right) over a stoma on the adaxial surface of *Phaseolus vulgaris*. (B) Appressorium formed in response to 0.5-μm-high ridges on polystyrene membranes. SEM preparation; glutaraldehyde-fixed, critical-point-dried, gold-palladium-coated. Bars both equal 20 μm. From Hoch et al. (1987), copyright 1987 by the AAAS.

and *P. sorghi* germlings were induced to form appressoria, while germlings of *P. graminis* did not (Allen et al., 1991). *P. graminis* will, however, form appressoria on scratches of undefined dimensions on aluminum and polyethylene (Staples et al., 1983), and on polystyrene membranes, suggesting that a more complex topography may be necessary for differentiation of this rust. A recent survey of 27 species of rust fungi indicated that approximately one-third of the species tested did not differentiate on

Table 5.1 Response of rust species to artificial topography for the formation of appressoria[a]

Thigmosensitive species	Nonthigmosensitive species
Melampsora medusae	*Coleosporium asterum*
Physopella zeae	*Coleosporium tussilaginis*
Puccinia antirrhini	*Phragmidium potentillae*
Puccinia arachidis	*Puccinia graminis* f. sp. *avenae*
Puccinia calcitrapae	*Puccinia graminis* f. sp. *tritici*
Puccinia carduorum	*Tranzchelia discolor*
Puccinia hieracii	
Puccinia melanocephala	
Puccinia polysora	
Puccinia sorghi	
Puccinia substriata	
Uromyces appendiculatus	
Uromyces vignae	

[a] Rust germlings were grown on polystyrene membranes with 0.5-μm-high ridges. Thigmosensitive rust species differentiated to form appressoria on the ridges.

artificial topographies with defined heights, although the majority were thigmosensitive (Table 5.1) (Allen et al., 1991). It was not determined whether the species that failed to form appressoria on the defined topographies require a more complex signal or whether some other stimulus, e.g., chemical, is necessary.

The parameters of the topographic signal that induce appressorium formation in *U. appendiculatus* were accurately quantified by Hoch et al. (1987). Using electron beam lithography, grooves of varying depths and widths were microfabricated into silicon wafers from which polystyrene replicas were made. The resulting plastic membranes, with ridges ranging from 0.03 μm to 5.0 μm high, and 2.0 μm to 100 μm wide, served as the test surface for the induction of appressoria. The rust germlings were found to be very sensitive to changes in ridge height but not ridge width. Appressorium formation was highest (over 75%) on 0.5-μm-high ridges, whereas ridge heights below 0.25 μm or above 1.0 μm were much less inductive. They concluded that the inductive signal was an elevation change of approximately 0.5 μm (Figure 5.8B), and that it involved germ tube contact with two acute angles. A critical examination of the morphology of bean stomata showed a cuticular lip extending from the guard cell ledge, with a mean height of 0.5 μm (Figure 5.3B). They therefore suggested that this physical parameter of the stoma, the guard cell lip, represented the inductive signal on the leaf surface. This finding supports Wynn's previous report (1976).

In most rust fungi, after an appressorium has formed, a series of infection structures develops, including the infection peg, substomatal vesicle, and infection hyphae. In bean rust, a single topographical signal provided by the stomatal lip, or in vitro by an artificial topography of appropriate height, is sufficient to induce the sequential development of all of these infection structures (Wynn and Staples, 1981; Hoch and Staples, 1987, 1990). In contrast, wheat rust will differentiate to form appressoria in response to scratches in artificial surfaces, but apparently requires other factors to induce vesicle formation. Appressoria formed by heat shock generally give rise to these later infection structures (Staples et al., 1983).

Close contact with the leaf surface is as important in the process of signaling appressorium formation as it is for germ tube growth orientation. On natural or artificial surfaces on which the fungus does not adhere properly, marked reductions in appressorium formation are observed (Wynn, 1976). Again, adhesion is related to the hydrophobicity of the surface. Uchiyama and Okuyama (1990) reported differences in appressorium formation of *Pyricularia oryzae* that were grown on various extracts of rice leaf wax that differed in wettability. Similarly, Wynn (1976) observed low efficiency in appressorium formation of *U. appendiculatus* grown on wheat and oat leaves, and attributed this to poor adherence.

5.7 Conclusions

All fungi that land and germinate on a leaf are affected in some way by the surface features on the leaf. The success of these fungi at colonizing the leaf surface and, for pathogens, at gaining entry to the leaf, depends to a large extent on the nature of the physical and chemical characteristics of the surface and in many cases, on the ability of the fungi to sense these signals. The examples from recent studies cited here illustrate some of these fungus-leaf interactions. It is unadvisable, however, to draw generalized conclusions from these studies regarding epiphytic fungal growth, especially in view of the variety of leaf surfaces present in nature, the numerous strategies of colonization and growth employed by fungal organisms, and the vast numbers of fungi and host-plant surfaces about which little or nothing is known. Many questions concerning the evolutionary significance and practical application of fungal-phylloplane interactions remain unanswered. For instance, what role have these interactions played in guiding the evolution of fungal-host preferences? The high levels of host specificity demonstrated by obligate parasites could be due in part to prepenetration interactions. Rust fungi that require the specific thigmotropic signal provided by protruding stomatal lips to induce appressorium formation would be incapable of infecting a host without this feature. Likewise, leaf surfaces possessing features that interfere with germ tube contact, misdirect germ tube growth, or initiate appressorium formation at sites away from stomata, would prevent infection. This selection pressure could conceivably direct the coevolution of the host-parasite relationship. Other questions from an applied agricultural perspective are also raised. Can our growing knowledge of pre-penetration interactions be used to develop novel approaches in the design of disease-resistant plants? An example of the latter might use conventional breeding techniques or genetic engineering technology to alter leaf surface characteristics, such as cuticle hydrophobicity or the topographic signals involved in the formation of infection structures, to reduce the efficiency of infection by plant pathogens. Another approach could involve spraying susceptible plant surfaces with inert materials that could mimic the natural thigmotropic signals resulting in misplaced appressorium formation. Such strategies could lead to the establishment of more stable, nonspecific resistance in a variety of agricultural and forest crops.

Support for the research and preparation of this chapter was funded in part from grants provided by the Whitehall Foundation and from the USDA/CRGO. The authors thank Dr. R.C. Staples and B.T. Terhune for their assistance in the preparation of this manuscript.

References

Aist, J.R. 1981. Development of parasitic conidial fungi in plants. pp. 75–110 in Cole, G.T. and Kendrick, B. (editors), *Biology of Conidial Fungi, vol. 2.* Academic Press, New York.

Allen, E.A., Hazen, B.E., Hoch, H.C., Kwon, Y., Leinhos, G.M.E., Staples, R.C., Stumpf, M.A., and Terhune, B.T. 1991. Appressorium formation in response to topographical signals by 27 rust species. *Phytopathology* 81:323–331.

Anderson, J.L. and Walker, J.C. 1962. Histology of watermelon anthracnose. *Phytopathology* 52:650–653.

Armentrout, V.N. and Downer, A.J. 1987. Infection cushion development by *Rhizoctonia solani* on cotton. *Phytopathology* 77:619–623.

Blakeman, J.P. and Atkinson, P. 1981. Antimicrobial substances associated with the aerial surfaces of plants. pp. 245–263 in Blakeman, J.P. (editor), *Microbial Ecology of the Phylloplane.* Academic Press, New York.

Bonde, M.R., Melching, J.S., and Bromfield, K.R. 1976. Histology of the suscept- pathogen relationship between *Glycine max* and *Phakopsora pachyrhizi*, the cause of soybean rust. *Phytopathology* 66:1290–1294.

Bonde, M.R., Melching, J.S., and Bromfield, K.R. 1982. Morphological development of *Physopella zeae* on corn. *Phytopathology* 72:1489–1491.

Burrage, S.W. 1969. Dew and the growth of the uredospore germ tube of *Puccinia graminis* on the wheat leaf. *Annals of Applied Biology* 64:495–501.

Carter, M.V. 1965. Ascospore deposition in *Euphyta armeniacae. Australian Journal of Agricultural Research* 65:825–836.

Christiansen, S.K. and Smedgaard, V. 1990. Microscopic studies of the interaction between barley and the saprophytic fungus, *Cladosporium macrocarpum. Journal of Phytopathology* 128:209–219.

Clark, C.A. and Lorbeer, J.W. 1976. The development of *Botrytis squamosa* and *B. cinerea* on surface of onion leaves as affected by exogenous nutrients and epiphytic bacteria. pp. 607–625 in Dickinson, C.H. and Preece, T.F. (editors), *Microbiology of Aerial Surfaces of Plants*. Academic Press, London.

Cook, M. 1980. Host- parasite relations in uredinial infections on peanut by *Puccinia arachidis*. *Phytopathology* 70:822–826.

Cutter, E.G. 1976. Aspects of the structure and development of the aerial surfaces of higher plants. pp. 1–40 in Dickinson, C.H. and Preece, T.F. (editors), *Microbiology of Aerial Plant Surfaces*. Academic Press, London.

Davies, R.R. 1961. Wettability and the capture, carriage and deposition of particles by raindrops *Nature (London)* 191:616.

Dickinson, S. 1949a. Studies in the physiology of obligate parasitism. I. The stimuli determining the direction of growth of the germ-tubes of rust and mildew spores. *Annals of Botany (London)* 13:89–104.

Dickinson, S. 1949b. Studies in the physiology of obligate parasitism. II. The behavior of the germ-tubes of certain rusts in contact with various membranes. *Annals of Botany (London)*. 13:219–236.

Dickinson, S. 1949c. Studies in the physiology of obligate parasitism. IV. The formation on membranes of haustoria by rust hyphae and powdery mildew germ-tubes. *Annals of Botany (London)*. 13:219–236.

Dickinson, S. 1969. Studies in the physiology of obligate parasitism. VI. Directed growth. *Phytopathologische Zeitschrift* 66:38–49.

Dickinson, S. 1970. Studies in the physiology of obligate parasitism. VII. The effect of a curved thigmotropic stimulus. *Phytopathologische Zeitschrift* 69: 115–124.

Dickinson, S. 1971. Studies in the physiology of obligate parasitism. VIII. An analysis of fungal responses to thigmotropic stimuli. *Phytopathologische Zeitschrift* 70:62–70.

Dickinson, S. 1972. Studies in the physiology of obligate parasitism. IX. The measurement of a thigmotropic stimulus. *Phytopathologische Zeitschrift* 73: 347–358.

Dickinson, S. 1977. Studies in the physiology of obligate parasitism. X. Induction of response to a thigmotropic stimulus. *Phytopathologische Zeitschrift* 89:97–115.

Dickinson, S. 1979. Growth of *Erysiphe graminis* on artificial membranes. *Physiological Plant Pathology* 15:219–221.

Dodman, R.L. 1979. How the defenses are breached. pp. 135–153 in Horsfall, J.G. and Cowling, E.B. (editors), *Plant Disease. An Advanced Treatise, vol. IV*. Academic Press, New York.

Edwards, M.C., and Bowling, D.J.F. 1986. The growth of rust germ tubes towards stomata in relation to pH gradients. *Physiological and Molecular Plant Pathology* 29:185–196.

Emmett, R.W. and Parbery, D.G. 1975. Appressoria. *Annual Review of Phytopathology* 13:147–167.

Endo, R.M. and Amacher, R.H. 1964. Influence of guttation fluid on infection structures of *Helminthosporium sorokinianum*. *Phytopathology* 54:1327–1334.

Epstein, L., Laccetti. L.B., Staples, R.C., and Hoch, H.C. 1987. Cell-substratum adhesive protein involved in surface contact responses of the bean rust fungus. *Physiological and Molecular Plant Pathology* 30:373–388.

Epstein, L., Laccetti. L.B., Staples, R.C., Hoch, H.C., and Hoose, W.A. 1985. Extracellular proteins associated with induction of differentiation in bean rust uredospore germlings. *Phytopathology* 75:1073–1076.

Esau, K. 1977. *Anatomy of Seed Plants*. John Wiley and Sons, New York. 550 pp.

Fisher, R.W. and Corke, A.T.K. 1971. Infection of Yarlington Mill fruit by the apple scab fungus. *Canadian Journal of Plant Science* 51:535–542.

Frossard, R. 1981. Effect of guttation fluids on growth of micro-organisms on leaves. pp. 213–226 in Blakeman, J.P. (editor), *Microbial Ecology of the Phyllo-*

plane. Academic Press, New York.

Gold, R.E. and Mendgen, K. 1984. Cytology of basidiospore germination, penetration, and early colonization of *Phaseolus vulgaris* by *Uromyces appendiculatus* var. *appendiculatus*. *Canadian Journal of Botany* 62:1989–2002.

Gregory, P.H. 1951. Deposition of air-borne *Lycopodium* spores on cylinders. *Annals of Applied Biology* 38:357–376.

Gregory, P.H. 1957. Electrostatic charges on spores of fungi in air. *Nature (London)* 180:330.

Gregory, P.H. 1971. The leaf as a spore trap. pp. 239–243 in Preece, T.F. and Dickinson, C.H. (editors), *Ecology of Leaf Surface Micro-organisms*. Academic Press, London.

Gregory, P.H. 1973. *The Microbiology of the Atmosphere*. Wiley, New York. 377 pp.

Gregory, P.H. and Stedman, O.J. 1953. Deposition of air-borne *Lycopodium* spores on plane surfaces. *Annals of Applied Biology* 40:651–674.

Hallam, N.D. and Juniper, B.E. 1971. The anatomy of the leaf surface. pp. 3–38 in Preece, T.F. and Dickinson, C.H. (editors), *Ecology of Leaf Surface Micro-organisms*. Academic Press, London.

Hamer, N.D., Howard, R.J., Chumley, F.G., and Valent, B. 1988. A mechanism for surface attachment in spores of a plant pathogenic fungus. *Science* 239:288–290.

Hau, F.C. and Rush, M.C. 1979. Leaf surface interactions between *Cochliobolus miyabeanus* and susceptible and resistant rice cultivars. *Phytopathology* 69:527.

Heath, M.C. 1974. Light and electron microscope studies of the interactions of host and non-host plants with cowpea rust *Uromyces phaseoli* var. *vignae*. *Physiological Plant Pathology* 4:403–414.

Heather, W.A. 1967a. Susceptibility of the juvenile leaves of *Eucalyptus bicostata* Maiden et al. to infection by *Phaeoseptoria eucalypti* (Hansf.) Walker. *Australian Journal of Biological Science* 20:769–775.

Heather, W.A. 1967b. Leaf characteristics of *Eucalyptus bicostata* Maiden et al. seedlings affecting the deposition and germination of spores of *Phaeoseptoria eucalypti* (Hansf.) Walker. *Australian Journal of Biological Science* 20:1155–1160.

Hickman, M.J. and Epstein, L. 1988. *Nectria haematococca* macroconidia attach to plant surfaces. *Phytopathology* 78:1523 (Abstr.).

Hoch, H.C. and Staples, R.C. 1987. Structural and chemical changes among the rust fungi during appressorium formation. *Annual Review of Phytopathology* 25:231–247.

Hoch, H.C. and Staples, R.C. 1991. Signaling for infection structure formation in fungi. pp. 25–46 in Cole, G.T. and Hoch, H.C. (editors), *The Fungal Spore and Disease Initiation in Plants and Animals*. Plenum Press, New York.

Hoch, H.C., Staples, R.C., Whitehead, B., Comeau, J., and Wolf, E.D. 1987. Signaling for growth orientation and cell differentiation by surface topography in *Uromyces*. *Science* 235:1659–1662.

Hopkin, A.A., Reid, J., Hiratsuka, Y., and Allen, E. 1988. Initial infection and early colonization of *Pinus contorta* by Endocronartium harknessii (western gall rust). *Canadian Journal of Plant Pathology* 10:221–227.

Hunt, P. 1968. Cuticular penetration by germinating uredospores. *Transactions of the British Mycological Society* 51:103–112.

Ingold, C.T. 1971. *Fungal Spores: Their Liberation and Dispersal*. Clarendon Press, Oxford. 302 pp.

Ingold, C.T. 1978. Role of mucilage in dispersal of certain fungi. *Transactions of the British Mycological Society* 70:137–140.

Jacobi, W.R., Amerson, H.V., and Mott, R.L. 1982. Microscopy of cultured loblolly pine seedlings and callus inoculated with *Cronartium fusiforme*. *Phytopathology* 72:138–143.

Jeffree, C.E., Baker, E.A., and Holloway, P.J. 1976. Origins of the fine structure of plant epicuticular waxes. pp. 119–158 in Dickinson, C.H. and Preece, T.F. (editors), *Microbiology of Aerial Plant Surfaces*. Academic Press, London.

Johnson, T. 1934. A tropic response in germ tubes of urediniospores of *Puccinia graminis tritici*. *Phytopathology* 24:80–82.

Johnstone, K.H. 1931. Observations on the varietal resistance of the apple to scab (*Venturia inequalis*, Aderh.) with special reference to its physiological aspects. *Journal of Pomology* 9:30–52.

Jones, M.J. and Epstein, L. 1989. Adhesion of *Nectria haematococca* macroconidia. *Physiological and Molecular Plant Pathology* 35:453–461.

Kitani, K. and Inoue, Y. 1960. Studies on the soybean rust and its control measure. (Part I) Studies on the soybean rust. *Shikoku Agricultural Experiment Station*. (*Zentsuji*, Japan), Bulletin. 5:319–342. (Japanese text, English summary).

Koch, E. and Hoppe, H.H. 1988. Development of infection structures by the direct-penetrating soybean rust fungus (*Phakopsora pachyrhizi* Syd.) on artificial membranes. *Journal of Phytopathology* 122:232–244.

Kolattukudy, P.E. 1985. Enzymatic penetration of the plant cuticle by fungal pathogens. *Annual Review Phytopathology* 23:223–250.

Kolattukudy, P.E. and Soliday, C.L. 1985. Effects of stress on the defensive barriers of plants. pp. 381–400 in Key, J.L. and Kosuge, T. (editors), *Cellular and Molecular Biology of Plant Stress*. Alan R. Liss, New York.

Köller, W. 1990. The plant cuticle: A barrier to be overcome by fungal plant pathogens. pp. (in press) in Cole, G.T. and Hoch, H.C. (editors), *Funaal Spore and Disease Initiation in Plants and Animals*. Plenum Press, New York.

Konar, R.N. and Linskens, H.F. 1966. The morphology and anatomy of the stigma of *Petunia hybrida*. *Planta* 71:356–371.

Kunoh, H., Yamaoka, N., Yoshioka, H., and Nicholson, R.L. 1988. Preparation of the infection court by *Erysiphe graminis* I. Contact-mediated changes in morphology of the conidium surface. *Experimental Mycology* 12:325–335.

Lapp, M.S. and Skoropad, W.P. 1978. Location of appressoria of *Colletotrichum graminicola* on natural and artificial barley leaf surfaces. *Transactions of the British Mycological Society* 70:225–228.

Law, S.E. and Lane, M.D. 1981 Electrostatic deposition of pesticide sprays onto foliar targets of varying morphology (broccoli, maize, cotton, cabbage). *Transactions of the American Society of Agricultural Engineers* 24:1441–1445.

Leach, C.M. 1976. An electrostatic theory to explain violent spore liberation by *Drechslera turcica* and other fungi. *Mycologia* 68:63–86.

Lewis, B.G. and Day, J.R. 1972. Behaviour of uredospore germ-tubes of *Puccinia graminis tritici* in relation to the fine structure of wheat leaf surfaces. *Transactions of he British Mycological Society* 58:139–145.

Maheshwari, R., Hildebrandt, A.C., and Allen, P.J. 1967. The cytology of infection structure development in uredospore germ tubes of *Uromyces phaseoli* var. *typica* (Pers.) Wint. *Canadian Journal of Botany* 45:447–450.

Martin, J.T. and Juniper, B.E. 1970. *The Cuticles of Plants*. Edward Arnold, Inc., London. 347 pp.

Mason, D.L. 1973. Host-parasite relations in the *Gibberidea* disease of *Helianthus strumosus*. *Mycologia* 65:1158–1170.

McKeen, W.E. 1974. Mode of penetration of epidermal cell walls of *Vicia faba* by *Botrytis cinerea*. *Phytopathology* 64:461–467.

Mercer, P.C., Wood, R.K.S., and Greenwood, A.D. 1971. Initial infection of *Phaseolus vulgaris* by *Colletotrichum lindemuthianum*. pp. 381–390 in Preece, T.F. and Dickinson, C.H. (editors), *Ecology of Leaf Surface Micro-organisms*. Academic Press, London.

Meredith, D.S. 1965. Violent spore release in *Helminthosporium turicum*. *Phytopathology* 55:1099–1102.

Mims, C.W. and Richardson, E.A. 1989. Ultrastructure of appressorium development by basidiospore germlings of the rust fungus *Gymnosporangium juniperi-virginianae*. *Protoplasma* 148:111–119.
Nicholson, R.L. and Epstein, L. 1991. Adhesion of fungi to the plant surface: prerequisite for pathogenesis. pp. 3–23 in Cole, G.T. and Hoch, H. C. (editors), *Fungal Spore and Disease Initiation in Plants and Animals*. Plenum Press, New York.
Nicholson, R.L. and Morales, W.B.C. 1980. Survival of *Colletotrichum graminicola*: Importance of the spore matrix. *Phytopathology* 70:255–261.
Nicholson, R.L., Yoshioka, H., Yamaoka, N., and Kunoh, H.1988: Preparation of the infection court by *Erysiphe graminis*. II. Release of esterase enzyme from spores in response to a contact stimulus. *Experimental Mycology* 12:336–349.
Niks, R.E. 1981. Appressorium formation of *Puccinia hordei* on partially resistant barley and two non-host species. *Netherlands Journal of Plant Pathology* 87:201–207.
Onyile, A.B., Edwards, H.H., and Gessner, R.V. 1982. Adhesive material of the hyphopodia of *Buergenerula spartinae*. *Mycologia* 74:777–784.
Palevitz, B.A. 1981. The structure and development of stomatal cells. pp. 1–23 in Jarvis, P.J. and Mansfield, T.A. (editors), *Stomatal Physiology*. Cambridge University Press, Cambridge, UK.
Patton, R.F. and Spear, R.N. 1989. Histopathology of colonization in leaf tissue of *Castilleja, Pedicularis, Phaseolus,* and *Ribes* species by *Cronartium ribicola*. *Phytopathology* 79:539–547.
Paus, F. and Raa, J. 1973. An electron microscope study of infection and disease development in cucumber hypocotyls inoculated with *Cladosporium cucumerinum*. *Physiological Plant Pathology* 3:461–464.
Preece, T.F. 1976. Some observations on leaf surfaces during the early stages of infection by fungi. pp. 1–10 in Friend, J. and Threlfall, D.R. (editors), *Biochemical Aspects of Plant-Parasite Relationships*. Academic Press, London.
Preece, T.F., Barnes, G., and Bayley, J.M. 1967. Junction between epidermal cells as sites of appressorium formation by plant pathogenic fungi. *Plant Pathology* 16:117–118.
Pring, R.J. 1980. A fine-structural study of the infection of leaves of *Phaseolus vulgaris* by uredospores of *Uromyces phaseoli*. *Physiological Plant Pathology* 17:269–276.
Pringle, R.B. 1981. Nonspecific adhesion of *Bipolaris sorokiniana* sporelings. *Canadian Journal of Plant Pathology* 3:9–11.
Ramadoss, C.S., Uhlig, J., Carlson, D.M., Butler, L.G., and Nicholson, R.L. 1985. Composition of the mucilaginous spore matrix of *Colletotrichum graminicola*, a pathogen of corn, sorghum, and other grasses. *Journal of Agricultural Food Chemistry* 33:728–732.
Rathaiah, Y. 1977. Stomatal tropism of *Cercospora beticola* in sugarbeet. *Phytopathology* 67:358–362.
Reissig, J.L, Lai, W.H., and Glasgow, J.E. 1975. An endogalactosaminidase from *Streptomyces griseus*. *Canadian Journal of Biochemistry* 53:1237–1249.
Robinson, P.M., Park, D., and Graham, T.A. 1968. Autotropism in fungal spores. *Journal of Experimental Botany* 19:125–134.
Royle, D.J. 1976. Structural features of resistance to plant-diseases. pp. 161–193 in Friend, J. and Threlfall, D.R (editors), *Biochemical Aspects of Plant-Parasite Relationships*. Academic Press, London.
Schönherr, J. and Bukovac, M.J. 1970. Preferential polar pathways in the cuticle and their relationship to ectodesmata. *Planta* 92:189–201.
Shaik, M. 1985. Race-nonspecific resistance in bean cultivars to races of *Uromyces appendiculatus* var. *appendiculatus* and its correlation with leaf epidermal characteristics. *Phytopathology* 75:481 (Abstr.).

Silva Fernandas A.M.S. 1965. Studies on plant cuticle VIII. Surface waxes in relation to water-repellency. *Annals of Applied Biology* 56:297–304.

Southwood, R. 1986. Plant surfaces and insects—an overview. pp. 1–22 in Juniper, B. and Southwood, R. (editors), *Insects and the Plant Surface*. Arnold, London.

Staples, R.C. and Macko, V. 1984. Germination of urediospores and differentiation of infection structures. pp. 255–289 in Bushnell, W.R. and Roelfs, A.P. (editors), *The Cereal Rusts*, vol. 1. Academic Press, New York.

Staples, R.C., Grambow, H.J., Hoch, H.C., and Wynn, W.K. 1983. Contact with membrane grooves induces wheat stem rust uredospore germlings to differentiate appressoria but not vesicles. *Phytopathology* 73:1436–1439.

Staples, R.C., Macko, V., Wynn, W.K., and Hoch, H.C. 1984. Terminology to describe the differentiation response by germlings of fungal spores. *Phytopathology* 74:380 (Abstr.).

Staples, R.C., Hoch, H.C., and Epstein, L. 1985a. The development of infection structures by the rusts and other fungi. *Microbiological Sciences* 2:193–198.

Staples, R.C., Hoch, H.C., Epstein, L., Laccetti, L., and Hassouna, S. 1985b. Recognition of host morphology by rust fungi: responses and mechanisms. *Canadian Journal of Plant Pathology* 7:314–322.

Staub, T., Dahmen, H., and Schwinn, F.J. 1974. Light- and scanning electron microscopy of cucumber and barley powdery mildew on host and non-host plants. *Phytopathology* 64:364–372.

Steadman, J.R. and Shaik, M. 1988. Leaf pubescence confers apparent race-nonspecific rust resistance in bean (*Phaseolus vulgaris*). *Phytopathology* 78:1566 (Abstr.).

Uchiyama, T. and Okuyama, K. 1990. Participation of *Oryza sativa* leaf wax in appressorium formation by *Pyricularia oryzae*. *Phytochemistry* 29:191–92.

von Alten, H. 1983. The effect of temperature, light, and leaf age on the frequency of appressoria formation and infection with *Uromyces phaseoli* (Pers.) Wint. *Phytopathologische Zeitschrift* 107:327–335.

Webster, J.A. 1975. *Association of Plant Hairs and Insect Resistance: An Annotated Bibliography*. United States Department of Agriculture. Agricultural Research Service. Miscellaneous Publication No. 1297. 16 pp.

Wheeler, H. 1977. Ultrastructure of penetration by *Helminthosporium maydis*. *Physiological Plant Pathology* 11:171–178.

Wilkinson, H.P. 1979. The plant surface (mainly leaf). pp. 97–165 in Metcalf. C. R. and Chalk, L. (editors), *Anatomy of the Dicotyledons*. Oxford University Press, New York.

Wynn, W.K. 1976. Appressorium formation over stomates by the bean rust fungus: response to a surface contact stimulus. *Phytopathology* 66:136–146.

Wynn, W.K. 1981. Tropic and taxic responses of pathogens to plants. *Annual Review of Phytopathology* 19:237–255.

Wynn, W.K. and Staples, R.C. 1981. Tropisms of fungi in host recognition. pp. 45–69 in Staples, R.C. and Toenniessen, G.H. (editors), *Plant Disease Control*. John Wiley and Sons, Inc., New York.

Young, D.H. and Kauss, H. 1984. Adhesion of *Colletotrichum lindemuthianum* spores to *Phaseolus vulgaris* hypocotyls and to polystyrene. *Applied Environmental Microbiology* 47:616–619.

Zaiter, H.Z., Coyne, D.P., Steadman, J.R., and Beaver, J.S. 1990. Inheritance of abaxial leaf pubescence in beans. *Journal of the American Society of Horticultural Science* 115:1158–1160.

6

Analysis of Spatial Patterns in the Phyllosphere

Murray K. Clayton and Brian D. Hudelson

6.1 Introduction

Microbes in the phyllosphere occupy a dynamic ecosystem where their activity can be affected by the interactions of a variety of factors, both physical and biological. Microbial activity can be directly affected by the environment, by the plant host, by anthropogenic influences (e.g., fungicide or bactericide applications), by introduction of competing or cooperating microbes, and by macroorganisms such as insects. Indirect influences on microbes abound: for example, soil type can influence leaf wetness, leaf surface temperature, surrounding insect population levels, and so on. Since the many direct and indirect influences on microbes can exhibit spatial variation, it is not surprising that the compound effect of these factors will result in the occurrence of spatial patterns in microbial activity at multiple scales. Indeed, the typical goal in studying spatial patterns is to better understand the relative importance of these influencing factors, and thus to better understand the phyllosphere as an ecosystem.

A successful understanding of the complex, multiple spatial patterns that can arise in the phyllosphere requires the collection of data of sufficient quality. However, the ability to collect such data can be severely restricted by time and labor constraints. Once data have been obtained, adequate tools for detecting and describing multiple spatial patterns are required. The development of such tools, as well as adequate sampling strategies, represents a task that is conceptually daunting and mathematically challenging. Even given adequate sampling and analytical techniques, the extent to which it is possible to make inferences from the patterns of microbial data to the causes of such patterns must also be addressed.

Studies of spatial patterns of microbial activity in the phyllosphere often focus on patterns of disease. Disease symptoms are usually relatively easy to observe and quantify, in contrast to the microbes that cause the disease. In commercial settings, quantification of mean or total disease within a field may be of primary interest. Generally, it is hoped that disease pattern information can lead to inferences regarding the epidemiological significance of the factors influencing disease development.

In this chapter we too shall focus on spatial patterns of disease. We shall provide a general review of approaches to spatial pattern analysis, focusing specifically on autocorrelation and autoregressive moving average (ARMA) approaches, a collection of powerful techniques for describing spatial patterns. Such approaches are useful in developing adaptive methods, whereby patterns determined at one scale are used to design sampling plans for elucidating patterns at a larger scale. They can also lead to sampling plans that provide more accurate assessment of mean disease levels for the benefit of crop management.

We shall illustrate several aspects of these ideas with our work on bacterial brown spot disease of snap beans caused by *Pseudomonas syringae* pv. *syringae*—a system that consistently exhibits multiple patterns of disease occurring at different scales within a given field. These patterns range from plant-to-plant patterns, to those that occur at distances greater than 60 m. We shall also draw on other examples of spatial pattern analysis, but we shall focus attention only on a few of the many studies of disease patterns in the phyllosphere. It should be noted that there have been numerous plant pathological studies in the rhizosphere, cf. Campbell and Noe (1985), and in some cases, the techniques that we review have only been employed in that setting. Also, many of the methods discussed here originally arose and continue to find application in the study of spatial patterns in areas other than plant pathology: examples include forestry, demographics, anatomy, astronomy, and geology.

6.2 Methods of Spatial Pattern Analysis

Numerous techniques exist for analyzing spatial data (cf. Campbell and Noe, 1985; Cliff and Ord, 1981; Diggle, 1983; Madden, 1989; Ripley, 1981; and Upton and Fingleton, 1985). The methods used depend on the type of data gathered and the goals of the analysis. We shall focus primarily on three groups of pattern analysis methods: quadrat methods, distance methods, and autocorrelation methods. (In this last group we also include the fitting of ARMA and ARMA-like models.) To illustrate these methods of pattern analysis, we begin by considering situations wherein each plant in a region of interest is either diseased or not, and the issue of concern is whether the diseased plants are randomly located, clustered at some scale, or regularly arranged. Later in this section, we shall discuss the case where disease on each plant can be quantified.

Quadrat Methods By far the most common method used in studying disease data of this type is to divide the study region into equal-sized, regularly shaped quadrats and to determine, for each quadrat, the number of diseased plants in the quadrat. An advantage of this method is that the data are relatively easy to collect. Quadrat data are then analyzed by determining the frequency distribution of quadrats with given numbers of diseased plants. If the observed frequency distribution is not significantly different from the Poisson distribution, the disease or inoculum is said to be randomly distributed (Upton and Fingleton, 1985). If the pattern is determined to be nonrandom, then the extent of nonrandomness might be described by an applicable index; Upton and Fingleton (1985) discuss a variety of indices of clumping, patchiness, crowding, etc. In addition, a probability distribution other than the Poisson distribution might be fit in an effort to describe further the pattern that exists.

Unfortunately, most efforts to analyze quadrat data using these methods fail. Upton and Fingleton (1985, p. 52) remark: "The use of quadrat counts as indicators of pattern is very popular. However . . . all the methods that have been suggested are beset by problems. Not one of the methods appears to take proper account of the spatial nature of the data—which is, of course, paradoxical."

To illustrate one aspect of the weakness of quadrat methods, we borrow from Nicot et al. (1984): Figure 6.1A shows a random distribution of points with a grid of quadrats superimposed. The quadrat counts fit a Poisson distribution quite well ($P \doteq 0.5$ based on a chi-squared test). In Figure 6.1B, the quadrats of Figure 6.1A have been rearranged without altering the arrangement of points *within* each quadrat. For example, the quadrat in Figure 6.1A, column 6, six rows down from the top, appears in Figure 6.1B, column 1, two rows down. Since the arrangement of points within quad-

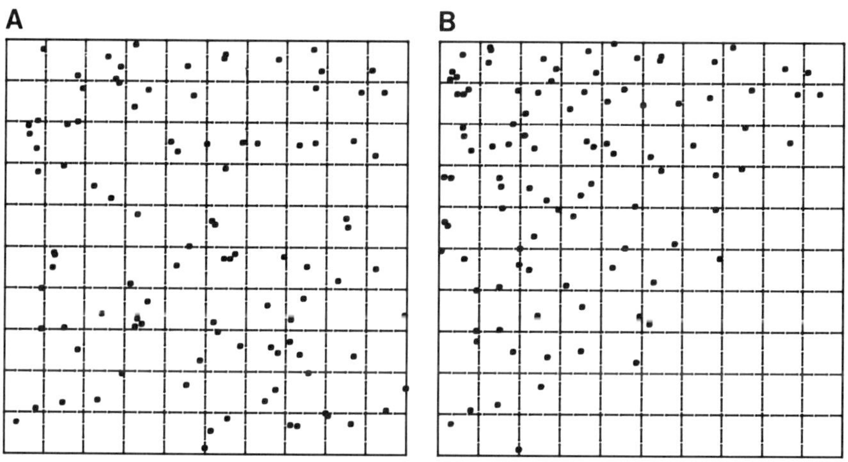

Figure 6.1 (A) Random spatial pattern of dots with quadrat grid superimposed. (B) The pattern within quadrats is retained, but the quadrats have been rearranged.

rats is unchanged, the fit of the quadrat counts of Figure 6.1B to a Poisson distribution is the same as the fit of the quadrat counts of Figure 6.1A. The points of Figure 6.1B clearly display a nonrandom pattern, however. The difficulty here is that tests based on fitting quadrat counts to a Poisson distribution fail to take into account the location of each quadrat, and therefore they do not provide any information on the spatial pattern of the disease at a scale other than the quadrat (Nicot et al., 1984).

A response to this criticism can be found in the formulations of Greig-Smith (1952) and others who look at quadrats of different sizes (cf. Upton and Fingleton), usually multiples of a given size. For example, in Figure 6.1B, if we were to use square quadrats with sides of length two then the pattern shown there would be apparent in a subsequent Poisson test ($P < 0.005$). Although this would seem to solve the problem, Upton and Fingleton (1985) show persuasively that this is not the case: this and other similar modifications of quadrat methods can be unsuccessful in detecting either single or multiple levels of spatial pattern. A similar point has been made by Ripley (1981). Shaw (1990) discussed an approach for detecting nonrandomness within quadrats, although the proposed method is essentially a variation on the distance methods to which we now turn.

Distance Methods In distance methods, the distance between each pair of diseased plants is calculated, and tests are based upon these interplant distances. For example, a test of randomness is based on a comparison of the distances between observed diseased plants with the distances that would be expected from randomly located points. Should the data prove to be nonrandom, the degree and type of patterns can then be quantified. A complete discussion is provided by Diggle (1983), Upton and Fingleton (1985), and Ripley (1981). Distance methods have been used to study spatial patterns of plant diseases although such use is relatively rare. Gray et al. (1986) used such methods to show that there was a tendency for watermelon mosaic virus 2 infections of muskmelon to occur in short runs within agronomic rows. Poushinsky and Basu (1984) showed that bacterial blight of soybean caused by *Pseudomonas syringae* pv. *glycinea* tended to be clustered when disease incidence was low, consistent with nonrandom distribution of primary inoculum in soil, but as incidence increased the pattern of disease became random or regular. In a study of tobacco diseases, Proctor (1984) showed that tobacco mosaic virus clusters were aligned more with rows than across rows, perhaps as a result of workers moving along field rows. On the other hand, potato virus Y on tobacco showed clustering without directional asymmetry, a pattern consistent with aphid transmission of the virus. Although distance methods can help to determine the existence of a pattern, they provide limited information that can be used in formulating sampling plans for estimating mean levels of disease, for example.

Although not strictly a distance method, a runs test is similar in principle: it too begins with information about the location of each diseased plant, in addition to the location of healthy plants. The number of groups (runs) of diseased and healthy plants along a line is recorded (Upton and Fingleton, 1985). The observed number of runs is compared to the expected number of runs that would arise from randomly arranged data. This method has been used, for example, by Gottwald et al. (1989) to study citrus greening disease in sweet orange groves; by Hudelson et al. (1989a) in their work on bacterial brown spot disease of snap beans; and in studies by Madden et al. (1982, 1987) of sweet corn plants infected by maize dwarf mosaic virus. Each of these studies showed that disease is often clustered within rows. As in these examples, runs tests are primarily restricted to patterns occurring in rows, and, similar to at least some of the distance methods, they provide little information beyond telling whether the existing pattern is clustered, random, or regular, and they are difficult to apply in the search for superimposed patterns. On the other hand, runs tests can be applied to a broad class of data types. For example, runs tests can be applied to disease severity data, to other quantitative measures of interest, and to quadrat data (Freund, 1981). Unfortunately, Upton and Fingleton's (1985) general criticism of quadrat methods could apply equally well to the use of quadrats in a runs test.

Autocorrelation Approaches In the remainder of this chapter we shall focus on methods that are based on the determination of autocorrelations or on the fitting of ARMA models. We believe these possess great potential for yielding useful information about spatial patterns of disease. These methods have been applied to data gathered on individual plants and to quadrat data. We begin by discussing the former; the extension to quadrats is straightforward. To apply autocorrelation methods, we must be able to quantify disease on each plant. Figure 6.2 illustrates this for data on brown spot disease on snap beans: the proportion of diseased leaflets (per plant) was recorded for every plant in a 5-m row segment. An autocorrelation analysis is based on the premise that the amount of disease on a given plant may be correlated with the amount of disease on neighboring plants. For example, depending on how disease spreads, a correlation might exist between immediately adjacent plants. In general, correlation information is summarized by an autocorrelation function ρ, whose value $\rho(s)$ at the "lag" s is the correlation of disease amounts associated with all pairs of plants that are s plants apart (Box and Jenkins, 1976). The estimated autocorrelation function, denoted $r(s)$, for the data in Figure 6.2 is given in Figure 6.3. (An arcsine square-root transformation was applied to the disease data before the correlations were calculated.) As might be expected, plants close to each other have amounts of disease that are positively correlated. This is apparent in Figure 6.2. Overall, the amounts of disease show a slowly undulating pattern moving along the row.

There are several points to be made here: first, the correlations that

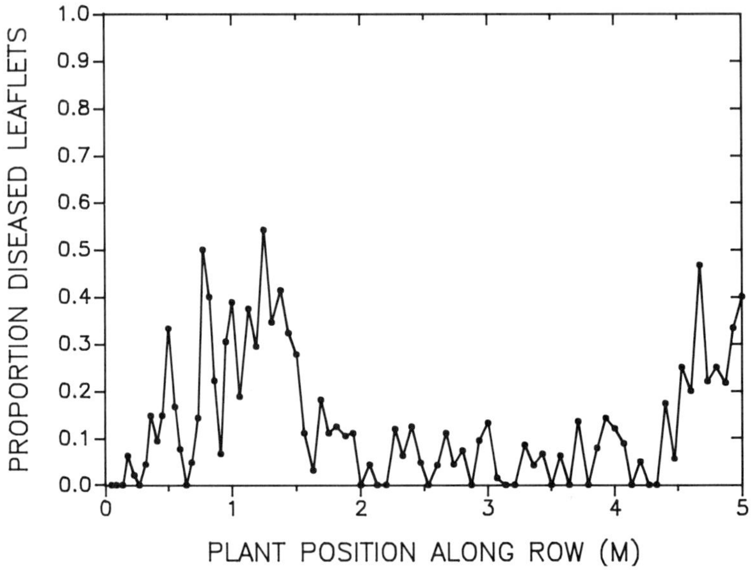

Figure 6.2 Bacterial brown spot disease incidence values for individual snap bean plants within a 5-m row segment. Disease incidence values are expressed as the proportion of diseased leaflets per plant. The number of leaflets per plant ranged from 4 to 66 with a median of 23 leaflets.

extend beyond the dashed lines in Figure 6.3 are statistically significant at the 5% level and thus suggest that the pattern shown in Figure 6.2 is not simply random. This provides a formal test of randomness in the data. It is important to add, however, that the overall shape of the autocorrelation function conveys information that is useful in its interpretation (cf. Box and Jenkins, 1976). Second, autocorrelations can readily be calculated using a statistical analysis package such as Minitab (3081 Enterprise Drive, State College, PA 16801, USA) or SAS (SAS Institute Inc., Box 8000, Cary, NC 27511-8000, USA). Finally, discussion of correlation of disease levels lends itself to hypothesizing about mechanisms. For example, the correlation pattern of Figure 6.3 that describes the smooth undulating pattern exhibited in Figure 6.2 would be consistent with a mechanism of spread of disease from a few foci to adjacent plants.

A second pattern is apparent in the data of Figure 6.2—a jagged sawtooth pattern in addition to the smooth undulating pattern described thus far. This pattern can also be explored through the use of correlation analyses, although the appropriate tool is a partial autocorrelation function. This device will not be discussed in detail here—a useful reference is Box and Jenkins (1976). The models to which we now turn will provide a description of both of the patterns discussed thus far.

To obtain a clearer and more concise description of the spatial patterns

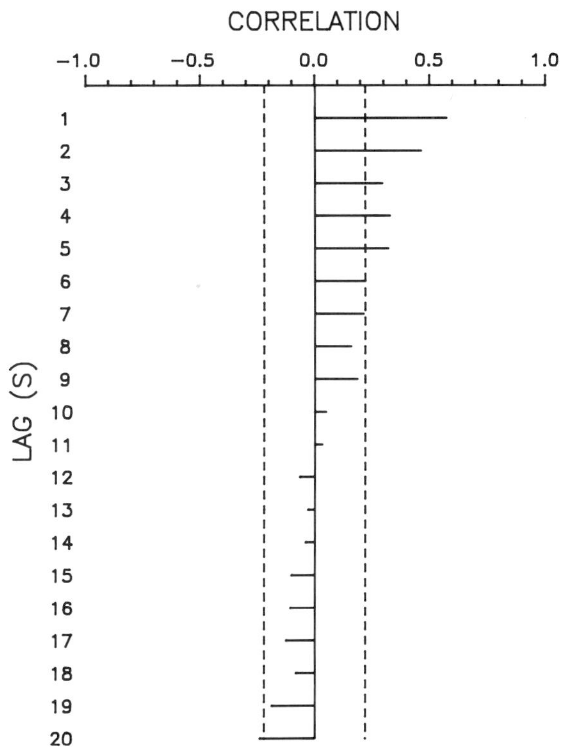

Figure 6.3 Autocorrelation function for the data shown in Figure 6.2 (after an arcsine square-foot transformation has been applied). The value of the autocorrelation function at lag *s* is the correlation of the transformed disease incidence values for all pairs of plants that are *s* plants apart. Dashed lines delimit approximate 95% confidence intervals assuming that the disease pattern is a random pattern.

of the data in Figure 6.2, we take advantage of the autoregressive moving average (ARMA) models developed for time series analysis (Box and Jenkins, 1976). These models provide a functional relationship between disease levels of adjacent plants. To describe this in more detail, suppose that the plants shown in Figure 6.2 have been numbered—let Y_t correspond to the amount of disease (or an appropriate transformation of this value) for the plant at position t. An ARMA(p,q) model has the form:

$$Y_t = \phi_1 Y_{t-1} + \ldots + \phi_p Y_{t-p} + \epsilon_t - \theta_1 \epsilon_{t-1} - \ldots - \theta_q \epsilon_{t-q} + \delta.$$

Here, the ϵs represent random "noise" components associated with each plant and are assumed to be independently and identically distributed with mean 0 and variance σ^2. To obtain some insight into the model, it is useful to examine its components. The piece: $Y_t = \phi_1 Y_{t-1} + \ldots + \phi_p Y_{t-p} + \epsilon_t + \delta$ is the autoregressive component of the model; in a sense Y_t is regressed on the p disease values preceding it in the row. In this context, the ϕs are similar to regression coefficients and δ is a constant related to the mean level of disease in the sample. The moving average portion of the model consists of $\theta_1\epsilon_{t-1} - \ldots - \theta_q\epsilon_{t-q}$, which, in a sense, represents a weighted average of the noise terms associated with the q plants that precede the plant at position t in the row. The θs represent the weights in this average. Although it appears that the ARMA model has "direction" insofar as numbering plants in the opposite direction in a row might appear to have an effect on the fitting of the ARMA model, in fact this will not generally occur, provided the ARMA model is the correct model for the data (Box and Jenkins, 1976).

Although the autoregressive portion of the ARMA model resembles a regression model, because there is a correlation between the Y_t terms, it is incorrect to use the usual regression methods to derive estimates of the ϕ terms. Again, Minitab, SAS, or other time-series software can be used to address this. Box and Jenkins (1976) or Chatfield (1984) should be consulted for more details on the model-fitting process. One step involves a detailed examination of the autocorrelation and partial autocorrelation functions. This step can often give useful clues regarding the number of parameters (p and q) needed in the model, although a definitive determination of these values can only be obtained by actually fitting the model.

The data in Figure 6.2 were fit using Minitab. The fitted model is an ARMA (1,1) model with model equation:

$$Y_t = 0.86Y_{t-1} + \epsilon_t - 0.41\epsilon_{t-1} + 0.04.$$

Essentially this model describes two levels of nonrandomness. We interpret the autoregressive portion of the model as describing the smooth undulating pattern in Figure 6.2 and interpret the moving average portion as describing the jagged pattern.

The data in Figure 6.2 give one example of a data set similar to those considered by Hudelson et al. (1989a). In 52 of 55 cases, a generalized ARMA (1,1) model provided a good fit to the disease incidence data for row segments about 5 m in length (Hudelson et al., 1989a; Hudelson, 1990). The generalized ARMA (1,1) model is of the form:

$$Y_t = \phi_1 Y_{t-1} + \epsilon_t - \theta_1\epsilon_{t-1} + \delta$$

where $-1.0 \leq \phi_1 \leq 1.0$ and $-1.0 \leq \theta_1 \leq 1.0$. This model is "generalized" in the sense that it includes: the pure autoregressive [ARMA (1,0)] model, $Y_t = \phi_1 Y_{t-1} + \epsilon_t + \delta$ (12 of the 52 data sets described above); the pure moving average [ARMA (0,1)] model, $Y_t = \epsilon_t - \theta_1\epsilon_{t-1} + \delta$ (0 of the 52 data sets described above); and the totally random model, $Y_t = \epsilon_t + \delta$ (20 of the 52 data sets described above). From this standpoint, the generalized ARMA (1,1) model consists of the class of those relatively simple ARMA models with $p \leq 1$ and $q \leq 1$, and excludes all ARMA models that require more parameters (i.e., $p > 1$ or $q > 1$) to describe the pattern in the Y values.

6.3 Interpreting ARMA Models

What are we to make of the fact that a high percentage of data sets discussed by Hudelson et al. (1989a) can be fit by a relatively small class of models? These data sets represent disease ratings collected over five years, in four different areas of Wisconsin (each with a distinct soil type), and from seven different snap bean cultivars. The very high frequency with which this model fits data from short row segments over a wide array of soil types, weather and bean cultivars argues quite strongly that these patterns are a common, integral part of the development of this disease. As the authors remark, this consistency suggests: " . . . that the biological processes that generated these data sets are not extremely diverse but are simply variations of a single underlying process or group of processes" (Hudelson et al., 1989a).

Unfortunately, it is not a trivial matter to deduce the mechanisms that might generate the patterns we observe. One reason for this is that there can be several ways in which an ARMA pattern can arise. For example, since incidence of brown spot disease is related to pathogen population size (Rouse et al., 1985), if population levels of *Pseudomonas syringae* pv. *syringae* follow an ARMA (1,1) pattern, then we would not be surprised to see the ARMA (1,1) pattern of disease described above. However, it is also possible for an ARMA (1,1) pattern to arise from the additive effects of an ARMA (1,0) pattern and a random pattern. Put more formally, if X_t is ARMA (1,0) and if W_t is random, then $Y_t = X_t + W_t$ is ARMA (1,1) (cf. Box and Jenkins, 1976). The following is a hypothetical biological model of this phenomenon: X_t represents the disease susceptibility of the plant at position t, and this varies along the row according to an ARMA (1,0) model due to smoothly varying soil conditions; W_t represents pathogen population size and it follows a completely random pattern, due to random dispersal of the pathogen. Finally, the amount of disease on the plant is Y_t, which represents the "sum" of the effects of plant susceptibility and pathogen dispersal. Thus constructed, Y_t will follow an ARMA (1,1) model. Conversely, it may be that plant susceptibility to disease is random (W_t) and pathogen populations (X_t) follow an ARMA (1,0) pattern. This too will result in an ARMA (1,1) model for disease. Even more complex situations can easily arise: an ARMA (1,1) model will result if, in fact, W_t is the sum of several random processes, or if X_t is ARMA (1,1) and W_t represents a random pattern. Because many different "constructions" can lead to the same overall model, some caution is necessary in attempting to relate the order of the model (i.e., p and q) or its components (the moving average or autoregressive portions of the model) to specific causes. At this point, it might not be surprising that a clear relationship has not been found in the relatively few studies in which patterns of pathogen populations were compared with patterns of disease (cf. Boland and Hall, 1988; Chellemi et al., 1988; Ferrin and Mitchell, 1986; Mihail, 1989).

The above discussion suggests that observing the spatial pattern of disease, or more generally, of microbial activity, will likely not lead to a definitive description of the causative mechanisms. Of course, disease pattern information can be used to speculate about possible mechanisms and to rule out certain mechanisms. For example, if a mechanistic model for brown spot disease leads to a spatial pattern that is not generalized ARMA (1,1) for short (i.e., 5 m) row segments, then the mechanistic model would be suspect.

We have argued that ARMA modeling facilitates a description of multiple spatial patterns, and it leads to inferences about mechanisms causing those patterns. Additional merits of this approach are discussed in the next two sections. However, some of its drawbacks need to be recognized. The most immediately apparent drawback is that it requires the quantification of disease on every plant. In addition, a relatively large number of plants must be evaluated for disease—Box and Jenkins suggest at least 50. More technically, the data should be approximately normally distributed with constant variance, or transformed to have such a characteristic, and there should be no missing data. Regarding this latter issue, the usual assumption of time series analysis, put in the current context, is that plants are evenly spaced. This assumption can in fact be relaxed somewhat: the approach will be valid as long as the interplant spacing is approximately even. More broadly, it should be noted that the fitting of ARMA-like models when the above assumptions fail is a topic of current statistical research, but it is beyond the scope of this review to delve into the details of these efforts.

A further potential difficulty of ARMA (p,q) models is that they are unidirectional, while disease probably operates in many directions. This may not be a serious drawback to their application since, mathematically, bidirectional models of disease spread in the row can be expressed in an ARMA form. This suggests caution, however, in interpreting the significance of a positive ϕ_1 term in an ARMA model as suggesting that Y_t *depends on* Y_{t-1}. Finally, ARMA models are designed to model correlation among observations. If the data should follow a trend (e.g., a linear trend), as well, then a more elaborate model is needed such as an Autoregressive Integrated Moving Average (ARIMA) model or a trend surface model. Box and Jenkins (1976) discuss details of the former, and Upton and Fingleton (1985) and Ripley (1981) discuss details for the latter.

6.4 Sampling Plans for Estimating Mean Disease Levels

The existence of spatial patterns in a population being sampled can have an effect on the efficiency of a sampling strategy, insofar as estimates of the mean disease level in a particular region may be excessively variable if an inappropriate sampling plan is used. (See Delp et al., 1986; Lin et al., 1979; Mihail and Alcorn, 1987; Poushinsky and Basu, 1984; Thal and Campbell, 1987; also the many references cited in Campbell and Noe, 1985; Cochran,

1977; and Iachan, 1982.) Consequently, when a sample (in lieu of a complete census) is to be used to estimate mean disease levels for a given region in which nonrandom spatial patterns occur, it will be desirable to identify sampling strategies that will give estimates with lower variance than given by simple random sampling. The strategies obtained should also be easy to implement.

Most studies on mean disease estimation in the realm of phytopathology have examined sampling plans of the region of interest where regularly spaced quadrats lying along a path in the shape of a W, an X, a diagonal stripe, etc., are observed (e.g., Basu et al., 1977; Delp et al., 1986; Hau et al., 1982; Lin et al., 1979; Mihail and Alcorn, 1987; Poushinsky and Basu, 1984). The general conclusion reached, typically based on computer simulations, is that this method of sampling is superior to simple random sampling when the disease pattern is clustered. Comparisons with other sampling methods such as stratified sampling have also been made. It is interesting to explore the link between these results and the work of sampling theorists.

The greatest theoretical efforts to define sampling strategies useful in the presence of spatial patterns have been made in the situation for which sampling is confined to a single dimension, for example, along a row (cf. Iachan, 1982). Most of the work in this direction has been based on knowledge of the form of the autocorrelation function, or equivalently, the form of ARMA model that fits the data. If the theoretical autocorrelation function satisfies certain conditions, then random start systematic sampling, within rows, will result in estimates of the mean level of disease that are less variable than estimates based on simple random samples (Cochran, 1977; Iachan, 1982, 1983). Data fitting the generalized ARMA (1,1) model considered by Hudelson et al. (1989a) satisfy the required conditions as long as $0 \leq \theta_1 \leq \phi_1$. A random start systematic sample, in this context, involves choosing a starting point at random in a given row, and then sampling every k^{th} plant along the row thereafter. Modifications meant to deal with the presence of linear, quadratic, and periodic trends have also been proposed, and variations useful for estimating the standard error of the mean are available (cf. Iachan, 1982; Wu, 1984).

As an example, suppose we wish to estimate the mean disease level for a 5-m row segment of data which follow an ARMA (1,1) model with $\phi_1 = 0.86$ and $\theta_1 = 0.41$ (the model for the data shown in Figure 6.2). Instead of taking a complete census, suppose we are limited to taking only 10 plants. Using the formulas of Iachan (1983), it is possible to show that a simple random sample would yield a standard error of the sample mean roughly twice the standard error that would be obtained from a systematic sample with $k = 10$. Clearly the systematic sample provides more precise information and is simpler to implement than a simple random sample. Depending on the size of k and the total sample size, the improvement can be even more dramatic.

To return to the discussion of sample plans using W, X, or other similarly shaped patterns, note that these can be treated effectively as systematic sampling plans of quadrats lying in a row. If the disease pattern is clustered, then neighboring quadrats along the sample path will show a positive autocorrelation, and thus random start systematic sampling will provide better estimates of mean disease than simple random sampling. Moreover, there will be no essential difference among sampling plans that follow a W, an X, or other path if the sample size and spacing is the same, and if the type of pattern is the same across the region of study. The effect of changes in the spacing between sampled quadrats, in the degree of aggregation, or in the sample size can be predicted, again using the formulas of Iachan (1983). Thus, given a fixed total sample size and with some foreknowledge of the spatial pattern present, we can determine good choices for k. As discussed in Delp et al. (1986) or Thal and Campbell (1987), sampling costs could also be taken into account.

We should note that this discussion has slightly blurred the distinction between one- and two-dimensional sampling: we have taken an optimality result regarding one-dimensional systematic sampling and applied it to what should properly be a problem of sampling a two-dimensional region, the original problem of Basu et al. (1977) and others. What was said remains correct, but we defer a discussion of further details regarding the sampling of two-dimensional regions to Section 6.6.

Systematic sampling is not a panacea: Ripley (1981), and Cochran (1977) discuss how inaccuracies can arise if the sampled pattern is periodic with period equal to the sampling gap k. We also note that, as with all sampling plans, the effectiveness of systematic sampling depends on the spatial pattern of the data: if the data follow an ARMA (1,0) pattern with $\phi_1 < 0$ then the best sampling plan is not systematic but clustered (Blight, 1973). The main point remains: by using knowledge of spatial patterns, and by successful modeling of those patterns, statistical theory can be invoked to construct and compare various sampling plans. Delp et al. (1986), Goodell and Ferris (1981), and many others have suggested adaptive applications of this notion: a pilot sample in a field is used to determine spatial patterns, and then a sampling plan is designed using that information. A particularly compact version of this has been discussed by Lin et al. (1979).

6.5 Sampling Plans for Detecting Spatial Patterns

Just as knowledge of spatial patterns can lead to improved sampling plans for estimating mean disease levels, spatial pattern information can be used in developing sampling schemes for detecting spatial patterns at other scales. The need for this is clear: if we wish to describe patterns in regions of substantial size, labor and time constraints force us to abandon the notion of taking a complete census, as described above for 5-m row segments. The approach taken by nearly all workers to date to deal with these constraints has been to use quadrat sampling. The use of quadrats can be prone to errors, however, patterns within quadrats might be missed, choice of quadrat size might influence the ability to detect patterns, and presence of spatial patterns can affect estimation of quadrat means if they are based on within-quadrat samples. Two non-quadrat-based approaches that we have found useful in detecting patterns within rows are outlined below. Both are modifications of systematic sampling.

The first approach is a sampling plan involving evaluating two adjacent plants for disease, skipping four adjacent plants, evaluating two adjacent plants, etc. (Hudelson et al., 1988). Within each pair, disease values are averaged, and these averages form the data to be analyzed. Using theoretical considerations, it is possible to show that the data from this modified sampling plan should follow a generalized ARMA (1,1) model if the generalized ARMA (1,1) patterns seen in the 5-m row segments are the only spatial patterns present in the longer row segment. Deviations from a generalized ARMA (1,1) model for the modified sample provide evidence of additional patterns. The advantage of this modified sampling plan is that it permits predictions to be made from 5-m data and provides a clear path toward the discovery of patterns at larger scales. In addition, it permits covering roughly three times the distance with the same effort as involved in the 5-m segment data collection. A large class of additional sampling plans with these characteristics has been outlined (Hudelson, 1990). The reader is referred to that work for more details of these calculations and of the resulting application.

A more elaborate modification of a systematic sampling plan, the cyclic sampling plan of Clinger and Van Ness (1976), can be used to aid in the detection of patterns over even greater distances. The data from one implementation of a cyclic sampling plan are shown in Figure 6.4. That sampling plan involved first locating a row segment consisting of 1550 plants. This was divided into 50 subsegments each consisting of 31 contiguous plants. Within each subsegment, plants were numbered from 1 to 31, and plants at positions 2, 5, 7, 11, 25, and 26 were sampled. Note that less than 20% of the plants in the entire row segment of 1550 plants were evaluated for disease. Also note, due to the cyclic nature of the sample, that it is possible to find numerous pairs of plants that are exactly one plant apart, numerous pairs that are exactly two plants apart, etc. It is therefore possible to construct an autocorrelation function from data collected in this manner.

The sample autocorrelation for the (arcsine square-root transformed) data of Figure 6.4 is shown in Figure 6.5. The dashed lines provide a test of significance similar to those of Figure 6.3—details for determining these appear elsewhere (Hudelson, 1990). (The periodic indentations of these lines every 31 lags as seen in Figure 6.5 correspond to the cyclic nature of the sampling scheme.) The particular lines displayed were determined under the assumption that spatial patterns over short distances are random. It is obvious that patterns exist extending over quite large distances.

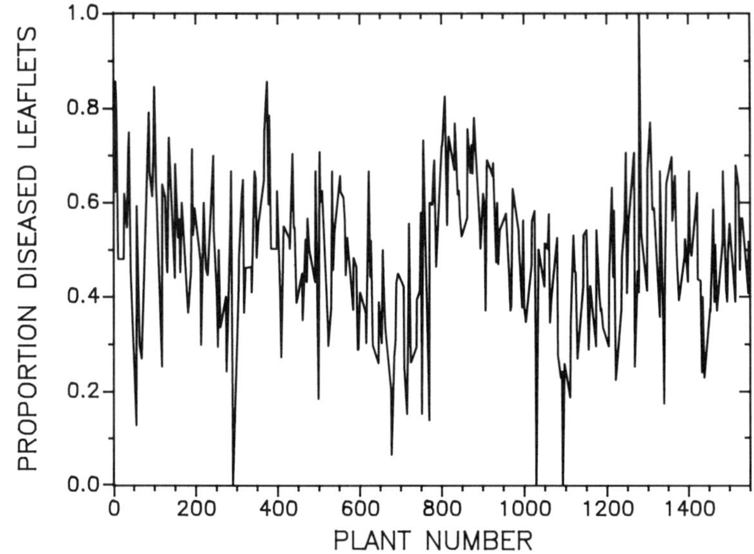

Figure 6.4 Bacterial brown spot disease incidence values for individual snap bean plants sampled using a cyclic sampling plan. In a cyclic sampling plan, sampled plants are spaced such that pairs of plants can be found that are any arbitrary number of spacings apart. For the cyclic sample used to collect the bacterial brown spot data from a total of 1550 plants, pairs of plants that are one plant apart, two plants apart, etc., up to 1525 plants apart can be found.

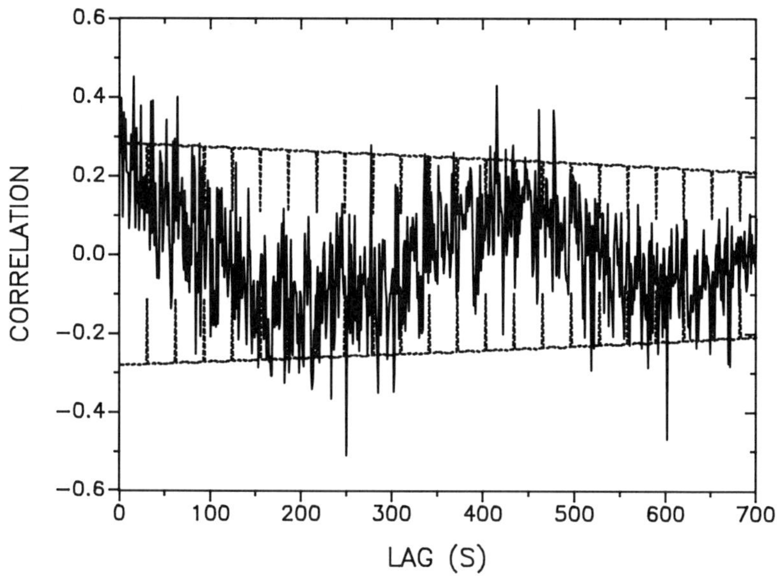

Figure 6.5 Autocorrelation function for the data shown in Figure 6.4 (after an arcsine square-root transformation has been applied). The value of the autocorrelation function at lag *s* is the correlation of the transformed disease incidence values for all pairs of plants that are *s* plants apart. Dashed lines delimit approximate 95% confidence intervals assuming that the disease pattern is a random pattern.

Indeed, the fitting of only three sinusoids of periods 255, 419, and 655 plants produces a periodic curve that explains roughly 23% of the variation in the data (Hudelson et al., 1989b). Although details are not shown here, note that the dashed lines in Figure 6.5 can also be constructed assuming that short-distance patterns follow a specific ARMA (1,1) model. These lead to the same conclusion regarding the existence of long-distance patterns. We speculate that these patterns may be linked to cultural practice. For example, the 655 plant pattern is roughly the same size as the width of the equipment used to spray fungicides on potatoes planted in that field in the previous year (Hudelson, 1990).

A key thrust in the two sampling designs described above is that they were designed in an adaptive fashion: spatial pattern information at a small scale was used to design a sampling plan at a larger scale and to interpret the resulting data. The strength of an ARMA model-based approach to spatial pattern analysis is that it facilitates an adaptive approach of this sort.

6.6 Higher-Dimensional Patterns

Theoretically, all that has been said thus far in applying ARMA modeling extends directly to two-dimensional pattern analysis or, if change over time is also considered, to a three-dimensional analysis. However, the numerical problems entailed in studying spatial patterns of observations even modestly irregularly arranged in two dimensions can be prohibitive. Greater progress can be made if the data represent observations on a rectangular grid—quadrat data, data from a sampling plan that uses a grid, or data from plants grown in a grid pattern will suffice, although again, unless care is taken in determining the sampling plan in the first two cases, there is a risk of missing patterns.

Assuming that grid data can be obtained, let $Y_{i,j}$ represent the observation in column i, row j of the grid. By direct extension of the one-dimensional situation, the spatial autocorrelation $\rho(u,v)$ is given by the correlation between all plant pairs which are exactly u columns and v rows apart, i.e., between $Y_{i,j}$ and $Y_{i+u,j+v}$. In their work on citrus greening disease Gottwald et al. (1989) used such autocorrelations to provide evidence of disease clustering and to investigate cluster shape and size. The use of spatial autocorrelations led Shew et al. (1984) to demonstrate the variability of spatial patterns of southern stem rot in peanut. Another example of the use of spatial autocorrelations is the study of Thal and Campbell (1987) who showed that autocorrelations of leaf spot disease in neighboring quadrats of alfalfa were usually low. Analogously to the one-dimensional situation, tests of significance can be applied to the spatial autocorrelation function $\rho(u, v)$, and aspects of the character of the spatial pattern can be deduced.

A measure related to the spatial autocorrelation function is Moran's I statistic, which is essentially a correlation between $Y_{i,j}$ and a prechosen weighted average of the neighboring Y values. The weighting scheme is flexible enough to include spatial autocorrelation as a special case although a more typical weighting scheme relates $Y_{i,j}$ to a simple average of neighboring values (cf. Upton and Fingleton, 1985; Cliff and Ord, 1981). Examples of the use of Moran's I statistic include the work of Nicot et al. (1984) and Mihail (1989). Another measure of spatial relationship is the semivariogram, which is based on the variance of the difference between $Y_{i,j}$ and $Y_{i+u,j+v}$ instead of the correlation between these two values. Although sometimes presented in the literature as competing concepts, there is in fact a close relationship between the semivariogram and the spatial autocorrelation function: if the variances of $Y_{i,j}$ and of $Y_{i+u,j+v}$ are the same, then the semivariogram is directly related to the spatial autocorrelation function. Sometimes a transformation of the data will suffice to ensure this equivalence. Cliff and Ord (1981) and Upton and Fingleton (1985), among others, give a more detailed discussion of these and other related measures of spatial relationship. Examples of phytopathological applications of semivariograms are given by Chellemi et al. (1988) and Lecoustre et al. (1989).

Just as the spatial autocorrelation function extends the autocorrelation function to two dimensions, so too can the one-dimensional ARMA model be extended to a two-dimensional analog, a spatial ARMA (SARMA) model. A general formulation of this model is complex, to say the least. The SARMA analog of an ARMA (1,1) model has the form:

$$\begin{aligned} Y_{i,j} = {} & \phi_{1,0}Y_{i+1,j} + \phi_{1,1}Y_{i+1,j+1} + \phi_{0,1}\, Y_{i,j+1} + \phi_{-1,1}Y_{i-1,j+1} \\ & + \phi_{-1,0}Y_{i-1,j} + \phi_{-1,-1}Y_{i-1,j-1} + \phi_{0,-1}Y_{i,j-1} + \phi_{1,-1}Y_{i+1,j-1} \\ & + \epsilon_{i,j} - \theta_{1,0}\epsilon_{i+1,j} - \theta_{1,1}\epsilon_{i+1,j+1} - \theta_{0,1}\epsilon_{i,j+1} \\ & - \theta_{-1,1}\epsilon_{i-1,j+1} - \theta_{-1,0}\epsilon_{i-1,j} - \theta_{-1,-1}\epsilon_{i-1,j-1} \\ & - \theta_{0,-1}\epsilon_{i,j-1} - \theta_{1,-1}\epsilon_{i+1,j-1} + \delta. \end{aligned} \tag{1}$$

Here, $Y_{i,j}$ is related to disease values and noise terms for all of the immediately adjacent quadrats, including those on the diagonal (see Figure 6.6). Obviously, the number of parameters would increase if $Y_{i,j}$ were to be related to other more distant Y or ϵ values. Unfortunately, even the SARMA model in equation (1) can be extremely difficult to fit. One method used to deal with this is to restrict how the parameters might vary. For example, with respect to the SARMA model, the restriction $\phi_{i,j} = \gamma W_{i,j}$ might be applied, where the $W_{i,j}$ are predetermined weights. This reduces the problem of estimating all eight ϕ parameters to a problem of estimating a single γ parameter. The θ parameters can be similarly restricted. Unfortunately, it is not always clear how the weights $W_{i,j}$ should be determined (cf. Upton and Fingleton, 1985; Reynolds and Madden, 1988; Reynolds et al., 1988). Many choices exist and, ideally, the application of a variety of weighting schemes combined with additional exploratory analyses would cast light on the choice of weights. To date such an approach has not been pursued extensively.

Recently, Basu and Reinsel (1990) have proposed an alternative approach, following Whittle (1954). In the context of the SARMA model of

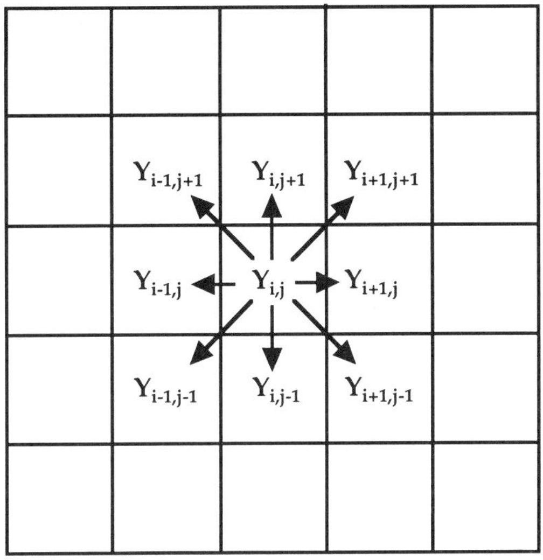

Figure 6.6 Illustration of dependence of an observation $Y_{i,j}$ on its adjacent neighbors for a spatial autoregressive moving average model analogous to an ARMA (1,1) model. $Y_{i,j}$ represents disease value for the quadrat in column i, row j in a grid of quadrats.

equation (1), they propose setting all the $\phi_{i,j} = 0$ except those corresponding to $Y_{i-1,j}$, $Y_{i-1,j-1}$, and $Y_{i,j-1}$, which are then left free to vary. A similar restriction is made on the $\theta_{i,j}$. More generally, to incorporate information from more distance observations in their model, $Y_{i,j}$ is related only to those neighboring values $Y_{u,v}$ for which $u \leq i$ and $v \leq j$. (Equivalently, either of the $\leq$ signs can be replaced by a $\geq$ sign.) Although this model is directionally asymmetric, this may not be a serious problem. For example, by fitting the model in different orientations, isotropy (directional invariance) can be examined. Moreover, the general (unrestricted) SARMA model can be reexpressed in this restricted form of "bilateral" ARMA model (Whittle, 1954). For certain forms of this bilateral model, Basu and Reinsel (1990) have conducted the requisite statistical calculations to obtain an algorithm for parameter estimation and model fitting. The appeal of this approach, in addition to its tractability, is that it permits explicit parameter estimation, combined with the aforementioned ability to examine hypotheses of isotropy.

Two-dimensional spatial modeling provides a description of spatial patterns for a "snapshot" in time. To study the progress of a disease epidemic through both time and space, Mihail (1989), Reynolds and Madden (1988), and Reynolds et al. (1988) discuss "spatio-temporal" autoregressive moving average models. Roughly speaking, in such models, each value $Y_{i,j}$ is related to neighboring values from the previous time cycle. Not surprisingly, the model-fitting exercise here is potentially very complex, and so most authors have confined themselves to calculating spatial autocorrelations and testing these for significance. The results of this effort are then used to guess the likely form of spatio-temporal ARMA model. It remains to be seen whether the approach of Basu and Reinsel (1990) can be applied to complete the difficult task of actually fitting an ARMA model to spatio-temporal data. (See Bennett, 1979; Cliff and Ord, 1981; Madden, 1989; and Upton and Fingleton, 1985, for a discussion of other variants of spatio-temporal models.)

A somewhat distinct class of two-dimensional pattern methods that we shall only discuss briefly are those that are used to model the spatial data with a surface. These include regression models, whereby $Y_{i,j}$ is modeled directly as a function of i and j (Upton and Fingleton, 1985; Cliff and Ord, 1981; Ripley, 1981); kriging (Ripley, 1981; Upton and Fingleton, 1985; Lecoustre et al., 1989); smoothing and interpolating splines (Wahba, 1990; Ripley, 1981); and two-dimensional spectral analysis (Ripley, 1981). The presence of spatial autocorrelation adds complexity to the application of these models. For example, the fitting of a regression model without taking into account spatial autocorrelation tends to lead to a false sense of precision in the model (cf. Upton and Fingleton, 1985; Basu, 1990). Interestingly, this fact seems to be ignored in epidemiological work where disease progress or gradient curves are fit. Loosely connected to this is the fact that spatial autocorrelation can also have an impact on studies of relationships between epidemiological factors. For example, in comparing disease levels in quadrats with inoculum levels in the same quadrats, it is generally incorrect to simply look at the correlation between these two factors without taking into account their spatial variability (Upton and Fingleton, 1985).

As might be expected based on the one-dimensional situation, the presence of a two- (or higher-) dimensional spatial pattern can have an impact on sampling design for determining mean disease level (cf. Iachan, 1982). Methods for estimating the standard error of mean disease estimates have been reported (Leu, 1988). Also, Matérn (1986) showed that for some types of two-dimensional autocorrelation function, a systematic sampling plan using a square or triangular grid is superior to random sampling. Presumably systematic sampling can also be of value in an adaptive context as outlined above for the one-dimensional situation; knowledge of two-dimensional spatial patterns at one scale might be useful in designing sampling plans for uncovering patterns at larger scales. We are currently implementing such an approach.

6.7 Concluding Remarks

In our work on modeling spatial patterns of bacterial brown spot disease, autocorrelation and ARMA models were found to be of great value for describing and quantifying complex, multiple spatial patterns. There is good reason to believe that these techniques will be similarly useful in exploring spatial patterns of other systems, not only from the standpoint of describing and quantifying spatial patterns, but also in terms of designing sampling plans for estimating disease or other microbial activities and for describing spatial patterns in adaptive approaches.

There is much theoretical work to be done, and while some of this is outlined in the previous sections, we might reiterate here the need for better understanding of the role spatial patterns play in designing sampling plans, and the need for adequate algorithms and software for fitting complex spatial and spatio-temporal models. The link between mechanisms and the observed patterns is not always clear, and this too needs further study. Probably a major part of this will be the development of methods for studying microbes more directly, in addition to the more common study of disease—in some sense an end product of microbial activity. Finally, our enthusiasm for ARMA-like models should not be interpreted as a sanction for their use exclusive of the other statistical and biological tools available; ARMA models play but one part in the complex task of understanding the microbial ecosystem, and attaining this understanding will undoubtably require the successful integration of these many tools.

Studies of the spatial variability of microbial activity are in their infancy, and exciting results undoubtedly will result from research of this type. We would like to close with the thought that the spatial analysis tools outlined above belong in the same class as the other high-technology tools used in biology, ranging from the now-commonplace electron microscope to more recently developed molecular tools. The analogy might be extended further: none of these tools is trivial to apply; they require skill, time, and resources. An understanding of their strengths and weaknesses is critical to their successful application. However, they possess the potential to reveal a wealth of information, to aid scientific research, and to lead to new insights and further discoveries.

The authors thank Sabyasachi Basu, Doug Rouse, Brian Yandell, and the editors, John Andrews and Susan Hirano, for their comments. This research was supported in part by the United States Department of Agriculture Competitive Grants 85-CRCR-1-1662 and 88-37151-3898.

References

Basu, P.K., Lin, C.S., and Binns, M.R. 1977. A comparison of sampling methods for surveying alfalfa foliage diseases. *Canadian Journal of Plant Science* 57:1091–1097.

Basu, S. 1990. Analysis of First-Order Spatial Bilateral ARMA Models. Ph.D. thesis, University of Wisconsin, Madison, WI. 168 pp.

Basu, S. and Reinsel, G.C. 1990. Analysis of first-order spatial bilateral ARMA models. University of Wisconsin-Madison, Department of Statistics Technical Report No. 862.

Bennett, R.J. 1979. *Spatial Time Series*. Pion Limited, London. 674 pp.

Blight, B.J.N. 1973. Sampling from an autocorrelated finite population. *Biometrika* 60:375–385.

Boland, G.J. and Hall, R. 1988. Relationships between the spatial pattern and

number of apothecia of *Sclerotinia sclerotiorum* and stem rot of soybean. *Plant Pathology* 37:329–336.
Box, G.E.P. and Jenkins, G.M. 1976. *Time Series Analysis: Forecasting and Control, revised ed.* Holden-Day, Oakland, CA. 575 pp.
Campbell, C.L. and Noe, J.P. 1985. The spatial analysis of soilborne pathogens and root diseases. *Annual Review of Phytopathology* 23:129–148.
Chatfield, C. 1984. *The Analysis of Time Series, An Introduction.* Chapman and Hall, London. 286 pp.
Chellemi, D.O., Rohrbach, K.G., Yost, R.S., and Sonoda, R.M. 1988. Analysis of the spatial pattern of plant pathogens and diseased plants using geostatistics. *Phytopathology* 78:221–226.
Cliff, A.D. and Ord, J.K. 1981. *Spatial Processes. Models and Applications.* Pion Limited, London. 266 pp.
Clinger, W. and Van Ness, J.W. 1976. On unequally spaced time points in time series. *Annals of Statistics* 4:736–745.
Cochran, W.G. 1977. *Sampling Techniques*, 3rd ed. John Wiley and Sons, New York. 428 pp.
Delp, B.R., Stowell, L.J., and Marois, J.J. 1986. Evaluation of field sampling techniques for estimation of disease incidence. *Phytopathology* 76:1299–1305.
Diggle, P.J. 1983. *Statistical Analysis of Spatial Point Patterns.* Academic Press, London. 148 pp.
Ferrin, D.M. and Mitchell, D.J. 1986. Influence of initial density and distribution of inoculum on the epidemiology of tobacco black shank. *Phytopathology* 76:1153–1158.
Freund, J.E. 1981. *Statistics, A First Course*, 3rd ed. Prentice-Hall, Englewood Cliffs, NJ. 466 pp.
Goodell, P.B. and Ferris, H. 1981. Sample optimization for five plant parasitic nematodes in an alfalfa field. *Journal of Nematology* 13:304–313.
Gottwald, T.R., Aubert, B., and Xue-Yuan, Z. 1989. Preliminary analysis of citrus greening (Huanglungbin) epidemics in the People's Republic of China and French Reunion Island. *Phytopathology* 79:687–693.
Gray, S.M., Moyer, J.W., and Bloomfield, P. 1986. Two-dimensional distance class model for quantitative description of virus-infected plant distribution lattices. *Phytopathology* 76:243–248.
Greig-Smith, P. 1952. The use of random and contiguous quadrats in the study of the structure of plant communities. *Annals of Botany* 16:293–316.
Hau, F.C., Campbell, C.L., and Beute, M.K. 1982. Inoculum distribution and sampling methods for *Cylindrocladium crotalariae* in a peanut field. *Plant Disease* 66:568–571.
Hudelson, B.D. 1990. Studies of the Spatial Patterns of Bacterial Brown Spot of Snap Beans. Ph.D. thesis, University of Wisconsin, Madison, WI. 429 pp.
Hudelson, B.D., Clayton, M.K., Smith, K.P., Rouse, D.I., and Upper, C.D. 1988. Use of adaptive sampling for the identification of nonrandom spatial patterns of bacterial brown spot. *Phytopathology* 78:1570 (Abstract).
Hudelson, B.D., Clayton, M.K., Smith, K.P., Rouse, D.I., and Upper, C.D. 1989a. Nonrandom patterns of bacterial brown spot in snap bean row segments. *Phytopathology* 79:674–681.
Hudelson, B.D., Clayton, M.K., and Upper, C.D. 1989b. Periodic spatial patterns of bacterial brown spot in a 90 meter snap bean row segment. *Phytopathology* 79:1162 (Abstract).
Iachan, R. 1982. Systematic sampling: a critical review. *International Statistical Review* 50:293–303.
Iachan, R. 1983. Asymptotic theory of systematic sampling. *Annals of Statistics* 11:959–969.

Lecoustre, R., Fargette, D., Fauquet, C., and de Reffye, P. 1989. Analysis and mapping of the spatial spread of African cassava mosaic virus using geostatistics and the Kriging technique. *Phytopathology* 79:913–920.
Leu, C.-H. 1988. Discriminant and Categorical Data Analysis and Variance Estimation in Complex Sample Surveys. Ph.D. thesis, University of Wisconsin, Madison, WI. 132 pp.
Lin, C.S., Poushinsky, G., and Mauer, M. 1979. An examination of five sampling methods under random and clustered disease distribution using simulation. *Canadian Journal of Plant Science* 59:121–130.
Madden, L.V. 1989. Dynamic nature of within-field disease and pathogen distributions. pp. 96–126 in Jeger, M.J. (editor), *The Spatial Component of Plant Disease Epidemics*. Prentice-Hall, Englewood Cliffs, NJ.
Madden, L.V., Louie, R., Abt, J.J., and Knoke, J.K. 1982. Evaluation of tests for randomness of infected plants. *Phytopathology* 72:195–198.
Madden, L.V., Louie, R., and Knoke, J.K. 1987. Temporal and spatial analysis of maize dwarf mosaic epidemics. *Phytopathology* 77:148–156.
Matérn, B. 1986. *Spatial Variation*, 2nd ed. Springer-Verlag, Berlin. 151 pp.
Mihail, J.D. 1989. *Macrophomina phaseolina*: Spatio-temporal dynamics of inoculum and of disease in a highly susceptible crop. *Phytopathology* 79:848–855.
Mihail, J.D. and Alcorn, S.M. 1987. *Macrophomina phaseolina*: Spatial patterns in a cultivated soil and sampling strategies. *Phytopathology* 77:1126–1131.
Nicot, P.C., Rouse, D.I., and Yandell, B.S. 1984. Comparison of statistical methods for the study of spatial patterns of soilborne plant pathogens in the field. *Phytopathology* 74:1399–1402.
Poushinsky, G. and Basu, P.K. 1984. A study of distribution and sampling of soybean plants infected with *Pseudomonas syringae* pv. *glycinea*. *Phytopathology* 74:319–326.
Proctor, C.H. 1984. On the detection of clustering and anisotropy using binary data from a lattice patch. *Communications in Statistics—Theory and Methods* 13:617–638.
Reynolds, K.M. and Madden, L.V. 1988. Analysis of epidemics using spatio-temporal autocorrelation. *Phytopathology* 78:240–246.
Reynolds, K.M., Madden, L.V., and Ellis, M.A. 1988. Spatio-temporal analysis of epidemic development of leather rot of strawberry. *Phytopathology* 78:246–252.
Ripley, B.D. 1981. *Spatial Statistics*. John Wiley and Sons, New York. 252 pp.
Rouse, D.I., Nordheim, E.V., Hirano, S.S., and Upper, C.D. 1985. A model relating the probability of foliar disease incidence to the population frequencies of bacterial plant pathogens. *Phytopathology* 75:505–509.
Shaw, M.W. 1990. A test of spatial randomness on small scales, combining information from mapped locations within several quadrats. *Biometrics* 46:447–458.
Shew, B.B., Beute, M.K., and Campbell, C. L. 1984. Spatial pattern of Southern stem rot caused by *Sclerotium rolfsii* in six North Carolina peanut fields. *Phytopathology* 74:730–735.
Thal, W.M. and Campbell, C.L. 1987. Sampling procedures for determining severity of alfalfa leaf spot disease. *Phytopathology* 77:157–162.
Upton, G. and Fingleton, B. 1985. *Spatial Data Analysis by Example*. John Wiley and Sons, Chichester, England. 410 pp.
Wahba, G. 1990. *Spline Models for Observational Data*. Society for Industrial and Applied Mathematics, Philadelphia. 169 pp.
Whittle, P. 1954. On stationary processes in the plane. *Biometrika* 41:434–449.
Wu, C.F.J. 1984. Estimation in systematic sampling with supplementary observations. *Sankhya, Series B* 46:306–315.

Part 3
Microbial Community Components

7

Methods for the Detection, Identification, and Enumeration of Microbes

David J. Drahos

7.1 Introduction

The greatest challenge a researcher usually faces when designing definitive experiments to measure the dynamics of microbial populations in the phyllosphere is to develop reliable and practical means to detect, identify, or enumerate a particular strain or group of interest in an environmental sample. While a number of sensitive, and often quite sophisticated and elegant, tools now exist for the task, each has advantages and disadvantages which must be fully appreciated and considered prior to settling on a particular strategy. Failure to take this step early can readily result in erroneous conclusions or an unexpected drain on resources and time. This chapter provides an overview of the current methodologies which have been or could be applied to phyllosphere microbiology, and it includes sufficient key references to allow a more comprehensive assessment of applicability to a particular research plan.

7.2 Detection vs. Enumeration vs. Identification

There is as yet no perfect detection or enumeration method. Each has advantages and drawbacks which require the user to have a clear understanding of the limitations, as well as the potential, of a particular technique. For example, under some conditions, a procedure which provides high detection sensitivity by selectable growth of a microbe may also give a reasonably accurate reflection of the actual viable population present. This is often the case for soilborne pseudomonad bacteria, which appear to be largely reculturable on laboratory media, and hence can be enumerated by this method. However, a significant percentage of the population of a similar organism in an aquatic environment is often not easily recovered, yet remains quite viable. Moreover, many methods effective in microbial detection may be poorly suited to enumeration. A good example of where caution is required is in the use of the polymerase chain reaction (PCR) method, described later, as an enumeration technique. The extreme sensitivity of this procedure renders enumeration by target dilution exceedingly difficult, if not virtually impossible. Nevertheless, one of the great attributes of the better detection/enumeration methods is their ability to function in a powerful complementary manner when several dissimilar strategies are combined. Reliable and efficient means to identify specific microorganisms in the complex environment of the phyllosphere are essential for any useful comparison of derived information with previous or ongoing related studies. Significant advances have been made toward extending the accuracy and speed of microbial identification capabilities. For example, extensions of the more classical biochemical characterization procedures, based on rapid and practical chromogenic enzymatic reactions and automated gas chromatographic analysis of characteristic fatty acids, have become commercially available. Also, the use of highly definitive nucleic acid fingerprinting procedures is providing an entirely new dimension to identification, and promises to be quite complementary to more standard classification practices. In the following sections, the capabilities, deficiencies, and potential synergy of the more promising methods for application to phyllosphere studies is described.

7.3 Detection and Enumeration: Antibiotic Resistance Factors

A standard practice in the detection of microorganisms from environmental samples has been the use of antibiotic resistance factors, which are either inherent to, or selected for, in the target organism (Kloepper et al., 1980; Kloepper and Schroth, 1981; Cook and Baker, 1983; Andrews, 1986). Certain of these resistance factors can be readily moved to other similar microbes, by plasmid conjugation or phage transduction; upon entry in the new host, an enzyme or gene product is expressed which is capable of conferring resistance to a particular set of antibiotics. These factors are referred to as transferable markers. Other organisms acquire resistance to an antibiotic by a mutation in a enzyme or gene product which is already expressed in the cell and which often plays a key role in critical cellular processes. This latter type of resistance is not typically transferred to other organisms (i.e., is nontransferable), either due to a much lower frequency of mobilization of genes on the host chromosome, or because the conjugal recipient already expresses the native, sensitive form of the gene product, and would not gain adequate antibiotic resistance from the mutant gene for selection.

Nontransferable Markers Most commonly, bacterial isolates are first selected in the laboratory for resistance to the antibiotics rifampicin (Rif^r) or nalidixic acid (Nal^r) by culturing large numbers of cells (10^5–10^8 CFU) on solid media containing 50–100 μg/ml of either rifampicin or nalidixic acid. Resistance to these antibiotics involves specific mutations in enzymes involved in nucleic acid replication. Rifampicin resistance is due to a specific change in the β subunit of RNA polymerase, and nalidixic acid resistance is due to a modification of the DNA gyrase enzyme system. Both of these changes result in enzymes which no longer interact with, and hence are no longer inhibited by the antibiotic. Reliance on Rif^r or Nal^r markers can provide significant ability to detect and track the target strain by reculturing soil and leaf wash samples on media containing these antibiotics, usually at a concentration of 50–100 μg/ml (Murray et al., 1990). However, this practice suffers from the inability of many such "marked" bacteria, particularly those with both Rif^r and Nal^r phenotypes, to maintain their pathogenicity, environmental persistence, or both (Lindstrom et al., 1990; Stotzky and Babich, 1984; Goodwin and Slater, 1979). Furthermore, the background level of resistance among soil microbes to either rifampicin or nalidixic acid alone is highly variable across different geographic regions. In many areas of the midwest United States, for example, it is not atypical to encounter a naturally occurring Rif^r population of 10^3–10^5 cells/g soil (Drahos et al., 1986). Such levels can appreciably limit detection capabilities, particularly when the number of bacteria to be monitored is low ($<10^3$/g soil) relative to the total bacterial population (often $>10^8$/g soil). However, this high background resistance to rifampicin has not been seen for samples taken from leaf surfaces (Weller and Saettler, 1978). For example, populations of Rif^r *Xanthomonas* strains could be detected even at densities less than 2×10^3 CFU/cm^2 of leaf surface. From the same samples, strains of bacteria resistant to other antibiotics, such as streptomycin and ampicillin, can be readily isolated.

Transferable Markers This type of resistance depends on the expression of enzymes which detoxify or inhibit the cross-membrane transport of certain antibiotics, such as ampicillin, kanamycin, streptomycin, tetracycline, and gentamicin. Typically these antibiotic resistance markers are introduced into the target bacterium by plasmid conjugation or transposon mutagenesis (Trevors et al., 1990). Similarly, the markers can be transferred to other strains by another conjugation, transduction, or transformation event occurring in the environment. The main advantage of the use of these markers is the high degree of selection they can provide under laboratory conditions using defined media and growth conditions. However, these resistance enzymes can have an adverse effect on the host, such as causing reduced growth rates or altering membrane characteristics and permeability (Moyed et al., 1983; Levy, 1984; Chopra et al., 1981). Further, resistance to a number of these antibiotics, such as kanamycin, streptomycin, and tetracycline, has been observed in bacteria from soil, leaf, and water samples, occasionally at appreciable levels (Schroth et al., 1979; Cole and Elkan, 1979; Kelch and Lee, 1978; Weller and Saettler, 1978). Moreover, intrinsic resistance to antibiotics is not necessarily stable over time. Brunel et al. (1988) found that while most serological and physiological properties of *Bradyrhizobium japonicum* inoculants remained constant over an eight-year period of study, intrinsic antibiotic resistance factors, including both transferable and nontransferable type factors, changed significantly.

7.4 Detection and Enumeration: Nutritional and Enzymatic Markers

The *xyl*E Gene Several reports have demonstrated the utility of using catabolic enzymes which are rare in certain environments. One of the best examples of these is catechol 2,3-dioxygenase, a product of the *xyl*E gene of the *Pseudomonas putida* TOL plasmid (Zukowski et al., 1983; Nakai et al., 1983; Ngo, 1985), which catalyzes the conversion of catechol to the yellow-pigmented 2-hydroxymuconic semialdehyde. Bacterial colonies expressing *xyl*E turn yellow following an overspray with an aqueous solution of 0.5 M catechol. Since this is not a selective method, however, some preselection on antibiotic medium is usually necessary when natural samples are analyzed due to the very low levels of the bacteria to be monitored. The development of plasmids capable of high-level expression of *xyl*E in a broad array of host bacteria (Winstanley et al., 1989) and of sensitive spectrophotometric detection, ELISA analysis, and immunoblotting procedures (Sala-Trepat and Evans, 1971; Morgan et al., 1989) has provided a more practical means to study a recombinant pseudomonad population in a lake-water microcosm (Morgan et al., 1989). Direct detection of the *xyl*E gene product by ELISA assessment was demonstrated even for population levels as low as 10^3 cells/ml lake water. In addition to *Pseudomonas* and *Escherichia coli*, the *xyl*E marker has been introduced as an efficient marker into a broad variety of microorganisms, including *Bacillus* (Zukowski et al., 1983; Ray et al., 1988) and *Streptomyces* species (Ingram et al., 1989).

The *lux* Gene Luminescent reporter genes have become more useful following the development of practical vector systems carrying genes coding for certain bacterial luciferase enzymes (Schauer, 1988). Such bioluminescent systems are based on the expression of a luciferase enzyme and its action on the substrate luciferin. In bacteria, luciferin is an aldehyde synthesized by the action of a reductase on a fatty acid precursor (Wall et al., 1984). The *lux* genes of *Vibrio harveyi* and *Vibrio fischeri* are among the better known of these systems (Schmetterer et al., 1986), the more versatile of which are contained on relatively small (7.5 kb) cassettes (Shaw et al., 1987). These *lux* cassettes carry genes for synthesis of both luciferase and luciferin bounded by convenient restriction sites, and have been expressed in a variety of bacteria (Shaw and Kado, 1986; Legocki et al., 1987), cyanobacteria (Schmetterer et al., 1986), yeast (Kirchner et al., 1989), and plant protoplasts (Koncz et al., 1987).

The *lux* marker system has been used very effectively as a noninvasive reporter in monitoring the ongoing process of infection by phytopathogenic bacteria, such as *Xanthomonas campestris, Agrobacterium tumefaciens, Erwinia carotovora, Erwinia amylovora, and Pseudomonas syringae* (Shaw and Kado, 1986). In many instances the progression of the marked strains can be routinely followed in plant tissue in a noninvasive and nondestructive maner using autoradiography or adapted photon detectors. In the past a major drawback of this technique has been its low sensitivity, as well as background problems due to the nonuniform quenching of photons by surrounding material. However, advances in instrument efficiency and cost effectiveness are rapidly diminishing these difficulties (Schauer, 1988). Shaw and Kado (1986) successfully used the *lux* marker to directly monitor the ongoing process of infection by plant-pathogenic and symbiotic bacteria in cauliflower leaves, potato tubers, and carrot roots. In these studies, light emissions were also quantified by a modified luminometer and shown to directly reflect the growth of the pathogen population in the host tissue.

The Ice Gene Certain strains of *Pseudomonas, Erwinia,* and *Xanthomonas* are capable of producing a gene product which acts to nucleate supercooled water to form ice (Maki et al., 1974; Lindow et al., 1978a, 1978b, 1978c). Ice+ bacteria can be followed by observing ice crystal formation with various techniques. For example, samples can be spotted onto the surface of a paraffin-coated sheet of aluminum foil floating on a glycol (60%) bath cooled to −5°C (Vali, 1971). Positive responses are clearly seen, again, in the immediate freezing of the droplets (Goodnow et al., 1990; Lindow et al., 1978b). Alternatively, a tube nucleation test has been successfully applied (Hirano et al., 1985; Hirano and Upper, 1986; Hirano et al., 1987), where leaflet samples are totally immersed in a buffer, and placed in a series of three refrigerated baths (−2°C, −2.5°C, and −3°C). The cumulative percent of leaflets frozen at each temperature can be used to estimate the relative population densities of certain Ice+ plant-pathogenic bacteria and to predict the relative intensity of foliar disease incidence (Hirano et al., 1987).

An interesting application of the ice nucleation method has been the genetic modification of certain bacteriophage genomes to carry and express the *inaZ* gene during the normal lytic replication cycle (T. Suslow, personal communication). The bacterial viruses seek out the host cells, and, only upon infection, produce amplified amounts of the ice nucleation gene product, which is detected by supercooling the sample solutions. Since these phages often have a high degree of specificity for their target host (Makkar and Casida, 1987), only a few molecules of the *ina*Z product would be required for ice formation. However, for this reason, these phages would be most effective for samples in which Ice+ strains are not typically found, such as soil or sewage. Nevertheless, this concept might conceivably be applied to phyllosphere samples using an alternative gene marker such as the *lux* cassette (see above).

An overall disadvantage in applying ice nucleation methods to phyllosphere samples is the presence of a relatively high background number of indigenous ice-nucleating bacteria. Measurement of an introduced population using an *ina*Z-based approach alone would depend on a substractive analysis, which may be highly variable, or impossible, when the target population density is low relative to native Ice+ strains.

The *lac*ZY Gene The *lac*ZY marker system relies on the introduction and expression of the *E. coli lac* operon genes, *lac*Z and *lac*Y, into bacterial species which are deficient in lactose metabolism. The product of the *lac*Z gene is β-galactosidase, which cleaves the lactose molecule into glucose and galactose. β-Galactosidase also cleaves the chromogenic dye "X-Gal" (5-bromo-4-chloro-3-indolyl-β-D-galactopyranoside), producing a brilliant blue color indicative of the presence of the active *lac*Z gene. The *lac*Y gene, which codes for the enzyme lactose permease, allows bacteria to take up the lactose sugar efficiently. Bacteria genetically modified to carry *lac*ZY can often be recovered by preselection on minimal media where lactose is the sole carbon source. The *lac*ZY marker was originally developed for fluorescent pseudomonad bacteria, common soil organisms, and efficient root colonizers. Fluorescent pseudomonads do not make β-galactosidase, and subsequently cannot efficiently use lactose as a sole energy source (Hemming and Drahos, 1984; Drahos et al., 1986). Thus, coupling recovery on lactose-minimal media with the scorable fluorescent phenotype, or with an inherently resistant phenotype to an antibiotic such as ampicillin, provides a practical and sensitive means to follow certain bacterial populations. Levels of marked strains as low as 100 CFU/g plant tissue can usually be detected (Cook et al., 1991; Drahos et al., 1988).

Further, effective delivery systems have been developed based on disarmed Tn7 transposon vectors. These permit efficient insertion of the *lac*ZY marker directly onto the chromosome of the target bacterium in a permanent, one-way manner (Barry, 1986, 1988a). In several field studies, this method has proven rapid and effective in monitoring the presence of both the *lac*ZY-marked strain as well as genes closely linked to the inserted *lac*ZY element (Drahos et al., 1988, 1990; Barry, 1986). Moreover, the U.S. EPA Office of Pesticide Products is considering providing broad latitude in the environmental release of strains marked with *lac*ZY (*Federal Register*, May 1990, 55:21094). Importantly, as discussed below, this system lends itself well to the application of direct detection methods, such as nucleic acid hybridization and fluorescent antibody probes. In addition, the Tn7-*lac*ZY vectors provide a means to genetically link other monitoring elements or cryptic genes to the *lac*ZY marker. For example, coupling the *xyl*E or the *lux* markers within the disarmed Tn7 arms (Barry, 1986) will effect insertion of the entire dual element into the target strain. This provides for multiple and mutually verifiable methods for microbe detection following reintroduction into the environment. The direct, *in situ* measurement of *lux* gene expression, for instance, could be correlated with either the level of microbes measured by recovery on lactose-selective media, or the population density estimated by probe hybridization against the inserted Tn7 elements (see Section 7.5).

The *uid*A Gene Detecting specific fungal organisms in the phyllosphere has usually meant reliance on standard morphological and biochemical characterizations, which are often tedious and difficult to apply on a large scale. However, the molecular techniques that have been developed for many filamentous fungi are now broad and practical enough to provide reliable transformation with practical genetic marker elements (Van Gorcom et al., 1986; Van Hartingsveldt et al., 1987; Oliver et al., 1987). Among the most promising of these marker elements is the *E. coli uid*A gene, which codes for the β-glucuronidase enzyme (GUS). Originally developed for expression in nematodes (Jefferson et al., 1986), and effective for the study of plant gene expression (Jefferson et al., 1987), GUS has now been adapted for use in several fungi, including *Aspergillus* and *Fulvia* species (Roberts et al., 1989), as a potential tool for the analysis of gene expression, protein sorting, and secretion. GUS has a very low background level in most plant tissues and microorganisms, and it is easily assayed by fluorimetry, spectrophotometry, microscopic observation, and colorimetry (Adams et al., 1990). Unlike certain other marker enzymes such as β-galactosidase, β-glucuronidase is readily translocated across cell membranes (Jefferson et al., 1987). This may aid significantly in the localization of small numbers of a marked fungal organism without requiring reculturing or enrichment. Furthermore, since GUS activity appears to have no discernable effect on pathogenicity (*F. fulva*; Roberts et al., 1989), the progress of the introduced organism might be followed using minimal amounts of leaf or cotyledon material.

7.5 Detection and Enumeration: Microscopy and Targeted Probes

Nutritional, enzymatic, and antibiotic marker genes have often been found to be effective and practical. However, by their nature, these methods rely on the ability of the target organism to sufficiently express these genes during a selection or reculturing process. Sufficient expression may require growth. In certain instances, such as survival under severe environmental stress, enzymes essential for other selection or detection procedures may not be adequately produced, or the organism itself may become quiescent (Colwell et al., 1985; Rollins and Colwell, 1986). Direct methods of detecting a specific microbe can bypass these problems. However, direct monitoring procedures must cope with large populations of similar organisms, high background interference by plant tissue and soil particles, and nonspecific binding of target strains to environmental surfaces and microniches. However, two methodologies, antibody probes and nucleic acid hybridization and amplification procedures, may prove capable of circumventing most of these problems. Importantly, they might provide a basis for calculating the relative efficiency of detection for the more rapid and practical selective markers, or for the more traditional and well-studied methods, such as microscopic visualization.

Microscopic Visualization Quantitative assessment by direct electron microscopy alone (Rovira et al., 1974; Børsheim et al., 1990) or with fluorescent stains (Zimmermann, 1977; Newell, 1991) has provided some useful information on rhizosphere microflora, primarily as a broad estimation of the total number of organisms present. In some cases estimates of direct viable counts (DVC) have been made for bacteria present in water samples (Kogure et al., 1979, 1987), by using epifluorescent microscopy to distinguish between growing (elongating) and presumably dead (not elongating) bacteria. The microscopic detection of specific species of bacteria has been demonstrated by first adding a fluorescent dye bound to a DNA probe specific for unique ribosomal RNA sequences (Giovianni et al., 1988). Yet procedures involving microscopic visualization are often quite labor intensive, particularly for broader scale field assessments, or require significant skill and expertise.

Antibody Probes Bohlool and Schmidt (1980) discussed the utility of applying polyclonal antibody methods in microbial ecology. Since then, there have been numerous examples describing the use of immunological methods to detect individual bacterial and fungal species in natural and artificial habitats (Muyzer, 1988). The most successful of these are based on the antigenic determinants in the outer membrane or exposed segments of the target strain. Antibodies have been coupled with a fluorescent dye such as fluorescein isothiocyanate (Brayton and Colwell, 1987; Brayton et al., 1987), and have been linked with a specific enzyme (for ELISA plate reading) such as lactoperoxidase and other enzymes (Chaicumpa et al., 1988; Robinson et al., 1983) to provide a moderately sensitive means to detect specific microbes in the general population. However, accurate enumeration or detection of low levels (less than 10^3 microbes/g plant tissue) from relatively complex leaf washings or soil samples is often difficult to achieve due to nonspecific staining, autofluorescence of nontargeted organisms, or visual interference from background material. Nevertheless, the advent of monoclonal antibodies, which are identical antibody clones specific to a single antigenic determinant (Macario and Conway de Macario, 1985; Zola, 1987), has provided a much greater sensitivity and accuracy in a complex environmental sample (Wright et al., 1986). The combination of direct viable microscopy with these sensitive detection assays has further enhanced the capability to detect viable but nonculturable microbes from native environments (Brayton and Colwell, 1987; Desmonts et al., 1990).

However, even these methods are not always effective for distinguishing a newly introduced microbial strain from the native community containing similar indigenous relatives. Linking diverse techniques can often solve such problems. For example, the availability of monoclonal antibodies against the *E. coli lac*Y gene product, lactose permease, allows a combination of the selective strength of the *lac*ZY marker system (see above) with the direct detection capacity of antibodies. This is based on the demonstrated presence of this lactose permease in the outer membrane of bacterial hosts expressing the *lac*Y gene (Drahos et al., 1986).

Nucleic Acid Hybridization With the advent of molecular techniques, we have gained the means to determine the precise nucleic acid sequences of certain genes and related genetic material contained within numerous native microbes. Moreover, such detailed information has typically been obtained for genes which have been inserted into genetically modified organisms destined for environmental introduction and testing. Such knowledge allows the construction of complementary, and often highly specific, DNA or RNA probe segments, which provide a means to detect these microbes even from very complex environmental samples by various hybridization procedures.

Typically, these probes are labeled with radionucleotides or linked enzyme substrates and used to detect nucleic acid material bound to a solid substrate (Ogram et al., 1988; Torsvik, 1980; Atlas and Sayler, 1988). For example, microbes in a leaf washing would be concentrated either by centrifugation or by passage through a high-capacity micropore filter (Millipore SVGS 01015; Colwell et al., 1988), then lysed enzymatically or physically (i.e., French press; grinding in liquid nitrogen). For soil sample analysis, the addition of polyvinylpyrrolidone to the sample prior to centrifugation will remove humic contaminants, which adsorb to the polymer (Holben et al., 1988). The sample is then bound to a nitrocellulose or nylon-based membrane hybridization filter by either direct application and absorption (Steffan and Atlas, 1988; Steffan et al., 1989), or following electrophoretic separation (Southern, 1975). The probe is usually either a small (less than

1000 base pairs), well-defined, native fragment, isolated and purified, or a DNA segment specifically synthesized chemically, which is then labeled by incorporating ^{32}P-radionucleotides or enzyme substrates by various means (Maniatis et al., 1982; Leary et al., 1983). The labeled probe is then incubated with the filter-bound sample, washed, and analyzed by autoradiography. Typically, this analysis is quantitated by densitometric scanning and compared with samples to which known numbers of the target microbe have been added, which provides the standard curve. In some cases, it is estimated that one target organism in 10^4 or 10^5 may be detected; in the rhizosphere, this would provide for detection of an organism having a population density of approximately 10^4 cells per gram dry weight of soil (Holben et al., 1988). However, this level may be inadaquate for many phyllosphere studies. While useful as a confirmation tool, this level of detection is much below that attainable by certain enzymatic and selection methods described previously.

The choice of the probe and the method for sample isolation are often the most critical limiting factors in sensitivity and specificity. Probes based on rRNA genes are perhaps the most definitive for assessment of a specific group of microorganisms. For example, a 360-base-pair DNA fragment from the 23S rRNA gene of *Pseudomonas aeruginosa* has been shown to hybridize specifically to bacteria belonging to the Group 1 pseudomonads (Festl et al., 1986; Barry, 1988b).

A serious limitation in nucleic acid hybridization approaches, however, has been the difficulty in quantitative extraction of target DNA material from complex environmental samples. This is particularly true when studying soil or phyllosphere habitats, where removal of tightly associated background material from root washings or leaf homogenates occludes the presence of small levels of the target strain (Holben et al., 1988; Ogram et al., 1988; Torsvik, 1980). There have been advances in purifying DNA from the soil (Holben et al., 1988) and from seawater (Colwell et al., 1988). Some methods provide novel extraction procedures based on density gradient centrifugation in percol clay resins and large-scale lysis in modified French press mechanisms. Yet none provide the sensitivity and speed afforded by the polymerase chain reaction (PCR) procedure.

Polymerase Chain Reaction (PCR) Method Conceptually simple and elegant, the PCR method is based on the biochemical amplification of even a single target DNA segment a millionfold or more (Mullis and Faloona, 1987). Typically, two short single-stranded DNA primers, usually 20–30 bases in length, are synthesized complementary to adjacent segments of the target DNA separated by 400–1000 bases (Figure 7.1). Importantly, these probes are complementary to opposite stands of the target DNA. The probes are then added to the test sample (Figure 7.1a) in a buffered solution containing a balanced mix of the four deoxynucleotides, ATP, and a heat-resistant DNA polymerase enzyme (*Taq* polymerase from *Thermus aquati-*

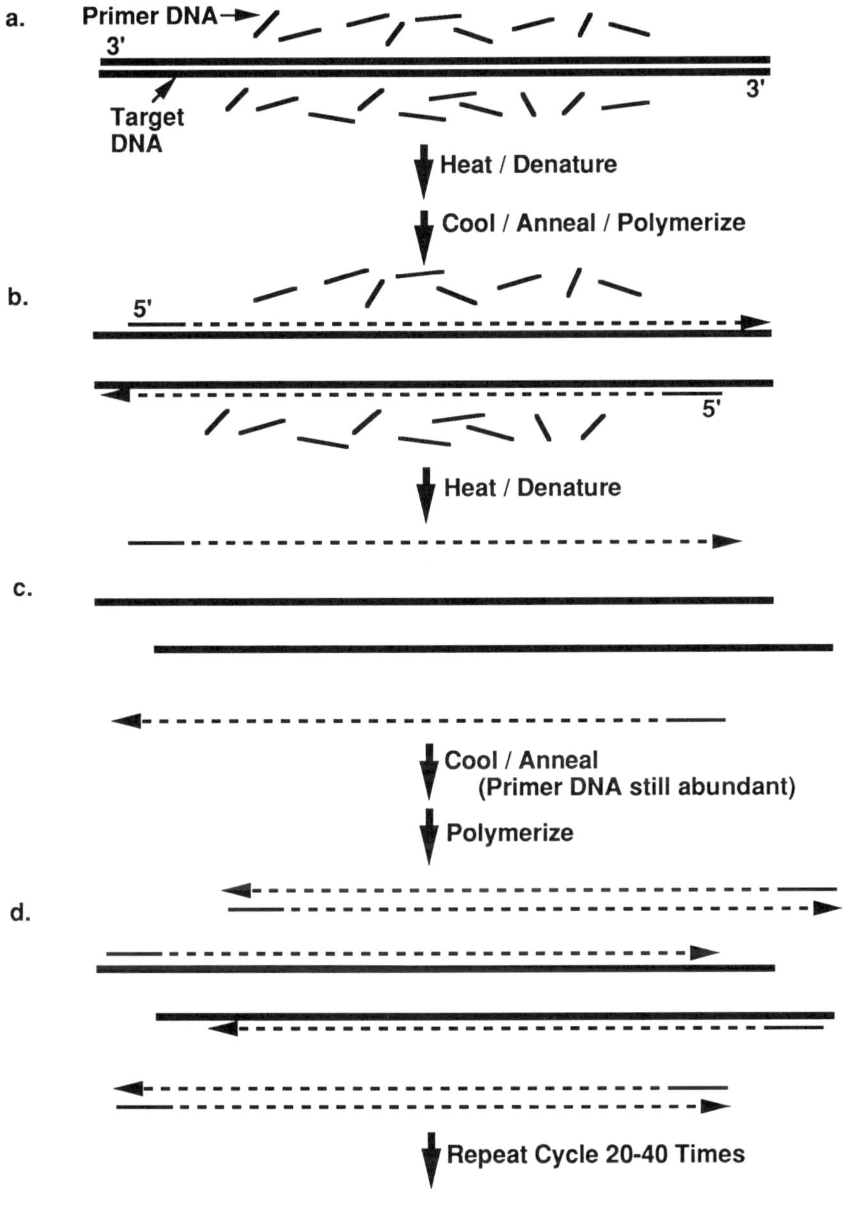

Figure 7.1 Polymerase chain reaction (PCR) procedure. (a) Two types of DNA primers are added to a sample containing target sequence to be amplified. (b) This is first heated to denature target DNA, then cooled to allow some of each primer to anneal to their complementary sequences which flank the target segment. Heat-stable DNA polymerase, also in the sample, synthesizes nascent complementary strands of the target sequence. (c) Nascent strands are denatured from the templates in another heating step, then cooled to allow hybridization with additional primers (added originally in excess) and (d) another round of synthesis. The entire process is repeated 20–40 times, providing an exponential increase of the specific flanked target DNA segment.

cus). The temperature is first raised to melt the target DNA present in the test sample, then lowered to allow complementary sequences to reanneal. During this process, some of the probe DNA also anneals to their corresponding segments of the target material contained in the test sample. A further short period of incubation at this lower temperature allows the DNA polymerase to catalyze a reaction which duplicates the complementary nucleotide region between the probe segments. This polymerase step is allowed to continue such that the newly synthesized strands will overlap with the region complementary to the opposite probe (Figure 7.1b). The sample is then cycled through another heating step, which releases the nascent segments (Figure 7.1c). A second round of cooling allows more of the short DNA probes, which had not annealed to the target DNA template in the first round, to now anneal with either the original target material, or with the newly synthesized complementary segments. Lowering the temperature once again repeats the synthesis process (Figure 7.1d). In this manner, an exponential amplification of the original target DNA contained in the test sample will take place. Usually the process is controlled through an automatic temperature cycling device which precisely regulates the timing of the step changes, and generally provides for 20–50 cycles during a 4- to 8-hr period. The level of amplification is often so great that for some simple qualitative assessments no radionucleotide labeling is required, and the large amount of nascent fragment can be visualized following standard agarose gel electrophoresis and ethidium bromide staining (Maniatis et al., 1982). For more accurate quantitation, ^{32}P-labeled synthetic probes are easily prepared prior to initiating the cycled steps. Numerous reports have described the merits and power of the PCR method as applied to analysis of environmental samples (Atlas and Steffan, 1988; Chaudhry et al., 1989; Steffan and Atlas, 1988; Bej et al., 1990).

While most direct detection methods, including those based on the PCR method, are highly sensitive and eliminate the problems associated with culturing some organisms from certain environments (Atlas, 1982; Roszak and Colwell, 1987), they may not accommodate large-scale analysis or less sophisticated field research laboratories. In such cases, a combination of methods may offer the best solution. For example, applying PCR methods to cells carrying the Tn*7-lac*ZY system provides a means to determine the relative efficiency of selective recovery based on selection for *lac*ZY gene expression (see Section 7.4). In this case, the abundance of a phyllosphere-colonizing fluorescent *Pseudomonas* strain is first determined by assessing an environmental sample for selective recovery on minimal lactose media (Drahos et al., 1986, 1988). The result can then be compared with that obtained from an identical sample analyzed by PCR amplification using primers homologous to defined regions in the end segments of the disarmed Tn7 transposon, which are present in all strains marked with the Tn*7-lac*ZY system (Barry, 1988a; E. Kline and D. Kluepfel, personal communication). Analyzing a small number of such samples over the course of the experiment allows great confidence in the fidelity of the data obtained from a much larger-scale simultaneous assessment based on minimal lactose selection alone. In this way, a PCR check provides a means to estimate the relative efficiency of lactose-dependent recovery that is relatively practical and applicable to large-scale assessments.

However, there are certain precautions which must be considered in successfully applying the PCR technique. It is generally acknowledged that great care must be exercised to insure the purity of the amplified target sample. Typically, a completely separate set of mechanical pipetting devices, and an isolated and dedicated section (or hood) in the laboratory are necessary to minimize false positive detection. Also, strictly defined and regimented experimental procedures are essential to maintain reproducibility, particularly over a long experimental period. This should include regular calibration of temperature cycling equipment down to the tube wells used in the heat block. However, even this extra care is not too difficult, and it can quickly become a routine part of the experimental program. The cost of establishing a PCR capacity in most microbiology and plant pathology laboratories is relatively affordable, the major expense being an investment in the temperature cycling apparatus.

7.6 Identification

Biochemical Characterization Along with application of various detection methods, it remains critical to at least occasionally confirm the identity of the target microbe by alternative, more standardized methods. Numerous standard biochemical and morphological methodologies for the identification of bacterial isolates are summarized extensively (see Holt, 1989). Many of these are practical and rapid, often based on chromogenic enzymatic reactions (Bochner and Savageau, 1977) and are marketed as kits by several companies (BioLog Inc., Hayward, CA, USA; API20E and API ZYM systems by Vitek, Inc., St. Louis, MO, USA). Fungal identification, on the other hand, can often require intensive analysis of characteristic morphological phenotypes, usually associated with sporulation events. While these methods can meet basic characterization needs, they do not provide for more definitive requirements of confirming the identity of a target microbe among a large number of very similar endogenous strains.

Combinations of several effective detection methods have been successfully applied to verify the identity of environmental isolates. These methods include identification by serological properties (Demezas and Bottomley, 1986; Ellis et al., 1984; Martensson and Gustafsson, 1985; Renwick and Gareth-Jones, 1985; Diatloff, 1977), and typing by bacteriophage specificity (Bromfield et al., 1986). Lindstrom et al. (1990) were able to use lipopolysaccharide typing to clearly differentiate and identify two closely related *Rhizobium* strains, which were recovered from the field five years after introduction. A combination of lipopolysaccharide analysis with protein profiling was quite effective in differentiating similar field isolates (Lipsanen and Lindstrom, 1989). Two-dimensional gel electrophoresis of protein extracts from environmental isolates is now becoming a far more reliable identification technique than in the past, with the advent of computer-assisted image analysis and large-capacity data storage systems (from Ambis Systems, Inc., San Diego, CA).

Gas Chromatographic/Fatty Acid Analysis The free fatty acids of most bacterial strains, as well as of some fungi and nematodes, are highly characteristic of specific genus and species. Saponification (methyl ester extraction) of these fatty acids, followed by gas chromatographic analysis, yields a distinct and reproducible profile (Miller, 1982; Sasser and Miller, 1984). The method has recently been automated (Microbial ID, Inc., Newark, DE), and a substantial data base developed for both clinical and environmental microbes. Often 100 samples can be prepared in one day and evaluated automatically overnight. The system includes a powerful and effective statistical analysis package as part of the dedicated computer control system. While expensive relative to most standard biochemical tests, the system has proven invaluable in providing a rapid, accurate, and surprisingly uncomplicated identification tool, which can easily function as a shared work station for a number of laboratories. Upgraded analysis programs allow for addition of new or previously uncharacterized strains, whose profiles then become a permanent part of the comparative data base. If care is taken in normalizing the microbial growth conditions prior to fatty acid extraction, this method usually provides accurate identification to the species level in most bacteria.

DNA Fingerprinting While nucleic acid hybridization procedures for microbial classification have been in use for some time, until recently they have been limited to broad identification and linkage of large classes of microbes. Now, the application of newer DNA isolation and probe hybridization techniques is opening the way to a very powerful and complementary identification scheme based on restriction profile analyses of total genomic DNA. Two of the more common procedures are chromosomal fingerprinting and restriction-fragment-length-polymorphism (RFLP) analysis. Chromosomal fingerprinting involves the isolation of the full complement of the microbial DNA, followed by enzymatic restriction and agarose gel electrophoresis (Drahos et al., 1985; Malvick and Moore, 1988; Lindstrom et al., 1990). This procedure will generate a complex, striated pattern upon staining with ethidium bromide and ultraviolet-light visualization. The pattern is due to specific variations in the length of DNA fragments formed following digestion with a restriction enzyme, usually EcoRI. This method provides data which is exceptionally specific to the target microbe and will usually allow differentiation at the biovar or even sub-biovar level (Drahos et al., 1985). However, this method does require that the target microbe first be isolated and grown in pure culture. Tests of environmental samples containing a *lac*ZY-marked pseudomonad demonstrated that a marked strain recovered from the soil 19 months after introduction bore the identical fingerprint pattern to a sample of the same strain stored frozen (−80°C) in the laboratory over the same period (E. Kline, personal communication). Chromosomal fingerprinting is generally rapid (20–30 samples can be readily processed per day) and requires little special equipment. If the profile of a particular organism is known, or its reference DNA is included as a control in the analysis, then this method provides a rapid and highly definitive means to confirm microbial identity. However, due to the complexity of the fingerprint pattern, it has not yet been possible to use this method to identify an unknown strain. This requires gel scanning devices capable of processing and digitizing the pattern information, in order to build a data base using known microbial standards. This may soon change, though, as high-resolution, computer-based scanning and image analysis has advanced greatly in the past several years.

RFLP analysis is an extension of the chromosomal fingerprint procedure, where a labeled nucleotide probe is hybridized to certain restriction fragments of the target strain (Botstein et al., 1980). Here the fragments are again separated by agarose gel electrophoresis and transferred onto a nitrocellulose membrane (or its equivalent). A probe based on either a defined sequence, or a sequence (or gene) known or suspected to be present in the target genome, is then hybridized to the membrane. Following autoradiography, specific banding patterns appear which are generally characteristic of the particular organism being analyzed. As with chromosomal fingerprinting, however, at present, the RFLP patterns allow identification only by comparison with known standards. There is a danger in basing an identification on RFLP analysis alone; the fewer different probes that are used, the greater the tendency toward false positive identification at the species or subspecies level. A combination of RFLP with chromosomal fingerprinting, gas chromatographic (GC) fatty acid analysis, or other biochemical tests helps to substantiate the identity, although this increases time and effort.

Since these methods assess nucleic acid structure rather than products of gene expression, they are independent of cell physiology, culture history, and apparent morphology. This makes them ideal as identification schemes, at least on a comparative basis. Nevertheless, DNA fingerprinting does require reculturing, which makes the procedure laborious, if not untenable, as a detection method in the field. However, combining DNA fingerprinting with other selectable measures has provided strong evidence for the source of the inoculum for foliar disease on fruit trees (Malvick and

Moore, 1988) and allowed for accurate typing of diverse *Xanthomonas campestris* isolates (Hartung and Civerolo, 1987).

7.7 Conclusion

The advent of powerful and diverse biological tools, as well as the more common availability of highly sensitive detection and characterization devices, has opened new doors toward understanding phyllosphere biology. Computer-assisted identification methodologies, such as GC analysis of bacterial fatty acids, DNA fingerprinting, and two-dimensional protein profiling are providing rapid, accurate, and often highly reproducible means to identify and catalog a variety of both prokaryotic and eukaryotic organisms. Sensitive and selectable genetic tracking tools, such as the *lac*ZY, *lux*, *ina*Z, *xyl*E, and *uid*A systems are offering effective, practical, yet inexpensive alternatives to the former reliance on the more standard antibiotic resistance elements or complex selection media. Many of these genetic systems have a proven field record for monitoring microbe longevity and dissemination, as well as for evaluating potential genetic exchange between an introduced strain and endogenous organisms. Advances in detection and enumeration methods based on fluorescent antibody probes and direct viable electron microscopy have significantly enhanced our capacity to detect microorganisms which have become dormant as part of their normal life cycle or under certain environmental conditions. Among the most promising of the new technologies are those based on the practical application of nucleic acid hybridization procedures. These include probes specific to ribosomal RNA species, which are often very characteristic of given microbe classes, or probes used in the polymerase chain reaction (PCR) method. Extremely sensitive microbe detection capabilities using PCR reactions have been demonstrated from a variety of environmental samples, including soil, marine, and leaf surface washings. However, this procedure requires extreme care in setting stringent control measures, as even very slight contamination can cause false positive results. Nevertheless, PCR detection may become an almost indispensible, and relatively inexpensive, tool if coupled with secondary selection or enumeration techniques. Overall, our greatest challenge may not lie in expanding the array of analytical tools, but in our resolve to attempt novel and sometimes unfamiliar technology and our wisdom to choose those methods that are appropriate to the questions asked.

The author extends great appreciation to G. Barry, B. Hemming, E.J. Brandt, E. Kline, H. Skipper, D. Kluepfel, D. Weller, and P. Mukerji, who played crucial roles in the development and field verification of certain monitoring methods described in this chapter, and in providing invaluable input used here for describing various detection and identification strategies.

References

Adams, M.R., Grubb, S.M., Hamer, A., and Clifford, M.N. 1990. Colorimetric enumeration of *Escherichia coli* based on β-glucuronidase activity. *Applied and Environmental Microbiology* 56:2021–2024.

Andrews, J.H. 1986. How to track a microbe. pp. 14–34 in Fokkema, N.J. and van den Henvel, J. (editors), *Microbiology of the Phyllosphere*. Cambridge University Press, Cambridge.

Atlas, R.M. 1982. Enumeration and estimation of biomass of microbial components in the biosphere. pp. 84–102 in Burns, R.G., and Slater, J.H. (editors), *Experimental Microbial Ecology*. Blackwell Scientific Publishers, Oxford.

Atlas, R.M. and Sayler, G.S. 1988. Tracking microorganisms and genes in the environment. pp. 31–45 in Omenn, G.S. (editor), *Environmental Biotechnology: Reducing Risks from Environmental Chemicals through Biotechnology*. Plenum Press, New York.

Atlas, R.M. and Steffan, R.J. 1988. Target DNA amplification. pp. 224–226 in Sussman, M., Collins, C.H., Skinner, F.A., and Stewart-Tull, D.E. (editors), *The Release of Genetically-engineered Micro-organisms*. Academic Press, London.

Barry, G. F. 1986. Permanent insertion of foreign genes into the chromosomes of soil bacteria. *Bio/Technology* 4:446–449.

Barry, G. F. 1988a. A broad-host-range shuttle system for gene insertion into the chromosome of Gram-negative bacteria. *Gene* 71:75–84.

Barry, G.F. 1988b. Construction of reporter strains. pp. 211–219 in Sussman, M., Collins, C.H., Skinner F.A., and Stewart-Tull, D.E. (editors), *The Release of Genetically-engineered Micro-organisms*. Academic Press, London.

Bej, A.K., Steffan, R.J., DiCesare, J., Haff, L., and Atlas, R.M. 1990. Detection of coliform bacteria in water by polymerase chain reaction. *Applied and Environmental Microbiology* 56:307–314.

Bochner, B.R. and Savageau, M.A. 1977. Generalized indicator plate for genetic, metabolic, and taxonomic studies with microorganisms. *Applied and Environmental Microbiology* 33:434–444.

Bohlool, B.B. and Schmidt, E.L. 1980. The immunofluorescence approach in microbial ecology. *Advances in Microbial Ecology* 4:203–241.

Børsheim, K.Y., Bratbak, G., and Heldal, M. 1990. Enumeration and biomass estimation of planktonic bacteria and viruses by transmission electron microscopy. *Applied and Environmental Microbiology* 56:352–356.

Botstein, D., White, R., Skolnick, M., and Davis, R.W. 1980. Genetic mapping using restriction fragment length polymorphisms. *American Journal of Human Genetics* 32:314–331.

Brayton, P.R. and Colwell, R.R. 1987. Fluorescent antibody staining method for enumeration of viable environmental *Vibrio cholerae* O1. *Journal of Microbiological Methods* 6:309–314.

Brayton, P.R., Tamplin, M.L., Huq, A., and Colwell, R.R. 1987. Enumeration of *Vibrio cholerae* O1 in Bangladesh waters by fluorescent-antibody direct viable count. *Applied and Environmental Microbiology* 53:2862–2865.

Bromfield, E.S.P., Sinha, I.B., and Wolynetz, M.S. 1986. Influence of location, host cultivar, and inoculation on the composition of naturalized populations of *Rhizobium meliloti* in *Medicago sativa* nodules. *Applied and Environmental Microbiology* 51:1077–1084.

Brunel, B., Cleyet-Marel, J.-C., Normand, P., and Bardin, R. 1988. Stability of *Bradyrhizobium japonicum* inoculants after introduction into soil. *Applied and Environmental Microbiology* 54:2636–2642.

Chaicumpa, W., Thin-Intra, W., Khusmith, S., Tapchaisri, P., Echeverria, P., Kalambaheti, T., and Chongsa-Nguan, M. 1988. Detection with monoclonal antibody of *Salmonella typhi* antigen 9 in specimens from patients. *Journal of Clinical Microbiology* 26:1824–1830.

Chaudhry, G.R., Topranzos, G.A., and Bhatti, A.R. 1989. Novel method for monitoring genetically engineered microorganisms in the environment. *Applied and Environmental Microbiology* 55:1301–1304.

Chopra, I., Shales, S.W., Ward, M.W., and Wallace, L.J. 1981. Reduced expression of Tn10-mediated tetracycline resistance in *Escherichia coli* containing more than one copy of the transposon. *Journal of General Microbiology* 126:45–54.

Cole, M.A. and Elkan, G.H. 1979. Multiple antibiotic resistance in *Rhizobium japonicum*. *Applied and Environmental Microbiology* 37:867–870.

Colwell R.R., Brayton, P.R., Grimes, D.J., Roszak, D.B., Huq, S.A., and Palmer, L.M. 1985. Viable but non-culturable *Vibrio cholerae* and related pathogens in the environment: Implications for release of genetically engineered organisms. *Bio/Technology* 3:817–820.

Colwell, R.R., Somerville, C., Knight, I., and Straube, W.L. 1988. Detection and monitoring of genetically engineered micro-organisms. pp. 47–60 in Sussman, M., Collins, C.H., Skinner F.A., and Stewart-Tull, D.E. (editors), *The Release of Genetically-engineered Micro-organisms*. Academic Press, London.

Cook R.J. and Baker, K.F. 1983. *The Nature and Practice of Biological Control of Plant Pathogens*. American Phytopathological Society, St. Paul, MN.

Cook, R.J., Weller, D.M., Drahos, D.J., Kovacevich, P.A., Hemming, B.C., Barnes, G., and Peirson, E.L. 1991. Establishment, monitoring, and termination of field tests with genetically altered bacteria applied to wheat for biological control of take-all. In MacKenzie, D.R. (editor), *Biotechnology Field Test Results*. Academic Press, New York. In press.

Demezas, D.H. and Bottomley, P.J. 1986. Autecology in rhizospheres and nodulating behavior of indigenous *Rhizobium trifolii*. *Applied and Environmental Microbiology* 52:1014–1019.

Desmonts, C., Minet, J., Colwell, R., and Cormier, M. 1990. Fluorescent-antibody method useful for detecting viable but nonculturable *Salmonella* spp. in chlorinated wastewater. *Applied and Environmental Microbiology* 56:1448–1452.

Diatloff, A. 1977. Ecological studies of root-nodule bacteria introduced into field environments. 6. Antigenic and symbiotic stability in *Lotononis* rhizobia over a 12-year period. *Soil Biology and Biochemistry* 9:85–88.

Drahos, D., Brackin, J., and Barry, G. 1985. Bacterial strain identification by comparative analysis of chromosomal DNA restriction patterns. *Phytopathology* 75:1381 (Abstract).

Drahos, D.J., Hemming, B.C., and McPherson, S. 1986. Tracking recombinant organisms in the environment: β-galactosidase as a selectable, non-antibiotic marker for fluorescent pseudomonads. *Bio/Technology* 4:439–443.

Drahos, D.J., Barry, G.F., Hemming, B.C., Brandt, E.J., Skipper, H.D., Kline, E.L., Kluepfel, D.A., Hughes, T.A., and Gooden, D.T. 1988. Pre-release testing procedures: US field test of a *lac*ZY-engineered soil bacterium. pp. 181–191 in Sussman, M., Collins, C.H., Skinner F.A., and Stewart-Tull D.E. (editors), *The Release of Genetically-engineered Micro-organisms*. Academic Press, London.

Drahos, D.J., Barry, G.F., Hemming, B.C., Brandt, E. J., Kline, E.L., Skipper, H.D., and Kluepfel, D.A. 1991. Spread and survival of genetically marked bacteria in soil. In Day, J.J. and Fry, J.C. (editors), *Environmental Release of Genetically Engineered and Other Microorganisms*. Academic Press, London. In press.

Ellis, W.R., Ham, C.E., and Schmidt, E.L. 1984. Persistence and recovery of

Rhizobium japonicum inoculum in a field soil. *Agronomy Journal* 76:573–576.

Festl, H., Lugwig, W., and Schleifer, K.H. 1986. DNA hybridization probe for the *Pseudomonas fluorescens* group. *Applied and Environmental Microbiology* 52:1190–1194.

Giovannoni, S.J., Delong, E.F., Olsen, G.J., and Pace, N.R. 1988. Phylogenetic group-specific oligodeoxynucleotide probes for identification of single microbial cells. *Journal of Bacteriology* 170:720–726.

Goodnow, R.A., Harrison, M.D., Morris, J.D., Sweeting, K.B., and Laduca, R.J. 1990. Fate of ice nucleation-active *Pseudomonas syringae* strains in alpine soils and waters and in synthetic snow samples. *Applied and Environmental Microbiology* 56:2223–2227.

Goodwin, D. and Slater, J.H. 1979. The influence of the growth environment on the stability of a drug resistance plasmid in Escherichia coli K-12. *Journal General Microbiology* 111:201–210.

Gross, D.C., Cody, Y.S., Proebsting, E.L., Jr., Radamaker, G., and Spotts, R.A. 1984. Ecotypes and pathogenicity of ice nucleation active *Pseudomonas syringae* isolated from deciduous fruit tree orchards. *Phytopathology* 74:243–269.

Hartung, J.S. and Civerolo, E.L. 1987. Genomic fingerprints of *Xanthomonas campestris* pv. *citri* strains from Asia, South America, and Florida. *Phytopathology* 77:282–285.

Hemming, B.C. and Drahos, D.J. 1984. β-Galactosidase, a selectable non-antibiotic marker for fluorescent pseudomonads. *Journal of Cellular Biochemistry, Supplement* 8b:252.

Hirano, S.S. and Upper, C.D. 1986. Bacterial nucleation of ice in plant leaves. pp. 730–738 in Parker, I. (editor), *Methods in Enzymology. Protons and Water: Structure and Translocation*, vol. 127, Academic Press, New York.

Hirano, S.S., Baker, L.S., and Upper, C.D. 1985. Ice nucleation temperature of individual leaves in relation to population sizes of ice nucleation active bacteria and frost injury. *Plant Physiology* 77:259–265.

Hirano, S.S., Rouse, D.I., Arny, D.C., Nordheim, E.V., and Upper, C.D. 1981. Epiphytic ice nucleation active (INA) bacterial populations in relation to halo blight incidence in oats. *Phytopathology* 71:881.

Hirano, S.S., Rouse, D.I., and Upper, C.D. 1987. Bacteial ice nucleation as a predictor of bacterial brown spot disease on snap beans. *Phytopathology* 77:1078–1084.

Holben, W.E., Jansson, J.K., Chelm. G.K., and Tiedje, J.M. 1988. DNA probe method for the detection of specific microorganisms in the soil bacterial community. *Applied and Environmental Microbiology* 54:703–711.

Holt, J.G. (editor). 1989. *Bergey's Manual of Systematic Bacteriology*. Williams & Wilkins, Baltimore, MD. 2648 pp.

Ingram, C., Brawner, M., Youngman, P., and Westpheling, J. 1989. *xyl*E functions as an efficient reporter gene in *Streptomyces* spp.: Use for the study of *gal*P1, a catabolite-controlled promoter. *Journal of Bacteriology* 171:6617–6624.

Jefferson, R.A., Burgess, S.M., and Hirsch, D. 1986. β-Glucuronidase from *Escherichia coli* as a gene-fusion marker. *Proceedings of the National Academy of Sciences USA* 83:8447–8451.

Jefferson, R.A., Kavanagh, T.A., and Bevan, M.W. 1987. GUS fusions: β-glucuronidase as a sensitive and versatile gene fusion marker in higher plants. *EMBO Journal* 6:3901–3907.

Kelch, W.J. and Lee, J.S. 1978. Antibiotic resistance patterns of gram-negative bacteria isolated from environmental sources. *Applied and Environmental Microbiology* 36:450–456.

Kloepper, J.W. and Schroth, M.N. 1981. Relationship in in vitro antibiosis of plant growth-promoting rhizobacteria to plant growth and the displacement of root microflora. *Phytopathology* 71:1020–1024.

Kloepper, J.W., Leong, L., Teintze, M., and Schroth, M.N. 1980. *Pseudomonas* siderophores: A mechanism explaining disease-suppressive soils. *Current Microbiology* 4:317–320.

Kirchner, G., Roberts, J.L., Gustafson, G.D., and Ingolia, T.D. 1989. Active bacterial luciferase from a fused gene: expression of a Vibrio Harveyi lux AB translational fusion in bacteria, yeast and plant cells. *Gene* 81:349–354.

Kogure, K., Simidu, U., and Taga, N. 1979. A tentative direct microscopic method for counting living marine bacterai. *Canadian Journal of Microbiology* 25:415–420.

Kogure, K., Simidu, U., Taga, N., and Colwell, R.R. 1987. Correlation of direct viable counts with heterotrophic activity for marine bacteria. *Applied and Environmental Microbiology* 53:2332–2337.

Koncz, C., Olsson, O., Langridge, W.R., Schell, J., and Szalay, A. 1987. Expression and assembly of functional bacterial luciferase in plants. *Proceedings of the National Academy of Sciences USA* 84:131–135.

Leary, J.J., Brigati, D.J., and Ward, D.C. 1983. Rapid and sensitive colorimetric method for visualizing biotin-labeled DNA probes hybridized to DNA or RNA immobilized on nitrocellulose: Bio-blots. *Proceedings of the National Academy of Sciences USA* 80:4045–4049.

Legocki, R.P., Legocki, M., Baldwin, T.O., and Szalay, A.A. 1987. Bioluminescence in root nodules of soybean controlled by nitrogenase *nif*D and *nif*H promoters. pp. 282–287 in Verma, D.P.S. and Bisson, N. (editors), *Molecular Genetics of Plant-Microbe Interactions*. Martinus Nijhoff Publications, Dordrecht, The Netherlands.

Levy, S.B. 1984. Resistance to the tetracyclines. pp. 191–240 in Bryan, L.E. (editor), *Antimicrobial Drug Resistance*. Academic Press, Orlando, FL.

Lindgren, P.B., Frederick, R., Govindarajan, A.G., Panopoulos, N.J., Staskawicz, B.J., and Lindow, S.E. 1989. An ice nucleation reporter gene system: identification of inducible pathogenicity genes in *Pseudomonas syringae* pv. *phaseolicola*. *The EMBO Journal* 8:1291–1301.

Lindow, S.E., Arny, D.C., Barchet, W.R., and Upper, C.D. 1978a. The role of bacterial ice nuclei in frost injury to sensitive plants. pp. 249–261 in Li, P.H. and Sakai, A. (editors), *Plant Cold Hardiness and Freezing Stress*. Academic Press, New York.

Lindow, S.E., Arny, D.C., and Upper, C.D. 1978b. *Erwinia herbicola*: an active ice nucleus incites frost damage to maize. *Phytopathology* 68:523–527.

Lindow, S.E., Arny, D.C., and Upper, C.D. 1978c. Distribution of ice nucleation active bacteria on plants in nature. *Applied and Environmental Microbiology* 36:831–838.

Lindstrom, K., Lipsanen, P., and Kaifalainen, S. 1990. Stability of markers used for identification of two *Rhizobium galegae* inoculant strains after five years in the field. *Applied and Environmental Microbiology* 56:444–450.

Lipsanen, P. and Lindstrom, K. 1989. Lipopolysaccharide and protein profiles of *Rhizobium* sp. (*Galega*) strains. *FEMS Microbiological Letters* 58:323–328.

Macario, A.J.L. and Conway de Macario, E. (editors). 1985. *Monoclonal Antibodies Against Bacteria*. Academic Press, Orlando, FL.

Maki, R.L., Galyan, E.L., Chang-Chien, M., and Caldwell, D.R. 1974. Ice nucleation induced by *Pseudomonas syringae*. *Applied and Environmental Microbiology* 28:456–460.

Makkar, N.S. and Casida, L.E. 1987. Technique for estimating low numbers of a bacterial strain(s) in soil. *Applied and Environmental Microbiology* 53:887–888.

Malvick, D.K. and Moore, L.W. 1988. Population dynamics and diversity of *Pseudomonas syringae* on maple and pear trees and associated grasses. *Phytopathology* 78:1366–1370.

Maniatis, T., Fritsch E.F., and Sambrook, J. 1982. *Molecular Cloning, a Laboratory Manual*. Cold Spring Harbor Laboratory, Cold Spring Harbor, New York. 545 pp.

Martensson, A. and Gustafsson, J.G. 1985. Competition between *Rhizobium trifolii* strains for nodulation, during growth in a fermenter, and in soil-based inoculants, studied by ELISA. *Journal of General Microbiology* 131:3077–3082.

Miller, L.T. 1982. Simple derivatization method for routine analysis of bacterial whole cell fatty acid methyl esters, including hydroxy acids. *Journal of Clinical Microbiology* 16:584–586.

Morgan, J.A.W., Winstanley C., Pickup, R.W., Jones J.G., and Saunders, J.R. 1989. Direct phenotypic and genotypic detection of a recombinant pseudomonad population released into lake water. *Applied and Environmental Microbiology* 55:2537–2544.

Moyed, H.S., Nguyen, T.T., and Bertrand, K.P. 1983. Multicopy Tn10 *tet* plasmids confer sensitivity to induction of tet gene expression. *Journal of Bacteriology* 155:549–556.

Mullis, K.B. and Faloona, F.A. 1987. Specific synthesis of DNA in vitro via a polymerase-catalyzed chain reaction. *Methods in Enzymology* 155:335–350.

Murray, R.E., Parsons, L.L., and Smith, M.S. 1990. Aerobic and anaerobic growth of rifampin-resistant denitrifying bacteria in soil. *Applied and Environmental Microbiology* 56:323–328.

Muyzer, G. 1988. Applications of immunofluorescent techniques. pp. 219–222 in Sussman, M., Collins, C.H., Skinner F.A., and Stewart-Tull, D.E. (editors), *The Release of Genetically-engineered Micro-organisms*. Academic Press, London.

Nakai, C., Hori, K., Kagamiyama, H., Nakazawa, T., and Nozaki, M. 1983. Purification, subunit structure, and partial amino acid sequence of metapyrocatechase. *Journal of Biological Chemistry* 258:2916–2922.

Newell, S.Y. 1991. Estimating fungal biomass and productivity is decomposing litter. In Carroll, G.C. and Wicklow, D. (editors), *The Fungal Community*, 2nd ed. Dekker, New York. In press.

Ngo, T.T. 1985. Ultrasensitive chromogen system for horse radish-peroxidase using 3-methyl-2-benzothiazolinone hydrazone and n-phenyldiethanolamine. *Applied Biochemistry and Biotechnology* 11:257–268.

Ogram, A., Sayler, G.S., and Barkay, T. 1988. DNA extraction and purification from sediments. *Journal of Microbiological Methods* 7:57–66.

Oliver, R.P., Roberts, I.N., Harling, R., Kenyon, L., Punt, P.J., Dingemanse, M.A., and van den Hondel, C.A.M.J.J. 1987. Transformation of *Fulvia fulva*, a fungal pathogen of tomato, to hygromycin B resistance. *Current Genetics* 12:231–233.

Ray, C., Igo, M., Shafer, W., Losick, R., and Moran, C.P., Jr. 1988. Suppression of *ctc* promoter mutations in *Bacillus subtilis*. *Journal of Bacteriology* 170:900–907.

Renwick, A. and Gareth-Jones, D. 1985. A comparison of the fluorescent ELISA and antibiotic resistance identification techniques for use in ecological experiments. *Journal of Applied Bacteriology* 58:199–206.

Roberts, I.N., Oliver, R.P., Punt, P.J., and van den Hondel, C.A.M.J.J. 1989. Expression of the *Escherichia coli* β-glucuronidase gene in industrial and phytopathogenic filamentous fungi. *Current Genetics* 15:177–180.

Robinson, B.J., Pretzman, C.I., and Mattingly, J.A. 1983. Enzyme immunoassay in which a myeloma protein is used for detection of salmonellae. *Applied and Environmental Microbiology* 45:1816–1821.

Rollins, D.M. and Colwell, R.R. 1986. Viable but nonculturable stage of *Campylobacter jejuni* and its role in survival in the natural aquatic environment. *Applied and Environmental Microbiology* 52:531–538.

Rossman, A.Y., Palm, M.E., and Spielman, L.J. 1987. *A Literature Guide for the*

Identification of Plant Pathogenic Fungi. American Phytopathological Society Press, St. Paul, MN. 252 pp.
Roszak, D.B. and Colwell, R.R. 1987. Survival strategies of bacteria in the natural environment. *Microbiological Reviews* 51:365–379.
Rovira, A.D., Newman, E.I., Bowen, H.J., and Campbell, R. 1974. Quantitative assessment of the rhizoplane microflora by direct microscopy. *Soil Biology and Biochemistry* 6:211–216.
Sala-Trepat, J.M. and Evans, W.C. 1971. The *meta* cleavage of catechol by Azotobacter species: 4-oxalocrotonate pathway. *European Journal of Biochemistry* 20:400–413.
Sasser, M. and Miller, L.T. 1984. Identification of pseudomonads by fatty acid profiling. pp. 45–46 in Psallidas, P.G. and Alivizatos, A.S. (editors), *Proceedings of the Second Working Group on Pseudomonas syringae Pathovars*. Hellenic Phytopathological Society Publishers, Athens, Greece.
Schauer, A.T. 1988. Visualizing gene expression with luciferase fusions. *Trends in Biotechnology* 6:23–27.
Schmetterer, G., Wolk, C.P., and Elhai, J. 1986. Expression of luciferases from *Vibrio harveyi* and *Vibrio fischeri* in filamentous cyanobacteria. *Journal of Bacteriology* 167:411–414.
Schroth, M.N., Thomson, S.V., and Moller, W.J. 1979. Streptomycin resistance in *Erwinia amylovora*. *Phytopathology* 69:565–568.
Shaw, J.J. and Kado, C.I. 1986. Development of a *Vibrio* bioluminescence gene-set to monitor phytopathogenic bacteria during the ongoing disease process in a non-disruptive manner. *Bio/Technology* 4:560–564.
Shaw, J.J., Settles, L.G., and Kado, C.I. 1987. Transposon Tn4431 mutagenesis of *Xanthomonas campestris* pv. *campestris*: Characterization of a nonpathogenic mutant and cloning of a locus for pathogenicity. *Molecular Plant-Microbe Interactions* 1:39–45.
Southern, E.M. 1975. Detection of specific sequences among DNA fragments separated by gel electrophoresis. *Journal of Molecular Biology* 98:503–517.
Steffan, R.J. and Atlas, R.M. 1988. DNA amplification to enhance detection of genetically engineered bacteria in environmental samples. *Applied and Environmental Microbiology* 54:2185–2191.
Steffan, R.J., Breen, A., Atlas, R.M., and Sayler, G.S. 1989. Monitoring genetically engineered microorganisms in freshwater microcosms. *Journal of Industrial Microbiology* 4:441–446.
Stotzky, G. and Babich, H. 1984. Fate of genetically engineered microbes in natural environments. *Recombinant DNA Technical Bulletin* 7:163–188.
Torsvik, V.L. 1980. Isolation of bacterial DNA from soil. *Soil Biology and Biochemistry* 12:15–21.
Trevors, J.T., van Elsas, J.D., van Overbeek, L.S., and Starodub, M.-E. 1990. Transport of a genetically engineered *Pseudomonas fluorescens* strain through a soil microcosm. *Applied and Environmental Microbiology* 56:401–408.
Vali, G. 1971. Quantitative evaluation of experimental results on the heterogeneous freezing nucleation of super cooled liquids. *Journal of Atmospheric Science* 28:402–409.
Van Gorcom, R.F.M., Punt, P.J., Pouwels, P.H., and van den Hondel, C.A.M.J.J. 1986. A system for the analysis of expression signals in *Aspergillus*. *Gene* 48:211–217.
Van Hartingsveldt, W., Mattern, I.E., van Zeijl, C.M.F., Pouwels P.H., and van den Hondel, C.A.M.J.J. 1987. Development of a homologous transformation system for *Aspergillus niger* based on the *pyr*G gene. *Molecular and General Genetics* 206:71–75.
Wall, L.E., Byers, D.M., and Meighen, E.A. 1984. In vivo and in vitro acylation

of polypeptides in *Vibrio harveyi*: Identification of proteins involved in aldehyde production for bioluminescence. *Journal of Bacteriology* 159:720–724.

Weller, D.M. and Saettler, A.W. 1978. Rifampin-resistant *Xanthomonas phaseoli* var. *fuscans and Xanthomonas phaseoli*: Tools for field study of bean blight bacteria. *Phytopathology* 68:778–781.

Winstanley, C., Morgan, J.A.W., Pickup, R.W., Jones, J.G., and Saunders, J.R. 1989. Differential regulation of lambda p_L and p_R promoters by a *c*I repressor in a broad-host-range thermoregulated plasmid marker system. *Applied and Environmental Microbiology* 65:771–777.

Wright, S.F., Foster, J.G., and Bennett, O.L. 1986. Production and use of monoclonal antibodies for identification of strains of *Rhizobium trifolii. Applied and Environmental Microbiology* 52:119–123.

Zimmermann, R. 1977. Estimation of bacterial number and biomass by epifluorescence microscopy and scanning microscopy. pp. 103–120 in Reinheimer, G. (editor), *Microbial Ecology of a Brackish Water Environment.* Springer-Verlag, Heidelberg, Federal Republic of Germany.

Zola, H. 1987. *Monoclonal Antibodies: A Manual of Techniques.* CRC Press, Boca Raton, FL.

Zukowski, M.M., Gaffney, D.F., Speck, D., Kauffmann, M., Findeli, A., Wisecup, A., and Lecocq, J. 1983. Chromogenic identification of genetic regulatory signals in *Bacillus subtilis* based on expression of a cloned *Pseudomonas* gene. *Proceedings of the National Academy of Sciences USA* 80:1101–1105.

8

The Yeast Community of Cacti

William T. Starmer, James C. Fogleman,
and Marc-Andre Lachance

8.1 Introduction

In this chapter, we review the work on the microbial communities associated with the decaying tissues of various cactus species. Most of the discussion centers around the determinants of yeast community organization. The biological system is composed of three main components: (1) host tissue such as stems or fruits of columnar and prickly pear cactus species; (2) insects such as drosophilids that feed on the decaying cactus tissue; and (3) microorganisms that initiate and/or participate in the rotting process. The three components of the system are biologically and evolutionarily interwoven. The insects rely on the tissue and microbes for nutrients, survival, and reproduction, while the microbes, especially the yeasts, are dependent on the insects for transportation to new necrotic plants. Thus, the insects and microorganisms have the potential for mutualistic relationships and coevolution. Primary herbivores and pathogenic microorganisms in the system are expected to affect the evolution of the host plant. However, the saprophytic community that is established after the tissue is dead is not expected to be important in the evolution of the host tissue characteristics. The saprophytic system including insects and microorganisms is, thus, viewed as a community that is tracking but not causing the evolution of their hosts, except possibly indirectly by affecting certain components of the herbivore fauna or by the vegetative propagation of the host plants.

Much of the work we discuss has been reviewed in several different contexts. Background information on the components of the cactus-microorganism-*Drosophila* system can be found in Barker and Starmer (1982). Overviews of the *Drosophila* species and on their origins, evolution, and host plant associations are given in Fellows and Heed (1972), Heed

(1978, 1982), Heed and Mangan (1986), and Fogleman and Heed (1989). The systematics and affiliations of the cactus species are discussed by Gibson and Horack (1978), Gibson (1982), and Gibson et al. (1986). Cactus chemistry and its relevance to *Drosophila* ecology has been recently reviewed by Fogleman and Abril (1990). The yeasts species associated with decaying cactus species have been classified and characterized as to host and geographic affinities; this work is reviewed by Lachance et al. (1988) and Starmer et al. (1990). The potential coadaptations between yeasts and drosophilids also have been reviewed by Starmer and Fogleman (1986).

8.2 The System

Overview Cactus tissue is valuable to desert inhabitants because it is a source of shelter, nutrients, and water. When the tissue is wounded or injured, infection from microorganisms can start a small pocket of decay, a "rot pocket," that attracts other larger organisms, such as adult drosophilids. The rot pocket can be small and only involve a few grams of tissue or can become very large and engulf an entire giant columnar cactus weighing hundreds of kilograms. Injury to the tissue is caused variously by frost damage, windfall, fire, systemic infection, herbivory, and human activities.

Females of various *Drosophila* species lay their eggs in the rotting tissue early in the rotting process. The larvae forage on the microbial growth, consuming host tissue and microorganisms in the process. As the larvae move about, they spread vegetative cells and spores of the microbes to other zones of tissue so that the rotting process is facilitated. Larvae pupate near the rotting tissue, usually on the interior of the rot pocket on dry skin but some larvae that use rot juices which infiltrate the soil from the rot pocket, pupate in the soil. Adults emerging from the pupae seek out shelter, water, and nutrients in their surroundings, and the rot pockets serve this purpose. Adults disperse to newly formed rot pockets when their original rot dries and becomes unusable (Johnston and Heed, 1976). The dispersal of adults starts a new community of microorganisms and completes the cycle.

Cactus Both columnar and platyopuntia (prickly pear) cacti are host to drosophilids and yeasts. The decaying trunks, arms (stems), and fruits provide tissue from the columnar plants, while trunks, cladodes (flattened stems), and fruits serve as habitats in the prickly pears. Most studies have been conducted on cacti of North America, northern South America, and islands in the Bahamas and the Caribbean. In addition, investigations on the introduced cacti and associated microflora in Hawaii and eastern Australia have been conducted. Five major groups of cactus/tissue have been studied. These include: (1) stem tissues of cactus species in the genera

Escontria, Myrtillocactus, and *Stenocereus,* which are interrelated genera in the cactus subtribe Stenocereinae; (2) stem tissues of cactus in the genera *Backebergia, Carnegiea, Cephalocereus* (most species), *Lophocereus, Neobuxbaumia,* and *Pachycereus,* which are interrelated genera in the cactus subtribe Pachycereinae; (3) cladode tissue of *Nopalea* and *Opuntia,* related prickly pears in the cactus tribe Opuntieae; (4) various other stem tissues of species not members of the Stenocereinae, Pachcereinae, or Opuntieae—these include species of the genera *Acanthocereus, Cephalocereus* (some species), *Cereus, Ferocactus, Melocactus,* and *Neoabbottia*; and (5) fruit tissues of all cacti. These five groups of hosts and tissues form a useful partition from a chemical perspective because the stems, but apparently not the fruits, of Stenocereinae and Pachycereinae cacti are chemically distinct.

The stem chemistry of the five major species of columnar cacti which are involved in the desert system includes several unusual compounds. For example, *Carnegiea gigantea, P. pringlei,* and *Lophocereus schottii* (subtribe Pachycereinae) possess isoquinoline alkaloids in their lipid fractions. The alkaloids in the first two species are simpler in structure and significantly lower in concentration (1–2% dry weight) than those in *L. schottii* (Brown et al., 1972; Kircher, 1982). The alkaloids in *L. schottii* can represent up to 20% of the dry weight of this plant and primarily consist of the compounds lophocereine and its trimer, pilocereine (Djerassi et al., 1962; Kircher, 1969). This species of cactus is further distinguished by its unusual sterol profile. The sterols in *L. schottii* are atypical due to the presence, of an interrupted biosynthetic pathway, which results in the accumulation of intermediate sterol structures (Campbell and Kircher, 1980).

Stenocereus gummosus and *S. thurberi* (subtribe Stenocereinae) are also characterized by some unusual plant compounds. Three chemical categories distinguish them from the other three species. First, they possess high concentrations (up to 36% dry weight) of pentacyclic triterpene glycosides (Kircher, 1977). Second, the fatty acids in their lipid fractions are predominantly shorter in chain length than the other species (Kircher, 1982). Finally, their sterol profiles consist primarily of dihydroxy forms (Kircher and Bird, 1982).

The presence of alkaloids in the Pachycereinae species and the triterpene glycosides in species of the subtribe Stenocereinae are a mutually exclusive pattern which extends to other members of the two respective subtribes. These chemical categories, as well as the other compounds mentioned above, have been demonstrated to play an important role in determining the pattern of host utilization among the desert *Drosophila.* This work has been recently reviewed by Fogleman and Abril (1990). Typically, the chemical effect is strong enough to exclude all but the resident *Drosophila* species for each cactus species. For this reason, in any given area, there is essentially a one-to-one relationship between cactus species and *Drosophila* species.

Drosophila *Drosophila* species associated with cactus necroses are mainly from a single lineage that radiated in the New World about 30 million years ago (Throckmorton, 1982). This radiation (New World Virilis Repleta Radiation) gave rise to the repleta group, which contains 70–80 extant species, most of which use cactus tissue as a breeding and feeding substrate. Host plant specialization for some of these flies has been extensively studied in the North American Sonoran Desert by W.B. Heed and coworkers (Fellows and Heed, 1972; Fogleman, 1982, 1984; Fogleman et al., 1981, 1982, 1986; Fogleman and Kircher, 1986; Heed, 1978, 1982; Heed and Mangan, 1986; Kircher, 1982; Kircher et al., 1967; Mangan, 1982; Starmer et al., 1986). These studies indicate a high degree of specialization in terms of host plant selection and host plant use by the resident species. In the Sonoran Desert, it is apparent that each of the year-round resident species has achieved its present host association pattern independently. The host specificities of the drosophilids could be an important factor in the realized distribution of yeasts in the system, especially since the yeasts rely, in part, on the drosophilids for transportation to new hosts and are not passively dispersed in the air.

Yeasts The importance of yeast to *Drosophila* has been known since the late 1800s (Delcourt and Guyenot, 1910; Baumberger, 1919). These early studies and subsequent investigations by Sang, Robertson, and colleagues (reviewed by Sang, 1978) showed the particular nutritional attributes of yeasts that are important for the development of larvae into adults in the laboratory. Wagner (1944, 1949) carried out the earliest investigation of naturally occurring yeasts and their nutritional importance to cactus-breeding drosophilids. Subsequent investigations into other habitats such as fruit-and tree exudate-breeding flies have been reviewed by Begon (1982). Begon's review clearly shows that different species groups and species groupings of the family Drosophilidae have distinct yeast communities associated with them. Physiological changes in these yeast communities parallel the evolutionary history of several independent *Drosophila* radiations (Starmer, 1981). This pattern is suggestive of coevolution between the *Drosophila* and yeasts. Evidence indicates that if coevolution has occurred, then it most likely involves a combination of diffuse mutualisms among members of the yeast communities, nutritional and detoxification benefits provided by the yeasts to the flies, and dispersal benefits provided by the flies to members of the yeast communities (Starmer and Fogleman, 1986; Starmer and Aberdeen, 1990).

Yeasts are recovered from samples of rotting cactus by suspending a sample of necrotic tissue in sterile water and then inoculating petri dishes of agar medium with dilutions of the suspension or by streaking a loopful of the suspension directly on the agar plate. Both general and selective media have been used for isolation of yeasts (Starmer, 1982a; Lachance et al.,

1988). Once the colonies have grown to a suitable size, they are picked and cultured on standard media for study and identification. Species have been defined by a number of different criteria with emphasis on reproductive isolation and genetic divergence as necessary conditions for recognizing and describing new taxa.

8.3 Patterns in Yeast Community Organization

Yeast communities can be viewed from several perspectives (evolutionary, ecological, host, space, and temporal) each emphasizing a different determinant aspect of community structure. In this section we review evidence from ecological, physiological, and biogeographical perspectives and highlight those studies that describe the quantitative and qualitative differences among yeast communities.

Spatial and Temporal Analysis on Individual Hosts An analysis of the community structure of yeasts associated with the decaying stems of *Stenocereus gummosus* was conducted on an increasing spatial scale for samples of plants in southern Baja California, Mexico (Starmer, 1982a). In this study, several noncontiguous samples of tissue in an individual rot pocket were taken from each of three to four different plants in one locality. This was repeated for three different localities about 100 km apart. Each sample was diluted and plated on a series of selective and nonselective media designed to reveal the yeasts known to inhabit *S. gummosus* rots. An average of 3.5 ± 0.4 species of yeasts were recovered from each sample with an average density of 5.47×10^7 cells per gram (wet weight) of tissue. A nested analysis of variance showed that most of the variation in the abundance of each species and their proportions relative to other yeasts was due to differences between plants as compared to samples within plants and to averages at each locality. This observation was interpreted to be a result of succession as follows: 1) differences between plants represent different stages in the rotting process; 2) similar composition within plants represent the same stage of a rot that had a single historical starting point; and 3) similar composition for the averages at each locality indicated that the yeast species present at the different localities were not very different.

This study prompted another investigation on the yeast communities associated with stem necroses of a related species of cactus *Stenocereus thurberi* (Fogleman and Starmer, 1985). The same three levels of organization (within plants, among plants within a locality, and among localities) were analyzed. The study resulted in the same conclusion: variability in the abundance and proportional representation of species is most likely due to samples among plants representing different successional stages of the rotting process. The yeast species recovered in the two studies were for the most part the same in name and number but the average density of yeast cells in *S. thurberi* tissue was significantly lower (1.5×10^7 cells per gram).

A similar but somewhat more detailed analysis of spatial and temporal variation in the community structure of yeasts associated with *Opuntia stricta* cactus in eastern Australia was conducted by Barker et al. (1987). In this study, samples were taken weekly for three to four consecutive weeks from the same rot for several plants. Two localities were studied. In one locality, the plants were older and more dispersed than in the other. The analysis showed that the effective number of yeasts species and species diversity increased as the rot aged to an observed maximum at four weeks. The two localities differed in terms of spatial and temporal influences. Spatial effects were more important at the locality that had older, more dispersed plants while temporal effects explained more variation for the locality with younger, less dispersed plants. There are several possible explanations for the observed differences including differences in: 1) the age of the tissue, 2) the size of the rots, 3) the longevity of the rots, 4) the heterogeneity of inocula available for the rots, and 4) the behavior of the primary rot initiator (the moth *Cactoblastis cactorum*). This work was consistent with an earlier finding by Starmer and Phaff (1983), who found that prickly pear cladodes in the desert southwest of the United States are more diverse in terms of their yeast florae and show greater differences between locality than do rots of *Stenocereus* species.

Generalists vs. Specialist Yeasts Yeasts from cactus necroses have been analyzed in terms of their fundamental physiological abilities, such as the ability to use single sources of carbon and nitrogen, ability to grow under high osmotic pressure, elevated temperatures, and anaerobic conditions, and their sensitivity to antibiotics. The overall physiological ability of a yeast or yeast community can then be summarized in a physiological profile which provides an estimate of the fundamental niche dimensions of that particular yeast (e.g., the effective number of positive physiological states). This measure of the fundamental niche can then be compared to the profile of habitats that the yeast actually occupies in nature or the realized niche (e.g., the effective number of cactus species used). This has been done for cactus-associated yeasts (Starmer, 1981). The comparison showed that there is a negative correlation between the fundamental and realized niche metrics. Those yeasts such as *Pichia cactophila* that are found in many different cactus species (i.e., a cactus generalist) are limited in their physiological abilities (i.e., a physiological specialist). Other yeasts such as *Pichia mexicana* have broad physiological abilities (physiological generalization) and use few host cacti (a cactus specialist).

This correlation has been interpreted as an evolutionary response to the different chemistry of the host cacti and stresses of the desert environment. In general, it is probable that the ancestral yeasts of the present cactus-inhabiting yeasts were more versatile in their physiological abilities. Adapting to toxins in the environment could occur by the evolution of new systems of detoxification or by excluding toxins from sites of toxic activity. Genetic studies with cactus-associated yeasts have shown that most changes in physiology are due to changes in the ability to transport compounds across the yeast plasmalemma. Furthermore, the genes responsible for resistance or sensitivity to known yeast growth inhibitors present in some cacti (such as triterpene glycosides) are most likely associated with membrane flexibility (Starmer et al., 1980). Thus, one possible explanation for observing a decrease in fundamental ability in conjunction with an increase in habitat use is that, by shutting down transport of rarely used substrates, the yeast may also eliminate transport or binding of growth-limiting toxins. Selective pressures of this sort could then lead to a yeast specialized on a limited number of predictable resources found in the cactus environment, as well as a yeast protected from toxins. In the cactus-rot environment, this specialization appears to be due to the use of fermentation products, i.e., alcohols and acids (Heed et al., 1976; Lachance and Starmer, 1982).

Biogeography Two facets of the biogeography of cactus-associated yeasts are of interest. One is the importance of geographic separation to the community organization of yeasts in the same or similar host plants. The second is the effect of moving the host plant to a region where cacti did not previously occur (i.e., cactus introductions to Australia).

Starmer et al. (1990) summarized the geographical distribution of yeasts living in cactus rots of North America, northern South America, and islands in the Bahamian and Caribbean seas. That study reported the pattern of yeast species distributions for 1885 plants in 121 different localities. Over 3000 strains of 92 species of yeast or yeastlike organisms were tabulated for five regional (southwestern USA, Sonoran Desert, southern Mexico, Caribbean, and Venezuela) and five cactus categories (Stems of Stenocereinae and Pachycereinae, stem and fruit of Opuntia and "other"). Table 8.1 summarizes the distribution of the most common cactus yeasts according to four specific host categories ("other" is not included). From the host and geographic distributions, several inferences were made about the origins of the different communities and the role of geographic separation in the differences or similarities observed for the communities. Three cactophilic yeasts (*Pichia cactophila*, *Candida sonorensis*, and *Cryptococcus cereanus*) are common in cactus stems of all types and are found in all regions where cacti grow. These species have no close relatives outside of the cactus-yeast-*Drosophila* system. They are restricted to cactus stems and, thus, are not found outside of cactus-growing habitats. *P. cactophila* and *C. sonorensis* are occasionally found in cactus fruits. Even in areas where cacti grow next to other yeast habitats, such as tree fluxes of oaks, cottonwoods, or mesquite, they are not found in these hosts (Ganter et al., 1986).

Table 8.1 The most common yeast species found in four host categories of cactus tissue[a]

	Columnar cactus		*Opuntia*	
	Pachycereinae	Stenocereinae	Stem	Fruit
Number of plants sampled	250	484	867	181
Number of yeast isolated	450	922	1882	276
Pichia cactophila	**24.9**	**33.1**	**19.6**	**4.8**
Candida sonorensis	**9.1**	**15.3**	**15.2**	2.6
Cryptococcus cereanus	**9.1**	**11.1**	**18.7**	
Candida ingens	**7.1**	8.3	0.5	
Pichia heedii	**14.2**	0.2		
Candida deserticola	0.4	**7.8**	0.1	0.4
Prototheca[b]	2.9	3.7	**7.8**	0.4
Candida mucilagina		1.7	**5.3**	
Clavispora opuntiae	0.7	1.5	**5.1**	3.7
Kloeckera spp.	0.3	0.4	0.3	**28.2**
Issatchenkia terricola				**8.3**
Pichia barkeri		0.6	0.6	**6.9**
Candida krusei			0.1	**5.4**

[a] The percentage of each species in the community (relative to all species recovered) of each host is given. The 5 or 6 dominant species in each host are indicated in bold.
[b] This genus is a yeastlike, colorless *Chlorella*.

Other yeasts in the cactus rot communities do show specificity in terms of regional and host plant distributions. Biogeographical analysis indicates that changes in the yeast communities of Stenocereinae species across regions usually involves replacements or additions of relatively unrelated species. However, changes in the yeast communities of Pachycereinae species are often replacements along lines of genetic relationship, and thus geography has apparently been important in fostering species splitting and thus speciation in lineages specific to Pachycereinae cactus species. Changes in the yeast species composition for communities associated with Opuntieae from region to region involve replacements by both related and unrelated species. Fruit yeast communities are distinct and radically different from the stem communities (Starmer et al., 1988b), and changes from region to region may be a function of the differences in yeast communities of other fruits that influence the available inocula carried by the insects, mammals, and birds that regularly visit fruits (Starmer et al., 1987b).

An overall analysis of the host and geographic effects on cactus yeast communities gives a greater weight to host than geographic effects, but individual cases exist where both appear equally important in determining the distribution of cactus-specific yeasts and yeast communities.

Prickly pear cactus, especially *Opuntia stricta,* was introduced to Australia in the last century, where it became an economic pest (Dodd, 1940). Measures to rid the infested areas resulted in the introduction of microorganisms normally associated with rotting cacti (Barker et al., 1983, 1984). It was thus possible to compare the yeast communities of *O. stricta* in its original environment and geographic region (pan-caribbean) with the yeast communities of *O. stricta* in the introduced region (eastern Australia). This comparison was conducted on yeasts of the cladode and fruit tissues (Starmer et al., 1987b) and showed that the yeasts of cladodes of the two regions were remarkably similar in terms of taxa present, while the fruit communities were different in this respect. The results were interpreted to mean that the yeast-cladode-insect associations were relatively "closed" to outside influences, while the yeast-fruit-vector associations were relatively "open" to outside influences (i.e., other reservoirs of yeasts). This finding is consistent with the view that host tissue similarities and differences are more important than geography in their influences on yeast community composition of stems and cladodes.

8.4 Causes of Yeast Community Composition

A sequence of causal factors that should be important to yeast community membership is diagrammed in Figure 8.1. In this figure, the first important facet is the dispersal of yeasts to the host plant. Given that yeasts are dispersed to new hosts, the chemical environment of the host tissue (especially secondary plant compounds that inhibit herbivory or infection) may exclude certain yeasts from growing on that host. Even if the yeasts that arrive are not inhibited by host chemicals, the yeasts that were already present may interfere with the growth of the new arrivals by producing toxic substances or by rapidly utilizing the available resources such as sugars, vitamins, and nitrogen. Finally, the nutritional benefits provided to potential vectors may be important because, if the vectors do not adequately mature and transport the yeast to a new host, the cycle is interrupted. Evidence on each of these four causal aspects of yeast community organization is reviewed in the next four sections.

Drosophila as Vectors of Yeasts Surveys of yeasts found associated with adult drosophilids have usually proceeded by capturing the adults in nature, aseptically dissecting the crop, and streaking the crop contents on a suitable yeast medium. Work of this sort has revealed the kinds and relative proportions of yeasts in the diets of the adults, and to some degree it has indicated the vectoring potential of the adults. General reviews that have considered this type of survey have been written by Heed et al. (1976) and

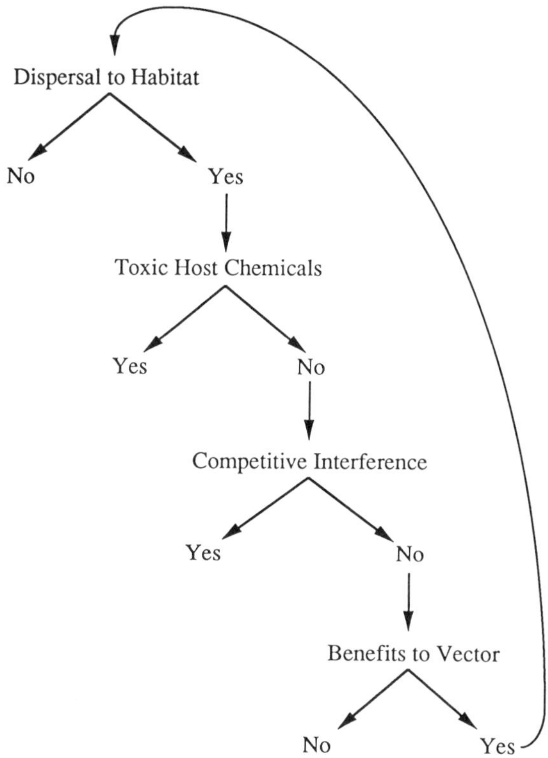

Figure 8.1 Levels and hierarchy of factors potentially affecting yeast community structure.

Begon (1982) while specific studies on cactus-breeding *Drosophila* have been presented by Starmer (1981).

Investigations into what is actually vectored by drosophilids in nature have been approached by allowing adult flies access to recently initiated rots in the field or by capturing adults and allowing them to inoculate petri dishes containing a complete yeast medium. Gilbert (1980) showed that woodland drosophilids are significant vectors of yeasts by exposing sterile substrates and observing insect visits in conjunction with subsequent observations of microbial growth on those substrates. Starmer et al. (1988a) also showed that cactophilic drosophilids and nereids visiting saguaro cactus are primary vectors of yeasts, while other insects, such as muscoids and dolichopodids, that visit saguaro cactus rots only carry incidental yeasts (Figure 8.2). In that study, the average number of yeast species vectored (i.e., transmitted to the test medium) by an individual fly was 2.4 ± 0.9. Ganter et al. (1986) and Ganter (1988), using similar methods, also found that *Drosophila* species were significant vectors of yeasts in other habitats such as tree fluxes and prickly pear cladode rots. Because most yeasts are

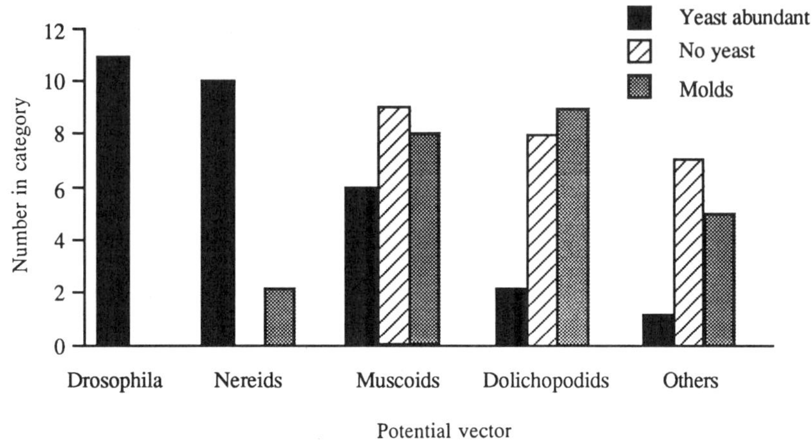

Figure 8.2 Number of adult insects carrying yeasts and filamentous fungi to rotting saguaro cactus by insect type.

not dispersed freely in air currents, these findings establish an important biological link between the yeasts and flies.

Chemical Determinants of Community Membership Habitat specificity for yeasts has been inferred from distributional surveys and by experimental testing. In general, yeast species have a high degree of specificity for the habitats in which they live (Phaff and Starmer, 1987). The cactus-associated yeasts provide several interesting examples in terms of the degree of specificity and possible determinants. Where possible, many cactus-specific yeasts have been tested for their ability to grow on various cactus tissues. Tests of this sort have shown that some cactus tissues inhibit yeast growth and, thus, exclude other cactus-specific yeasts from the yeast communities of these cacti. This has been demonstrated for yeast communities associated with the stem rots of *Stenocereus gummosus* and *S. thurberi*. In these cases, surface-active triterpene glycosides, which are abundantly present in the stems, inhibit the growth of other cactus-associated species (e.g., *Pichia heedii*, which is common in stem rots of *Carnegiea gigantea* and *Lophocereus schottii*). The triterpene glycosides also have activity against other yeasts not associated with cacti and thus may function to exclude these yeasts as well (Starmer et al., 1980).

Genetic studies on two varieties of *Pichia amethionina* provide insights, not only into the determinants of community membership, but also suggest a role of the host chemicals in genetic divergence and speciation. *Pichia amethionina* var. *amethionina* is commonly found in *Stenocereus gummosus* and *S. thurberi* rots while the related form *P. amethionina* var. *pachycereana* is found in *Carnegiea gigantea* and other cactus rots but only rarely in *Stenocereus* species. The latter variety is sensitive to triterpene glycosides and its growth is inhibited. Analysis of the haploid progeny from crosses between the two varieties indicates the differences are controlled by a single temperature-sensitive locus (Starmer et al., 1980). Thus, community membership and genetic isolation can be the result of fairly simple (one-gene) differences in the taxa.

Additional experiments and studies with cactus chemicals have not revealed other chemicals which have "all-or-none" exclusion properties. However, some chemicals such as short-chain fatty acids in *Stenocereus thurberi* appear to have differential inhibitory effects on yeast growth and could function to change the proportional representation of yeasts in necrosis of *S. thurberi* (Starmer, 1982b).

Experiments with alkaloids found in *Carnegiea gigantea* and *Lophocereus schottii* have shown that these compounds have no effect on the growth of yeasts and are not directly involved in shaping yeast communities in these cacti (Fogleman and Kircher, 1983).

Yeast-Yeast Interactions There has been one report on the yeast-yeast interactions within *Stenocereus gummosus* rots (Starmer and Fogleman, 1986). This paper reported the quantitative effects of yeasts on the growth of other yeasts in *S. gummosus* tissue with and without *Drosophila* larvae present. All statistically significant effects of one yeast on another during the initial growth phase were positive, and these positive effects did not depend on the presence or absence of larvae (Figure 8.3). Thus, yeasts in *S. gummosus* tissue facilitate the growth of other community members, and a system of diffuse mutualism is apparent. When yeast densities reached an asymptote, the yeast-yeast effects were sometimes positive and sometimes negative. Furthermore, the type of interaction was dependent on the presence or absence of larvae in the tissue (Figure 8.3).

This type of dependence is known for other systems as well. Insect grazing is known to affect plant succession (McBrien et al., 1983) and to mediate algal interactions in stream benthic communities (Hart, 1985). An overall analysis of the yeast-yeast interactions using May's (1974) criteria for qualitative stability of communities indicated that yeast communities with larvae were qualitatively stable while those without larvae were unstable. These findings were consistent with the results of Heithaus et al. (1980), which showed that the addition of a predator can lead to stable equilibrium for systems involving mutualism. Several models of mutualism indicate that mutualism occurring during initial growth enhances stability (Wolin and Lawlor, 1984; Addicott, 1979).

Killer Yeasts Killer yeasts produce extracellular toxins that kill other yeasts. The killer phenomenon in yeasts has been well studied in the laboratory (Bruenn, 1980; Bussey, 1981; Wickner, 1981; Tipper and Bostian, 1984). Killer toxins are usually active at relatively low pH and are produced in a nonagitated colonial state (i.e., as a stationary colony). These conditions

NO LARVAE

Initial Growth

	Pc	Cs	Pa	Cc	Ci
Pc					+
Cs	+		+		
Pa	+	+			
Cc	+				
Ci					

Asymptote

	Pc	Cs	Pa	Cc	Ci
Pc		+	+		+
Cs	+				
Pa					
Cc					
Ci		+			

25-75 LARVAE

Initial Growth

	Pc	Cs	Pa	Cc	Ci
Pc					
Cs	+		+		
Pa	+	+			
Cc	+				
Ci					

Asymptote

	Pc	Cs	Pa	Cc	Ci
Pc				+	+
Cs			+		
Pa	-	-			
Cc	-	-			
Ci	-				

Figure 8.3 Summary of yeast-yeast effects in decaying *Stenocereus gummosus* cactus with and without *Drosophila mojavensis* larvae present. The competitor yeast species is listed on top and the responding yeast on the side. + for the responding yeast, indicates an increased initial growth rate or higher asymptote; − indicates a decrease. Nonsignificant yeast-yeast interactions are left blank. Pc, *Pichia cactophila;* Cs, *Candida sonorensis;* Pa, *Pichia amethionina* var. *amethionina;* Cc, *Cryptococcus cereanus;* and Ci *Candida ingens*. Summarized from Starmer and Fogleman (1986).

implicate the toxins as potentially important to the killer yeast during colonization if pH and larval activity are low (in cactus tissue and fruit) and when interference with the growth of other forms is potentially valuable for future growth. We have surveyed (Starmer et al., 1987a) various habitats for killer yeasts and found that killer yeasts are common, with at least one killer species represented in each yeast community type. About 1 in 10 isolates from an average sample (e.g., a cactus stem yeast community) is a killer yeast, while 1 in 3 isolates from rotting fruit is a killer yeast. Of 82 species of yeast recovered in this survey, 23 were determined to be killers of at least one or more species of yeast. Cross-testing the most common yeasts of five host types (Table 8.2) that occur together in the desert southwest (i.e., saguaro cactus, opuntia cactus, opuntia fruit, mesquite flux, and oak flux) indicated that most killing activity occurs between yeasts from different hosts (e.g., yeasts from the fruit community are immune to the killers in fruit but the killers in this community kill most yeasts from the other host types). This generalization holds with the exception of the killer yeast *Pichia thermotolerans* (in saguaro), which kills most other yeasts found in saguaro. Some evidence from collections of rotting stems or cladodes of cactus show

Table 8.2 Distribution of dominant yeast species according to host in the desert southwest[a]

	Saguaro	Opuntia	Fruit	Mesquite	Oak
Number of plants	45	238	36	56	43
Pichia cactophila	0.47	0.36	0.03		0.02
Candida sonorensis	0.13	0.21	0.11		
Pichia amethionina	0.36	0.07			
Cryptococcus cereanus	0.24	0.66			
Candida ingens	0.22	0.01			
Candida mucilagina		0.21			
Pichia deserticola		0.06			
Pichia thermotolerans	**0.29**			0.04	
Pichia heedii	0.49				
Hansenula polymorpha		0.18			
Pichia kluyveri		**0.05**	**0.31**		
Pichia membraneafaciens			**0.11**	**0.26**	**0.02**
Kloeckera apiculata			**0.86**		
Debaryomyces hansenii				0.57	
Cryptococcus unigutulatus				0.21	
Candida sp. "PT"				0.43	
Kluyveromyces lactis var. *drosophilarum*					**0.14**
Hansenula anamola					0.23
Pichia pastoris					**0.19**
Saccharomyces dairensis					0.21

[a]The values given below the rule are the number recovered per number of plants sampled. Only taxa that occurred in at least 10% of any one host type are listed. The killer yeasts in each community are shown in bold.

patterns of mutually exclusive species distributions (Figure 8.4). In these cases, one of the two species is often a killer yeasts and the other is sensitive to that particular killer (as determined in the laboratory by standard killer assays). Therefore, there is circumstantial evidence that killer toxins are functioning to exclude other yeasts from particular yeast communities (Starmer et al., 1987a).

Larval-Yeast Interactions Because yeasts, in general, rely on insects such as *Drosophila* species as vectors to new habitats, the nutritional benefits of yeasts to *Drosophila* could be important to community organization. Studies on the nutritional sufficiency of cactus-inhabiting microorganisms for cactophilic *Drosophila* have shown that individual yeast species are, in general, sufficient for larval growth (Wagner, 1944, 1949; Vacek, 1982; Starmer, 1982b). Studies with combinations of yeasts have shown that even though single species of yeasts on cactus tissue are nutritionally sufficient, mixed cultures of different yeast species provide additional benefits to lar-

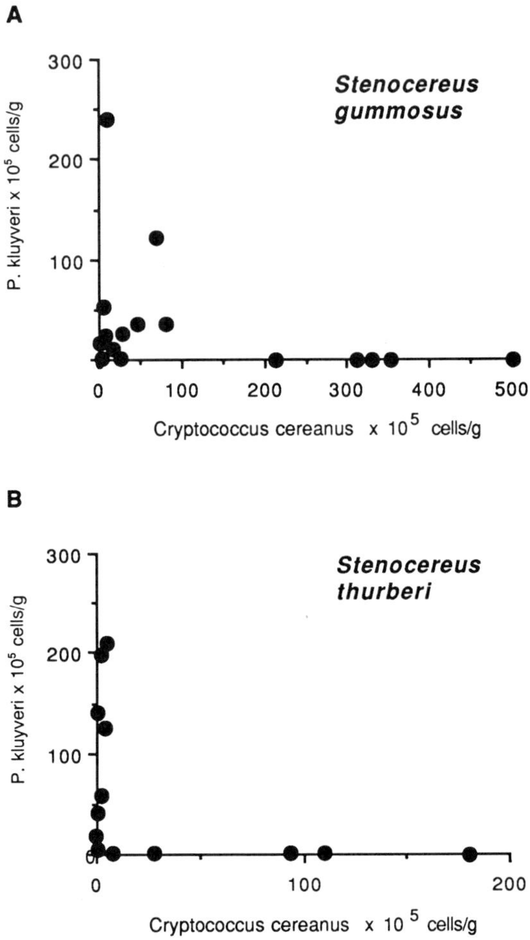

Figure 8.4 Co-occurrence of *Pichia kluyverii* and *Cryptococcus cereanus* in (A) *Stenocereus gummosus* and (B) *S. thurberi* rots. Data from Fogleman and Starmer (1985); *Candida valida* is equated with *P. kluyveri*. *P. kluyveri* is a killer of *Cryptococcus cereanus* in laboratory tests.

vae, in terms of faster development, larger adult size, or increased viability (Starmer and Fogleman, 1986; Starmer and Barker, 1986; Starmer and Aberdeen, 1990). Figure 8.5 shows an example of such an effect for *Drosophila buzzatii* larvae growing on monocultures and bicultures of yeasts on *Opuntia stricta* cladode tissue. This mixed culture effect appears to be host dependent since larval benefits are not experienced in mixed cultures of yeasts from one host tissue (Opuntia cladode) growing in a different tissue (fruit). The community of yeasts that provides the best nutrition for its vector is

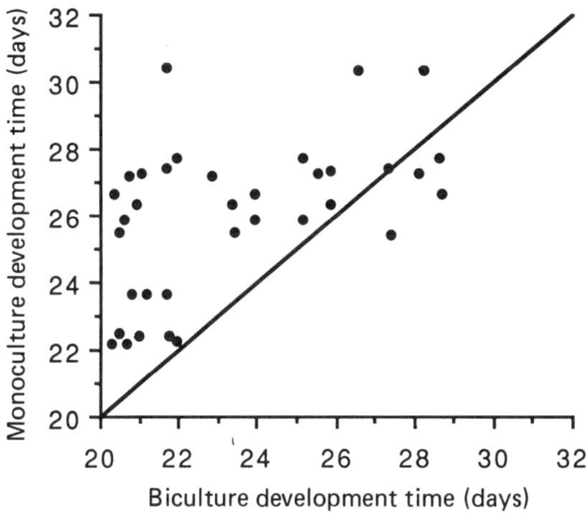

Figure 8.5 Development time of *Drosophila buzzatii* larvae grown on *Opuntia stricta* cladode tissue inoculated with monocultures and all possible biculture combinations of four different yeast species. Each point represents a monoculture and the biculture combination of which it is a member for three different strains of the fly averaged over five replicates. Modified from Starmer and Barker (1986).

thus host specific (Starmer and Aberdeen, 1990). Coadaptation of this type could lead to a closed system and greater specificity of yeasts for particular communities. The interactions of yeasts within a community could then be expected to become more specialized.

8.5 Conclusions

We have reviewed the distribution of yeast species that live in the decaying tissues of cactus species. Cactus yeasts are, in general, specific for the cactus habitat and do not occur elsewhere. They also appear to have evolved within the framework of the cactus-microorganism-*Drosophila* system, and the closest relatives of most species are within the system. Host differences appear to be important determinants of yeast community structure. Yeast vectors such as drosophilids and their host affinities may also contribute significantly to observed yeast distributions. The nutritional benefits provided by the yeasts to their vectors are potentially potent organizing factors for the stability and continuation of particular community types. Yeast-yeast interactions range from mutualism to interference competition. This latter form of competition is achieved in some cases by production of killer toxins. These toxins appear to be most important for between community exclusions. Yeast communities of cactus might be viewed as communities with diffuse mutualisms occurring among members of the same community, which, as a "unit community" (Swift, 1976), is mutualistic with its vectors.

The authors acknowledge funding by the U.S. National Science Foundation (W.T.S.) and by the Natural Science and Engineering Research Council of Canada (M.A.L.). Much of the work presented here was done in collaboration with Drs. H.J. Phaff, W.B. Heed, and H.W. Kircher. The authors thank Virginia Aberdeen, Mary Miranda, and Phil Ganter for their technical assistance and support.

References

Addicott, J.F. 1979. A multispecies aphid-ant association: Density dependence and species-specific effects. *Canadian Journal of Zoology* 55:558–569.

Barker, J.S.F. and Starmer, W.T. 1982. *Ecological Genetics and Evolution: The Cactus-Yeast-*Drosophila *Model System*. Academic Press, Sydney, Australia.

Barker, J.S.F., Toll, G.L., and East, P.D. 1983. Heterogeneity of the yeast flora in the breeding sites of cactophilic *Drosophila*. *Canadian Journal of Microbiology* 29:6–14.

Barker, J.S.F., East, P.D., Phaff, H.J., and Miranda, M. 1984. The ecology of the yeast flora in necrotic *Opuntia* cacti and of associated *Drosophila* in Australia. *Microbial Ecology* 10:379–399.

Barker, J.S.F., Starmer, W.T., and Vacek, D.C. 1987. Analysis of spatial and temporal variation in the community structure of yeasts associated with decaying *Opuntia* cactus. *Microbial Ecology* 14:267–276.

Baumberger, J.D. 1919. A nutritional study of insects, with special reference to microorganisms and their substrata. *Journal of Experimental Zoology* 28:1–81.

Begon, M. 1982. Yeasts and Drosophila. pp. 345–384 in Ashburner, M., Carson, H.L., and Thompson, J. (editors), *The Genetics and Biology of* Drosophila, vol. 3. Academic Press, New York.

Brown, S.D., Hodgkins, J.E., Massingill, J.L.Jr., Reinecke, M.G. 1972. The isolation, structure, synthesis and absolute configuration of the cactus alkaloid gigantine. *Journal of Organic Chemistry* 37:1825–1828.

Bruenn, J.A. 1980. Virus-like particles of yeast. *Annual Review of Microbiology* 34:49–68.

Bussey, H. 1981. Physiology of killer factor in yeasts. *Advances in Microbial Physiology* 22:93–122.

Campbell, C.E. and Kircher, H.W. 1980. Senita cactus: a plant with interrupted sterol biosynthetic pathways. *Phytochemistry* 19:2777–2779.

Delcourt, A. and Guyenot, E. 1910. De la possibilité d'étudier certains Diptères en milieu défini. *Comptes Rendus Hebdomadaires des Séances de l'Academie des Sciences. Serie D. Sciences Naturelles* 151:255–257.

Djerassi, C., Brewer, H.W., Clark, C., Durham, L.C. 1962. Alkaloid studies. XXXVIII. Pilocereine—a trimeric cactus alkaloid. *Journal of the American Chemical Society* 79:2203–2210.

Dodd, A.P. 1940. *The Biological Campaign Against Prickly-Pear*. Government Printer. Brisbane, Australia.

Fellows, D.P. and Heed, W.B. 1972. Factors affecting host plant selection in desert-adapted cactiphilic *Drosophila*. *Ecology* 53:850–858.

Fogleman, J.C. 1982. The role of volatiles in the ecology of cactophilic *Drosophila*. pp. 191–206 in Barker, J.S.F. and Starmer, W.T. (editors), *Ecological Genetics and Evolution: the Cactus-Yeast-*Drosophila *Model System*. Academic Press, Sydney, Australia.

Fogleman, J.C. 1984. The ability of cactophilic Drosophila to utilize soaked soil as larval substrates. *Drosophila Information Service* 60:105–107.

Fogleman, J.C. and Abril, J.R. 1990. Ecological and evolutionary importance of host plant chemistry. pp. 121–143 in Barker, J.S.F., MacIntyre, R.J., and Starmer, W.T. (editors), *Ecological and Evolutionary Genetics of Drosophila*. Plenum Press, New York.

Fogleman, J.C. and Heed, W.B. 1989. Columnar cacti and desert *Drosophila*: The chemistry of host plant specificity. pp. 1–24 in Schmidt, J.O. (editor), *Special Biotic Relationships in the Arid Southwest*. University of New Mexico Press, Albuquerque, NM.

Fogleman, J.C. and Kircher, H.W. 1983. Senita alkaloids: No inhibition of sterol biosynthesis in yeasts or cacti. *Journal of Natural Products* 46:279–282.

Fogleman, J.C. and Kircher, H.W. 1986. Differential effects of fatty acid chain length on the viability of two species of cactophilic *Drosophila*. *Comparative Biochemistry and Physiology* 83A:761–764.

Fogleman, J.C. and Starmer, W.T. 1985. Analysis of the community structure of yeasts associated with the decaying stems of cactus. III. *Stenocereus thurberi*. *Microbial Ecology* 11:165–173.

Fogleman, J.C., Starmer, W.T., and Heed, W.B. 1981. Larval selectivity for yeast species by *Drosophila mojavensis* in natural substrates. *Proceedings of the National Academy of Sciences. USA*. 78:4435–4439.

Fogleman, J.C., Starmer, W.T., and Heed, W.B. 1982. Comparisons of yeast florae from natural substrates and larval guts of southwestern *Drosophila*. *Oecologia* 52:187–191.

Fogleman, J.C., Duperret, S.M., and Kircher, H.W. 1986. The role of phytosterols in host plant utilization by cactophilic *Drosophila*. *Lipids* 21:92–96.

Ganter, P.F. 1988. The vectoring of cactophilic yeasts by *Drosophila*. *Oecologia* 75:400–404.

Ganter, P., Starmer, W.T., Lachance, M.A., and Phaff, H.J. 1986. Yeast communities from host plants and associated *Drosophila* in southern Arizona: new isolations and analysis of the relative importance of hosts and vectors on community composition. *Oecologia* 70:386–392.

Gibson, A.C. 1982. Phylogenetic relationships of Pachycereeae. pp. 3–13 in Barker, J.S.F. and Starmer, W.T. (editors), *Ecological Genetics and Evolution: The Cactus-Yeast* Drosophila *Model System*. Academic Press, Sydney, Australia.

Gibson, A.C. and Horak, K.E. 1978. Systematic anatomy and phylogeny of Mexican columnar cacti. *Annals of the Missouri Botanical Garden* 65:999–1057.

Gibson, A.C., Spencer, K.C., Bajaj, R., and McLaughlin, J.L. 1986. The ever-changing landscape of cactus systematics. *Annals of the Missouri Botanical Garden* 73:532–555.

Gilbert, D.G. 1980. Dispersal of yeasts and bacteria by *Drosophila* in a temperate forest. *Oecologia* 46:135–137.

Hart, D.D. 1985. Grazing insects mediate algal interactions in a stream benthic community. *Oikos* 44:40–46.

Heed, W.B. 1978. Ecology and genetics of Sonoran Desert *Drosophila*. pp. 109–126 in Brussard, P.F. (editor), *Ecological Genetics: the Interface*. Springer-Verlag, New York.

Heed, W.B. 1982. The origin of *Drosophila* in the Sonoran Desert. pp. 65–80 in Barker, J.S.F. and Starmer, W.T. (editors), *Ecological Genetics and Evolution:*

The Cactus-Yeast-Drosophila *Model System*. Academic Press, Sydney, Australia.

Heed, W.B. and Mangan, R.L. 1986. Community ecology of the Sonoran Desert *Drosophila*. pp. 311–345 in Ashburner, M., Carson, H.L., and Thompson, J.N., Jr. (editors), *The Genetics and Biology of* Drosophila, vol. 3e. Academic Press, New York.

Heed, W.B., Starmer, W.T., Miranda, M., Miller, M.W., and Phaff, H.J. 1976. An analysis of the yeast flora associated with cactiphilic *Drosophila* and their host plants in the Sonoran desert and its relation to temperate and tropical associations. *Ecology* 57:151–160.

Heithaus, E.R., Culver, D.C., and Beattie, A.J. 1980. Models of some ant-plant mutualisms. *American Naturalist* 116:347–361.

Johnston, J.S. and Heed, W.B. 1976. Dispersal of desert-adapted *Drosophila*: The saguaro-breeding *D. nigrospiracula*. *American Naturalist* 110:629–651.

Kircher, H.W. 1969. The distribution of sterols, alkaloids, and fatty acids in senita cactus, *Lophocereus schottii*, over its range in Sonora, Mexico. *Phytochemistry* 8:1481–1488.

Kircher, H.W. 1977. Triterpene glycosides and queretaroic acid in organ pipe cactus. *Phytochemistry* 16:1078–1080.

Kircher, H.W. 1982. Chemical composition of cacti and its relationship to Sonoran Desert *Drosophila*. pp. 143–158 in Barker, J.S.F. and Starmer, W.T. (editors), *Ecological Genetics and Evolution: The Cactus-Yeast*-Drosophila *Model System*. Academic Press, Sydney, Australia.

Kircher, H.W., and Bird, H.L. 1982. Five 3(β), 6(α)-dihydroxysterols in organ pipe cactus. *Phytochemistry* 21:1705–1710.

Kircher, H.W., Heed, W.B., Russell, J.S., and Grove, J. 1967. Senita cactus alkaloids: Their significance to Sonoran Desert ecology. *Journal of Insect Physiology* 13:1869–1874.

Lachance, M.A. and Starmer, W.T. 1982. Evolutionary significance of physiological relationships among yeast communities associated with trees. *Canadian Journal of Botany* 60:285–293.

Lachance, M.A., Starmer, W.T., and Phaff, H.J. 1988. Identification of yeasts found in decaying cactus tissue. *Canadian Journal of Microbiology* 34:1025–1036.

Mangan, R.L. 1982. Adaptations to competition in cactus breeding *Drosophila*. pp. 257–272 in Barker, J.S.F. and Starmer, W.T. (editors), *Ecological Genetics and Evolution: the Cactus-Yeast*-Drosophila *Model System*. Academic Press, Sydney, Australia.

May, R.M. 1974. *Stability and Complexity in Model Ecosystems*. Princeton, New Jersey. 265 pp.

McBrien, H., Harmsen, R., and Crowder, A. 1983. A case of insect grazing affecting plant succession. *Ecology* 64:1035–1039.

Phaff, H.J. and Starmer, W.T. 1987. Yeasts associated with plants, insects and soil. pp. 123–179 in Rose, A.H. and Harrison, J.S. (editors), *The Yeasts*, vol. 1, 2nd ed. Academic Press, London.

Sang, J.H. 1978. The nutritional requirements of *Drosophila*. pp. 159–192 in Ashburner, M. and Wright, T.R.F. (editors), *The Genetics and Biology of* Drosophila, vol. 2, Academic Press, New York.

Starmer, W.T. 1981. A comparison of *Drosophila* habitats according to the physiological attributes of the associated yeast communities. *Evolution* 35:35–52.

Starmer, W.T. 1982a. Analysis of the community structure of yeasts associated with the decaying stems of cactus. I. *Stenocereus gummosus*. *Microbial Ecology* 8:71–81.

Starmer, W.T. 1982b. Associations and interactions among yeasts, *Drosophila* and their habitats. pp. 159–174 in Barker, J.S.F. and Starmer, W.T. (editors),

*Ecological Genetics and Evolution: The Cactus-Yeast-*Drosophila *Model System*. Academic Press, Sydney, Australia.

Starmer, W.T. and Aberdeen, V. 1990. The nutritional importance of pure and mixed cultures of yeasts in the development of *Drosophila mulleri* larvae in *Opuntia* tissues and its relationship to host plant shifts. pp. 145–160 in Barker, J.S.F., Starmer, W.T., and MacIntyre, R.J. (editors), *The Ecological and Evolutionary Genetics of* Drosophila. Plenum Publishing Company, New York.

Starmer, W.T. and Barker, J.S.F. 1986. Ecological genetics of the *Adh*-1 locus of *Drosophila buzzatii*. *Biological Journal of the Linnean Society* 28:373–385.

Starmer, W.T., and Fogleman, J.C. 1986. Coadaptation of *Drosophila* and yeasts in their natural habitat. *Journal of Chemical Ecology* 12:1037–1055.

Starmer W.T., and Phaff, H. J. 1983. Analysis of the community structure of yeasts associated with the decaying stems of cactus. II. *Opuntia* species. *Microbial Ecology* 9:247–259.

Starmer, W.T., Kircher, H.W., and Phaff, H.J. 1980. Evolution and speciation of host plant specific yeasts. *Evolution* 34:137–146.

Starmer, W.T., Ganter, P., and Phaff, H.J. 1986. Quantum and continuous evolution of the DNA base composition in the yeast genus *Pichia*. *Evolution* 40:1263–1274.

Starmer, W.T., Ganter, P., Aberdeen, V., Lachance, M.A., and Phaff, H.J. 1987a. The ecological role of killer yeasts in natural communities of yeasts. *Canadian Journal of Microbiology* 33:783–796.

Starmer, W.T., Lachance, M.A., and Phaff, H.J. 1987b. A comparison of yeast communities found in necrotic tissue of cladodes and fruits of *Opuntia stricta* on islands in the Caribbean Sea and where introduced into Australia. *Microbial Ecology* 14:179–192.

Starmer, W.T., Phaff, H.J., Bowles, J.M., and Lachance, M.A. 1988a. Yeasts vectored by insects feeding on decaying saguaro cactus. *Southwestern Naturalist* 33:362–363.

Starmer, W.T., Aberdeen, V., and Lachance, M.A. 1988b. The yeast community associated with decaying *Opuntia stricta* (Haworth) in Florida with regard to the moth, *Cactoblastis cactorum* (Berg). *Florida Scientist* 51:7–11.

Starmer, W.T., Lachance, M.A., Phaff, H.J., and Heed, W.B. 1990. The biogeography of yeasts associated with decaying cactus tissue in North America, the Caribbean, and Northern Venezuela. *Evolutionary Biology* 24:253–296.

Swift, M.J. 1976. Species diversity and the structure of microbial communities in terrestrial habitats. pp. 185–222 in Anderson, J.M. and Macfadyen, A. (editors), *The Role of Terrestrial and Aquatic Organisms in Decomposition Processes*. Blackwell, Oxford.

Throckmorton, L.H. 1982. Pathways of evolution in the genus *Drosophila* and the founding of the Repleta group. pp. 33–47 in Barker, J.S.F. and Starmer, W.T. (editors), *Ecological Genetics and Evolution: The Cactus-Yeast-*Drosophila *Model System*. Academic Press, Sydney, Australia.

Tipper, D.J., and Bostian, K.A. 1984. Double-stranded ribonucleic acid killer systems in yeasts. *Microbiological Reviews* 48:125–156.

Vacek, D.C. 1982. Interactions between microorganisms and cactophilic *Drosophila* in Australia. pp. 175–190 in Barker, J.S.F. and Starmer, W.T. (editors), *Ecological Genetics and Evolution: The Cactus-Yeast-*Drosophila *Model System*. Academic Press, Sydney, Australia.

Wagner, R.P. 1944. The nutrition of *Drosophila mulleri* and *D. aldrichi*. Growth of the larvae on a cactus extract and the microorganisms found in the cactus. *University of Texas Publications* 4445:104–128.

Wagner, R.P. 1949. Nutritional differences in the Mulleri group. *University of*

Texas Publications 4920:39–41.

Wickner, R.B. 1981. Killer systems in *Saccharomyces cerevisiae*. pp. 415–444 in Strathen, J.N. (editor), *Molecular Biology of the Yeast* Saccharomyces cerevisiae: *Life Cycle and Inheritance*. Cold Spring Harbor Press, Cold Spring Harbor Laboratory, New York.

Wolin, C.L. and Lawlor, L.R. 1984. Models of facultative mutualism: Density effects. *American Naturalist* 124:843–862.

9

Fungal Endophytes of Tree Leaves

Orlando Petrini

9.1 Introduction

The term epiphyte is used in general to characterize organisms that subsist only on plant surfaces. De Bary's (1866) definition of endophytes—all the organisms that colonize internal plant tissues—was also used by Petrini (1986). In the same volume, Carroll (1986) restricted the use of the term endophyte to organisms that cause asymptomatic infections within plant tissues, excluding pathogenic fungi and mutualists such as mycorrhizal fungi. In view of the additional findings discussed in this chapter, however, I feel that this definition is no longer tenable. I propose, therefore, that Carroll's definition be expanded to include all organisms inhabiting plant organs that at some time in their life, can colonize internal plant tissues without causing apparent harm to their host. This would account for those endophytic organisms that have a more or less lengthy epiphytic phase and also for latent pathogens that may live symptomless in their hosts for some time in their life.

In the last 30 years, the ecology of phyllosphere-inhabiting microorganisms has been extensively studied. All these investigations, however, were centered for a long time on the epiphytic components of the phyllosphere community, and extensive reviews on fungal epiphytes of plant leaves have been published (Dickinson, 1976, 1981; see also Fokkema and van den Heuvel, 1986).

Reports on the presence of fungal endophytes in grass leaves were published more than 60 years ago (Lewis, 1924). Sampson (1935) and Neill (1940) described endophytic fungi in *Festuca rubra* and *Lolium* spp. Schüepp (1961) later gave evidence of endophytism for *Guignardia citricarpa* Kiely in *Citrus* spp. However, this relationship between fungi and plants was not clear until 1977 when Bernstein and Carroll showed conclusively that needles of *Pseudotsuga menziesii* (Mirb.) Franco can harbor endophytes that apparently cause no effect to their host. Since then, symptomless fungal endophytes have been discovered in the aerial plant tissues of at least 300 plant species, including ferns and mosses (summarized in Petrini, 1986; see also other references in this chapter) and have attracted the interest of taxonomists and ecologists. The presence of known pathogens among endophytes has prompted plant pathologists and epidemiologists to seriously reconsider the possibility that some plant pathogens are latent within their host tissues for a considerable period. Nevertheless, latency by some fungal pathogens had already been described in the early 1970s (Verhoeff, 1974). Epiphytes and, especially, endophytes, have also met with increasing interest by biotechnologists because of the intrinsic potential of genetically engineered endophytic organisms to serve as vectors of resistance and other genes to be introduced into economically important plant species (Wood, 1990).

In this review, I describe the recent developments of research in the ecology of endophytic fungi of tree leaves, their spatial and temporal relationship with some epiphytes, and certain physiological aspects of the host-endophyte symbiosis. I then discuss the histological evidence for tissue colonization by endophytic fungi and some issues related to their host and tissue specificity. Finally, I summarize the possible significance of epiphytes and endophytes for their hosts, and the potential application of these microbes in different areas of basic and applied research.

9.2 Ecology of Fungal Endophytes

Initial research on endophytic fungi concentrated mainly on their taxonomy. Detailed census work was published by several authors who isolated endophytes from a large number of phanerogams, ferns, and mosses (Petrini, 1986). Studies on ecological aspects of fungal endophytism have been published, and some authors have tried to formulate evolutionary hypotheses on the host-fungus symbiosis (Carroll, 1988; Clay, 1988).

Research on species composition of endophyte communities of several hosts has shown that generally a large number of fungal taxa can be recovered from the tissues of a single host species after surface sterilization (Butin, 1986; Petrini, 1986; Sieber-Canavesi and Sieber, 1988; Bertoni and Cabral, 1988; Sieber et al., 1988; Petrini et al., 1989). In all plant species investigated, however, a distinct pattern of dominance by only one or few fungal species can be detected. For instance, *Lophodermium piceae* (Fuckel) Höhn. is the most frequent endophyte of *Picea abies* Karst. (Butin, 1986; Sieber, 1988). Petrini et al. (1989) found that *Phyllosticta multicorniculata* Bissett et Palm, a fungus that Bissett and Palm (1989) associated with needle blight of its hosts, is the most important symptomless colonizer of balsam fir needles [*Abies balsamea* (L.) Mill.]. Leaves of beech (*Fagus sylvatica* L.) in Europe are mainly colonized by *Apiognomonia errabunda* (Rob.) Morelet

(Sieber and Hugentobler, 1987; Toti and Petrini, unpublished). Generally, species composition of endophyte communities varies in relation to the collection site (Sieber-Canavesi and Sieber, 1988; Petrini, 1987). Within each site, grouping of individual trees or shrubs can often be made according to their topographical position. For instance, Sieber-Canavesi and Sieber (1988) demonstrated that European white fir (*Abies alba* Mill.) trees growing at the same place can be grouped by their endophyte communities in distinct clusters correlated with their relative position in the field. Dominance of different fungal species in geographically separated locations is occasionally reported (Sieber-Canavesi and Sieber, 1988), although, as a rule, only one or two fungal taxa dominate a single host species, regardless of the geographical provenance of the samples (Petrini, 1986). Petrini (1987) observed that the degree of fungal colonization of alpine plants by endophytes is very much dependent on ecological conditions. High colonization rates by endophytes are positively correlated with long-lasting snow cover and poor nutrient supply for the host plant, and negatively with increasing wind exposure. Correspondingly, the number of endophyte species would be expected to be rather low in extreme climates and to follow a pattern of site specificity. Other studies dealing with endophyte communities of twigs have yielded similar results (Sieber, 1989a; Sieber et al., 1991a).

Endophytic communities of trees of the same species growing at the same location are generally similar, but marked differences in species richness and distribution of selected fungal species can be detected between young and old individuals of the same host (Espinosa-Garcia and Langenheim, 1990). Higher colonization rates by endophytic fungi can be observed for samples from homogeneous stands with a closed canopy (Petrini and Carroll, 1981; Petrini et al., 1982; Legault et al., 1989a). There may be a correlation between elevation above sea level and/or humidity and moisture and infection by a given endophyte. Sieber (1988) showed a tendency by *Tiarosporella parca* (Berk. et Broome) Whitney et al. to colonize needles of Norway spruce (*Picea abies*) growing at low elevation in rather humid sites, while *Lophodermium piceae* is apparently more frequent at drier sites situated at higher elevations.

Several studies have focused on the spatial and temporal patterns of endophytes within the same tree. In general, a correlation between increase in species richness and/or frequency of colonization and the age of the tissues has been demonstrated (Petrini and Carroll, 1981; Espinosa-Garcia and Langenheim, 1990). Usually, distinct endophyte communities characterize leaf samples of different age classes; changes in diversity of the endophytic associations are mostly correlated with the differential distribution of some species in young versus old leaves (Sieber and Hugentobler, 1987; Sieber-Canavesi and Sieber, 1988; Espinosa-Garcia and Langenheim, 1990). There is a tendency for a marked increase of overall colonization rates and changes in species diversity with increasing height of the canopy; no significant differences, on the other hand, have been found for leaf and twig samples taken from different compass directions on the same tree (Johnson and Whitney, 1989a; Legault et al., 1989a; Fisher and Petrini, 1990a; Petrini and Cuny, unpublished).

Studies on seasonal changes of the endophyte populations and communities have yielded partially contrasting results. Widler and Müller (1984) described a seasonal pattern in the occurrence of some endophytes of *Arctostaphylos uva-ursi* (L.) Sprengel, but the species undergoing seasonal variation were never host-specific, frequently observed taxa. Sieber and Hugentobler (1987) detected no distinct seasonal patterns in the species composition of endophyte communities in beech leaves; they have reported, however, marked seasonal variations in colonization rates by *Apiogomonia errabunda*, *Diaporthe eres* Nitschke, and *Bisporella* sp., the three most frequent endophytes.

Canavesi (1987) demonstrated a succession of microorganisms within European white fir needles. Following the comparatively stable and uniform endophyte community characteristic of symptomless green needles, a second group of fungal taxa, including some species known to be also epiphytic, colonize aging and brown needles still attached to the twigs. Within four to six weeks after needle fall, this community is replaced completely by saprobes typical of the fungal associations described by Gourbière (1974, 1980) for conifer litter. The fungal communities of green, symptomless needles can thus be considered in an ecological species equilibrium as long as no senescence process starts. Senescing tissues gradually allow the establishment of a new community, composed mainly of saprobes. Results of subsequent investigations on needles of *Sequoia sempervirens* (Endl.) Del Don (Espinosa-Garcia and Langenheim, 1990) confirm, at least partially, the results of Canavesi (1987).

The effect of anthropogenic modifications of the ecosystem on endophyte populations and communities has not yet been fully established. Sieber-Canavesi and Sieber (1988) reported that species composition of the endophyte population of white fir needles and the relative occurrence of each fungal species is dependent on the forest management type applied to the site studied. Clear-cutting and plantations tend to reduce the transmission of endophytic fungi: In one site, where European white fir is propagating spontaneously, colonization of needles by endophytes was more frequent than in plantations. Air pollutants apparently also affect species composition and frequency of colonization by a given species in endophyte communities. Sieber and Hugentobler (1987) and Sieber (1988, 1989a) detected differences in endophyte associations between healthy and diseased trees in beech and Norway spruce, at least partly confirming the findings by Barklund and Rowe (1983) and supporting the hypothesis that forests and fungal communities are equally affected by pollutants. Riesen and Close (1987) showed that endophyte colonization of barley leaves is significantly reduced after propiconazole treatment of the barley fields. In their opinion, the elimination by this treatment of latent pathogens growing symptomless as endophytes could explain unexpected yield increase after treatment of an apparently healthy crop with a fungicide.

Epiphytes and endophytes occupy two distinct niches in the plant and are involved in different physiological and ecological events peculiar to the phyllosphere ecosystem. Epiphytic fungi are exposed to rapid changes of the external environment, whilst endophytes have to protect themselves against defense reactions by the living plant cells. Both groups of organisms play probably an equally important role in the phyllosphere ecosystem. So far, investigations on phyllosphere organisms have been exclusively centered either on the epiphytic (e.g., Kinkel et al., 1987,1989; Helander and Rantio-Lehtimäki, 1990) or the endophytic communities. To my knowledge, there is only one published study carried out on both communities with the same model and at the same time (Legault et al., 1989a, 1989b). For this reason, only limited comparisons of the taxonomy, physiology, and ecology of both communities are possible.

9.3 Evidence for Endophytic Colonization

The efficacy of the sterilization procedures used to eliminate spores or mycelial fragments attached to the surface of the plant material investigated has been assessed several times (Petrini, 1986). Thus, it is likely that only in rare cases will spores or hyphal fragments attached to the surface survive surface sterilization. Among the fungi reported as endophytes by several authors, however, a number of taxa known to be frequent epiphytes are able to live endophytically within plant tissues (Petrini, 1986). O'Donnell and Dickinson (1980) have already demonstrated that some epiphytes may, under appropriate ecological and climatical conditions, penetrate into the host tissue. For instance, *Alternaria alternata* (Fr.) Keissler, *Cladosporium cladosporioides* (Fr.) De Vries, and *C. herbarum* (Pers.) Link ex S.F. Gray are phylloplane colonizers able to penetrate into living leaf tissues at the onset of the senescence process. *Epicoccum purpurascens* Ehrenb. ex Schlecht. may behave in a similar way (Dickinson, 1981). The senescence process apparently modifies the ecological niche provided by the plant tissues to "true" endophytes and allows the development of organisms that are usually better adapted to saprotrophic life. Switching from epiphytic to endophytic life may enable epiphytic fungi to protect themselves from adverse ecological conditions (desiccation, irradiation) and at the same time avoid antagonistic activities by other more competitive surface organisms. It could also allow these primary saprobes to rapidly colonize substantial volumes of tissue (Dickinson, 1976). Verhoeff (1974) points out that some fungi living epiphytically are sometimes able to form subcuticular appressoria or hyphae which could survive surface sterilization. This, too, could explain the occasional occurrence of some common epiphytes in the inner tissues of plants.

The work carried out by Legault et al. (1989a, 1989b) demonstrated that the epiphytic and the endophytic communities of needles of *Pinus banksiana* Lamb. and *P. resinosa* Ait. differ in species composition. The most frequently isolated epiphytic species are common, primary saprobes found on the surface of a large number of higher plants. Endophytes of the same hosts, on the other hand, are rather host specific: it is possible that most species which usually live epiphytically but can at times penetrate into their host tissues have developed some kind of host specificity at the strain level and that only these strains will be able to successfully colonize the interior of their hosts. Some preliminary experiments with *Cladosporium cladosporioides* (Petrini, unpublished) point in this direction, but too little information is available to confirm the validity of this hypothesis.

The results of investigations by Canavesi (1987), Legault et al. (1989a, 1989b), and by Cabral (1985) support the hypothesis that certain phylloplane fungi will colonize the interior of plant organs only when the frequency of true endophytic species decreases after the onset of tissue senescence. Facultative endophytes, thus, are probably fungal species mainly involved in primary litter decomposition and are likely to be found more frequently in older tissues.

Colonization of leaves by epiphytic organisms is generally thought to occur rapidly by fungal air spora. Kinkel et al. (1987, 1989), for instance, consider air spora to be the major determinant of the composition and quantity of fungal immigrants on apple leaves. Air spora, on the other hand, cannot be considered the only mode of fungal endophyte immigration in tree leaves. Thus far, only limited information is available on the colonization of plant tissues by endophytes. However, at least three different mechanisms can be proposed: (1) infection of aerial plant tissues by air spora is certainly a widespread mechanism of infection for a number of endophytes that sporulate readily on their hosts or host remains. This may apply, for instance, to *Apiognomonia errabunda*, although this fungus is also likely to be transmitted with seeds (L. Toti, O. Viret, and O. Petrini, unpublished). (2) Bose (1947) and Bloomberg (1966) showed that some endophytes of *Casuarina equisetifolia* and *Pseudotsuga menziesii* are seedborne; the same has been demonstrated for a number of grass endophytes (Clay, 1986). (3) Insects can also be carriers of endophyte inoculum. Such insect-endophyte associations have not yet been demonstrated conclusively, but the results of some investigations (Carroll, 1986; Petrini et al., 1989) strongly suggest close fungus-insect relationships, similar to those known for some tree pathogens, e.g., *Ceratocystis ulmi* (Sinclair et al., 1987).

Asymptomatic fungal colonization of tree tissues has been demonstrated by histological investigations only in a limited number of cases (Stone, 1987, 1988; Johnson and Whitney, 1989b; Suske and Acker, 1989). Stone (1987, 1988) showed that *Rhabdocline parkeri* Sherwood forms intracellular infections, which are apparently confined to a single epidermal cell until the onset of needle senescence. The fungus infects healthy foliage by direct penetration of the host cell wall with a very thin penetration hypha. Haustoria are produced from intracellular hyphae only at the onset of needle senescence. Endophytic infection of Norway spruce by *Lophodermium piceae* has been studied in detail by Suske and Acker (1989) using immunological methods. Colonization is carried out by only a few hyphae, often containing endocells, located in the subepidermal needle tissues. Thus, confinement of infection loci to small tissue areas or the presence of subcuticular analogues of the structures described by Muirhead and Deverall (1981) for some plant pathogens may be the reason why endophytic infections are frequently symptomless.

Recently I.H. Chapela (unpublished) observed that endophytic stages of the wood-inhabiting endophyte *Hypoxylon fragiforme* (Pers.: Fr.) Kickx are initiated from unicellular, multinucleate bodies within the ray parenchyma cells. The hyphal initials are extremely thin, extending throughout ray tissues before invading the vessel lumina where they produce irregularly swollen hyphae, which eventually give rise to regular hyphae. Chapela postulates that the fungal initials observed in beech ray parenchyma could be composites of endophyte and plant material, possibly, fungal protoplasts included within plant cells, from which nuclei can migrate into hyphae as soon as mycelial growth has been triggered by the appropriate ecological conditions. Plant organelles or even protoplasts are frequently found in hyphal cells of *Colletotrichum lindemuthianum* (Sacc. et Magn.) Briosi et Cav. (O'Connell et al., 1984; O'Connell, 1987). Recently, Lynch et al. (1989) succeeded in inducing the uptake of fungal protoplasts by celery (*Apium graveolens* L.) protoplasts. The formation of such fungus-plant chimeras may also occur in nature, probably at low frequency. This kind of tissue colonization by endophytic fungi may be more widespread than so far realized and may explain why sometimes it is so difficult to visualize endophyte infections in plant tissues.

Stone (1987) provided estimates of infection frequencies by *Rhabdocline parkeri*, which indicate that more than 15 infection loci per mm^2 can be present in a Douglas fir needle, and has postulated that several genotypes of the same species may be present within the same needle. Analysis of vegetative-compatibility-group systems with isolates of *Apiognomonia errabunda* from the same and from different beech leaves (Hämmerli and Petrini, unpublished) fully supports this hypothesis. The significance of this kind of distribution gives some clues toward the understanding of the fungus-fungus and fungus-host interaction. Endophytes may avoid substrate competition or antagonism by occupying tight compartments within the same tissue; on the other hand, the host plant does not react to its endophytes in the same way as it would react to infections by pathogenic organisms, where the activation of defense mechanisms would tend to prevent subsequent infections. This implies a complex host-fungus recognition mechanism that allows only the symbiont partners to penetrate into the host tissues without activating the defense barriers usually erected by the host against pathogens (Darvill and Albersheim, 1984). Such a mechanism, indicative of a close coevolution between the endophyte and its host, would further support the hypothesis of the long-term existence of endophyte-host symbioses (Carroll, 1988; Clay, 1988).

9.4 Physiology of Endophytes

The production of extracellular enzymes involved in infection and penetration processes (Collmer and Keen, 1986) necessary for the degradation of selected plant cell components is a common trait of endophytic fungi. Amylases, cellulases, esterases, lipases, ligninases, pectinases, and proteases have been detected in many endophytic species by substrate-utilization tests and isozyme analysis (Carroll and Petrini, 1983; Sieber, 1989b; Petrini et al., 1991; Sieber et al., 1991b). Substrate-utilization capacity may vary among conspecific isolates but, in general, consistent patterns are seen within species. Conspecific isolates derived from the same host are remarkably homogeneous with respect to their enzymatic activities (A. Leuchtmann, O. Petrini, L. Petrini and G.C. Carroll, unpublished; L. Toti and O. Petrini, unpublished). This observation supports the possibility that host-specific strain formation within the same fungal species may take place (Petrini et al., 1982; Petrini, 1986). At the same time, the lack of variability may be indicative of the age-old nature of the symbiosis.

Production of indole-3-acetic acid (IAA) and indole-3-acetonitrile has been demonstrated in vitro in a number of fungal species, including *Aureobasidium pullulans* (De Bary) Arnaud, *Epicoccum purpurascens* (Pugh, 1972), and endophytic strains of *Balansia* spp. (Porter et al., 1985). Strains of endophytes tested by Bergamin-Strotz (1988) have the ability to produce IAA and other auxins in vitro; in some cases, production of cytokinins has been also assessed. The influence of selected strains of *E. purpurascens* and *A. pullulans* on seed germination has been tested in separate experiments; in both cases, culture filtrates of these organisms have affected seed germination either positively (in ivy; Luginbühl and Müller, 1982) or negatively (in wheat; Sieber, 1985). Investigations by Bergamin-Strotz (1988) showed that one endophytic strain of *Hypoxylon serpens* isolated from tobacco stimulated the flowering process of its host, while other strains of the same species induced wilting and inhibited the growth of tobacco seedlings.

These results provide several clues about the ecological significance of endophytic fungi. The potential of endophytic fungi to produce extracellular enzymes needed for cell-wall degradation supports the hypothesis by Carroll (1988) that fungal endophytes of aerial plant tissues represent a highly specialized group of organisms. The general tolerance of endophytes to gallic acid (Carroll and Petrini, 1983) suggests their widespread ability to cope with phenolic compounds normally present in the plant tissues. The generation of substances that could somehow influence the growth of the host plant points to the close symbiotic relationship which has evolved between selected endophytes and their hosts. The capacity to produce hormones involved in senescence processes suggests that endophytes may play a role in the plant senescence process, where epiphytic organisms evidently are involved (Jachmann and Fehrmann, 1989).

Production of extracellular enzymes responsible for plant infection and colonization as well as of auxin-like growth factors by endophytes indicates that these fungi are likely to have undergone a close co-evolution with their host and have adapted to their environment, a kind of situation already known in plant-pathogenic fungi (e.g., Albersheim and Valent, 1974). On the other hand, reports of symptomless infections by a number of fungal plant pathogens (Cerkauskas and Sinclair, 1980; Nathaniels and Taylor, 1983; Kulik, 1984; Cerkauskas, 1988; Huang, 1989; Petrini et al., 1989) suggest that the host-endophyte symbiosis cannot be represented conceptually as a static relationship. Carroll (1988) noticed that in some situations the distinction between latent pathogen and endophyte is not clear. Apparently, symptomless endophytes may become symptomatic when the host is stressed (Millar, 1980; Andrews et al., 1985). For instance, Minter and Ingham (1984) reported leaf spotting on *Chrysanthemum* associated with ejected ascospores of a species of *Podospora*, an otherwise nonpathogenic, saprobic fungus. *Apiognomonia errabunda* produces symptoms only on excised leaves of beech and, only rarely, minor symptoms on nondetached leaves (L. Toti, O. Viret, and O. Petrini, unpublished). It can be expected that in some cases the same individual undergoes one or more physiological alterations during its lifetime that switch it from one type of symbiosis to the other

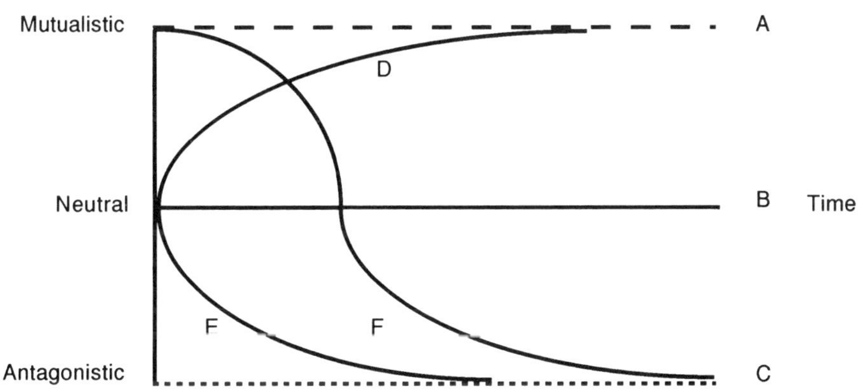

Figure 9.1 Hypothetical changes in physiological states of an endophytic fungus over time. A, The mutualistic state; B, neutral; C, antagonistic during the whole lifetime; D, the transition from neutral to mutualistic; E, from neutral to antagonistic; and F, from mutualistic to antagonistic. The detailed shapes of the curves and the choice of the time intervals are arbitrary.

(Figure 9.1). Changes in host susceptibility caused by the presence or absence of stress factors, such as drought and poor nutrient supply, will induce the transition from one state to the other (e.g., from mutualistic to antagonistic, or from neutral to mutualistic) and at the same time will determine the duration of the period during which an individual will remain in the same symbiotic state.

9.5 Host Organ and Tissue Specificity of Endophytic Fungi

Some endophytes may be specific only at the family level of the host plant (Petrini, 1986). If different species within the same plant family are present at the same site, colonization of all species by the same endophyte would occur. However, host specificity in endophytic fungi is probably much narrower than previously thought. In particular, the formation of host-specific strains in fungal endophytes that can infect different hosts, postulated by Petrini et al. (1982) for a species of *Septogloeum* and known for a number of plant-pathogenic species (e.g., *Fusarium*; see Booth, 1971), has now been proved.

Leuchtmann and Clay (1989) demonstrated, by isozyme electrophoresis, considerable variation within isolates of *Atkinsonella hypoxylon* (Peck) Diehl from different hosts, and suggested that this variation may be correlated with the ability of a given strain to infect a particular host. Similar results have been obtained in my laboratory with conifer-inhabiting *Phyllosticta* spp. and with strains of *Apiognomonia errabunda* isolated from different hosts. Analysis of wood-inhabiting *Melanconium* spp. (Sieber et al., 1991b) isolated from different species of Alder (*Alnus* spp.) in Europe and Canada and of *Leptostroma* spp. from conifer needles (Sieber-Canavesi et al., 1991) yielded the same results.

Early stages of fungal infection in any host-symbiont interaction include germination of the spores, attachment of the parasite to the host (prior to or after germination) and, frequently, penetration of the host cuticle by the infection hyphae. We have described a host-mediated recognition mechanism in *Hypoxylon fragiforme* (Chapela et al., 1990), where the activation of ascospores, a necessary initial step of germination resulting in the emergence of the cell body from a rigid outer shell formed by a differentiated transparent wall layer, is mediated by two monolignols, faguside and syringoside (I.H. Chapela and O. Petrini, unpublished). We looked for a similar mechanism in *A. errabunda*, but germination of *A. errabunda* conidia is apparently not dependent on any host-specific stimulus. Therefore, we looked for host-mediated recognition in this fungus at other stages of the infection process. We applied a fixed quantity of fresh *A. errabunda* conidia to the leaf surface of *Fagus sylvatica* L., to a *Hordeum* sp., and to teflon, and we tried to remove them by subsequent washing of the treated surface

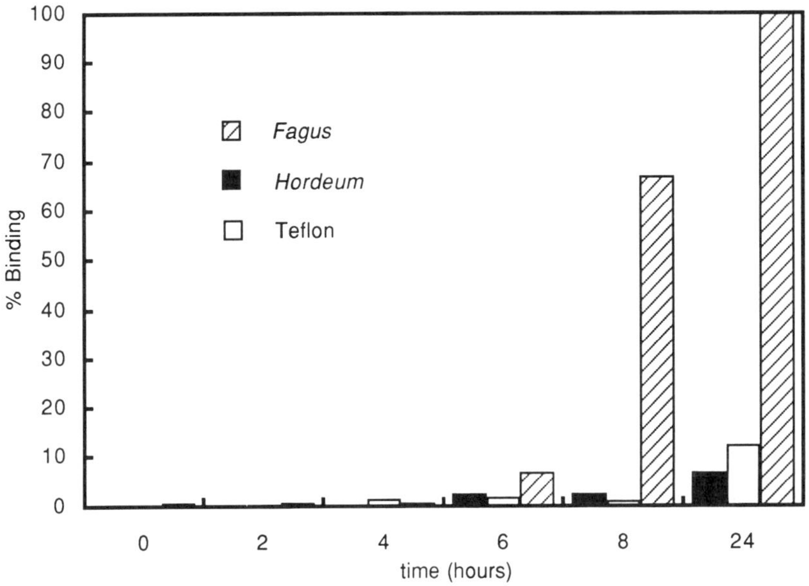

Figure 9.2 Results of preliminary binding experiments with conidia of *Apiognomonia errabunda,* carried out by spraying or applying fixed quantities of fresh *A. errabunda* conidia per leaf area to the leaf surface of *Fagus sylvatica* L., of *Hordeum* sp., and to teflon, with subsequent washing of the treated surface under running deionized water. The bars represent the percentage of conidia that have remained attached to the various surfaces. In the display, the number of conidia binding to *Fagus* leaves has been set to 100%. See text for details.

under running deionized water at a pressure of 0.5 Pa. After 8 hours of incubation, a high proportion of fresh conidia on beech leaves resists removal, whereas only a negligible amount remains attached to *Hordeum* or teflon (Figure 9.2). Furthermore, while conidia of strains isolated from beech germinate readily on a large number of solid substrates, only beech leaves can be infected by the fungus. Other known hosts such as *Castanea sativa* L. or nonhost plants (*Corylus avellana* L., *Carex* sp.) cannot be successfully infected (L. Toti, O. Viret, I.H. Chapela, and O. Petrini, in preparation). So far we cannot explain the reasons for this differential attachment and infection; electron microscopy observations of the germinating conidia suggest a mechanism similar to that described for ascospores of *Venturia inaequalis* (Cke) Winter (Smereka et al., 1987), the rice blast fungus, *Magnaporthe grisea* Barr (Hamer et al., 1988); and the endoparasitic nematophagous fungus *Drechmeria coniospora* W. Gams et Jansson (Dijksterhuis et al., 1990). This evidence, supported by similar examples from the phytopathological literature (Carver and Ingerson, 1987; Carver and Adaigbe, 1990), suggests that differential recognition mechanisms may be widespread among endophytic fungi. Organ specificity by endophytic fungi was first demonstrated by Sieber (1985) for wheat endophytes, and again by Sieber (1988, 1989a) for endophytes of Norway spruce. Unpublished results from my laboratory with beech leaves, twigs, and roots confirm that the fungal community structures of different plant organs are very specific.

Carroll et al. (1977) first reported some tissue specificity by endophytes. Many of the fungi isolated from the petiole segment of coniferous hosts in Europe were restricted to that part and were rarely detected in more distal portions of the needle. Carroll and Carroll (1978) and Petrini and Müller (1979) came to similar conclusions. Stone (1986) showed that *Rhabdocline parkeri* and *Phyllosticta* sp. coexist in needles of *Pseudotsuga menziesii*: *R. parkeri* is confined to epidermal and hypodermal cells and *Phyllosticta* sp. occurs intercellularly in the mesophyll. Subsequent investigations on endophytes of broad-leaved trees (Bertoni and Cabral, 1988) and of wood (Fisher and Petrini, 1987, 1988; Petrini and Fisher, 1988; Fisher and Petrini, 1990a, 1990b) confirmed that many endophytic fungi show a certain degree of tissue specificity.

Tissue and organ specificity, combined with the development of specialized food preferences (Carroll and Petrini, 1983; Sieber, 1989b), may be a mechanism to prevent niche overlap and competition among endophytes occupying the same physical habitat. Alternatively, the specificity may be a particular response of the fungus to its environment.

9.6 The Future of Endophyte and Epiphyte Research in Agriculture and Biology

The presence of well-established epiphytic and endophytic fungal communities on virtually all plant species has triggered investigations of their ecology in the fields of both basic and applied research. Our knowledge of the ecological and physiological aspects of these organisms, however, is still fragmentary, and, mainly, only the qualitative aspects of the symbiosis have been examined.

For instance, although much qualitative information is now available on the factors regulating the dynamics of phyllosphere organisms, almost nothing is known on the relative quantitative impact of each factor within the epiphytic and endophytic communities. Investigations aiming to elucidate the biochemical basis of the host-endophyte symbiosis are still lacking. Work is in progress in our laboratories (L. Toti, O. Viret, and O. Petrini) to investigate this aspect in the model system of *Fagus sylvatica* and *Apiognomonia errabunda*. Information about the novel, host-mediated, recognition mechanism described by Chapela et al. (1990), as well as about the host-specific binding of conidia of *A. errabunda* to beech leaves, is also very incomplete; additional research on the host specificity of endophytic fungi and other symbionts will add to our understanding of the regulation of infection mechanisms not only in mutualistic symbioses, but also in antagonistic ones. Work by Hamer et al. (1989), among many others, on *Magnaporthe grisea* showed that the breeding and cultivation of rice has favored host selection by specific pathogen genotypes; it can be expected, by analogy, that a similar situation occurs among endophytes and epiphytes.

Many endophytic fungi are able to produce antibiotic compounds in culture that are active against human- and plant-pathogenic bacteria (Fisher et al., 1984a, 1984b, 1986; Dreyfuss, 1987). Fisher et al. (1984a, 1984b) suggest that the primary function of antibiotic production may be competition against antagonists. If this is so, antibiotic substances should be produced by a large number of endophytes, thus making these organisms particularly suited for pharmacological screening (Dreyfuss, 1987) and for agricultural programs using fungi for the biological control of plant pests and diseases.

Barklund and Rowe (1983) found a negative correlation between endophyte infection and tree vitality. Sieber and Hugentobler (1987) and Sieber (1988, 1989a) presented results that are in partial contradiction with this. Helander and Rantio-Lehtimäki (1990), however, reported a significant decrease in the number of phylloplane organisms isolated from birch leaves after simulated acid rain treatment. In either case, the potential use of endophytic fungi as bioindicators of vitality is an issue that needs further investigations.

Clay (1989) discussed the potential of Clavicipitaceous endophytes of grasses as biocontrol agents. Some nonpathogenic endophytes from conifer needles, however, show features that could be used in the biocontrol of insect pests (Clark et al., 1989; Petrini et al., 1989) or as deterrents against subsequent invasion of plant tissues by pathogens, as suggested by Wicklow (1988) for maize kernel saprobes. Alternatively, highly host-specific and virulent strains of endophytes could be used as mycoherbicides in the biological control of weed (Cerkauskas, 1988; Chiang et al., 1989; Sieber et al., 1990).

Endophytic associations appear to offer the greatest potential for biocontrol, because the fungi are integrated into the host system (Clay, 1989). For the same reason, they also are ideally suited to be used for the introduction of foreign genes into plant tissues. Host-specific endophytes could be genetically manipulated to produce well-defined substances within their hosts—a kind of indirect genetic engineering of plants. The recent progresses in fungal transformation (Fincham, 1989; Rambosek and Leach, 1987) make the production of genetically engineered "endophytic bioinsecticides" an exciting, economically important, and, at the same time, realistic proposition. Dewan and Sivasithamparam (1988, 1989a, 1989b) reported that a sterile red fungus (SRF) isolated from wheat roots provides significant protection to the host from infection by the "take-all" fungus (*Gaeumannomyces graminis* Arx et Olivier var. *tritici* Walker). Their work has stimulated intensive research toward the use of endophytic microorganisms for biocontrol (Wood, 1990).

Drs. G.C. Carroll (Eugene, Oregon, USA), I.H. Chapela (Basel, Switzerland), P.J. Fisher (Exeter, England), A. Leuchtmann (Zürich, Switzerland), and L.E. Petrini (Zürich, Switzerland) offered useful criticism of preliminary versions of the manuscript. L. Toti and O. Viret disclosed some of their unpublished results and helped to perform some of the experiments described in this paper. The author thanks them all for their generous support.

References

Albersheim, P. and Valent, B. 1974. Host-pathogen interactions Vll. Plant pathogens secrete proteins which inhibit enzymes of the host capable of attacking the pathogen. *Plant Physiology* 53:684–687.

Andrews, J.H., Hecht, E.P., and Bashirian, S. 1985. Association between the fungus *Acremonium curvulum* and Eurasian water milfoil, *Myriophyllum spicatum*. *Canadian Journal of Botany* 60:1216–1221.

Barklund, P. and Rowe, J. 1983. Endophytic fungi in Norway spruce—possible use in bioindication of vitality. *Aquilo Series Botanicae* 19:228–232.

Bergamin-Strotz, L.M. 1988. *Floristische und ökophysiologische Aspekte im Zusammenleben von Tabakpflanzen (Nicotiana tabacum L. var. SR1) und endophytischen Pilzen*. Dissertation. ETH no. 8520, Swiss Federal Institute of Technology, Zürich, Switzerland.

Bernstein, M.E. and Carroll, G.C. 1977. Internal fungi in old-growth Douglas fir foliage. *Canadian Journal of Botany* 55:644–653.

Bertoni, M.D. and Cabral, D. 1988. Phyllosphere of *Eucalyptus viminalis*. II: distribution of endophytes. *Nova Hedwigia* 46:491–502.

Bissett, J. and Palm, M.E. 1989. Species of *Phyllosticta* on conifers. *Canadian Journal of Botany* 67:3378–3385.

Bloomberg, W. J. 1966. The occurrence of endophytic fungi in Douglas fir seedlings and seeds. *Canadian Journal of Botany* 44:413–420.

Booth, C. 1971. *The Genus Fusarium*. Commonwealth Mycological Institute, Kew, Surrey, 237 pp.

Bose, S.R. 1947. Hereditary (seed-borne) symbiosis in *Casuarina equisetifolia*. *Nature* (London) 159:512–514.

Butin, H. 1986. Endophytische Pilze in grünen Nadeln der Fichte (*Picea abies* Karst.). *Zeitschrift für Mykologie* 52:335–346.

Cabral, D. 1985. Phyllosphere of *Eucalyptus viminalis*. Dynamics of fungal populations. *Transactions of the British Mycological Society* 85:501–511.

Canavesi, F. 1987. *Beziehungen zwischen Endophytischen Pilzen von Abies alba Mill. und den Pilzen der Nadelstreue*. Dissertation. ETH no. 8325, Swiss Federal Institute of Technology, Zürich, Switzerland.

Carroll, F.E., Müller, E., and Sutton, B.C. 1977. Preliminary studies on the incidence of needle endophytes in some European conifers. *Sydowia* 29:87–103.

Carroll, G.C. 1986. The biology of endophytism in plants with particular reference to woody perennials. pp. 205–222 in Fokkema, N.J. and van den Heuvel, J. (editors). *Microbiology of the Phyllosphere*. Cambridge University Press, Cambridge, UK.

Carroll, G.C. 1988. Fungal endophytes in stems and leaves: from latent pathogen to mutualistic symbiont. *Ecology* 69:2–9.
Carroll, G.C. and Carroll, F.E.. 1978. Studies on the incidence of coniferous needle endophytes in the Pacific Northwest. *Canadian Journal of Botany* 56:3034–3043.
Carroll, G.C. and Petrini, O. 1983. Patterns of substrate utilization by some endophytes from coniferous foliage. *Mycologia* 75:53–63.
Carver, T.L.W. and Adaigbe, M.E. 1990. Effect of oat host genotype, leaf age and position and incubation humidity on germination and germling development by *Erysiphe graminis* f. sp. *avenae*. *Mycological Research* 94:18–26.
Carver, T.L.W. and Ingerson, S.M. 1987. Responses of *Erysiphe graminis* germlings to contact with artificial and host surfaces. *Physiological and Molecular Plant Pathology* 30:359–372.
Cerkauskas, R.F. 1988. Latent colonization by *Colletotrichum* spp.: epidemiological considerations and implications for mycoherbicides. *Canadian Journal of Plant Pathology* 10:297–310.
Cerkauskas, R.F. and Sinclair, J.B. 1980. Use of paraquat to aid detection of fungi in soybean tissues. *Phytopathology* 70:1036–1038.
Chapela, I.H., Petrini, O., and Petrini, L.E. 1990. Unusual ascospore germination in *Hypoxylon fragiforme*: first steps in the establishment of an endophytic symbiosis. *Canadian Journal of Botany* 68:2571–2575.
Chiang, M.-Y., van Dike, C.G., and Leonard, K.J. 1989. Evaluation of endemic foliar fungi for potential biological control of johnsongrass (*Sorghum halepense*): Screening and host range tests. *Plant Disease* 73:459–464.
Clark, C.L., Miller, J.D., and Whitney, N.J. 1989. Toxicity of conifer needle endophytes to spruce budworm. *Mycological Research* 93:508–512.
Clay, K. 1986. Grass endophytes. pp. 188–204 in Fokkema, N.J. and van den Heuvel, J. (editors), *Microbiology of the Phyllosphere*. Cambridge University Press, Cambridge, UK.
Clay, K. 1988. Clavicipitaceous fungal endophytes of grasses: coevolution and the change from parasitism to mutualism. pp. 79–105 in Pirozynski, K.A. and Hawksworth, D.L. (editors), *Coevolution of Fungi with Plants and Animals*. Academic Press, London.
Clay, K. 1989. Clavicipitaceous endophytes of grasses: their potential as biocontrol agents. *Mycological Research* 92:1–12.
Collmer, A. and Keen, N.T. 1986. The role of pectic enzymes in plant pathogenesis. *Annual Review of Phytopathology* 24:383–409.
Darvill, A.G. and Albersheim, P. 1984. Phytoalexins and their elicitors: a defense against microbial infection in plants. *Annual Review of Plant Physiology* 35:243–275.
De Bary, A. 1866. *Morphologie und Physiologie der Pilze. Flechten und Myxomyceten*. Engelmann, Leipzig.
Dewan, M.M. and Sivasithamparam. 1988. A plant growth-promoting sterile fungus from wheat and rye-grass roots with potential for suppressing take-all. *Transactions of the British Mycological Society* 91:687–717.
Dewan, M.M. and Sivasithamparam. 1989a. Growth promotion of rotation crop species by a sterile fungus from wheat and effect of soil temperature and water potential on its suppression of take-all. *Mycological Research* 93:156–160.
Dewan, M.M. and Sivasithamparam. 1989b. Behaviour of a plant growth-promoting sterile fungus on agar and roots of rye-grass and wheat. *Mycological Research* 93:161–166.
Dickinson, C.H. 1976. Fungi on the aerial surfaces of higher plants. pp. 293–324

in Preece, T.F. and Dickinson, C.H. (editors), *Microbiology of Aerial Plant Surfaces*. Academic Press, London.

Dickinson, C.H. 1981. Biology of *Alternaria alternata, Cladosporium cladosporioides* and *C. herbarum* in respect of their activity on green plants. pp. 169–184 in Blakeman, J.P. (editor), *Microbial Ecology of the Phylloplane*. Academic Press, London.

Dijksterhuis, J., Veenhuis, M., and Harder, W. 1990. Ultrastructural study of adhesion and initial stages of infection of nematodes by conidia of *Drechmeria coniospora*. *Mycological Research* 94:1–8.

Dreyfuss, M.M. 1987. Neue Erkenntnisse aus einem pharmakologischen Pilzscreening. *Sydowia* 39:22–36.

Espinosa-Garcia, F.J. and Langenheim, J.H. 1990. The leaf fungal endophytic community of a coastal redwood population. Diversity and spatial patterns. *New Phytologist* 116:89–98.

Fincham, J.R.S. 1989. Transformation in Fungi. *Microbiological Reviews* 53:148–170.

Fisher, P.J. and Petrini, O. 1987. Location of fungal endophytes in tissues of *Suaeda fruticosa*: a preliminary study. *Transactions of the British Mycological Society* 89:246–249.

Fisher, P.J. and Petrini, O. 1988. Tissue specificity by fungi endophytic in *Ulex europaeus*. *Sydowia* 40:46–50.

Fisher, P.J. and Petrini, O. 1990a. Occurrence of fungal endophytes in twigs of *Salix fragilis* and *Quercus robur*. *Mycological Research* 94:1077–1080.

Fisher, P.J. and Petrini, O. 1990b. A comparative study of fungal endophytes in xylem and bark of *Alnus* species in England and Switzerland. *Mycological Research* 94:313–319.

Fisher, P.J., Anson, A.E., and Petrini, O. 1984a. Antibiotic activity of some endophytic fungi from ericaceous plants. *Botanica Helvetica* 94:249–253.

Fisher, P.J., Anson, A.E., and Petrini, O. 1984b. Novel antibiotic activity of an endophyte *Cryptosporiopsis* sp. isolated from *Vaccinium myrtillus*. *Transactions of the British Mycological Society* 83:145–148.

Fisher, P.J., Anson, A.E., and Petrini, O. 1986. Fungal endophytes in *Ulex europaeus* and *Ulex gallii*. *Transactions of the British Mycological Society* 86:153–156.

Fokkema, N.J. and van den Heuvel, J. 1986. *Microbiology of the Phyllosphere*. Cambridge University Press, Cambridge, UK. 392 pp.

Gourbière, F. 1974. Les champignons microscopiques liés aux aiguilles du sapin (*Abies alba* Mill.). II. Variation saisonnière de la microflore des aiguilles tombantes. *Bulletin de la Société Mycologique de France* 90:325–333.

Gourbière, F. 1980. Les champignons microscopiques liés aux aiguilles du sapin (*Abies alba* Mill.). V. Synthèse des études précédentes. *Bulletin de la Société Mycologique de France* 96:35–42.

Hamer, J.E., Howard, R.J., Chumley, F.G., and Valent, B. 1988. A mechanism for surface attachment in spores of a plant pathogenic fungus. *Science* 239:288–290.

Hamer, J.E., Farrall, L., Orbach, M.J., Valent, B., and Chumley, F.G. 1989. Host species-specific conservation of a family of repeated DNA sequences in the genome of a fungal plant pathogen. *Proceedings of the National Academy of Sciences USA* 86:9981–9985.

Helander, M.L. and Rantio-Lehtimäki. 1990. Effects of watering and simulated acid rain on quantity of phyllosphere fungi of Birch leaves. *Microbial Ecology* 19:119–125.

Huang, H.C. 1989. Distribution of *Verticillium albo-atrum* in symptomed and symptomless leaflets of alfalfa. *Canadian Journal of Plant Pathology* 11:235–241.

Jachmann, H.-Th. and Fehrmann, H. 1989. Effects of phyllosphere microorganisms on the senescence of wheat leaves. *Zeitschrift für Pflanzenkrankheiten und Pflanzenschutz* 96:124–133.

Johnson, J.A. and Whitney, N.J. 1989a. An investigation of needle endophyte colonization patterns with respect to height and compass direction in a single crown of balsam fir (*Abies balsamea*). *Canadian Journal of Botany* 67:723–725.

Johnson, J.A. and Whitney, N.J. 1989b. A study of fungal endophytes of needles of balsam fir (*Abies balsamea*) and red spruce (*Picea rubens*) in New Brunswick, Canada, using culture and electron microscope techniques. *Canadian Journal of Botany* 67:3513–3516.

Kinkel, L.L., Andrews, J.H., Berbee, F.M., and Nordheim, E.V. 1987. Leaves as island for microbes. *Oecologia* 71:405–408.

Kinkel, L.L., Andrews, J.H., and Nordheim, E.V. 1989. Fungal immigration dynamics and community development on apple leaves. *Microbial Ecology* 18:45–58.

Kulik, M.M. 1984. Symptomless infection, persistence, and production of pycnidia in host and non-host plants by *Phomopsis batatae*, *Phomopsis phaseoli*, and *Phomopsis sojae*, and the taxonomic implications. *Mycologia* 76:274–291.

Legault, D., Dessureault, M., and Laflamme, G. 1989a. Mycoflore des aiguilles de *Pinus banksiana* et *Pinus resinosa*. I. Champignons endophytes. *Canadian Journal of Botany* 67:2052–2060.

Legault. D., Dessureault, M., and Laflamme, G. 1989b. Mycoflora of *Pinus banksiana* and *Pinus resinosa* needles. II. Epiphytic fungi. *Canadian Journal of Botany* 67:2061–2065

Leuchtmann, A. and Clay, K. 1989. Isozyme variation in the fungus *Atkinsoniella hypoxylon* within and among populations of its host grasses. *Canadian Journal of Botany* 67:2600–2607.

Lewis, F.J. 1924. An endotrophic fungus in the Coniferae. *Nature* (London) 114:860.

Luginbühl, M. and Muller, E. 1982. Untersuchungen über endophytische Pilze. II. Förderung der Samenkeimung bei *Hedera helix* durch *Aureobasidium pullulans* und *Epicoccum purpurascens*. *Berichte der Schweizerischen Botanischen Gesellschaft* 90:262–267.

Lynch, P.T., Isaac, S., and Collin, H.A. 1989. Uptake of protoplasts from the filamentous fungi *Aspergillus nidulans* and *Fusarium oxysporum* into protoplasts from celery (*Apium graveolens* L.). *Protoplasma* 150:9–18.

Millar, C.S. 1980. Infection processes on conifer needles. pp. 185–209 in Blakeman, J.P. (editor), *Microbial Ecology of the Phylloplane*. Academic Press, London.

Minter, D.W. and Ingham, J.C. 1984. Leaf spotting on *Chrysanthemum* associated with ejected spores of a species of *Podospora*. *Bulletin of the British Mycological Society* 18:133–134.

Muirhead, I.F. and Deverall, B.J. 1981. Role of appressoria in latent infection of banana fruits by *Colletotrichum musae*. *Physiological Plant Pathology* 19:77–84.

Nathaniels, N.Q.R. and Taylor, G.S. 1983. Latent infection of winter oilseed rape by *Leptosphaeria maculans*. *Plant Pathology* 32:23–31.

Neill, J.C. 1940. The endophyte of Rye-grass (*Lolium perenne*). *New Zealand Journal of Science and Technology* 21(5a):280A–291A.

O'Connell, R.J. 1987. Absence of a specialized interface between intracellular hyphae of *Colletotrichum lindemuthianum* and cells of *Phaseolus vulgaris*. *New Phytologist* 107:725–734.

O'Connell, R.J., Bailey, J.A., and Richmond, D.V. 1984. A chloroplast and other organelles of *Phaseolus vulgaris* within a hypha of *Colletotrichum lindemuthianum*. *New Phytologist* 96:35–41.

O'Donnell, J. and Dickinson, C.H. 1980. Pathogenicity of *Alternaria* and *Clado-*

sporium isolates on *Phaseolus*. *Transactions of the British Mycological Society* 74:335–342.

Petrini, L.E., Petrini, O., and Laflamme, G. 1989. Recovery of endophytes of *Abies balsamea* from needles and galls of *Paradiplosis tumifex*. *Phytoprotection* 70:97–103.

Petrini, L.E., Petrini, O., Leuchtmann, A., and Carroll, G.C. 1991. Conifer-inhabiting species of *Phillosticta*. *Sydowia* 43: In press.

Petrini, O. 1986. Taxonomy of endophytic fungi of aerial plant tissues. pp. 175–187 in Fokkema, N.J. and van den Heuvel, J. (editors), *Microbiology of the Phyllosphere*. Cambridge University Press, Cambridge, UK.

Petrini, O. 1987. Endophytic fungi of alpine Ericaceae. The endophytes of *Loiseleuria procumbens*. pp. 71–77 in Laursen, G.A., Amirati, J.F., and Redhead, S.A. (editors), *Arctic and Alpine Mycology II*. Environmental Science Research, vol. 34. Plenum Press, New York and London.

Petrini, O. and Carroll, G.C. 1981. Endophytic fungi in foliage of some Cupressaceae in Oregon. *Canadian Journal of Botany* 59:629–636.

Petrini, O. and Fisher, P.J. 1988. A comparative study of fungal endophytes in xylem and whole stems of *Pinus sylvestris* and *Fagus sylvatica*. *Transactions of the British Mycological Society* 91:233–238.

Petrini, O. and Müller, E. 1979. Pilzliche Endophyten am Beispiel von *Juniperus communis* L. *Sydowia* 32:224–251.

Petrini, O., Stone, J., and Carroll, F.E. 1982. Endophytic fungi in evergreen shrubs in Western Oregon: a preliminary study. *Canadian Journal of Botany* 60:789–796.

Porter, J.K., Bacon, C.W., Cutler, H.G., Arrendale, R.F., and Robbins, J.D. 1985. In vitro auxin production by *Balansia epichloë*. *Phytochemistry* 24:1429–1431.

Pugh, G.J.F. 1972. Saprophytic fungi and seeds. pp. 337–345 in Heydecker, W. (editor), *Seed Ecology*. Butterworth, London.

Rambosek, J. and Leach, J. 1987. Recombinant DNA in filamentous fungi: progress and prospects. *CRC Critical Reviews in Biotechnology* 6:357–393.

Riesen, T.K. and Close, R.C. 1987. Endophytic fungi in propiconazole-treated and untreated barley leaves. *Mycologia* 79:546–552.

Sampson, K. 1935. The presence and absence of an endophytic fungus in *Lolium temulentum* and *Lolium perenne*. *Transactions of the British Mycological Society* 19:337–343.

Schüepp, H. 1961. Untersuchungen über *Guignardia citricarpa* Kiely, den Erreger der Schwarzfleckenkrankheit auf *Citrus*. *Phytopathologische Zeitschrift* 40:258–271.

Sieber, T.N. 1985. *Endophytische Pilze von Winterweizen (Triticum aestivum L.)*. Dissertation. ETH no. 7725, Swiss Federal Institute of Technology, Zürich, Switzerland.

Sieber, T.N. 1988. Endophytische Pilze in Nadeln von gesunden und geschädigten Fichten [*Picea abies* (L.) Karsten]. *European Journal of Forest Pathology* 18:321–342.

Sieber, T.N. 1989a. Endophytic fungi in twigs of healthy and diseased Norway spruce and white fir. *Mycological Research* 92:322–326.

Sieber, T.N. 1989b. Substratabbauvermögen endophytischer Pilze von Weizenkörnern. *Zeitschrift für Pflanzenkrankheiten und Pflanzenschutz* 96:627–632.

Sieber, T.N. and Hugentobler, C. 1987. Endophytische Pilze in Blättern und Ästen gesunder und geschädigter Buchen (*Fagus sylvatica* L.). *European Journal of Forest Pathology* 17:411–425.

Sieber, T., Riesen, T.K., Muller, E., and Fried, P.M. 1988. Endophytic fungi in four winter wheat cultivars (*Triticum aestivum* L.) differing in resistance against *Stagonospora nodorum* (Berk.) Cast. & Germ. = *Septoria nodorum* (Berk.) Berk. *Journal of Phytopathology* 122:289–306.

Sieber, T.N., Sieber-Canavesi, F., and Dorworth, C.E. 1991a. Endophytic fungi of red alder (*Alnus rubra* Bong.) leaves and twigs in British Columbia. *Canadian Journal of Botany*. In press.

Sieber, T.N., Sieber-Canavesi, F., Petrini, O., Ekrameddoullah, A.K.M., and Dorworth, C.E. 1991b. Partial Characterization of Canadian and European *Melanconium* (*Melanconis*) from some *Alnus, Fagus,* and *Quercus* species by morphological and biochemical studies. *Canadian Journal of Botany*: in press.

Sieber-Canavesi, F. and Sieber, T.N. 1988. Endophytische Pilze in Tanne (*Abies alba* Mill.)—Vergleich zweier Standorte im Schweizer Mittelland (Naturwald–Aufforstung). *Sydowia* 40:250–273.

Sieber-Canavesi, F., Petrini, O., and Sieber, T.N. 1991. Endophytic *Leptostroma* species on *Picea abies, Abies alba,* and *Abies balsamea*: a cultural, biochemical, and numerical study. *Mycologia*. 83:89–96.

Sinclair, W.A., Lyon, H.H., and Johnson, W.T. 1987. Diseases of Trees and Shrubs. Comstock Publishing Associates, Cornell University Press, Ithaca, NY, and London. 574 pp.

Smereka, K.J., Machardy, W.E., and Kausch, A.P. 1987. Cellular differentiation in *Venturia inaequalis* ascospores during germination and penetration of apple leaves. *Canadian Journal of Botany* 65:2549–2561.

Stone, J.K. 1986. *Foliar Endophytes of Pseudotsuga menziesii (Mirb.) Franco. Cytology and Physiology of the Host-Endophyte Relationship*. Dissertation, University of Oregon, Eugene.

Stone, J.K. 1987. Initiation and development of latent infections by *Rhabdocline parkeri* on Douglas fir. *Canadian Journal of Botany* 65:2614–2621.

Stone, J.K. 1988. Fine structure of latent infections by *Rhabdocline parkeri* on Douglas fir, with observations on uninfected epidermal cells. *Canadian Journal of Botany* 66:45–54.

Suske, J. and Acker, G. 1989. Endophytic needle fungi: culture, unltrastructural and immunocytochemical studies. pp. 121–136 in Schulze, E.D., Lange, O.L., and Oren, R. (editors), *Forest Decline and Air Pollution: A study of spruce (Picea abies) on Acid Soils Ecological Studies*, vol. 77. Springer-Verlag, New York.

Verhoeff, K. 1974. Latent infections by fungi. *Annual Review of Phytopathology* 12:99–110.

Wicklow, D.T. 1988. Patterns of fungal association within maize kernels harvested in North Carolina. *Plant Disease* 72:113–115.

Widler, B. and Muller, E. 1984. Untersuchungen über endophytische Pilze von *Arctostaphylos uva-ursi* (L.) Spreng. (Ericaceae). *Botanica Helvetica* 94:307–337.

Wood, I. 1990. Plant fungus keeps its host in good health. *New Scientist*, March 1990:27.

10

Fungal Endophytes of Grasses: Detrimental and Beneficial Associations

Malcolm R. Siegel and Christopher L. Schardl

10.1 Introduction

Relationships between grasses and endophytic fungi have recently received considerable attention because of the impact of certain of these associations on agricultural production, and because of unique features that make these relationships excellent models of the ecology and evolution of mutualistic associations. Fungal species of interest include those belonging to the tribe Balansieae, family Clavicipitaceae (Ascomycetes), species of which exhibit a range of host interactions that span a continuum from pathogenic to non-pathogenic. The fungal genera involved include *Balansia, Myriogenospora, Atkinsonella* and, *Epichloe*. The asexual state of *Epichloe* and one of the two asexual states of *Atkinsonella* are classified in the form genus *Acremonium* section *Albo-lanosa,* which also includes a number of imperfect (nonsexual), seed-disseminated endophytes. Many *Acremonium* fungi produce no disease on their hosts and are referred to as "true endophytes" (Siegel et al., 1987b).

Species of *Balansia, Atkinsonella,* and *Epichloe* are typically maintained within vegetative plant tissues for months or even years before causing disease. The pathogenic states are most frequently initiated with flowering and can result in sterilization of the grass plants due to interference with flower maturation. Should any infected inflorescences "escape" the disease, these fungi then grow in the seed and are thus maintained in succeeding host generations. Because the true endophytes exist in high frequency in certain grass species and are disseminated only in seed and vegetative propagules, their fates are inextricably tied to those of their host lineages. By implication, these latter associations must be beneficial to the plant hosts; otherwise, they will not be maintained on an evolutionary time scale. Antiherbivore alkaloids are characteristic of many of the host-fungus associations and are of major benefit to the hosts. Other important benefits conferred by the mycosymbionts to the plants can include improved growth and drought tolerance. It must be noted that the host benefits are not restricted to associations with the true endophytes, but are also characteristic of some of the pathogenic associations.

In reference to the terminology used to describe the interactions of plants and microorganisms, Lewis (1985) stated, "as in the appreciation of beauty, interpretation[s] of technical terms and their definitions depend on the eyes of the beholder." This is especially true when discussing concepts involving beneficial and detrimental host-endophyte associations. In this chapter, we use the term symbiosis without connotation of harm or benefit to the symbionts. The perennial, constitutive symbioses that characterize grass-endophyte systems approach the nature of individual units of selection and development. In order to describe pathogenic relationships as individual entities, Loegering (1978) coined the term aegricorpus ("disease body"), which is inappropriate for nonpathogenic interactions. Here we use the term "symbiotum" (plural, "symbiota"), with the individual components (symbionts) stated explicitly or in context. The symbiotic continuum includes parasitism as well as mutualism. In a typical parasitic relationship, a heterotrophic fungal symbiont is nutritionally dependent upon its autotrophic plant host and acts as a pathogen in that it causes overt destructive symptoms to at least some host tissues. Generally, the disease state itself is considered to cause a decreased fitness of the host. In mutualism, the heterotrophic partner also receives nutrition and other enhancements but confers benefits by producing or participating in production of beneficial metabolites, or by causing other physiological effects that may enhance the survival or fecundity of its host.

That grass-endophyte associations confer benefits and detriments for humans is of critical importance in this discussion because the characteristics that provide enhanced host fitness in nature may or may not be regarded as desirable in agriculture or other settings. For this reason, for example, there have been considerable efforts to manipulate the *Acremonium* endophytes. They have been removed from forage grasses where they adversely affect grazing animals and introduced into turf grasses, where their toxic alkaloids provide control of insects and improve grass persistence. As will be shown, the success of these manipulations is dependent on the clear understanding of the symbiota at the levels of ecology, physiology, biochemistry, and genetics.

Numerous reviews cover many aspects of the grass-endophyte associations (Bacon, 1990; Bacon and De Battista, 1990; Bacon and Siegel, 1988; Bacon et al., 1986; Clay, 1986, 1988a, 1988b, 1989, 1991; Carroll, 1988; Dahlman et al., 1991; Siegel et al., 1985, 1987b, 1989). In referring to aspects which these reviews cover in depth, they will be cited instead of the primary literature sources. This chapter will summarize our current state of knowledge of the field and will include information which furthers our understanding of the relationships. Particular emphasis will be placed on the true endophytes and their impacts on grasses used for pasture, turf, and conservation. In addition, endophyte antagonisms of plant pests will be discussed by Clay (Chapter 17, this volume).

10.2 Fungal Taxonomy and Biology

Tribe Balansieae Systemic infections of grasses by endophytic fungi were known prior to the discovery that these grasses caused adverse effects on herbivores (Siegel et al., 1987b; Bacon and De Battista, 1990). Clay (1986, 1989) and Bacon and De Battista (1990) have summarized the taxonomy, life cycle, and distribution of the clavicipitaceous endophytes of grasses (also see Deihl, 1950). It has been estimated that over 80 genera and several hundred species of graminoids worldwide are hosts for Balansieae endophytes. Further estimates are that only 15% of new world species are hosts for endophytes (Bacon and De Battista, 1990), although this number may increase substantially as research intensifies. Most Balansieae infect grasses (family Poaceae) although infections of species in the related families Cyperaceae and Juncaceae are also known. The teleomorphic states (giving rise to meiotic spores) of the Balansieae are classified in the genera *Atkinsonella, Myriogenospora, Balansia,* and *Epichloe*. *Balansia* contains over half of the known species of the tribe. The anamorphic (mitotic spore) states of the Balansieae are *Acremonium* and *Ephelis*. The genus *Balansiopsis* was originally described without an anamorphic state by Diehl (1950), but Rykard et al. (1984) demonstrated that the conidial state exists and is closely related to those of *Balansia* spp. Members of the genera *Atkinsonella* and *Epichloe* infect C3 (cool season) grasses, while the other three genera infect C4 (warm season) grasses. Since C3 and C4 grasses are distributed from temperate to tropical regions, potential hosts exist in a wide range of habitats. Of interest is that host grasses in different regions can be infected with the same species (Diehl, 1950). While several species of Balansieae can grow primarily as epiphytes (Leuchtmann and Clay, 1988a) on meristems, young leaves, and inflorescences, most grow systemically. The hyphae growing in the plant are strictly intercellular, neither producing haustoria nor entering host cells (Siegel et al., 1987a, 1987b). The anamorphic and teleomorphic stages usually appear after a latent period as external stromata on leaves and/or emerging flower panicles. Mating requirements have been determined for species of *Atkinsonella* and *Epichloe*, both of which are heterothallic (Leuchtmann and Clay, 1989a; Morgan-Jones and White, 1989; White and Bultman, 1987). Due to expression of the stromatal state, most Balansieae cause their host plants to loose the ability to flower and produce seed.

Genus Epichloe The anamorph of *Epichloe typhina, Acremonium typhinum* Morgan-Jones et Gams, and related asexual *Acremonium* spp. are included in the section *Albo-lanosa*. Members of this genus infect many grass species, primarily of the subfamily Pooideae. Approximately 90% of the species in this subfamily are infected with *E. typhina*. Fewer endophytic anamorphs have been detected (ca. 25 grass species), but further evaluations should increase this number (Bacon and De Battista, 1990). Many endophyte-infected, cool season grasses (e.g., tall fescue, perennial and annual ryegrass, and the fine leaf fescues) are of considerable economic importance and account for the interest in this group of fungi. The life cycles of *E. typhina* and related endophytes are illustrated in Figure 10.1. *E. typhina* causes "choke disease" by forming sporogenous stromata around emerging flower panicles, resulting in abortion of the inflorescences (White and Morgan-Jones, 1987). Hyaline, nonseptated, ellipsoid conidia, which

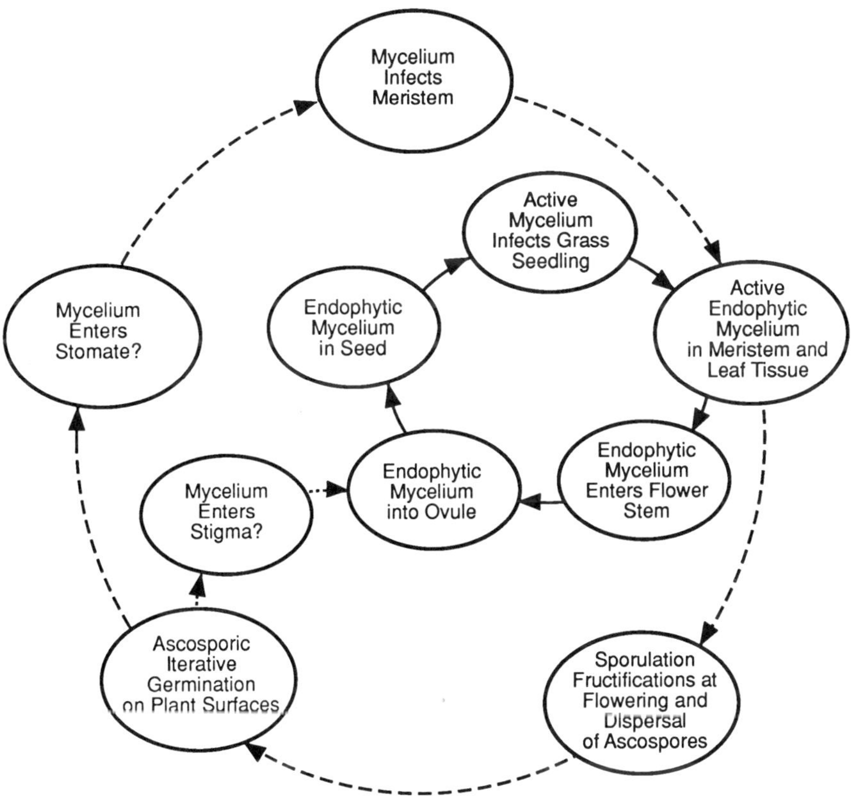

Figure 10.1 Life cycles of *Epichloe typhina* and related *Acremonium* endophytes. Solid lines represent life cycle of nonpathogenic *Acremonium* seedborne endophytes; dashed lines represent the sexual cycle of *E. typhina*. ? = route of infection by ascospores not determined.

are borne on short conidiophores in a T-shaped configuration, act as spermatia in cross-fertilization of mating types (simple bipolar heterothallism). This gives rise to the orange-tan perithecial stage (White and Bultman, 1987). Bultman and White (1988) have shown that heterothallism is facilitated by a fly, *Phorbia phrenione* Seguy, which moves conidia of one mating type to stromata of the opposite mating type, where nuclear transfer, dikaryotization, karyogamy, meiosis, and ascospore production occur. The fly had been considered to be a fungal parasite, but it is now thought that the interaction of fungus and fly is highly coevolved and mutualistic.

Hyaline fusiform ascospores are forcibly ejected from mature perithecia. They then germinate to produce phialides and conidia, which, in turn, germinate to produce hyphae and more conidiophores (ascosporic iterative germination) (Bacon and Hinton, 1988). During this microcyclic conidiation, each cycle gives rise to smaller conidia. After the last cycle (usually, the third), the mycelia presumably infect the host via the stigmata, leaf-stem stomata (Bacon, personal communication), and/or ancillary buds (White, personal communication). Whatever the infection route, meristematic tissue must become infected for survival of the fungus. The related species, *Claviceps purpurea,* is known to infect through stigmata, but unlike Balansieae, *Claviceps* spp. destroy the host ovule (Diehl, 1950).

Whereas almost all species of the Balansieae produce an external pathogenic (ectophytic) stage, the associations of *E. typhina* and anamorphs with grasses of the subfamily Pooideae exhibit considerable variation in disease development. In some grass-fungus associations, nearly all flowering panicles are choked. In other associations, only a few panicles are diseased, and the healthy flowers give rise to *E. typhina*-infected seeds (Clay and Leuchtman, 1989; Sampson, 1933; White, 1988). In fact, many pooid grasses contain *Acremonium* spp. or *Acremonium*-like endophytes that cause no choke disease and are strictly seed disseminated. Serology (Siegel et al., 1987b), isozyme analysis (Leuchtmann and Clay, 1990), and nuclear DNA sequences (Schardl and Siegel, unpublished) indicate that these are anamorphs of *E. typhina* or closely related species.

Phylogenetic Relationships and Taxonomic Differentiation The basis for the taxonomic relationship of genera in the tribe Balansieae was established by Diehl (1950), with the addition of the genus *Myriogenospora* by Luttrell and Bacon in 1977, and appears to have withstood the scrutiny caused by recent interest in grass endophytes. However, some questions have arisen concerning the diversity of species that have no teleomorphic state. These species, which primarily appear to be allied with the genus *Epichloe,* have been found in many cultivated and wild pooid grass species. While some species designations have been assigned (six as of now), Clay and Leuchtmann (1989) have pointed out that it may be difficult to accommodate many *Acremonium* anamorphs in established taxa.

Various techniques are now available to determine genetic diversity, elucidate phylogenetic relationships, and assist in taxonomic differentiation. These biological and biochemical techniques include cross-inoculation studies, isozyme electrophoresis, restriction-fragment-length-polymorphism (RFLP) studies (Miller and Tanksley, 1990), and nucleic acid sequence determination (Bruns et al., 1990).

Isozyme analysis has been used to study genetic variation in *Atkinsonella* spp. (Leuchtman and Clay, 1989a). There is a high degree of genetic variability among isolates. Genetic distances correlate with taxonomic distances among the host grasses. Coevolutionary relationships are also suggested by the complete inability of isolates from *Danthonia* spp. or *Stipa* spp. to infect seedlings of the other (Leuchtmann and Clay, 1989b). This, and morphological characteristics, provide the basis for classifying the *Stipa leucotricha* isolates as a new species, *At. texensis* (Morgan-Jones and White, 1989; Leuchtmann and Clay, 1989c). The possible coevolutionary relationships of *E. typhina* and/or anamorphic endophytes with their hosts are not as apparent. When isozyme analysis was applied to *E. typhina* and *Acremonium* spp. endophytes, Leuchtmann and Clay (1990) confirmed that the true endophytes are related to the choke pathogens. However, although *E. typhina* exhibits a high degree of host specificity (Schardl and Siegel, unpublished), cluster analysis of isozyme data (Leuchtman and Clay, 1990) does not provide any clear indications regarding coevolutionary relationships.

Schardl and Siegel (unpublished) used sequences of nuclear ribosomal RNA genes (Bruns et al., 1990), subject to maximum parsimony analysis (Swofford, 1990; Li and Gouy, 1990) to better elucidate the interrelationships of *E. typhina* and the anamorphic endophytes, and to establish the degree to which they may have coevolved with their hosts. Two of the isolates were from perennial ryegrass; one a choke pathogen (*E. typhina*), and the other a nonpathogenic endophyte (*A. lolii*). Comparison of their DNA sequences with those of *A. coenophialum* and *C. purpurea* (as an outlier) shows that the two endophytes are much more closely related to each other than to the perennial ryegrass pathogen (Figure 10.2). In fact, the nonpathogenic endophytes of tall fescue and perennial ryegrass appear to be very closely related to a choke pathogen of red fescue (Schardl and Siegel, unpublished). Therefore, although isolates from the same grass species are

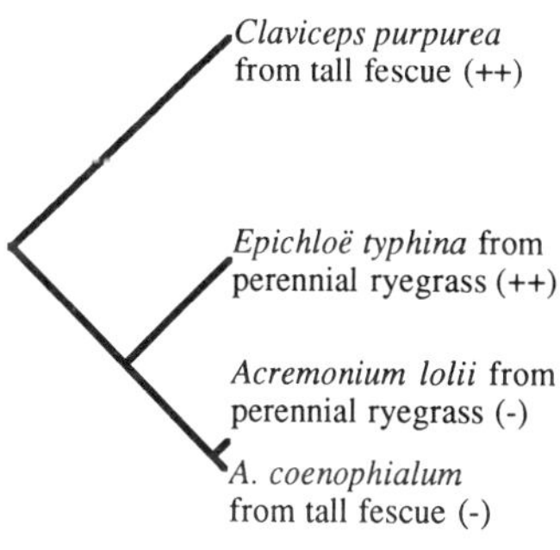

Figure 10.2 Parsimonious tree of clavicipitaceous isolates based upon sequence of the internal transcribed spacer regions of the nuclear rRNA genes (*rrnITS*). The branching order was determined, by the method of Li and Gouy (1990), to be significant ($P < 0.0005$). *Claviceps purpurea* was assumed to establish the root. The degree of fungal pathogenicity is indicated: ++ = highly pathogenic, − = nonpathogenic.

not necessarily close relatives, the choke pathogens and endophytes are generally of the same phylogenetic group.

Sequence analysis of the conserved β-tubulin genes has also been used to elucidate phylogenetic relationships (Byrd et al., 1990). As expected, the *E. typhina* and *A. coenophialum* genes share a high degree of homology in comparison with other fungal genes (Mogen and Liu, unpublished). Also, they both lack two introns that are characteristic of the β-tubulin genes of other Euascomycetes. Sequence analysis and intron presence or absence in other Clavicipitaceae may provide useful indicators of the interrelationships of the genera in the family, and the relationship of the family with other fungi.

Nonclavicipitaceous Endophytes Asymptomatic infections of grasses by nonclavicipitaceous fungi appear to be rare. Only three such endophytic species have been described: *Gliocladium*-like and *Phialophora*-like species have been isolated from perennial ryegrass and tall fescue, respectively (Latch et al., 1984), and *Pseudocercosporella trichachnicola* has been recovered from *Trichachne insularis* (sourgrass) (White and Morrow, 1990). Since some or all of these endophytes grow intercellularly and are seed disseminated, it is assumed that they are also in mutualistic association with their hosts. However, only indirect evidence supports this assumption. No biological activity has been noted for the *Gliocladium*-like endophyte. The *Phialophora*-like fungus has extensive antifungal activity against 11 grass pathogens, as determined by in vitro agar plate assays (Siegel and Latch, unpublished). However, it has not been determined if the antifungal activity is significant in nature. Sourgrass is frequently infected with *P. trichachnicola* and is avoided by cattle, suggesting the presence of antiherbivore compounds.

10.3 Detrimental and Beneficial Associations

Symbiosis Survival in nature is based, in part, on symbiotic associations. Considerable research and discussion have concerned symbiotic relationships and have resulted in a clearer understanding of the components of this important aspect of the biosphere (see Boucher, 1985; Pirozynski and Hawksworth, 1988; Ahmadjian and Paracer, 1986). Myriad symbioses are evident and, when defined on the basis of potential fitness of the interacting organisms, they can be arranged in a continuum. Lewis (1985) describes this continuum such that antagonism (parasitism) includes competitive, amensalistic, and agonistic interactions (fitness of one or more of the symbionts is reduced). In beneficial associations, either of the symbionts (commensalism) or both (mutualism) have enhanced fitness over individuals that do not participate in the relationship. The symbionts are maintained within their ecological niches by coevolutionary forces. Coevolution of interactive partners is through reciprocal selection pressures (Clay, 1988b).

How do the grass-fungal endophyte associations fit into the symbiotic continuum? The Balansieae are obligate biotrophs. Throughout most of their life cycles they grow as nonpathogenic endophytes or epiphytes without penetrating or damaging host cells. Most species are potential pathogens, with the most common symptoms involving sterilization of the flower panicles. Therefore, we might regard them as antagonists although, even in the disease state, the grass hosts may sometimes derive benefits such as protective alkaloids and improved vegetative growth (see below). In contrast, the symptomless symbiota are clearly commensal or mutualistic (beneficial) relationships.

Clay (1988b) describes a continuum for grass-endophyte associations based on integration of the reproductive modes of the symbionts. In this scheme the sexual capacity of the grass host—i.e., ability to produce seed—is impeded by formation of fungal stromata which are, themselves, necessary for expression of fungal sexuality (ascospore production). This "choice" between fungal and host sexuality is far from absolute since a single symbiotum may express both diseased and symptomless flowers (White, 1988; Clay, 1988b). In some strains of *E. typhina* (or nonsexual allies) production of stromata is very rare (10% or less) (White, 1988). It must also be noted that stroma formation does not necessarily lead to the fungal sexual cycle, since many or all Balansieae are heterothallic (White and Bultman, 1987; Leuchtmann and Clay, 1989b), requiring the availability of spermatia from distinct mating types. Therefore, the loss of sexuality may occur by natural or artificial isolation of a single mating type. Within the confines of Clay's model, such nonsexual pathogens may either be doomed to extinction or their lineages may evolve to nonpathogenic forms (true endophytes). It is intriguing to speculate that this process has occasionally resulted from agricultural use of grasses. Both geographical translocation and selection for high seed production may have selected some of the true endophytes of cultivated grasses.

It is difficult to place many grass-endophyte symbiota in the continuum described by Lewis (1985). Regardless of whether a symbiotum manifests disease, the fungus and/or symbiotum may nevertheless produce secondary metabolites that cause the relationship to be beneficial to the host. The range and degree of benefit varies. For example, several types of protective alkaloids have been detected in both diseased and asymptomatic associations of grasses with *E. typhina* or related endophytes (Siegel et al., 1990; also see below).

In species of the other Balansieae genera, pathogenesis is more the rule. However, some of these also produce ergot alkaloids (Bacon et al., 1986) and plant auxins (Porter et al., 1985), and their associations manifest increased tolerance to insects and grazing animals (Bacon et al., 1986; Clay, 1988a), and enhanced plant growth (Clay, 1990, 1991). Clearly, the prevention of flowering with loss of seed as a means of genetic recombination and spread can be a detriment to the plant. However, there is also benefit in increased plant biomass and longevity over noninfected plants (Clay, 1990; Bradshaw, 1959). As Bacon and De Battista (1990) have stated, the short-term gain of increased vegetative growth, reproduction, and persistence by infected grasses is essentially mutualistic only if seed production (sexuality and spread) is not a prerequisite for survival of a population over the long term; otherwise, the association would be antagonistic. Examples of antagonistic, pathogenic relationships have been noted (Diehl, 1950; Rykard et al., 1985; Clay, 1990). If all of the costs associated with symptomatic and asymptomatic infections are brought together, then the balance between benefits and detriments may become more apparent. As discussed by Carroll (Chapter 18), these costs can be hidden and have probably been greatly underestimated. They include both continuous and episodic outlays by the plant (loss of host metabolites required to maintain and produce fungal biomass, and alteration of the allocation of plant biomass). It will require considerable research into all aspects of grass-endophyte associations to determine, with any degree of certainty, what kind of cost/benefit relationship exists between the symbionts and how associations should be evaluated.

Human or societal perception adds an additional perspective on grass-endophyte associations and on what are perceived as detriments or benefits. Understandably, those concerned with animal husbandry tend to view the toxicity to grazing mammals of endophyte-infected pasture grasses as detrimental, although the antiherbivore activity is an ecological benefit to the plant. In fact, the alkaloids that cause mammalian toxicoses generally also contribute antiinsect activity (Clay, 1988a; Dahlman et al., 1991; Siegel et al., 1987b). Since the persistence of the grasses is required for their agricultural usefulness, particularly in marginal lands where forage and pasture grasses are so crucial, humans must seek the appropriate balance of these societal detriments and ecological benefits. Manipulation of cultivar characteristics will need to be undertaken only with careful consideration of particular uses and environments for which they are intended. This will continue to be true even as genetic engineering techniques become more refined and applicable to these systems, because the forage grasses are the primary producers in a multitrophic system, which includes humans, other mammals, insects, nematodes, microbial parasites, and, of course, the symbionts.

Biologically Active Compounds Table 10.1 lists the secondary metabolites produced in endophyte-infected plants and their known biological activities. The peramine, lolitrem, loline, and ergot alkaloids have antiherbivore activity and consequently have been studied most intensively (for details on their chemistry and biological activity, see Bacon et al., 1986; Siegel et al., 1987b; Dahlman et al., 1991; Bacon and De Battista, 1990; Clay, 1988a; Stuedemann and Hoveland, 1988; Yates, 1983). Most of the compounds listed or their precursors are of fungal origin, having also been

Table 10.1 Secondary metabolites and their biological activity in grass-endophyte symbiota

Chemical	Source[a]			Biological activity
	Fungus	Grass/fungus	Grass[b]	
Ergot alkaloids	+ (E, A, B)	+ (E, A, B)		Insecticidal and mammalian toxic
Peramine	+ (E, A)	+ (E, A)		Insecticidal
Paxilline	+ (A)	+ (A)		Insecticidal
Lolitrem		+ (E, A)		Insecticidal and mammalian toxic
Lolines		+ (A) + (A)		Insecticidal and mammalian toxic
Indole acetic acid	+ (E, A, B)			Improved plant growth
Indole ethanol	+ (B)			
Indole acetamide	+ (B)			
Ergosterol, ergosterol epoxide, and other sterol derivatives	+ (A)			Antibacterial and animal toxic
Sesquiterpenes, phenolic glycerides, fatty acids, and ring aromatic sterol	+ (E)			Antifungal
Perloline		+	+	Antimicrobial, synergistic, and animal toxic

[a] + = detected; E = *Epichloe typhina;* A = *Acremonium* spp.; B = other *Balansieae* spp.
[b] Compound found in noninfected plants.

observed in culture. It is still uncertain whether those compounds found in grass-*Acremonium* spp. associations are present in other Balansieae-infected plants. The ergot alkaloids have been detected in cultures of *E. typhina* and *Balansia* spp. (Bacon et al., 1986; Bacon and De Battista, 1990), but the spectrum of compounds differs considerably. The primary ergot alkaloids recovered from grass-*Acremonium* spp. associations are the ergopeptine types, ergovaline and its isomer ergovalinine, with the clavine types being minor constituents (Lyons et al., 1986). Ergovaline and ergovalinine can account for 10–180% of the total ergot alkaloids present in grasses infected with *A. coenophialum* (Lyons et al., 1986; Belesky et al., 1988) and *A. lolii* (Rowan and Shaw, 1987). In symbiota with *Balansia* species, the clavines predominate, and ergopeptines are minor constituents (Bacon et al., 1986).

Peramine (Rowan and Gaynor, 1986) and paxilline, a precursor of lolitrem B (Weedon and Mantle, 1987), are of fungal origin. The pyrrolizidine alkaloids (N-formyl and N-acetyl loline) have been detected in tall fescue plants only when *A. coenophialum* is present (Kennedy and Bush, 1983). Little is known about loline biosynthesis, so the possiblity of the endophyte producing an intermediate in culture has not been determined. However, the high concentration of lolines detected in plants—up to 0.8% dry weight—suggests that some or, conceivably, all of the steps are catalyzed by plant enzymes. The other fungal products occur in infected plants at 100- to 500-fold-lower levels than the lolines.

As might be expected, abiotic factors play an important role in determining levels and biological activity of secondary metabolites in fungal culture as well as in the plant. For example, factors such as growth medium, culture age, repeated subcultures, and temperature affect ergot alkaloid synthesis in culture (Bacon, 1990). Soil moisture and soil nutrition affect the levels of ergot (Belesky et al., 1988, 1989b) and loline alkaloids (Kennedy and Bush, 1983; Belesky et al., 1987b, 1989b) in infected plants. In addition, repeated clipping of greenhouse-grown tall fescue plants has been found to increase loline alkaloid levels over nonclipped plants (Bush, personal communication). Obviously, more information is needed concerning the effects of abiotic factors during the growing season on the levels of all secondary metabolites produced in infected plants.

Siegel et al. (1990) have shown that the number of symbiosis-specific antiherbivore alkaloids can vary substantially in *E. typhina*- and *Acremonium* spp.-infected grasses. This variation is influenced by the fungal isolate and, rarely, by the host grass species or variety. In 34 grass-endophyte associations, the most commonly detected alkaloids were peramine and ergovaline (found in 66% and 60% of the associations, respectively); the least common were the lolines and lolitrem B (each found in 17% of the associations). Various combinations of the four classes of alkaloids were found in most of the infected grasses, but none of the alkaloids were apparent in five of the associations. One or combinations of two or three types of alkaloids were detected. In artificial plant-fungus combinations, the alkaloids are usually characteristic of the natural symbiota from which the fungi, rather than the grasses, were derived. For example, the natural tall fescue-*A. coenophialum* association produces lolines and not lolitrems, whereas perennial ryegrass-*A. lolii* produces lolitrems but no lolines (both mycosymbionts also produce ergovaline and peramine). The reciprocal, artificial associations were perennial ryegrass-*A. coenophialum* and tall fescue-*A. lolii*. The former produces lolines but no lolitrems, and the latter produces lolitrems and not lolines.

Although loline production correlates with the fungal strain, DNA sequence analysis indicates no particular phylogenetic relationships between endophytes (or their host plants) from loline-producing symbiota in comparison to those from symbiota without detectable loline (Schardl and Siegel, unpublished). Loline biosynthesis appears to be a plesiomorphic (ancestral) trait. Perhaps there are considerable differences in the regulation of alkaloid synthesis in different symbiota.

Some data suggest that biotypes or genotypes of endophytic *Acremonium* spp. exist which have limited production of various alkaloids. Latch (1989) has observed that a few of the *A. lolii*-infected perennial ryegrass plants from a large cultivar collection do not produce appreciable lolitrem. The lolitrem-minus phenotype is evident even when the endophytes are isolated and introduced into other cultivars (Latch, personal communication). Hill et al. (1990) have found one *A. coenophialum*-tall fescue symbiotum which produces sixfold less ergovaline than another, but it is not known whether the low-ergot-alkaloid level is due to the fungal genotype, host genotype, or interaction of the two. These observations hold promise that plant and fungal manipulations may provide well-tailored cultivars for agriculture. The potential for such manipulations is discussed later in this chapter (see Section 10.4).

Certain Balansieae species produce the auxin, indoleacetic acid (IAA), and related compounds in culture (Porter et al., 1985; De Battista et al., 1990a). The concentration of IAA produced in culture varies according to the isolate, culture conditions, and age. However, no differences in IAA concentrations are apparent in infected versus noninfected tall fescue, so unbound IAA does not appear to be related to changes in growth expression (De Battista et al., 1990a). However, levels of biologically active bound IAA have not been determined, so the contribution of the endophyte to the total auxin potential may still be significant.

Ergosterol, ergosterol epoxide, and related sterols have been recovered from *A. coenophialum* cultures (Davis et al., 1986). Ergosterol epoxide was shown to have significant toxicity to brine shrimp, two bacterial species, and chick embryos. However, there are no data concerning the levels of these compounds in infected plants, so their roles as toxic agents are unknown. Antifungal chemicals—sesquiterpenes, phenolic glycerides, unsaturated fatty acids, and a ring sterol compound—have been isolated from stromata of *E. typhina* on *Phleum pratense* (timothy grass) (Koshino et al., 1989), and exhibit biological activity against *Cladosporium pheli*.

An important plant product recovered from both infected and noninfected tall fescue is perloline. This compound can be recovered at concentrations of up to 2,000 ppm (Hovin and Buckner, 1983) and has been reported to be toxic to fermentative rumen microflora (Bush et al., 1976). Of more importance is that perloline exhibits synergism, increasing ergocryptine toxicity to the large milkweed bug, *Oncopeltus faciatus* (Yates et al., 1989). In this respect, other natural products found in grasses (e.g., perlolidine, harmane, norharmane, and halostachine) could also play an important role in the physiological activity of fungal and symbiotum-specific metabolites.

The Endophytic Mycosymbionts The nonpathogenic *Acremonium* spp. endophytes live in a protected environment within the host and are seed-disseminated. This aspect of survival is beneficial for organisms that have lost external sexual and asexual reproductive stages and means of infection. As Clay (1991) points out, the success of such an existence is dependent on the well-being of the host. Therefore, it is not surprising that the nonpathogenic endophytes often produce protective substances. It must be remembered, however, that many pathogenic isolates also produce such alkaloids, can actually promote vegetative growth, and may also be seed-disseminated when they are incompletely effective in causing disease. These capacities make them excellent candidates for ancestors of strict mutualists. Those that exhibit mild pathogenicity either may be intermediates in the evolutionary process or may have achieved a balance of seed dissemination and spore dissemination that is nearly optimal in their particular ecological contexts.

The intercellular growth and lack of host cell destruction by fungal hyphae suggest that plant nutrients are released from the host's cells to support the growth of the mycosymbiont. Little is known either concerning endophyte nutrition in the host or the release of secondary metabolites by the fungus. The mycosymbionts would require products of photosynthesis such as simple sugars, fixed nitrogen, and inorganic elements. Unless degraded by the plant in response to infection or changes in the plant's physiology, complex organic molecules within cells are probably not available to the mycosymbiont.

Translocation of a simple carbohydrate (sucrose) from host to ectophytic stromata, and fungal synthesis of sugar alcohols have been demonstrated (Smith et al., 1985; Thrower and Lewis, 1973). The vitamin thiamin has been shown to be required for growth of *A. coenophialum* in culture (Kulkarni and Neilsen, 1986) suggesting that the plant supplies it. Lyons et al. (1990) have indicated that *A. coenophialum*-infected tall fescue has a greater rate of nitrogen accumulation and metabolism than noninfected plants. A number of amino acids and ammonium accumulate in the sheaths and blades, along with an increase in NO_3^{2-}. The nitrogen compounds may be available to the fungus and the nitrogenous products of fungal biosynthesis (e.g., alkaloids) could be released to the host via reciprocal translocation.

The Phytosymbionts As previously stated, endophyte infection of grasses results in improved host fitness by reducing biotic and abiotic stresses. The first to be described and the most visible effect of endophyte infection was tolerance of cultivated grasses to grazing herbivores. The symptoms of fescue toxicosis (Bush et al., 1979; Stuedemann and Hoveland, 1988) and sheep staggers (Mortimer and di Menna, 1985) were described prior to the discovery that these maladies were caused by infection of tall fescue and perennial ryegrass by nonpathogenic *Acremonium* spp. (Siegel et al., 1987b). Cattle losses in the USA due to reduced weight gains and fecundity have been estimated at 600 million dollars annually (Hoveland, personal communication). Not included in this estimate are losses to the horse industry due to agalactia and reduced live birth rates. These effects on animal production must be viewed from a societal standpoint as detrimental. However, this is a host benefit because reduced grazing, especially during seed production or drought, increases grass survival. Both ergot (Garner and Cornell, 1978) and loline (Lauriault et al., 1990; Dew et al., 1990) alkaloids have been implicated in fescue toxicosis, while the lolitrem neurotoxins in perennial ryegrass cause sheep staggers (Gallagher and Hawkes, 1985). Other Balansieae-infected perennial weed grasses have also been implicated in animal toxicoses in locations where they have become part of established pastures (Bacon et al., 1986).

Endophyte-infected plants also greatly affect insect herbivores, by reducing predation rates. Numerous studies have indicated that natural and introduced pests of grasses suffer direct toxicosis or are deterred by the presence of various endophytes (see Siegel et al., 1987b; Bacon and De Battista, 1990; Clay, 1989, 1991; Dahlman et al., 1991; also see Chapter 17, this volume, for a review of the spectrum of biological activity and ecological implications). Four classes of chemicals (lolines, lolitrems, peramine, and ergot alkaloids) have been implicated as insect-active, but, as stated earlier, the number of these chemicals present in infected grasses varies. Consequently, the spectrum of biological activity depends on the specific grass-endophyte association (Siegel et al., 1990).

Enhanced resistance by *Acremonium spp.*-infected grasses to other pests such as nematodes (West et al., 1988, Pedersen et al., 1988; Kimmons et al., 1990) and fungi (White and Cole, 1985; Koshino et al., 1989) have been reported. The importance of fungicidal and nematocidal activity of infected plants needs further elucidation in order to determine what role these activities have on host fitness.

Enhanced growth and persistence of infected grasses have been noted presumably in the absence of biotic stresses. In addition, studies indicate that infected grasses are more resistant to abiotic stresses, such as drought, than are noninfected plants. Symptomatic infections of perennial grasses can cause host sterility, but often increase vegetative growth and induce other morphological changes (Clay, 1988b). For example, *Danthonia spicata* infected with *At. hypoxylon* (Clay, 1984) and *Panicum agrostoides* infected with *Balansia henningsiana* (Clay et al., 1989) produced fewer inflorescences but more tillers and biomass than noninfected plants. Enhanced growth of infected *D. spicata* plants was confirmed at 10 weeks after artificial inoculation of seedlings when the plants had 50% more tillers than noninfected plants (Leutchmann and Clay, 1988b). Clay (1988b, 1990) argues that these vigorous, but generally sterile, infected hosts are equally or more ecologically fit than the corresponding noninfected plants.

Most studies of changes in growth and persistence have involved cultivated grasses infected with nonpathogenic *Acremonium* spp. In various studies, increased yield has been correlated with greater numbers of tillers in the infected grasses (Belesky et al., 1989a; Bouton et al., 1989; West et al., 1988; Arachevaleta et al., 1989; Cheplick et al., 1989; De Battista et al., 1990b; Latch et al., 1985). In addition, greater root growth, plant regrowth (De Battista et al., 1990b), and seedling vigor and germination are observed (Clay, 1987). Increased persistence and greater growth of rhizomatous-infected fescue genotypes over noninfected rhizomatous and nonrhizomatous plants in the presence and absence of competing bermudagrass have been demonstrated (Bouton et al., 1989). Under drought conditions, one clone of infected tall fescue had narrow and thicker leaf blades and increased leaf roll, which would favorably affect the plants' response to drought (Arachevaleta et al., 1989). Some of these studies were carried out under controlled conditions involving soil nutrition and moisture, and in the field where conditions were less easy to control. Increased biomass (yield) was correlated in some studies with increased drought tolerance (e.g., Arachevaleta et al., 1989; Cheplick et al., 1989; Hill et al., 1990; West et al., 1988). If resistance of infected grasses to drought stress is to be evaluated, then the interaction of light, temperature, and soil nutrition during stress must also be studied. While this is easier to do in the greenhouse or growth chamber, properly monitored field studies involving competitive grasses and grazing animals are equally desirable.

As pointed out by Bacon and De Battista (1990), it is difficult to determine whether the observed differences between infected and noninfected grasses are due to either the presence of the endophyte, the grass genotypic variation, or a combination of both. In the previously mentioned studies, grass genotypes and effects of fungal infection were confounded. The use of clonal material can be facilitated by fungicide treatment of individual infected tillers (De Battista et al., 1990b). Detailed studies using infected and noninfected plants of the same genotype have been undertaken. Hill et al. (1990) have demonstrated in greenhouse studies that differences in a series of plant responses (i.e., forage yields, tiller number, dry matter production, crown depth, leaf area, and ergovaline concentration) of various infected and noninfected clones were related to differences in the grass genotypes, endophyte genotypes, or both. It was concluded that endophyte infection increased plant-phenotypic variation, and this could increase the ability of mixed stands of infected and noninfected tall fescue to adapt to diverse environments. De Battista et al. (1990b) determined the effect of endophytes on rhizome and herbage production of rhizomatous clones of tall fescue in greenhouse experiments. No differences in rhizome production occurred between infected and noninfected clones and populations. However, the presence of the endophyte did increase tillering, herbage, and root growth in all infected clones. Competitiveness (persistence) of infected and noninfected clones of tall fescue in mixed and pure stands was determined by Hill et al. (1991). The presence of endophytes increased tiller numbers in both stand types. However, when competitiveness was determined, one infected genotype outcompeted the noninfected at all host population densities, and the other infected genotype was more competitive at the highest population densities. From these studies it would appear that, depending on the genotype, infected tall fescue may show greater persistence over noninfected plants. It is obvious that a number of interacting physiological and biochemical processes account for improved host fitness of infected tall fescue.

There is not much information on the physiological and biochemical mechanisms that control enhanced growth, persistence, and drought tolerance of infected grasses. As previously stated, plant and fungal auxin production may account for changes in growth. It has been suggested that tall fescue produces allelochemicals that can inhibit other species of plants in mixed stands (Bacon and De Battista, 1990). However, other than one report that N-formyl loline inhibits germination of annual ryegrass seed (Petroski et al., 1990), no other allelopathic compounds have been identified. Studies have indicated that infected tall fescue exhibits physiological responses related to improved drought tolerance, including lower net photosynthetic rate, increased stomatal resistance (decreased stomatal conductance) (Belesky et al., 1987a), and osmotic adjustment (West et al., 1990). Increased stomatal resistance and osmotic adjustment would enhance turgor pressure. Stomate closure and leaf rolling during drought stress would improve water use efficiency. Reduced tiller mortality, greater regrowth, and increased biomass during recovery after drought have been associated with endophyte infection (Arachevaleta et al., 1989). Stands of infected tall fescue have been noted to have a lower canopy-to-air temperature differential than noninfected stands (West et al., 1988). This suggests that infected plants can retain more water in the canopy, and thus they are better able to utilize available water for cooling.

The biochemical mechanisms that control the physiology of drought tolerance may involve the accumulation of soluble sugars that occur in plants under stress. The role of the endophyte could involve the release by the host of simple sugars, which are required for the growth of the fungus. If release is stimulated and the fungus acts as a sink (Belesky et al., 1987a), then apoplastic redistribution of sugars to leaf blades and roots could occur. These sugars could then produce the changes noted in osmotic adjustment. One source of plant-soluble sugars is nonstructural carbohydrates. However, the concentration of nonstructural carbohydrates decreased in infected tall fescue under drought conditions in all parts of the plant except the roots, where they increased irrespective of endophyte status (Belesky et al., 1989a). Hill et al. (1990) determined that carbohydrate levels were either the same or lower in various clones of infected tall fescue than in noninfected plants. They suggested that the endophyte provides a mechanism by which regrowth after drought is enhanced through efficient utilization of carbohydrate reserves.

10.4 Manipulation of Endophytes

Concepts and Relationships We have discussed the observation that grass-endophyte associations can be perceived as either detrimental or beneficial, based on ecological fitness or from a societal viewpoint. The initial societal view that endophyte infection of tall fescue and perennial ryegrass was detrimental led to the assumption that removal of the fungi would be beneficial for livestock husbandry. This assumption has often proved erroneous. The removal of *A. lolii* from pasture perennial ryegrass in New Zealand was not successful because the endophyte-free cultivar did not survive overgrazing by sheep and attack by the Argentine stem weevil (Siegel et al., 1987b). In the USA, removal of *A. coenophialum* from tall fescue pasture grass appears, under certain circumstances, to be detrimental for survival of the grass. Apparently, this is not based on insect predation, but on overgrazing and a lack of drought tolerance, stand establishment, and persistence (Siegel et al., 1989). The inability of certain endophyte-cured pasture grass cultivars to survive biotic and abiotic stresses indicates that the obligatory nature of the mutualistic symbiosis between these hosts and their *Acremonium* endophytes has been poorly understood.

It is important to understand that, for native-endophyte-infected grasses, the ultimate, long-term outcome primarily involves enhanced host fitness. However, with cultivated grasses, society intervenes by plant breeding, monoculture, and animal management practices, and judges the outcome on the short-term basis of production and economics. Cultivated grasses infected with *E. typhina* and *Acremonium* spp. were selected for improved agronomic character. Inadvertently, this probably involved selection of endophyte genotypes whose character was directly linked to improved host survival and a lack of pathogenicity (permitting seed production). The existence of certain noninfected turf and pasture grass cultivars (Funk et al., 1985, 1989; Siegel et al., 1984) that have lost their fungal mutualists due to long-term seed storage (Siegel et al., 1984; Welty et al., 1987), might suggest that they exist, at least for a time, either because of a lack of severe biotic or abiotic stresses or because of high inputs of pesticides, fertilizers, and water, along with other management practices. Conventional plant breeding may well produce new adapted host genotypes that do not require, on a long-term basis, an associated endophyte. However, the potential for biological control of insects, improved growth, and persistence of endophyte-infected grasses suggests that the manipulation of the mycosymbionts and their introduction into certain noninfected cultivars are desirable.

Introduction of Naturally Occurring Endophytes Success in producing turf cultivars of *Festuca* and *Lolium* spp. has been achieved primarily where weakly pathogenic or nonpathogenic endophytes have been introduced by maternal line selection and backcrossing (Funk et al., 1985, 1989;

Siegel et al., 1989). Here, mammalian toxicity is not considered and control of insect predation is desired, along with other factors which improve the survival of the grass. Endophytes can also be introduced into grasses by artificial inoculations of one- to three-week-old seedlings (Latch and Christensen, 1985; Leuchtmann and Clay, 1988b). However, the introduced endophyte must be compatible with the host, seed-disseminated, and maintain its beneficial nature to the host. The production of new synthetic grass-fungus associations by artificial inoculation of naturally occurring endophytes offers great potential in eliminating specific toxic alkaloids that affect grazing livestock. This approach has been used by Latch (personal communication) who discovered a strain of *A. lolii* that does not produce lolitrem, but still maintains levels of peramine and ergot alkaloids when introduced into a pasture cultivar of perennial ryegrass. The new synthetic association should eliminate the incidence of sheep ryegrass staggers, the neurological disorder caused by lolitrems. While this approach holds promise, it should be noted that the loss of one or more toxic alkaloids also affects the spectrum of insect toxicity and increases the potential for insect damage.

In our studies (Siegel, unpublished), we have attempted to move a number of known endophytes with potentially desirable alkaloid spectra into tall fescue. As described previously, some of these cross-inoculations are stable; however, the majority are not. In the future, prior phylogenetic analysis, such as the one shown in Figure 10.2 is likely to be a useful guide as to which artificial combinations may prove most stable.

Genetically Modified Endophytes Another promising approach is the production of synthetic associations by the introduction of genetically engineered endophytes. Toward this end, we have developed a fungal transformation system (Schardl, unpublished). The selectable marker employed the 5′-end of the *sti35* gene of *Fusarium solani* f. sp. *phaseoli* (Choi et al., 1990) to promote transcription of the reading frame of the *Escherichia coli hph* gene for hygromycin B detoxification (Gritz and Davies, 1983). The previously adapted electroporation protocol (Richey et al., 1989) was employed, and the integration of the exogenous DNA into the *A. coenophialum* genome was confirmed by Southern-blot analysis (Ausubel et al., 1990).

There are three general purposes to which this transformation technology is applicable: the disruption of or interference with fungal genes; elevation of expression of fungal genes; and the introduction of new genetic capabilities from other sources. All three of these approaches have applications in cultivar development. For example, critical enzymes for synthesis of ergopeptines, lolines, or lolitrems may be targeted for gene disruption (Barnes and Thorner, 1985), or their expression may be inhibited by the introduction of chimeric genes expressing the corresponding antisense RNAs (Hamilton et al., 1990). Other genes, such as those necessary for peramine production, may have their expression enhanced by manipulation of *cis* or *trans* regulatory elements. Finally, the endophytes might provide convenient packages for foreign genes, enabling them to express proteins, enzymes, and secondary metabolites associated with resistance to insects (Ponz et al., 1988; Höfte and Whiteley, 1989), drought and diseases, herbicide tolerance, enhanced growth, and better nutritional quality. It must be reemphasized that these approaches are only useful in the context of much more extensive understanding of the symbiotic entities at the biochemical and ecological levels.

The authors wish to thank many colleagues for providing unpublished information and copies of manuscripts in press. Research in the authors' laboratories was supported, in part, by USDA Competitive Grant 88-37151-3860 and a McKnight Foundation Grant. This chapter is a Kentucky Agricultural Experiment Station Journal Series Paper 90-11-176.

References

Ahmadjian, V. and Paracer, S. 1986. *Symbiosis*. University Press of New England, Hanover and London. 212 pp.

Arachevaleta, M., Bacon, W.C., Hoveland, C.S., and Radcliffe, D.E. 1989. Effect of the tall fescue endophyte on plant response to environmental stress. *Agronomy Journal* 81:83–90.

Ausubel, F.M., Brent, R., Kingston, R.E., Moore, D.D., Seidman, J.G., Smith, J.A., and Struhl, K. (editors). 1990. *Current Protocols in Molecular Biology*. John Wiley and Sons, New York.

Bacon, C.W. 1990. Isolation, culture, and maintenance of endophytic fungi of grasses. pp. 259–282 in Labeda, D.P. (editor), *Isolation of Biotechnological Organisms from Nature*. McGraw-Hill Pub. Co., New York.

Bacon, C.W., and De Battista, J.D. 1990. Endophytic fungi of grasses. pp. 231–256 in Avora, D.K., Rai, B., Mukerji, K.G., and Knudsen, G.R. (editors), *Handbook of Applied Mycology, vol 1. Soils and Plants*. Marcel Dekker, Inc., New York & Basel.

Bacon, C.W. and Hinton, D.M. 1988. Ascoporic iterative germination in *Epichloe typhina*. *Transactions of the British Mycological Society* 90:563–569.

Bacon, C.W. and Siegel, M.R. 1988. Endophyte parasitism of tall fescue. *Journal of Production Agriculture* 1:45–55.

Bacon, C.W., Lyons, P.C., Poter, J.K., and Robbins, J.D. 1986. Ergot toxicity from endophyte-infected grasses: a review. *Agronomy Journal* 78:106–116.

Barnes, D.A. and Thorner, J. 1985. Use of the *LYS2* gene for gene disruption, gene replacement, and promoter analysis in *Saccharomyces cerevisiae*. pp. 197–226 in Bennett, J.W. and Lasure, L.L. (editors), *Gene manipulation in Fungi*. Academic Press, New York.

Belesky, D.P., Devine, O.J., Pallas, J.E., Jr., and Stringer, W.C. 1987a. Photosynthetic activity of tall fescue as influenced by a fungal endophyte. *Photosynthetica* 21:82–87.

Belesky, D.P., Robbins, J.D., Stuedemann, J.A., Wilkinson, S.R., and Devine, O.J. 1987b. Fungal endophyte infection—loline derivative alkaloid concentration of grazed tall fescue. *Agronomy Journal* 79:217–220.

Belesky, D.P., Stuedemann, J.A. Plattner, R. D., and Wilkinson, S.R. 1988. Ergopeptine alkaloids in grazed tall fescue. *Agronomy Journal* 80:209–212.

Belesky, D.P., Stringer, W.C., and Hill, N.S. 1989a. Influence of endophyte and water regime upon tall fescue accessions. I. Growth characteristics. *Annals of Botany* 63:495–503.

Belesky, D.P., Stringer, W.C., and Plattner, R.D. 1989b. Influence of endophyte and water regime upon tall fescue accessions. II. Pyrrolizidine and ergopeptine alkaloids. *Annals of Botany* 64:343–349.

Boucher, D.H. (editor). 1985. *The Biology of Mutualism*. Oxford Press, New York. 388 pp.

Bouton, J.H., Whitehead, F.C., and De Battista, J.P. 1989. Tall fescue rhizome production as influenced by bermudagrass competition and cutting frequency. *Agronomy Journal* 81:220–223.

Bradshaw, A.D. 1959. Population differentiation in *Agrostis tenuis* Sibth. II. The incidence and significance of infection by *Epichloe typhina*. *New Phytologist* 58:310–315.

Bultman, T.L. and White J.W. 1988. "Pollination" of a fungus by a fly. *Oecologia* 75:317–319.

Bruns, T.D., Fogel, R., and Taylor, J.W. 1990. Amplification and sequencing of DNA from fungal herbarium specimens. *Mycologia* 82:175–184.

Bush, L.P., Burton, H., and Boling, J.A. 1976. Activity of tall fescue alkaloids and analogues in in vitro rumen fermentation. *Journal of Agricultural and Food Chemistry* 24:869–872.

Bush, L., Boling, J., and Yates, S. 1979. Animal disorders. pp. 247–292 in Buckner, R.C. and Bush, L.P. (editors), *Tall Fescue*. American Society of Agronomy, Madison, WI.

Byrd, A.D., Schardl, C.L., Songlin, P.J., Mogen, K.L., and Siegel, M.R. 1990. The β-tublin gene of *Epichloe typhina* from perennial ryegrass. *Current Genetics*. 18:347–354.

Carroll, G.C. 1988. Fungal endophytes in stems and leaves: from latent pathogen to mutualistic symbiont. *Ecology* 69:2–9.

Cheplick, G.P., Clay, K., and Marks, S. 1989. Interactions between infection by endophytic fungi and nutrient limitation in the grass *Lolium perenne* and *Festuca arundinacea*. *New Phytology* 111:89–97.

Choi, G.H., Marek, E.M., Richey, M.G., Chang, S., Schardl, C.L., and Smith, D.A. 1990. A stress responsive gene in *Fusarium* spp. *Journal of Bacteriology*. 172:4522–4528.

Clay, K. 1984. The effect of the fungus *Atkinsonella hypoxylon* (Clavicipitaceae) on the reproductive system and demography of the grass *Danthonia spicata*. *New Phytologist* 98:165–175.

Clay, K. 1986. Grass endophytes. pp. 190–201 in Fokkema, N.J. and van den Heuvel, J. (editors), *Microbiology of the Phyllosphere*. Cambridge University Press, Cambridge, UK.

Clay, K. 1987. Effect of fungal endophytes on seed and seedling biology of *Lolium perenne* and *Festuca arundinacea*. *Oecologia* 73:358–362.

Clay, K. 1988a. Fungal endophytes of grasses: a defensive mutualism between plants and fungi. *Ecology* 69:10–16.

Clay, K. 1988b. Clavicipitaceous fungal endophytes of grasses: coevolution and the change from parasitism to mutualism. pp. 79–105 in Pirozynski, K.A. and Hawksworth, D. (editors), *Coevolution of Fungi with Plants and Animals*. Academic Press, London.

Clay, K. 1989. Clavicipitaceous endophytes of grasses: their potential as biocontrol agents. *Mycological Research* 92:1–12.

Clay, K. 1990. Comparative demography of three graminoids infected by systemic, clavicipitaceous fungi. *Ecology* 7:558–570.

Clay, K. 1991. Fungal endophytes, grasses, and herbivores. pp. 199–226 in Barbosa, P., Kirschik, L., and Jones, E. (editors), *Microbial Mediation of Plant-Herbivore Interactions*. John Wiley and Sons, New York.

Clay, K. and Leuchtmann, A. 1989. Infection of woodland grasses by fungal endophytes. *Mycologia* 81:805–811.

Clay, K., Cheplick, G.P., and Marks, S. 1989. Impact of the fungus *Balansia henningsiana* on *Panicum agrostoides*: frequency of infection, plant growth and reproduction and resistance to pests. *Oecologia* 80:374–380.

Dahlman, D.L., Eichenseer, H., and Siegel, M.R. 1991. Chemical perspectives on endophytegrass interaction and their implications to insect herbivory. pp. 227–252 in Barbosa, P., Kirschik, L., and Jones, E. (editors), *Microbial Mediation of Plant-Herbivore Interactions*. John Wiley and Sons, New York.

Davis, N.D., Cole, R.J., Dorner, J.W., Weete, J.D., Backman, P.A., Clark, E.M., King, C.C., Schmidt, S.P., and Diener, U.L. 1986. Steroid metabolites of *Acremonium coenophialum*, an endophyte of tall fescue. *Journal of Agricultural and Food Chemistry* 34:105–108.

De Battista, J.P., Bacon, C.W., Severson, R., Plattner, R., and Bouton, J.H. 1990a. Indole-acetic acid production by the fungal endophyte of tall fescue. *Agronomy Journal* 82:1272–1279.

De Battista, J.P., Bouton, J.H., Bacon, C.W., and Siegel, M.R. 1990b. Rhizome and herbage production of endophyte-removed tall fescue clones and populations. *Agronomy Journal*. 82:651–654.

Dew, R.K., Boissonneault, G.A., Gay, N., Boling, J.A., Cross, R.J., and Cohen, D.A. 1990. The effect of the endophyte (*Acremonium coenophialum*) and associated toxin(s) of tall fescue on serum titer response to immunization and spleen cell flow cytometry analysis and response to mitogens. *Veterinary Immunology and Immunopathology* 26:285–295.

Diehl, W.W. 1950. Balansia and the Balansiae in America. *Agriculture Monograph 4*. U.S. Department of Agriculture, Washington, D.C.

Funk, C.R., Halisky, P.M., Ahmad, S., and Hurley, R.H. 1985. How endophytes modify turfgrass performance and response to insect pests in turfgrass breeding and evaluation trials. pp. 137–145 in Lemaire, F. (editor), *Proceedings of the 5th International Turf Conference* (Avignon). Institute National de la Recherche Agronomique Versailles.

Funk, C.R., Clark, B.B., and Johnson-Cicalese, J. M. 1989. Role of endophytes in enhancing the performance of grasses used for conservation and turf. pp. 203–210 in Leslie, A. R., and Metcalf, R. L. (editors), *Integrated Pest Management for Turfgrass and Ornamentals*. U.S. Environmental Protection Agency, Washington, D.C.

Gallagher, R.T. and Hawkes, A.D. 1985. Estimation of neurotoxin levels in perennial ryegrass by mouse bioassay. *New Zealand Journal of Agricultural Research* 28:427–431.

Garner, G.B. and Cornell, C.N. 1978. Fescue foot in cattle. pp. 45–46 in Wyllie, T.D. and Morehouse, L.G. (editors), *Mycotoxic Fungi, Mycotoxins, Mycotoxicoses: An Encyclopedic Handbook*, vol. 2. Marcel Dekker, New York.

Gritz, L. and Davies, J. 1983. Plasmid-encoded hygromycin B resistance: the sequence of hygromycin B phosphotransferase gene and its expression in *Escherichia coli* and *Saccharomyces cerevisiae*. *Gene* 25:179–188.

Hamilton, A.J., Lycett, G.W., and Grierson, D. 1990. Antisense gene that inhibits synthesis of the hormone ethylene in transgenic plants. *Nature* 346:284–287.

Hill, N.S., Stringer, W.C., Rottinghaus, G.E., Belesky, D.P., Parrot, W.A., and Pope, D.D. 1990. Growth, morpholological, and chemical component responses of tall fescue to *Acremonium coenophialum*. *Crop Science* 30:156–161.

Hill, N.S., Belesky, D.P., and Stringer, W.C. 1991. Competitiveness of tall fescue (*Festuca arundinacea* Schreb.) as influenced by endophyte (*Acremonium coenophialum* Morgan-Jones and Gams). *Crop Science*. 31:185–190.

Höfte, H. and Whiteley, H.R. 1989. Insecticidal crystal proteins of *Bacillus thuringiensis*. *Microbiological Reviews* 53:242–255.

Hovin, A.W. and Buckner, R.C. 1983. Alkaloids in tall fescue and reed canary grass. pp. 241–247 in Rechcigl, M. (editor), *Handbook of Naturally Occurring Food Toxicants*. CRC Press, Boca Raton, FL.

Kennedy, C.W. and Bush, L.P. 1983. Effect of environmental and management factors on the accumulation of N-acetyl and N-formyl loline alkaloids in tall fescue. *Crop Science* 23:547–552.

Kimmons, C.A., Gwinn, K.D., and Bernard, E.C. 1990. Nematode reproduction on endophyte-infected and endophyte-free tall fescue. *Plant Disease* 74:757–761.

Koshino, H., Yoshihara, T., Sakamura, S., Shimanuki, T., Sato, T., and Tajima, A. 1989. A ring B aromatic sterol from stromata of *Epichloe typhina*. *Phytochemistry* 28:771–772.

Kulkarni, R.K. and Nielsen, B.D. 1986. Nutritional requirements for growth of a fungal endophyte of tall fescue grass. *Mycologia* 78:781–786.

Latch, G.C.M. 1989. Plant improvement using endophytic fungi. pp. 345–346 in *XVI International Grasslands Congress*. French Grassland Society, Nice, France.

Latch, G.C.M. and Christensen, M.J. 1985. Artificial infections of grasses with endophytes. *Annals of Applied Biology* 107:17–24.

Latch, G.C.M., Christensen, M.J., and Samuels, G.J. 1984. Five endophytes of *Lolium* and *Festuca* in New Zealand. *Mycotaxon* 20:535–550.

Latch, G.C.M., Hunt, W.F., and Musgrave, D.R. 1985. Endophytic fungi affect growth of perennial ryegrass. *New Zealand Journal of Agricultural Research* 28:165–168.

Lauriault, L.M., Dougherty, C.T., Bradley, N.W., and Cornelius, P.L. 1990. Thiamin supplementation and the ingestive behavior of beef cattle grazing endophyte-infected tall fescue. *Journal of Animal Science* 68:1245–1253.

Leuchtmann, A. and Clay, K. 1988a. *Atkinsonella hypoxylon* and *Balansia cyperi*. epiphytic members of the Balansiae. *Mycologia* 80:192–199.

Leuchtmann, A. and Clay, K. 1988b. Experimental infection of host grasses and sedges with *Atkinsonella hypoxylon* and *Balansia cyperi* (Balansiae, Clavicipitaceae). *Mycologia* 80:291–297.

Leuchtmann, A. and Clay, K. 1989a. Isozyme variation in the fungus *Atkinsonella hypoxylon* within and among populations of its host grasses. *Canadian Journal of Botany* 67:2600–2607.

Leuchtmann, A. and Clay, K. 1989b. Experimental evidence for genetic variation in compatibility between the fungus *Atkinsonella hypoxylon* and its three host grasses. *Evolution* 43:825–834.

Leuchtmann, A. and Clay, K. 1989c. Morphological, cultural and mating studies on *Atkinsonella*, including *A. texensis*. *Mycologia* 81:692–701.

Leuchtmann, A. and Clay, K. 1990. Isozyme variation in the *Acremonium/Epichloe* fungal endophyte complex. *Phytopathology*, 80:1133–1139.

Lewis, D.H. 1985. Symbiosis and mutualism: crisp concepts and soggy semantics. pp. 29–39 in Boucher, D.H. (editor), *The Biology of Mutualism*. Oxford University Press, New York.

Li, W.H. and Gouy, M. 1990. Statistical tests of molecular phylogenies. *Methods in Enzymology* 183:645–659.

Loegering, W.Q. 1978. Current concepts in interorganismal genetics. *Annual Review of Phytopathology* 16:309–320.

Luttrell, E.S. and Bacon, C.W. 1977. Classification of *Myriogenospora* in the Clavicipitaceae. *Canadian Journal of Botany* 55:2090–2097.

Lyons, C.L., Plattner, R.C., and Bacon, C.W. 1986. Ocurrence of peptide and clavine ergot alkaloids in tall fescue. *Science* 232:487–489.

Lyons, C.P., Evans, J.J., and Bacon, C.W. 1990. Effects of the fungal endophyte *Acremonium coenophialum* on nitrogen accumulation and metabolism in tall fescue. *Plant Physiology* 92:726–732.

Miller, J.C. and Tanksley, S.D. 1990. RFLP Analysis of phylogenetic relationships and genetic variation in the genus *Lycopersicon*. *Theoretical and Applied Genetics* 80:437–448.

Morgan-Jones, G. and White, J.F., Jr. 1989. Concerning *Atkinsonella texensis*, a pathogen of the grass *Stipa leucotricha*: developmental morphology and mating system. *Mycotaxon* 35:455–467.

Mortimer, P.H. and di Menna, M.E. 1985. Interaction of *Lolium* endophyte on pasture production and perennial ryegrass staggers disease. pp. 149–158 in Lacey, J. (editor), *Trichothecenes and Other Mycotoxins*. John Wiley, Chichester/ New York.

Pedersen, J.F., Rodriguez-Kabana, R., and Shelby, R.A. 1988. Ryegrass cultivars and endophyte in tall fescue affect nematodes in grass and succeeding soybean. *Agronomy Journal* 80:811–814.

Petroski, R.J., Dornbos, D.L., Jr., and Powell, R.G. 1990. Germination and growth inhibition of annual ryegrass (*Lolium multiflorum* L.) and alalfa (*Medicago sativa* L.) by loline alkaloids and synthetic N-acylloline derivatives. *Journal of Agricultural and Food Chemistry*. 38:1716–1718.

Pirozynski, K,A. and Hawksworth, D.L. (editors). 1988. *Coevolution of Fungi with Plants and Animals*. Academic Press, London. 285 pp.

Ponz, F., Glascock, C.B, and Bruening, G. 1988. An inhibitor of polyprotein processing with the characteristics of a natural virus resistance factor. *Molecular Plant-Microbe Interactions* 1:25–31.

Porter, J.K., Bacon, C.W., Cutler, H.G., Arrendale, R.F., Robbins, J.D. 1985. In vitro auxin production by *Balansia epichloe*. *Phytochemistry* 24:1429–1431.

Richey, M.G., Marek, E.T., Schardl, C.L, and Smith, D.A. 1989. Transformation of filamentous fungi with plasmid DNA by electroporation. *Phytopathology* 79:844–847.

Rowan, D.D. and Gaynor, D.L. 1986. Isolation of feeding deterrents against Argentine stem weevil from ryegrass infected with the endophyte *Acremonium loliae*. *Journal of Chemical Ecology* 12:647–658.

Rowan, D.D. and Shaw, J.G. 1987. Detection of ergopeptine alkaloids in endophyte-infected perennial ryegrass by tandem mass spectrometry. *New Zealand Veterinary Journal* 35:197–198.

Rykard, D.M., Luttrell, E.S., and Bacon, C.W. 1984. Conidiogensis and conidiomata in the Clavicipitoideae. *Mycologia* 76:1095–1103.

Rykard, D.M., Bacon, C.W., and Luttrell, E.S. 1985. Host relations of *Myriogenospora atrementosa* and *Balansia epichloe* (Clavicipitaceae). *Phytopathology* 75:950–956.

Sampson, K. 1933. The systemic infection of grasses by *Epichloe typhina* (Pers.) Tul. *Transactions of the British Mycological Society* 18:30–47.

Siegel, M.R., Johnson, M.C., Varney, D.R., Nesmith, W.C., Buckner, R.C., Bush, L.P., Burrus II, P.B., Jones, T.A., and Boling, J.A. 1984. A fungal endophyte in tall fescue: incidence and dissemination. *Phytopathology* 74:932–937.

Siegel, M.R., Latch, G.C.M., and Johnson, M.C., 1985. *Acremonium* fungal endophytes of tall fescue and perennial ryegrass: significance and control. *Plant Disease* 69:179–183.

Siegel, M.R., Jarlfors, U., Latch, G.C.M., and Johnson, M.C. 1987a. Ultrastructure of *Acremonium coenophialum*, *Acremonium lolii*, and *Epichloe typhina* endophytes in host and nonhost *Festuca* and *Lolium* species of grasses. *Canadian Journal of Botany* 65:2357–2367.

Siegel, M.R., Latch, G.C.M., and Johnson, M.C. 1987b. Fungal endophytes of grasses. *Annual Review of Phytopathology* 25:293–315.

Siegel, M.R., Dahlman, D.L., and Bush, L.P. 1989. The role of endophytic fungi in grasses: new approchs to biological control of pests. pp. 169–186 in Leslie, A.R. and Metcalf, R.L. (editors), *Integrated Pest Management for Turfgrass and Ornamentals*. U. S. Environmental Protection Agency, Washington, D.C.

Siegel, M.R., Latch, G.C.M., Bush, L.P., Fannin, F.F., Rowan, D.D., Tapper, B.A., Bacon, C.W., and Johnson, M.C. 1990. Fungal endophyte-infected grasses: alkaloid accumulation and aphid response. *Journal of Chemical Ecology* 16:3301–3315.

Smith, K.T., Bacon, C.W., and Luttrell, E.S. 1985. Reciprocal translocation of carbohydrates between host and fungus in Bahiagrass infected with *Myriogenospora atramentosa*. *Phytopathology* 75:407–411.

Stuedemann, J.A. and Hoveland, C.S. 1988. Fescue endophyte: history and impact on animal agriculture. *Journal of Production Agriculture* 1:39–44.

Swofford, D.L. 1990. PAUP: phylogenetic analysis using parsimony, Version 3.0. Illinois Natural History Survey, Champaign, Illinois.

Thrower, L.B. and Lewis, D.H. 1973. Uptake of sugars by *Epichloe typhina* (Pers. ex Fr.) Tul. in culture and from its host, *Agrostis stolonifera* L. *New Phytologist* 72:501–508.

Weedon, C.M. and Mantle, P.G. 1987. Paxilline biosynthesis by *Acremonium loliae*; a step towards defining the origin of lolitrem neurotoxins. *Phytochemistry* 26:969–971.

Welty, R.E., Azevedo, M.D., and Copper, T.M. 1987. Influence of moisture content, temperature and length of storage on seed germination and survival of endophytic fungi in seeds of tall fescue and perennial ryegrass. *Phytopathology* 77:893–900.

West, C.P., Izekor, D.M., Oosterhuis, D.M., and Robbins, R.T. 1988. The effect of *Acremonium coenophialum* on growth and nematode infestation of tall fescue. *Plant and Soil* 112:3–6.

West, C.P., Oosterhuis, D.M., and Wullschleger, S.D. 1990. Osmotic adjustment in tissues of tall fescue in response to water deficit. *Environmental and Experimental Botany* 30:149–156.

White, J.F., Jr. 1988. Endophyte-host associations in forage grasses. XI. A proposal concerning origin and evolution. *Mycologia* 80:442–446.

White, J.F., Jr. and Bultman, T.L. 1987. Endophyte-host associations in forage grasses. VIII. Heterothallism in *Epichloe typhina*. *American Journal of Botany* 74:1716–1721.

White, J.F., Jr. and Cole, G.T. 1985. Endophyte-host associations in forage grasses. III. In vitro inhibition of fungi by *Acremonium coenophialum*. *Mycologia* 78:102–107.

White, J.F., Jr. and Morgan-Jones, G. 1987. Endophyte-host associations in forage grasses. IX. Concerning *Acremonium typhinum*, the anamorph of *Epichloe typhina*. *Mycotaxon* 29:489–500.

White, J.F., Jr. and Morrow, A.C. 1990. Endophyte-host associations in forage grasses. XII. A fungal endophyte of *Trichachne insularis* belonging to *Pseudocerosporella*. *Mycologia* 82:218–226.

Yates, S.G. 1983. Tall fescue toxins. pp. 249–273 in Recheigel, M. (editor), *Handbook of Naturally Occurring Food Toxins*. CRC Press, Boca Raton, FL.

Yates, S.G., Fenster, J.C., and Bartelt, R.J. 1989. Assay of tall fescue seed extracts, fractions, and alkaloids using the large milkweed bug. *Journal of Agricultural and Food Chemistry* 37:354–357.

11

Crop Plants as a Source of Fungus Spores of Medical Importance

Harriet A. Burge, Michael L. Muilenberg, and Jean A. Chapman

11.1 Introduction

Fungi are primarily either decay organisms obtaining their food saprophytically or plant pathogens. Many have developed enzyme systems (for example cellulases, ligninases, and pectinases) that facilitate the decomposition of virtually any nonliving plant material, if appropriate water and temperature conditions are provided. Fungi, in fact, are responsible for most aerobic decay of such materials (Kendrick, 1985).

Because the fungi play this essential role in the natural ecosystem, it is logical that a change in the attributes (such as source, moisture content, and species mix) of available nonliving plant material on a given land area will also change the fungal population emanating from or associated with that land area. Modern agriculture undeniably changes the natural plant ecosystem. Vast acreages are planted to a single plant or genotype (e.g., a particular cultivar of corn, soybeans, or wheat). The plants grow nearly synchronously, and they all mature and/or die within a short period. Harvest of many crops occurs after the plant has been dead for some time (often after late season rains have provided sufficient water for the growth of fungi on the plant material substrate). Then, the remaining dead plant material is left in the field (to prevent erosion) throughout the winter and is incorporated into the soil in the spring. After the plants have died and prior to spring plowing, fungal populations become well established on the dead and dying leaves and stems. The fungal mycelium initiates the decay process by secreting enzymes and (often toxic) metabolites. Each species of fungus produces spores that are almost always designed for aerial dispersal. Any activity that disturbs fungal colonies causes aerosolization of the spores and metabolites.

Extensive agricultural operations (including row crops and grain crops) are necessary to feed the billions of human beings in today's world. The changes that agriculture impose on the environment, however, are not generally beneficial for humans and may be detrimental. This paper is a brief review of the effects of the fungal saprophytes on human health; the evidence that agricultural practices contribute to these health problems; some speculations with respect to control of agricultural bioaerosols; and suggestions for future research directions.

11.2 Medical Effects of Common Fungal Saprophytes

Asthma and Hay Fever Asthma and hay fever (both known to be allergies) result from exposure to antigens that stimulate the production of antibodies (immunoglobulin E, or IgE) that recognize some facet of the antigen structure. Once the antigen has caused this specific IgE to be produced, further exposure to the same antigen causes the release of mediators (e.g., histamine) that induce the typical symptoms of asthma (chest tightness, wheezing) and hay fever (itchy and watery eyes and nose). The antigens to which individuals have become sensitized can be identified by means of skin tests, in which minute amounts of extracts of suspected substances are scratched onto or injected into the skin. In patients with IgE antibodies specific for the given substance, a wheal and an area of redness are produced at the scratch or injection site. Virtually every fungus for which suitable extracts have been prepared has elicited a positive skin test in some fraction of the patients tested.

To cause these allergic diseases, fungus spores need only enter the nose, eye, or mouth of a susceptible person, and release soluble antigen(s) which act on the mucous membranes and are absorbed and transported through the blood stream. This means that spores need not be small enough to enter the lower airways, and virtually any airborne spore with appropriate antigenic material can cause these diseases.

Apparently for genetic reasons, antigen-specific IgE antibodies are produced by only about 15–30% of the population. Risk for acquiring the allergies is related to both the genetic pattern of a person and exposure to the antigen. Researchers have yet to determine the level of spores (or, more accurately, the level of spore antigen) required to induce sensitization in a genetically predisposed person or symptoms in a sensitized person. For dust mite allergens, it has been hypothesized that long-term, low-level exposure is necessary to cause sensitization; more intense exposure, it is further hypothesized, is required for symptom development (Platts-Mills et al., 1990). It is important to note that excluding the antigen from the environment of the potentially sensitive person, or vice versa, will prevent sensitization. If sensitization occurs, prevention of future exposure will prevent symptoms of disease.

Allergic Bronchopulmonary Fungosis Asthma can cause the lungs to retain excessive mucous secretions in affected individuals. In some asthmatics, fungi can become established and grow in these sticky secretions to produce a disease called allergic bronchopulmonary fungosis, also known as allergic bronchopulmonary aspergillosis (ABPA) because *Aspergillus fumigatus* is the most frequently found causal agent. Reports of this disease are increasing, possibly due to improvement in diagnostic materials.

For a fungus to cause ABPA, its spores must be small enough ($< 5\ \mu m$) to penetrate into the lower airways. These spores must also be able to survive in the lung and be able to germinate and grow at body temperature (about 37°C). Fortunately for human beings, most fungi are not able to withstand conditions present in the human lung.

Risk factors for development of ABPA include the presence of asthma in the individual, and exposure of that person to a fungus that can live and grow in the lung. The number of fungus spores that must be inhaled to induce this disease is not known at present. A single viable spore may be sufficient, although it is probable that more intense exposures are usually necessary.

Hypersensitivity Pneumonitis Hypersensitivity pneumonitis (also called allergic alveolitis) results when an appropriate antigen stimulates local, cellular, immune responses in the human lung. Circulating antibodies (immunoglobulin G, or IgG) also are produced. Once such responses have been stimulated, further antigen exposure induces the release of mediators that cause the lungs to become inflamed, producing shortness of breath, cough, and, if antigen exposure continues, fibrosis of lung tissues.

To produce hypersensitivity pneumonitis (HP), spores must be small enough to enter the lower airways and exposure levels apparently must be very high ($> 10^9/m^3$ of air) (Malmberg et al., 1987). The agents that most commonly cause hypersensitivity pneumonitis are the thermophilic actinomycetes. These filamentous bacteria produce very small spores (about $1\ \mu m$) in very large quantities when moisture and temperature conditions are favorable. Temperatures in excess of 50°C are necessary for these bacteria to grow. Appropriate conditions occur in self-heating situations (stored wet hay or compost) and in mechanical systems such as humidifiers associated with heating units. It appears that the thermophilic actinomycetes contain both antigens and materials (adjuvants) that directly stimulate the immune system. Fungus-related HP may be more likely to occur when the fungus spores are present in air in association with high levels of Gram-negative bacterial endotoxin. Risk factors for development of HP include exposure to appropriate kinds and levels of spores and, probably, internal risk factors that are currently unknown.

Toxicoses Toxic compounds produced by the fungi cause diverse acute as well as chronic conditions that depend on the nature of the toxin and, probably, the route through which the toxin is absorbed. Ingested toxins usually produce systemic acute effects or diseases of specific target organs. For example, ingestion of the trichothecene toxins produces severe central nervous system symptoms, gastric hemorrhage, and immune suppression. The ingestion of aflatoxin produces primary liver cancer. Almost all of the mycotoxin research has focused on natural or experimental ingestion, or, rarely, injection (Rodericks et al., 1977; Shank, 1981). On the other hand, inhaled toxins are absorbed systemically and produce systemic effects. Inhaled toxin can, however, directly damage lung tissue. For example, inhalation of grain dust aerosols (with entrained fungus spores and toxins) can cause serious pulmonary inflammation; inhalation of aflatoxin-containing particles probably can result in lung cancer. Limited evidence exists indicating that inhalation of toxins (at least for aflatoxin) may be as much as 1000 times more dangerous than ingestion (Sorenson et al., 1984).

The fungi produce a variety of toxic substances, although only a few of the many thousands of fungi have been extensively studied with respect to toxin biosynthesis (Rodericks et al., 1977; Shank, 1981). The most familiar fungal toxins are the antibiotics (e.g., penicillin), cyclosporin (an immunosuppressive agent used in transplant procedures), the trichothecenes, and the aflatoxins. Most of the fungi that produce toxins are saprophytes and have broad nutritional requirements. Many of the saprophytic fungi commonly grow on dead or dying crop plants or on stored agricultural material. Some fungi (e.g., *Penicillium chrysogenum*) excrete toxin (penicillin) into the environment but do not store it within the fungus cell structure. Other fungi, including *Stachybotrys atra* and *Aspergillus flavus*, accumulate toxins (satratoxin and aflatoxins respectively) in the fungus spores and mycelium as well as in the substrate (Wicklow and Shotwell, 1983). When material contaminated with toxigenic fungi is disturbed, spores and fragments of both the substrate and the fungus mycelium become airborne, and the toxins are inhaled.

Human health risks associated with mycotoxins are directly related to exposure, although it is possible that other factors (such as smoking) may have a synergistic effect.

Opportunistic Infections Few fungi cause human infectious disease (see Chapter 12, this volume). The highly infectious pathogenic fungi (e.g., *Histoplasma, Blastomyces, Coccidioides*) are not known to be associated with crop plants although they do have natural reservoirs in the environment (Rippon, 1988). In order for a fungus to cause human invasive disease, it must be able to grow at body temperature (37°C) and to tolerate the tissue environment, including reduced oxidation-reduction potential and characteristic CO_2 and O_2 tensions (Rippon, 1988). Fungi that cause infectious disease do so only in people who are compromised with an underlying disease or other factor that has reduced immune system function. Acquired immunodeficiency syndrome (AIDS), some forms of cancer, and drug-induced immunosuppression (as for example with cyclosporin used to prevent transplant rejection, or high-dose steroid treatment) increase the risk for invasive fungus disease. The level of exposure necessary to cause disease depends on the immune status of the host. In severely immunodeficient people, a single spore of an appropriate fungus is probably sufficient.

11.3 Crop Plants as Sources for Disease-Causing Bioaerosols

Most of the literature on fungus aerosols and agriculture-associated disease concentrates on indoor environments where agricultural products are handled. It is clear, for example, that handling moldy hay presents a risk for hypersensitivity pneumonitis (farmer's lung disease). Note that the "mold" in these cases is usually one or more thermophilic actinomycetes. However, it seems obvious (at least intuitively) that crops in the field support large microbial populations, and that, as the plants begin to die, these populations include an abundance of fungi (Darke et al., 1976). This discussion, then, will be limited to parameters affecting exposure to fungal aerosols in the outdoor environment, and primarily to those environments associated with harvest of field crops.

Fungal Aerosols Associated with Crop Plants Measured levels of biological aerosols are strongly dependent on the method used to collect the sample and the method of enumeration. In general, suction samplers have been used for most farm-related studies (including filter cassettes, Andersen culture plate impactors, particulate cascade impactors, and slit samplers). Although these samplers have not been carefully compared, major disparities in sample collection efficiency over the size range of interest are unlikely. Enumeration methods, however, have varied from one investigator to the next. Some investigators use light or fluorescence microscopy for "total" particle counts; others use cultural methods. Other than in a very general way, methods relying on viability cannot be compared to total count methods.

Cultural methods select not only for viability, but for "culturability," and some fungus spores typically present in agricultural situations (e.g., *Cercospora*) do not readily grow in mixed cultures on broad spectrum media. On the other hand, small, colorless, fungus spores are extremely difficult to count on particulate collections and may be seriously underestimated if only direct counts are used. Thus not only are total numbers exceedingly difficult to compare, but cultural methods will produce a different list of fungal taxa than particulate counts (even when both are applied to the same sample, Burge et al., 1977). Palmgren et al. (1986), using filter cassette samples, compared counts of particles stained with acridine orange to counts obtained from dilution culture of particles washed from the filter membranes. Viable counts ranged from 7–91% of total counts estimated by

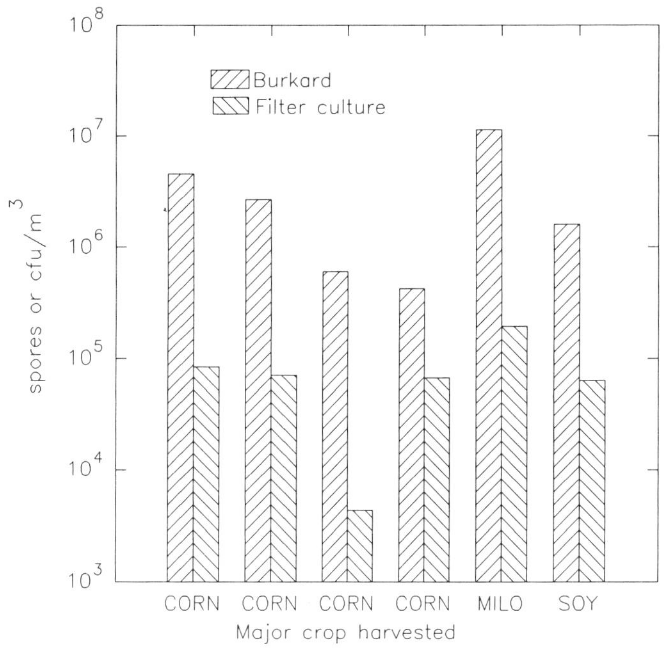

Figure 11.1 Data from a single Burkard personal spore trap (microscopic spore counts) and a single filter cassette sampler (cultural analysis) collected during a single fall harvest day, downwind of an operating harvester on each of six farms.

microscopy, depending on the type of aerosol present. The lowest viability percentages were for samples taken in a facility housing pigs, where one would expect high numbers of bacteria to dominate in the wet organic debris. The highest viability percentages were for samples taken in dryer grain-and wood fuel chip-handling situations, where fungus spores dominated the aerosol. In a similar study, Lacey et al. (1987) present data indicating a range in viability of 12–63% when light microscopy of unstained preparations and culture are compared, and 3 to 340% when fluorescence counts and culture are compared. In all situations except the pig house (where only bacteria were recovered), light microscopy produced higher counts than fluorescence.

We used filter cassettes for cultural counts and the Burkard spore trap for particulate counts (Figure 11.1). The filter sampler pulls air through a membrane with a known maximum pore diameter (0.8 μm). Only particles considerably smaller than 0.8 μm will penetrate the filter (i.e., the filter collects 100% of spores greater than 0.8 μm in diameter). The Burkard trap impacts particles on an adhesive-coated slide. Assuming the entrained particles hit an "empty" site, 50% of particles 5 μm in diameter are collected. Smaller spores are less efficiently trapped, while larger spores are

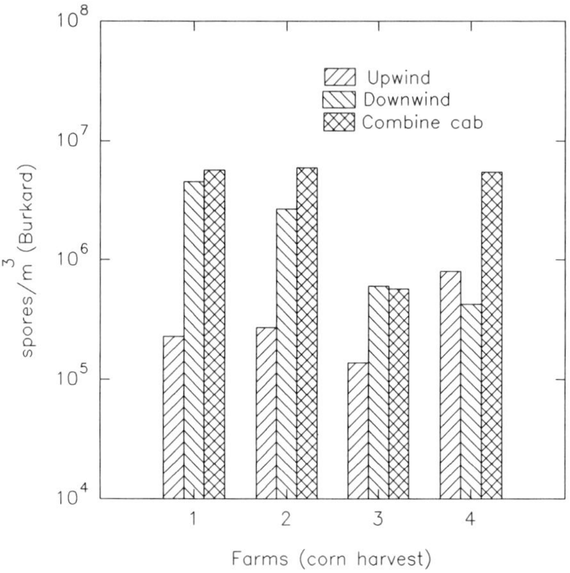

Figure 11.2 Single Burkard personal spore trap samples (during a single fall harvesting occasion on each farm) collected at four sites: upwind; in the spore cloud generated by the combine (downwind); and inside the combine cab.

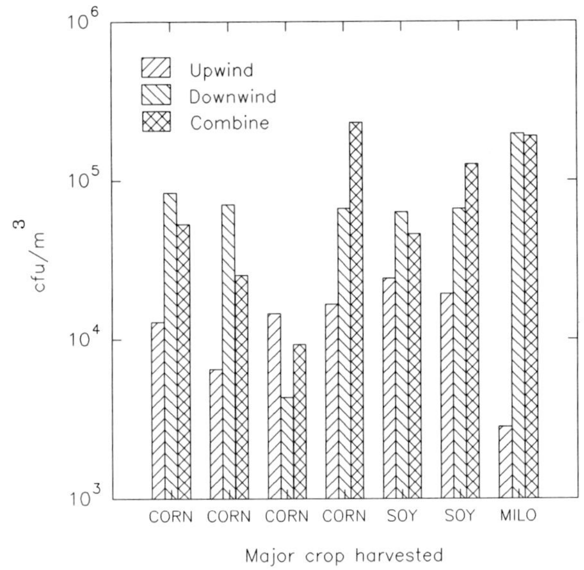

Figure 11.3 Single filter cassette samples related to harvesting operations collected on seven farms at three sites: upwind; in the spore cloud generated by the combine (downwind); and inside the combine cab.

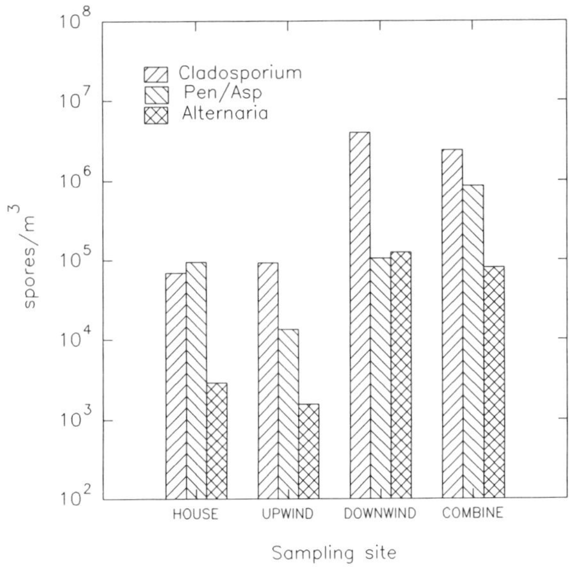

Figure 11.4 Fungus taxa recovered on single Burkard personal spore trap samples at four sites: inside the farm residence (house); upwind of harvesting operations; in the spore cloud generated by the combine (downwind); and inside the combine cab (combine) (average of seven farms; Pen/Asp = *Penicillium/Aspergillus*).

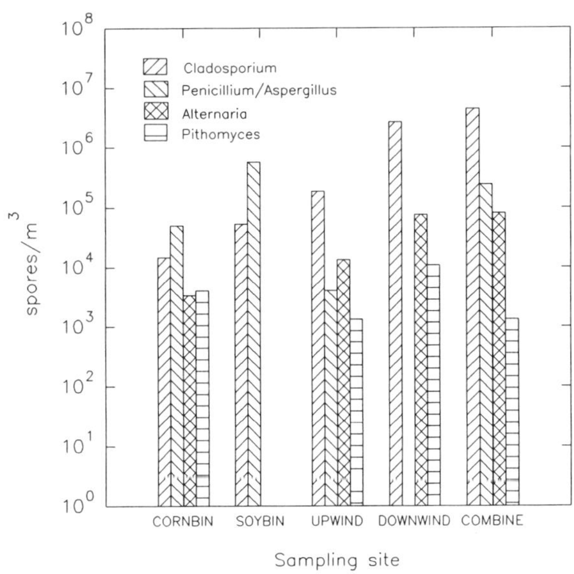

Figure 11.5 Fungus taxa recovered on a single occasion on a single Burkard personal spore trap sample at each of five sites: in the air stream exiting from a corn dryer and from a soybean dryer (cornbin, soybin); upwind of harvesting operations; in the spore cloud generated by the combine (downwind); and inside the combine cab (combine).

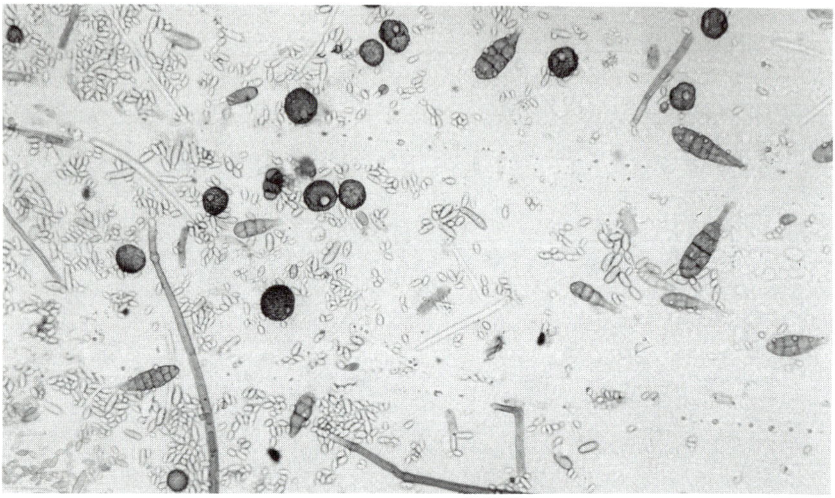

Figure 11.6 Photomicrograph of a spore aerosol from a Burkard personal spore trap sample collected in a closed, air-conditioned combine cab during a fall corn harvest.

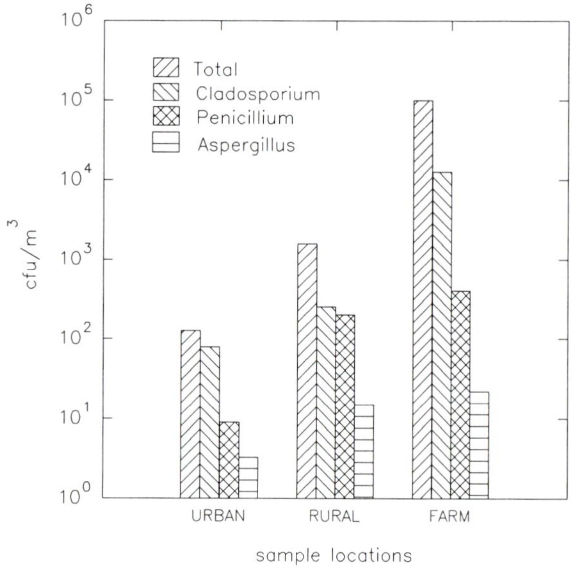

Figure 11.7 Cultural analysis of the major fungus taxa recovered from filter cassettes worn for 8 hours as personal samplers by urban dwellers (n = 8), farm wives (rural, n = 10), and farmers during harvesting operations (farm, n = 10).

more efficiently collected. As sampling proceeds, and the adhesive surface becomes covered with particles, overall efficiency of collection falls. In spite of these theoretical differences in efficiency, Burkard counts were 5–10 times or more higher than concomitantly collected filter cassette recoveries analyzed by cultural methods. The most likely factor causing these differences is culturability (including viability) of the aerosol.

In addition, sites chosen for sampling can have a dramatic impact on recoveries. Lacey (1987a) compared cascade impactor counts (using light microscopy) for a sampler located behind a combine and one worn by the operator in the cab. Spore levels behind the combine ranged from $3–7 \times 10^7$ per m^3 of air; those in the cab were an order of magnitude lower ($3–18 \times 10^6$ per m^3 of air). We compared Burkard total counts and filter cassette cultures upwind from combines, directly downwind (close to the rear of the machine), and in the cab (Figures 11.2 and 11.3). In four of the seven farms studied, levels in the combine cab were higher than those in the dust cloud behind the machine. In addition, on two farms where total Burkard counts in the cab were lower than those downwind, levels of *Penicillium/Aspergillus* type spores were higher in the cab than downwind (Figures 11.4 and 11.5) indicating that these fungi may have colonized substrates within the combine. Although all the combines we studied were air conditioned, the operators frequently opened the door while in dense spore clouds, many machines had large openings in the cab which allowed penetration of the aerosol, and none of the filters were properly maintained.

Thus, in spite of potential sources of error, the data indicate that farming activities (including harvesting of field crops) impose a significant risk of exposure to high levels of fungus spores. Figure 11.6 is a photograph of a Burkard spore trap sample collected during combine operations at a site with relatively low fungus spore recovery levels. Fungus spore levels in ambient urban air rarely reach levels above 10,000/m^3 (AAAI, 1989). On the other hand, the lowest total particle count levels reported by Lacey in air-conditioned combine cabs were 5×10^5 spores per m^3 of air and the average for the lowest year of sampling was 4.2×10^6 spores per m^3 of air (Lacey, 1987a). We compared personal exposure to viable fungi (using filter cassette cultures) of farmers, their wives, and people from a nearby town (Figure 11.7). Farmers were exposed to viable spore levels almost three orders of magnitude (1000-fold) higher than urban dwellers, and two orders of magnitude (100-fold) higher than their wives. These exposures may contribute to diseases such as grain fever, farmer's lung disease, and asthma.

While all fungi are probably allergenic (can cause asthma and/or hay fever in susceptible people), some taxa present the risk of other diseases as well. A number of studies have examined the aerosols resulting from handling moldy hay (e.g., Lacey, 1972), working in grain elevators (DoPico et al., 1977; Smalley et al., 1986), and other indoor farming activities. The thermophilic actinomycetes are usually dominant when moldy hay is handled, and aerosol levels can exceed 10^9 spores/m^3 of air. In grain elevators and other places where grain is handled and stored after harvest, species of *Penicillium, Aspergillus,* and yeasts are abundant (Smalley et al., 1986). Figure 11.5 compares Burkard spore counts from effluents of corn and soybean dryers to counts from combining operations. The dominant spore type at the dryers was *Aspergillus/Penicillium*. Cultural studies demonstrated that the soybeans were supporting active *A. flavus* growth and the corn bin contained a variety of *Aspergillus* and *Penicillium* species. On the other hand, aerosols from combines are nearly always dominated by *Cladosporium* species (Hill et al., 1984; Darke et al., 1976; Lacey, 1987, Figures 11.4, 11.5, 11.6, and 11.7). It has been suggested that the "field" fungi die after harvest to be replaced by the "storage" fungi (Baxter et al., 1981). It is more likely, however, that field fungi are leaf and stem saprophytes that are left in the field, while the storage fungi are those able to invade the seeds either before or after harvest. Crop-related differences in saprophytic fungus colonization and growth have not been adequately examined.

Of special concern are the risks of exposure to the toxins produced by some fungi. Although the potent trichothecene toxins are known to be produced by saprophytic fungi that occur on crop plants (e.g., *Fusarium, Stachybotrys*), attention with respect to field-crop exposures has been directed primarily toward the aflatoxins. *Aspergillus flavus* and *A. parasiticus* are known to invade stored grains (including peanuts) and to have caused severe acute toxic effects when such grain was eaten (Rodericks et al., 1977; Shank, 1981). Aflatoxin has been recovered from air during handling of contaminated peanuts (Sorenson et al., 1984; Van Nieuwenhuize et al., 1973) and corn (Burg et al., 1982). In addition, it is now clear that *A. flavus* contaminates corn in the field, as well as during storage (McMillian et al., 1978; Marsh et al., 1973) and aflatoxin becomes airborne during harvesting in heavily contaminated fields (Burg et al., 1982). It has also been demonstrated that high levels of aflatoxin (> 100,000 ng/gram) is contained in spores of *A. flavus* (Wicklow and Shotwell, 1983), so that the presence of *A. flavus* spores in a sample is presumptive evidence for the presence of aflatoxin.

The contribution of these agriculturally generated aerosols to the ambient air spora in adjacent or downwind urban areas remains to be documented. Our data (Figure 11.7) indicate that urban dwellers are exposed to essentially the same fungi as are farmers during harvest, and, further, that urban spore peaks tend to be associated with agricultural activities in adjacent land areas (AAAI, 1989). Long-term sampling studies combined with analysis of weather parameters and other factors known to influence airborne spore levels will be necessary to verify these observations.

Evidence for Human Disease Related to Crop Plant Aerosols As mentioned above, serious diseases have been linked to the handling of contaminated hay and grain in indoor situations. However, very little re-

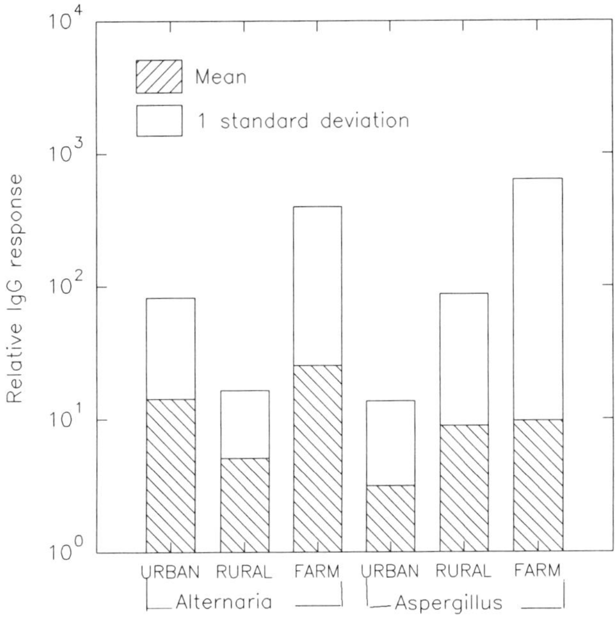

Figure 11.8 IgG levels specific for *Alternaria alternata* and *Aspergillus fumigatus* in urban dwellers (urban, n = 8), farm wives (rural, n = 10), and farmers (farm, n = 10).

search has focused on risks of diseases associated with harvest operations. Since the aerosol composition for outdoor/harvest conditions is so different from indoors, extrapolation from indoor environments is not useful. Darke et al. (1976) studied 78 workers employed in the harvesting and/or handling of grain. Of these, 23% reported symptoms of respiratory distress including coughing, wheezing, and (often severe) breathlessness. Complainants were more likely to react to skin test extracts and to respond to a greater variety of materials than noncomplainants. Incidence of precipitins to fungi recovered from dust was approximately equal in both complainants and noncomplainants, providing some evidence that development of antigen-specific IgG is related to exposure rather than a specific disease state. None of the workers tested had precipitating antibodies to the standard farmer's lung antigens (thermophilic actinomycetes, *Aspergillus fumigatus*).

We measured antigen-specific IgG (using an enzyme-linked immuno sorbent assay) to several species of *Aspergillus* and to *Alternaria alternata* in our personal exposure studies described above (Figure 11.8). Significant between-group differences were not found for any antigen, although farmers were more likely than members of other groups to have elevated specific IgG as determined by our clinical standards. We also examined reactivity to *Cladosporium* using an IgG immunoblot technique. In this technique, extracts are electrically separated in polyacrylamine gel on the basis of molecular weight, transferred to nitrocellulose paper, and incubated with human serum and enzyme-labelled anti-human IgG antibodies. Although all subjects had specific IgG for at least some *Cladosporium* antigens, farmers were positive for a greater number of allergens than either their wives or town dwellers. These data indicate that, although *Cladosporium*-specific IgG is produced in response to ambient exposures (which usually range from 1000–10,000/m^3), some *Cladosporium* antigens require higher exposure levels to stimulate the development of specific antibodies. The relevance of these findings to disease has not yet been evaluated.

Whether or not disease related to toxin exposure occurs as a result of harvesting field crops remains unclear. Exposure data are summarized above, but are too limited to allow adequate risk analysis. Van Nieuwenhuize et al. (1973) reported elevated cancer levels in a group of workers at a peanut and linseed oil processing plant. Aflatoxin exposure was estimated to be 0.03 to 2.5 μg of aflatoxins inhaled per 45-hour week. However, his sample size was too small for statistical conclusions. In areas where *Aspergillus flavus* is known to be a problem (e.g., in the hot, humid parts of the world), the potential risks for aflatoxin exposure should be recognized. Although other toxigenic fungi do grow on field crops (the dominant *Cladosporium* species produce toxins), health effects related to toxin exposure have yet to be examined.

11.4 Conclusions and Future Directions

Fungal aerosols are certainly produced during harvesting of field crops. These aerosols are composed primarily of fungal saprophytes, and *Cladosporium* is usually overwhelmingly dominant. Aerosols generated by grain harvest operations are qualitatively different from those associated with indoor farming activities, especially those involving the handling of stored organic material. Harvest-generated aerosols probably contribute significantly to the ambient fungal aerosol in the surrounding farm and nonfarm environments. These aerosols constitute a health hazard for atopic individuals in general and for farmers (atopic or not) who harvest crops infested with toxin-producing fungi.

It is difficult to imagine a means for controlling the production of such aerosols. Fungicides would probably decrease fungal levels, but at an unacceptable risk to the farmers and to the environment. Harvest timing (both with respect to time of day and length of time after leaf and stem senescence) may contribute to the composition of fungal aerosols, although this has not been studied. Even if a correlation can be shown, farmers are unlikely to schedule harvesting operations specifically for controlling the aerosols.

Means for protection of farmers are available. Respiratory protection can be worn for severely affected (asthmatic) individuals, or in high-risk situations (where aflatoxin is known to be a problem). Proper maintenance of combine cab ventilation systems would go far to reduce the farmer's exposure.

Given the dearth of published studies dealing with field-crop fungal aerosols, research in virtually any aspect of this interesting field will be valuable. Specifically, both qualitative and quantitative information is needed on the composition of aerosols as a function of such factors as the crop, weather, soil, humidity, and time of harvest. Epidemiological studies are necessary to establish the risks of disease associated with harvesting (and possibly other field operations) that are related to exposure to fungal aerosols. Such studies should include environmental investigations concurrent with health studies.

Agricultural aerosols should be seriously considered by those attempting to do modeling of airborne fungal prevalence. It is probable that harvesting, and perhaps also other farming activities, contribute significantly to the ambient aerosol, providing a major confounding factor in models based on "natural" environmental factors alone.

The authors thank the farmers and their families who generously allowed us to interrupt their harvesting operations. This work was supported in part by NIAID Grant Number AI10181 from the National Institutes of Health, USPHS.

References

American Academy of Allergy and Immunology (AAAI). 1989. *Pollen and Spore Counts*. American Academy of Allergy and Immunology, Milwaukee, WI.

Baxter, C.S., Wey, H.E., and Burg, W.R. 1981. A prospective analysis of the potential risk associated with inhalation of aflatoxin-contaminated grain dusts. *Food and Cosmetic Toxicology* 19:765–769.

Burg, W.R., Shotwell, O.L., and Saltzman, B.E. 1982. Measurements of airborne aflatoxins during the handling of 1979 contaminated corn. *American Industrial Hygiene Association Journal* 43:580–586.

Burge, H.P., Boise, S.R., Rutherford, J.A., and Solomon, W.R. 1977. Comparative recoveries of airborne fungus spores by viable and nonviable modes of volumetric collection. *Mycopathologia* 61:27–33.

Darke, C.S., Knowelden, J., Lacey, J., and Ward, A. 1976. Respiratory disease of workers harvesting grain. *Thorax* 31:294–302.

DoPico, G.A., Reddan, W., Flaherty, D., Tsiatis, A., Peters, M.E., Rao, P., and Rankin, J. 1977. Respiratory abnormalities among grain handlers. *American Review of Respiratory Disease* 115:915–927.

Hill, R.A., Wilson, D.M., Burg, W.R., and Shotwell, O. 1984. Viable fungi in corn dust. *Applied Environmental Microbiology* 47:84–87.

Kendrick, B. 1985. *The Fifth Kingdom*. Mycologue Publications, Waterloo, Ontario. 364 pp.

Lacey, J. 1972. Actinomycete and fungus spores in farm air. *Journal of Agricultural and Labour Science* 1:61–78.

Lacey, J. 1987a. Airborne spores in the farm environment and respiratory disease. pp. 117–120 in *Aerosols: Their Generation, Behaviour and Applications*. Proceedings of The Aerosol Society First Conference, Loughborough University of Technology, 31 March–1 April 1987. Aerosol Society, London.

Lacey, J. 1987b. Airborne fungus spores in outdoor air. pp. 125–128 in *Aerosols: Their Generation, Behaviour and Applications*. Proceedings of The Aerosol Society First Conference, Loughborough University of Technology, 31 March–1 April 1987. Aerosol Society, London.

Lacey, J., Eduard, W., Karlsson, K., Palmgren, U., Strom, G., and Blomquist, G. 1987. The assessment of filter samples from agricultural environments highly contaminated with microorganisms. pp. 77–82 in *Aerosols: Their Generation, Behaviour and Applications*. Proceedings of The Aerosol Society First Conference, Loughborough University of Technology, 31 March–1 April 1987. Aerosol Society, London.

Malmberg, P., Rask-Andersen, A., Palmgren, U., Haglund, S., and Kolmodin-Hedman, B. 1987. Respiratory problems among Swedish farmers, correlation between symptoms and environment. *European Journal of Respiratory Disease Supplement* 154:22–27.

Marsh, P.E., Simpson, M.E., Craig, G.O., Donaso, J., and Ramey, H.H. 1973. Occurrence of aflatoxin in cotton seed at harvest in relation to location of growth and field temperatures. *Journal of Environmental Quality* 2:276.

McMillian, W.A., Wilson, D.M., and Widstrom, N.W. 1978. Insect damage, *Aspergillus* ear mold, and aflatoxin contamination in south Georgia corn field in 1977. *Journal of Environmental Quality* 7:565.

Palmgren, U., Strom, G., Blomquist, G., and Malmberg, P. 1986. Collection of airborne micro-organisms on Nuclepore filters, estimation and analysis–CAMNEA method. *Journal of Applied Bacteriology* 61:401–406.

Platts-Mills, T.A.E., Chapman, M.D., Pollart, S.M., Heymann, P.W., and Luczynska, C. 1990. Establishing health standards for indoor foreign proteins related to asthma: dust mite, cat and cockroach. *Toxicology and Industrial Health* 6:197–208.

Rippon, J.W. 1988. *Medical Mycology*. 3rd edition. W.B. Saunders Co., Philadelphia.

Rodericks, J.V., Hesseltine, C.W., and Mehlman, M.A. (editors). 1977. *Mycotoxins in Human and Animal Health*. Pathotox Publishers, Inc., Park Forest South, IL. 807 pp.

Shank, R.C. 1981. *Mycotoxins and N-nitroso Compounds: Environmental Risks, vol. I*. CRC Press, Boca Raton, FL. 285 pp.

Smalley, E.B., Burkholder, W.E., Caldwell, R.W., Mai, S.H., Phillips, J.K., and Whidden, M.P. 1986. *Microbial Flora and Fauna of Respirable Grain Dust from Grain Elevators*. United States Department of Health and Human Service, National Institute of Occupational Safety and Health, Morgantown, WV. 95 pp.

Sorenson, W.G., Jones, W., Simpson, J., and Davidson, J.I. 1984. Aflatoxin in respirable airborne peanut dust. *Journal of Toxicology and Environmental Health* 14:525–533.

Van Nieuwenhuize, J.P., Herber, R.F.M., DeBruin, A., Meyer, I.P.B., and Duba, W.C. 1973. Aflatoxins—Epidemiological study on the carcinogenicity of prolonged exposure to low levels among workers of a plant. *Tijdschr. Soc. Geneesk* 51:754–769.

Wicklow, D.T. and Shotwell, O.L. 1983. Intrafungal distribution of aflatoxins among conidia and sclerotia of *Aspergillus flavus* and *Aspergillus parasiticus*. *Canadian Journal of Microbiology* 29:1–5.

12

Association Between the Human Pathogen *Sporothrix schenckii* and Sphagnum Moss

Dennis M. Dixon and Ira F. Salkin

12.1 Introduction

The discipline of medical mycology can be divided into a study of: 1) mycoses (fungal infections of man and other animals); 2) mycotoxicoses (toxic fungal metabolites); 3) mycetismus (mushroom poisoning); and 4) fungal allergies (Rippon, 1988; see also Chapter 11). The mycoses are subdivided on the basis of the extent of invasion of the host and on the basis of the innate virulence of the etiologic agents, i.e., primary pathogens or opportunistic pathogens (Table 12.1). The primary pathogens are those that readily cause infection in the normal (healthy) host, whereas the opportunistic pathogens such as *Candida albicans* and *Aspergillus fumigatus* typically only infect an immunocompromised or otherwise debilitated host. The medically important fungi are noted for demonstrating specific ecological habitats and geographical distributions. *Cryptococcus neoformans*, the etiological agent of cryptococcosis and a major pathogen in immunosuppressed patients, is associated with pigeon guano and is distributed throughout the world. Other examples include *Coccidioides immitis* in the soil of the San Juaquin Valley, California; *Histoplasma capsulatum* in starling, chicken, and bat guano, and *Sporothrix schenckii* in sphagnum moss. Epidemics have been caused by all of these fungi.

A complete understanding of an infectious disease requires a knowledge of the distribution of the causal fungus in the environment. This is not unique to sporotrichosis, but is a general consideration in any infectious disease. The purpose of this presentation is to discuss the significance of

Table 12.1 Clinical classification of the major mycoses

Classification	Characteristics	Examples[a]
Superficial	Restricted to nonliving portions of host; little or no pathology	Tinea versicolor
Cutaneous	Infection of the skin and its appendages; pathology usually results	Dermatophytoses, candidiasis
Sub-cutaneous	Infection of the skin and sub-cutaneous tissue, usually as a result of traumatic inoculation	Sporotrichosis, chromoblasto-mycosis, mycetoma, phaeo-hyphomycosis[b]
Systemic	Infection of multiple organ sys-tems in what can be a life-threatening disease.	Blastomycosis, coccidioido-mycosis, histoplasmosis, candidiasis, cryptococcosis

[a] A disease can be grouped into more than one category depending upon clinical presentation.
[b] Phaeohyphomycosis can be classified as superficial, cutaneous, subcutaneous, or systemic, depending upon the site of infection and etiology (McGinnis, 1983).

fungal ecology in medical mycology using as a paradigm the largest reported U.S. epidemic of sporotrichosis.

Sporotrichosis is a subcutaneous fungal infection caused by the primary pathogen, *Sporothrix schenckii*. Infection is thought to be initiated by traumatic implantation of the fungus into the skin or subcutaneous tissues. The resulting disease may be of the fixed cutaneous, lymphocutaneous, or extracutaneous type. Cutaneous lesions typically begin as nodules which enlarge, ulcerate, and drain. In fixed cutaneous sporotrichosis, the lesions do not extend beyond the primary inoculation site. In lymphocutaneous sporotrichosis, contiguous spread of the disease occurs along the lymphatics to result in a chain of lesions. Extracutaneous spread may occur, but it is rare; bone is the most commonly involved site after skin. Dissemination from a cutaneous site to visceral organs is exceedingly rare. However, dissemination with multiple organ involvement may be a consequence of primary pulmonary sporotrichosis, which is a disease usually encountered only in immunosuppressed patients (Rippon, 1988).

12.2 Medical Mycology, Epidemiology, and Ecology

Epidemiology and ecology occupy prominent places in the traditional study of the mycoses. For example, the dermatophytoses are grouped as to geophilic, zoophilic, and anthropophilic species. Knowledge of the source of the organisms is important in establishing the mode of infection and has implications in public health. Infection with a zoophilic species such as

Table 12.2 Habitats and geographical distribution of some representative systemic fungal pathogens of humans

Mycosis	Fungus	Habitat	Geographical distribution
Candidiasis	*Candida albicans*	Intestines, skin of humans	Worldwide
Coccidioido-mycosis	*Coccidioides im-mitis*	Sandy soil	California, Arizona, New Mexico, Central and South America
Crypto-coccosis	*Cryptococcus neoformans*	Guano of pigeons	Worldwide
Blastomycosis	*Blastomyces der-matitidis*	Woody plant mate-rial; association with river banks and beaver lodges	Eastern USA and Canada; especially Kentucky, Wis-consin, Michigan
Histoplas-mosis	*Histoplasma capsulatum*	Guano of starlings, chickens, and bats	Essentially worldwide, especially USA along the Mississippi River Valley

Trichophyton verrucosum is often quite inflammatory and suggests the need to seek out the animal serving as the source in order to prevent additional cases. Ecological and epidemiological considerations are particularly important for the systemic mycoses. For example, systemic diseases such as coccidioidomycosis, paracoccidioidomycosis, and blastomycosis are of such restricted geographical distribution that they are not usually included in the differential diagnosis of an infectious disease without the patient's residence in or travel to an endemic geographical region. Coccidioidomycosis is perhaps the best studied of the systemic mycoses in terms of the ecology of the etiologic agent, *Coccidioides immitis*. Although the ecological niche remains unknown, the fungus is known to inhabit soils of the Lower Sonoran Life Zone that are rich in $CaSO_4$ and borates (Rippon, 1988). The other well-documented associations between specific etiologic agents of systemic mycoses and ecological habitats are summarized in Table 12.2. *Candida albicans* is included as an example of a fungal pathogen that is characteristically endogenous in origin. The other infectious agents are of exogenous origin.

The subcutaneous mycoses include chromoblastomycosis, mycetoma, phaeohyphomycosis and sporotrichosis. The clinical diseases are heterogeneous, and the etiologic agents are equally diverse. Common factors are the mode of infection, which is usually traumatic implantation, the relative ubiquity of the etiological agents in nature, and the fact that the hyphae and/or conidia of most of the etiological agents are melanized by a common pigment known as dihydroxynaphthalene (DHN) melanin (Dixon et al., 1990). The presence of DHN melanin in the cell walls of these fungi is thought to confer a selective advantage on the organisms, allowing them to survive desiccation and ultraviolet (UV) irradiation (Wheeler and Bell, 1987).

Numerous studies have documented the recovery of the etiological agents of chromoblastomycosis, mycetoma, and phaeohyphomycosis from nature (Dixon et al., 1980; Gezuele et al., 1972; Iwatsu et al., 1981; Okeke and Gugnani, 1986). Common sample materials have been structural wood, decaying wood from dead trees, other plant debris, and soil. The role of the phyllosphere has not been systematically investigated. In a study where leaves were cultured to examine for the presence of pathogenic dematiaceous fungi, mostly saprophytic species were found (Dixon, 1982). In other studies based on examination of potted plants and the epidemiology of the mycoses, soil but not leaf surfaces has been the typical source of the pathogens (Staib, 1984; Staib et al., 1978; Kenyon, 1984; Summerbell et al., 1989). In one such study, *Sporothrix schenckii* was isolated from 2 of 12 samples of potting soil (Kenyon et al., 1984).

12.3 Sporotrichosis

Sporotrichosis occurs worldwide, and infection is commonly associated with decaying vegetation. Occupations or activities such as gardening, horticulture, forestry service, and gold mining have been associated with epidemics. The largest epidemic occurred in Witwaterstrand, South Africa, and involved over 3000 cases. *Sporothrix schenckii* was subsequently isolated from timbers in the mines; workers presumably acquired the infection when brushing against the contaminated timbers (Rippon, 1988). The fungus has also been isolated from such diverse sources as the floor of a swimming pool (Staib, 1983), from desiccated mushrooms (Kazanas, 1987), straw, armadillo burrows (Mackinnon et al., 1969), soil, and sphagnum moss (Rippon, 1988). Numerous interesting reports are available on the ecology and epidemiology of this fungus (Findlay et al., 1984; Findlay, 1970; Fukushiro, 1984; Howard and Orr, 1963; Mackinnon et al., 1969; Mariat, 1975; Rippon, 1988; Shippee, 1970).

At least eight epidemics of sporotrichosis related to sphagnum moss have occurred in the USA (Table 12.3). In seven of these, the moss originated from bogs in Wisconsin. *Sporothrix schenckii* was isolated from stored sphagnum moss at various sites in six of the eight epidemics, but not from any moss directly from bogs in Wisconsin.

12.4 Analysis of an Epidemic

The largest recorded outbreak of sporotrichosis occurred in 1988 (Coles, F.B., Schuchat, A., Hibbs, J.R., Kondracki, S.F., Salkin, I.F., Dixon, D.M.,

Table 12.3 Data on U.S. epidemics of sporotrichosis involving sphagnum moss

No. of cases	Location	Occupation	Moss source	Author	Year
6	Indiana[a]	Florist	Not reported	Gastineau et al.	1941
8	Florida	Nursery worker	Wisconsin	Crevasse and Ellner	1960
5	Michigan	Nursery worker	Wisconsin	Hayes	1960
14	Vermont[a]	Nursery worker	Wisconsin	D'Alessio et al.	1965
7	Oregon[a]	Botanist	Not reported	CDC	1969
17	Mississippi[a]	Forestry worker	Wisconsin	Powell et al.	1978
3	Minnesota[a]	Nursery worker	Wisconsin	Grotte and Younger	1981
4	Wisconsin[a]	Garden-shop worker	Wisconsin	CDC	1982

[a] *Sporothrix schenckii* was cultured from moss samples at the epidemic sites, but not from the bogs in Wisconsin.

Chang, H.G., Duncan, R.D., Hurd, N.J., and Morse, D.L., in preparation; Dixon et al., 1991). Eighty-four cases of primary cutaneous sporotrichosis were detected in 15 states (Table 12.4). Most of the cases were associated with sphagnum moss obtained from a single supplier in Wisconsin. The epidemic occurred in conjunction with the annual reforestation programs in Illinois and New York, and accordingly, most cases were from patients residing in those two states.

Table 12.4 Distribution by patient residence of cases of sporotrichosis in the 1988 epidemic

State of residence	No. of cases
New York	25
Illinois	24
Pennsylvania	15
Ohio	5
Wisconsin	3
Connecticut	2
Michigan	2
Vermont	2
Indiana	1
Iowa	1
Maryland	1
Massachusetts	1
New Hampshire	1
North Carolina	1
Virginia	1
Total	84

Samples and Sampling The Laboratories for Mycology of the Wadsworth Center for Laboratories and Research, New York State Department of Health, received samples of moss collected by other investigators. The samples originated from collections from five Pennsylvania nurseries, from the New York State Soil and Water Conservation (NYSWC) Extension Services in various New York counties, and from bogs in Wisconsin from which the moss was harvested. The purpose of the studies was to determine the source of the *S. schenckii* and to determine the point at which the moss was contaminated with the fungus. As will be shown, it was possible to isolate *S. schenckii* from eight samples of sphagnum moss at various epidemic sites, but not from fresh moss at the bogs in Wisconsin.

The samples consisted primarily of sphagnum moss, evergreen seedlings, soil, and water (Table 12.5). We received the water samples in glass jars and the other samples in plastic bags from June through August of 1988. The samples were processed by an aqueous extraction technique that was demonstrated previously to be successful in isolating dematiaceous fungal pathogens from nature (Dixon et al., 1980). In brief, the procedure entailed placing 25 ml loose sample volumes in flasks, adding antibacterial drugs (penicillin and streptomycin) in water to reduce bacterial contamination, preparing suspensions, and then making dilutions for direct plating (0.2 ml/plate) on Mycosel agar.

Characterization and Identification of *Sporothrix* Isolates All representative colonies were examined for macroscopic and microscopic characteristics consistent with *Sporothrix schenckii*. This required examination of all colonies that were white, light-brown to dark-brown, or black. Isolates suspected of being *Sporothrix schenckii* were deposited in a sterile water culture collection and stored at room temperature for critical evaluation (Dixon et al., 1991).

Selected isolates were compared on the basis of the following characteristics: macroscopic morphology was evaluated on Sabouraud dextrose agar (S+; Difco) supplemented with 40 μg gentamicin/ml, Mycosel agar (Myco; BBL), potato dextrose agar, malt extract agar, and cornmeal agar supplemented with 1% Tween 80 (PDA, malt, and CM; Difco). Plates were inoculated, incubated at 27°C, and colony morphology evaluated at 7 days. Microscopic morphology was evaluated with PDA slide cultures incubated at 30°C for 14 days. Exoantigen test, growth at 35 and 37°C, conversion to a yeast phase, and determination of virulence by injection of $\geq 10^6$ conidia intravenously in mice were determined as described elsewhere (Dixon et al., in preparation).

Sixty-three environmental samples associated with the epidemic were evaluated for the presence of *Sporothrix schenckii* (Table 12.5). The direct plating of dilutions of aqueous samples on Mycosel agar resulted in isolated fungal colonies for quantitative and qualitative study. Isolates producing one-celled conidia sympodially on terminal or lateral conidiophores were

Table 12.5 Summary of the properties of environmental samples and the culture results

Material	Condition	No. neg.[a]	No. pos.[a]	Source[b]	No. of isolates: Virulent[e]	Avirulent
Burlap	Dry	1		N–A		
Dirt	Dry	1	1	N–A	0	1
		1		N–E		
Dirt and grass	Wet	1		N–B		
Dirt and needles	Dry	1		N–C		
Hay	Dry	1		N–A		
Moss	Dry	4	7	N–A	1	11
		4		N–B		
		1	4	N–C	0	5
		4	1	N–D	0	2
		3	2	N–E	0	4
		2		N–F		
			4	BR[c]	2	6
		2		CH		
			2	DE[c,d]	0	3
			2	GE[d]	0	7
		3	1	HE[c,d]	0	2
		1		IL		
			2	MO[d]	1	1
			2	ON[d]	2	2
		10		WI		
Moss	"Mud"	1		WI	0	
Moss	Wet		1	N–A	1	2
		1		N–B		
		1	1	N–C	0	1
			1	N–D	0	1
			2	N–E	0	3
		2		WI		
Moss and dirt	Dry		1	N–A	0	1
Moss and needles	Dry		1	N–A	1	1
Moss and wood	Dry		1	N–A	0	2
Needles	Dry	1		N–F		
Packing tape	Dry	1		N–E		
Seedlings	Dry		2	IL	0	5
Water		8	1	N–A	0	1
		2		N–D		
	Dry	1		N–E		
Water and gel		2		N–A		
Wood		3		N–A		
Total		63	39		8	61

[a] For *Sporothrix* or *Ophiostoma*.
[b] Abbreviations. N–A through N–F = Pennsylvania (PA) nurseries A through F. IL = Illinois; WI = Wisconsin. NYS County Soil and Water Conservation Extension Service in the following counties: BR = Broome; CH = Chenango; DE = Delaware; GE = Genessee; HE = Herkimer; MO = Montgomery; ON = Onondaga.
[c] Moss and seedlings from Nursery A.
[d] Moss and seedlings from Nursery B.
[e] Virulence was determined by the ability to produce disseminated life-threatening infections in mice following intravenous injection of $\geq 10^6$ conidia/mouse.
Data from Dixon et al., 1991.

Table 12.6 Morphological classification of *Sporothrix* isolates associated with an epidemic of sporotrichosis[a]

	Group number					
Assessment	I	II	III	IV	V	VI
Morphology on S+						
Color						
Grey or taupe	+	−	−	−	−	−
Ivory to beige	−	+	+	−	+	+
Tan to brown	−	−	−	+	−	−
Appearance						
Small, slightly raised, smooth	+	−	−	−	−	−
Finely wrinkled with radial grooves	−	+	+	+	−	−
Flat, smooth, spreading	−	−	−	−	+	+
Texture						
Glabrous, waxy	−	+	+	+	−	−
Dull, granular	+	−	−	−	−	−
Coarse "ground glass"	−	−	−	−	+	+
Brown diffusible pigment	−	−	−	+	−	−
White aerial hyphae	−	−	−	−	−	+
Morphology on PDA slide						
Dematiaceous conidia	+	−	−	−	−	−
Perithecia	−	+	−	−	−	−

[a]The isolates were streaked on Sabouraud dextrose agar (S+) plates to check for purity and to compare morphology, and then inoculated on potato dextrose agar (PDA) slide cultures, which were incubated at 30°C and mounted at 14 days. The colony morphology on S+ and the presence of dematiaceous conidia or visible structures resembling perithecia on PDA slide cultures were used to divide the isolates into six general groups. Data from Dixon et al., 1991.

given a preliminary identification of *Sporothrix* sp. Thirty-nine of the environmental samples yielded 69 isolates meeting these criteria (Table 12.5). Additional studies were required to determine which were *S. schenckii*.

Twenty-one clinical isolates from the epidemic were included for comparative study. All were isolated from lesions on the upper extremities from the fingers to the forearms. Fourteen of the patient isolates were from New York, four were from Illinois, two were from Pennsylvania, and one was from Michigan. They were either reference cultures sent by other laboratories (14 isolates) or were isolated from primary clinical specimens sent to our laboratories (seven isolates).

The 21 patient isolates and the 69 environmental isolates were compared on the five different agar media described above. This enabled us to divide them into six arbitrary groups, numbers I–VI, on the basis of morphological observations (Table 12.6). The patient isolates were remarkably consistent. All produced dark brown to black colonies on Myco, CM, malt, and PDA, and white to grey colonies on S+. Microscopically, these isolates were characterized by the presence of oval (length-to-width ratio of ≤ 1.5), sympodial, hyaline conidia on conidiophores and dematiaceous conidia arranged in cylindrical layers about the hyphae. These characteristics were identical to those of environmental group I isolates. Environmental groups II–VI were distinguished by the absence of dark-brown to black pigment on all of the five media, by the absence of dematiaceous conidia borne in cylindrical layers along the main hyphal axis, and by the presence of clavate (length-to-width ratio >1.5; usually >2.0), hyaline, sympodial conidia. Environmental groups II-VI were differentiated from each other on the basis of the criteria listed in Table 12.6. Noteworthy was the production of perithecia and ascospores by group II. These isolates were identified as *Ophiostoma stenoceras* (Robak) Melin and Nannf.

The results from comparative studies of clinical versus environmental isolates are shown in Table 12.7. On the basis of all the studies, the 21 patient isolates and the eight isolates in environmental group I were identified as *Sporothrix schenckii* Hektoen and Perkins, group II isolates as *O. stenoceras*, and group III–VI isolates as *Sporothrix* species other than *S. schenckii*.

The two most important mycological characteristics for confirmatory identification were 1) the presence of melanized, oval, sympodial conidia on lateral conidiophores and 2) sleeves of melanized, oval conidia along the main hyphal axis. The production of melanized conidia corresponded to the production of dematiaceous colonies on permissive media such as Myco, malt, PDA, and CM. These characteristics also correlated with virulence in the mouse bioassay. Ability to convert to a yeastlike phase, ability to grow at 37°C, and the exoantigen test were not reliable characteristics for separating the species encountered in this study.

The virulent isolates in Table 12.5 represent *S. schenckii*. Thus, it can be seen that isolates of *S. schenckii* were recovered from the environment. Three of these were from moss from nursery A, and five were from moss at NYSWC sites that received moss from nursery A or other nurseries in the study. *S. schenckii* was not isolated from fresh moss from the bogs in Wisconsin.

Significance This is the ninth and largest American epidemic of sporotrichosis associated with sphagnum moss. In eight of these epidemics, the moss originated from bogs in Wisconsin. Although *S. schenckii* has been isolated from moss at the epidemic sites in this and in five of the previous epidemics, the fungus was not isolated from moss at the bogs in Wisconsin. This raises the question of where the moss becomes associated with *S. schenckii*. It is possible that the moss becomes contaminated after being harvested from the bogs and transported to distant sites. The situation may be analogous to the epidemiology of cryptococcosis. *Cryptococcus neoformans* can be recovered from the guano of pigeons, but normally not from the intestines of pigeons. Thus, it must originate from some as yet undetected source. In the case of *C. neoformans* var. *gattii*, the source was recently deter-

Table 12.7 Comparison of selected characters of clinical and environmental isolates

Group:	Number of isolates	No. showing growth at		Character			
		37°C	35°C	Melanized conidia	Conidial sleeves	Mouse mortality	ExoAg[a]
Clinical	21	21[b]/21[c]		+	+	+	+
Environmental							
Group I	10	8/8	2/2	+	+	+	+
Group II	5	0/0	0/4	−	−	−	+
Group III	13	2/4	7/9	−	−	−	+
Group IV	21	1/3	16/18	−	−	−	+
Group V	12	5/8	1/4	−	−	−	+
Group VI	8	4/6	2/2	−	−	−	+

[a]Exoantigen test: + indicates that one or more precipitin bands of identity were obtained in the double immunodiffusion assay reacting with positive control antiserum to *S. schenckii* with exoantigen extracts from the cultures in the study.
[b]Number of isolates that converted to a yeast phase.
[c]Total number of isolates capable of growth at given temperature.

mined to be *Eucalyptus camaldulensis* (Ellis and Pfeiffer, 1990). The authors speculated that an angiosperm may also be involved in the natural history of *C. neoformans* var. *neoformans*.

We know of no evidence to support the hypothesis that *S. schenckii* is present on moss in bogs in Wisconsin. In fact, sporotrichosis is not a risk factor for workers who harvest moss. Additionally, in a serological study of the 1988 sporotrichosis epidemic, there was no evidence of diagnostic agglutinating, precipitating, or complement-fixing antibodies in bog workers, including one worker with more than 30 years of occupational exposure (Salkin, I.F., Duncan, R.A., and Dixon, D.M., unpublished). It is possible, however, that *S. schenckii* is present in such bogs. Since bogs represent such a large sampling area, negative data from small sample numbers or volumes, such as in our study, may constitute false negative data. Also, it is possible that the fungus is present in the bogs in concentrations too small to be detected by the methods used and that these concentrations increase after the moss is harvested and conditions become more favorable. We plan to address such possibilities by using systematic sampling methods and detection methods independent of fungal growth.

The authors thank Dr. Jonathan Hibbs, Dr. Anne Schuchat, and Dr. Bruce Coles, Epidemic Intelligence Service Officers from the Centers for Disease Control stationed in Pennsylvania, Georgia, and New York, respectively, for their help in the environmental sample collections and epidemiological studies.

References

Centers for Disease Control (CDC). 1969. Sporotrichosis—Oregon. *Morbidity and Mortality Weekly Report* 18:170.

Centers for Disease Control (CDC). 1982. Sporotrichosis associated with Wisconsin sphagnum moss. *Morbidity and Mortality Weekly Report* 31:542–544.

Crevasse, L. and Ellner, P.D. 1960. An outbreak of sporotrichosis in Florida. *Journal of the American Medical Association* 173:29–33.

D'Alessio, D.J., Leavens, L.J., Strumpf, G.B., and Smith, C.D. 1965. An outbreak of sporotrichosis in Vermont associated with sphagnum moss as the source of infection. *New England Journal of Medicine* 272:1054–1058.

Dixon, D.M. 1982. Isolation of dematiaceous fungi from a college greenhouse. *Mycopathologia* 78:123–124.

Dixon, D.M., Shadomy, H.J., and Shadomy, S. 1980. Dematiaceous fungal pathogens isolated from nature. *Mycopathologia* 70:153–161.

Dixon, D.M., Szaniszlo, P.J., and Polak, A. 1990. Dihydroxynaphthalene (DHN) melanin and its relationship with virulence in the early stages of phaeohyphomycosis. pp. 297–318 in Cole, G.T., and H.C. Hoch (editors), *The Fungal Spore and Disease Initiation in Plants and Animals*. Plenum Press, New York.

Dixon, D.M., Salkin, I.F., Duncan, R.A., Hurd, N.J., Haines, J.H., Kemna,

M.E., and Coles, B. 1991. Isolation and characterization of *Sporothrix schenckii* from clinical and environmental sources associated with the largest U.S. epidemic of sporotrichosis. *Journal of Clinical Microbiology*. 29:1106–1113.

Ellis, D.H. and Pfeiffer, T.J. 1990. Natural habitat of *Cryptococcus neoformans* var. *gattii*. *Journal of Clinical Microbiology*. 28:1642–1644.

Findlay, G.H. 1970. The epidemiology of sporotrichosis in the Transvaal. *Sabouraudia* 7:231–236.

Findlay, G.H., Vismer, H.F., and Dreyer, L. 1984. Studies on sporotrichosis. Pathogenicity and morphogenesis in the Transvaal strains of *Sporothrix schenckii*. *Mycopathologia* 87:85–93.

Fukushiro, R. 1984. Epidemiology and ecology of sporotrichosis in Japan. *Zentralblatt fuer Bakteriologie Mikrobiologie und Hygiene* A257:228–233.

Gastineau, F.M., Spolyar, L.W., and Haynes, E. 1941. Sporotrichosis: report of six cases among florists. *Journal of the American Medical Association* 117:1074–1077.

Gezuele, E., Mackinnon, J.E., and Conti-Diaz, I.A. 1972. The frequent isolation of *Phialophora verrucosa* and *Phialophora pedrosoi* from natural sources. *Sabouraudia* 10:266–273.

Grotte, M. and Younger, B. 1981. Sporotrichosis associated with sphagnum moss exposure. *Archives of Pathology and Laboratory Medicine* 105:50–51.

Hayes, W.N. 1960. Sporotrichosis in employees of a tree nursery. *General Practice* 22:114.

Howard, D.H. and Orr, G.F. 1963. Comparison of strains of *Sporotrichum schenckii* isolated from nature. *Journal of Bacteriology* 85:816–821.

Iwatsu, T., Miyaji, M., and Okamoto, S. 1981. Isolation of *Phialophora verrucosa* and *Fonsecaea pedrosoi* from nature in Japan. *Mycopathologia* 75:149–158.

Kazanas, N. 1987. Pathogenic fungi isolated from desiccated mushrooms, seaweed, anchovies and rice sticks imported from the Orient. *Journal of Food Protection* 50:933–939.

Kenyon, E.M., Russell, L.H., and McMurray, D.N. 1984. Isolation of *Sporothrix schenckii* from potting soil. *Mycopathologia* 87:128.

Mackinnon, J.E., Conti-Diaz, I.A., Gezuele, E., Civila, E., and Da Luz, S. 1969. Isolation of *Sporothrix schenckii* from nature and considerations on its pathogenicity and ecology. *Sabouraudia* 7:38–45.

Mariat, F. 1975. Observations sur l'ecologie de *Sporothrix schenckii* et de *Ceratocystis stenoceras* en Corse et en Alsace, provinces Francaises indemnes de sporotrichose. *Sabouraudia* 13:217–225.

McGinnis, M.R. 1983. Chromoblastomycosis and phaeohyphomycosis: New concepts, diagnosis, and mycology. *Journal of the American Academy of Dermatology* 8:1–16.

Okeke, C.N. and Gugnani, H.C. 1986. Studies on pathogenic dematiaceous fungi. I. Isolation from natural sources. *Mycopathologia* 94:19–25.

Powell, K.E., Taylor, A., Phillips, B.J., Blakey, D.L., Campbell, G.D., Kaufman, L., and Kaplan, W. 1978. Cutaneous sporotrichosis in forestry workers: epidemic due to contaminated sphagnum moss. *Journal of the American Medical Association* 240:232–235.

Rippon, J.W. 1988. *Medical Mycology. The Pathogenic Fungi and the Pathogenic Actinomycetes* 3rd ed. W.B. Saunders, Philadelphia. 797 pp.

Shippee, J.K. 1970. A study of potential sources of *Sporotrichium schenckii* in the processing of sphagnum moss. M.S. thesis, University of Wisconsin, Madison.

Staib, F. 1983. Isolation of *Sporothrix schenckii* from the floor of an indoor swimming pool. *Zentralblatt fuer Bakteriologie Mikrobiologie und Hygiene* B177:499–506.

Staib, F. 1984. Ecological and epidemiological aspects of aspergilli pathogenic for man and animal in Berlin (West). *Zentralblatt fuer Bakteriologie Mikrobiologie und Hygiene* A257:240–245.

Staib, F., Tompak, B., Thiel, D., and Blisse, A. 1978. *Aspergillus fumigatus* and *Aspergillus niger* in two potted ornamental plants, cactus (*Epiphyllum truncatum*) and clivia (*Clivia niniata*). Biological and epidemiological aspects. *Mycopathologia* 66:1–2:27–30.

Summerbell, R.C., Krajden, S., and Kane, J. 1989. Potted plants in hospitals as reservoirs of pathogenic fungi. *Mycopathologia* 106:13–22.

Wheeler, M.H. and Bell, A.A. 1987. Melanins and their importance in pathogenic fungi. pp. 338–387 in McGinnis, M. (editor), *Current Topics in Medical Mycology* vol. 2. Springer-Verlag, New York.

Part 4
Microbial Community Processes and Evolution

13

Fungal Community Dynamics

Linda Kinkel

13.1 Introduction

Phylloplane fungal communities are dynamic in time and space. The patterns of spatial and temporal change in the complement of microbes on the plant surface reflect successive balances among the population processes of immigration (arrival of propagules), emigration (physical loss or removal of propagules), growth (increase in biomass or number of propagules), and death for individual species. The individual population processes are in turn driven by physical forces and biotic interactions. Community study may focus on description of patterns, quantifying the population processes that determine community dynamics, or on the manner in which biotic and abiotic factors influence microbial population processes and therefore community dynamics.

Traditional descriptive approaches to community dynamics, while providing important basic ecological information on what organisms are found where and when, afford little insight into the functioning of microbial communities. Many fundamental questions about the nature of phylloplane fungal communities and community dynamics remain either unresolved or uninvestigated. Is there a limit (carrying capacity), in practice, to the numbers of microorganisms that leaves can support? What factors may limit community size? Will similar communities develop on similar sites? How important are microbial interactions in phylloplane existence? Gaining insight into these issues will depend fundamentally on information on the relative importance of the four major microbial population processes (immigration, emigration, growth, and death) in community change through time.

Community ecologists, including those working on the phylloplane, have tended to focus on internal forces and within-community dynamics. Thus, the emphasis in describing communities has been on growth and death and the within-leaf biotic (microbial interactions: Fokkema et al.,

1979; van den Heuvel, 1971; McBride, 1971) and abiotic (ultraviolet radiation, desiccation stress, etc.: Bashi and Fokkema, 1977; Dickinson and O'Donnell, 1977) factors that influence these processes. The possible role of the air spora in influencing phylloplane populations has been recognized (Last, 1955a, 1955b; Collins and Hayes, 1976; Dickinson, 1976), as has the fact that leaves act as spore traps (Gregory, 1971). Additionally, work on the dynamics of removal of microbes from plant surfaces (Grace and Collins, 1976; Lindemann et al., 1982) indicates that leaves are a source of atmospheric microbes. However, these studies have not led to direct investigations of immigration and emigration as important processes in community dynamics. The essential issue is that phylloplane communities are open systems that continually experience both influx as well as outflow of individuals. The regulation of population and community dynamics in an open system may be fundamentally different from that in more closed systems such as the rhizosphere (Hughes, 1990). Open systems may have population sizes that are independent of local fecundity because of continuous recruitment from outside the system (Hughes, 1990; Roughgarden, 1986). Recognition of the open nature of phylloplane communities requires a consideration of population processes that provides an emphasis on immigration or emigration processes equal to that placed on within-community (growth and death) processes. Additionally, investigations of the influences of abiotic and biotic forces on phylloplane community dynamics must consider their effects on recruitment and loss as well as growth and death.

This chapter focuses on microbial population processes, primarily immigration, in relation to community dynamics, beginning with a brief review of descriptive studies of phylloplane fungal community dynamics. These studies provide information on process dynamics and motivation for pursuing a process-oriented approach. Next, approaches to the investigation of immigration in other systems (nonphylloplane) and resulting immigration models are considered. These studies provide useful approaches to conceptualizing immigration in relation to community dynamics. I present data from a study of the quantitative influence of immigration in fungal community development (Kinkel et al., 1989a, 1989b), and the insight these data may provide into the dynamics of phylloplane fungal communities is discussed. Finally, I consider some simple approaches to quantifying and modeling immigration in relation to other population processes and discuss possible future directions in process-oriented study of phylloplane fungal community dynamics.

13.2 Defining and Studying Communities: Sampling and Questions of Scale

Communities are usually defined or delimited based upon spatial or organismal criteria (Southwood, 1987; Giller and Gee, 1987). Definitions vary depending upon the nature of the organisms studied and sometimes on the investigator's perception of the importance of interactions (competition, predation) in community organization (Abele et al., 1984). Functionally, phylloplane fungal communities consist of groups of fungal species of varying abundances coexisting on individual leaves. This definition delimits the community both spatially (the individual leaf) and organismally (excluding bacteria, viruses, microarthropods, and other organisms that may be present on the leaf) and avoids the issue of interactions as a criterion for community bounds. The fungal community (and the fungal "individual") is defined in practice by the sampling procedures selected, including propagule removal methods, isolation media, and incubation methods (Dickinson, 1971; Lindsey and Pugh, 1976; Parbery et al., 1981).

An unfortunate fact of phylloplane community study is that most practical approaches to investigating the community are destructive. Therefore, field studies typically view individual leaves as replicates (Hurlbert, 1984). For instance, changes in community composition through time are investigated by sampling communities on different leaves at each of many sample dates. While it is not illogical to assume that similar processes are occurring concurrently on separate leaves over time, this has not been rigorously documented, and it may be difficult to do so. Thus, to investigate phylloplane community dynamics, the answer to a very basic ecological question—are similar processes of community development followed on similar but distinct sites?—is assumed to be yes (McCune and Allen, 1985). Destructive sampling approaches also necessitate logistic tradeoffs between the number of leaves sampled at a single time and the number of individuals sampled per leaf. These tradeoffs influence the ability to assess changes in fungal species number or species abundance distribution on leaves through time (Kinkel, 1988).

An additional issue in community study relates to the selection of spatial and temporal scales of investigation (Giller and Gee, 1987; O'Neill, 1989). Spatially, dynamics at scales ranging from microsites on individual leaves to large-scale patterns among leaves in a field or canopy or even a region may be considered. Temporally, dynamics of phylloplane microbes have been described at scales ranging from a few hours to seasonal or multiple-year intervals. Substantially different types of information may be obtained by investigation at different scales (Wiens, 1984; Hirano and Upper, 1989; Fenn et al., 1989), and for descriptive studies the selection of scale will depend largely on the questions being asked. Though ideally the study of fungal population processes in relation to community dynamics should be determined by the scales at which those processes occur, limited field data exist on the relevance of different scales to population processes. Additionally, Wiens (1984) notes that spatial and temporal scales may not be independent in relation to particular population processes.

13.3 Fungal Community Dynamics: A Brief Review

Community Composition Previous studies of phylloplane fungal community dynamics have provided a large body of information on the identities and predominance of microorganisms found on different host leaves through time (Pugh and Williams, 1968; Collins and Hayes, 1976; Carroll, 1979; Breeze and Dix, 1981; Mishra and Dickinson, 1981; Thomas and Shattock, 1986). These data have allowed a schematic classification of fungi identified on leaves into nonpathogenic epiphytes that are phylloplane inhabitants (able to grow and sporulate in fluctuating climatic and nutritional conditions; e.g., *Aureobasidium, Cladosporium*), phylloplane invaders (grow extensively only when conditions are favorable; *Alternaria*), and primary saprophytes (unable to grow extensively until onset of senescence; *Phoma*) (Dickinson, 1976). Phylloplane pathogens are grouped into those that are wholly restricted to the phylloplane (*Erysiphe*), those that grow epiphytically or survive for periods prior to penetration (*Alternaria, Botrytis, Septoria*), and those that are nonpathogenic to the particular plant on which they land (Dickinson, 1976). The groups in this classification scheme are not mutually exclusive, and not all possible phylloplane fungi (e.g., endophytes) are included. However, such a scheme provides a useful breakdown of the ecological strategies and roles of fungi that are common components of dynamic phylloplane communities.

Recognition of alternative ecological strategies has led to investigations of the environmental (physical and nutritional) requirements for growth and survival on the phylloplane for organisms in each of the groups (Bashi and Fokkema, 1977; Dickinson and O'Donnell, 1977; Hurst et al., 1984). Additional research has focused on the interactions among phylloplane microorganisms (McBride, 1971; Collins, 1976; Fokkema et al., 1979; Thomas and Shattock, 1986). Characterization of epiphytic fungi along an r-K strategy continuum (Andrews and Harris, 1986; Thomas and Shattock, 1986) may be possible using these types of data. However, these approaches have not provided substantial insight into the mechanisms of community change on leaves. The next step may be to consider groups or single species of fungi and to determine the population processes that may be limiting for those fungi on the phylloplane, i.e., is species x limited by its ability to grow, survive, or immigrate, and what processes are dominant for particular species at different times? This approach moves beyond the description of traits of individual organisms to an attempt to understand the functioning of phylloplane communities based upon an evaluation of the processes that determine community composition.

Successional Dynamics Apparent successional trends in fungal community composition on several plant hosts have been identified (Dickinson, 1967; Hogg and Hudson, 1966; Sherwood and Carroll, 1974; Lindsey and Pugh, 1976; Wildman and Parkinson, 1979). These assessments of the communities range from indirect (leaf wash or imprint) summaries of the taxa present on leaves to quantifications of hyphal growth and sporulation over time. General trends show small populations of microorganisms on young leaves, with both increasing numbers of propagules and increasing hyphal development with time and leaf age (Dickinson, 1976; McBride and Hayes, 1977; Breeze and Dix, 1981). Changes in community composition over time have also been described (Pugh and Buckley, 1971; Lindsey and Pugh, 1976; Wildman and Parkinson, 1979).

A difficult issue in analyzing and interpreting these data has been to attribute the observed changes in community composition and size to any specific factor (Frankland, 1981). Traditional models of succession have emphasized the roles of species interactions and organismal effects on the physical substratum (Connell and Slatyer, 1977). However, the phylloplane substratum itself is aging and changing with time independently of any fungal species interactions or influences. Also, in addition to seasonal changes in the physical environment, such as temperature, relative humidity, and solar radiation, longer periods of time may simply allow for greater cumulative immigration or for compositional changes in the identities of immigrants (air spora) (Gregory, 1973; Hirst, 1953). Determining the relative importance of species interactions, leaf age, seasonal environmental changes, and quantitative and qualitative differences in cumulative immigration in successional trends will require investigation of each factor independently. The quantitative and qualitative impacts of immigration, emigration, growth, and death on the components of the community through the growing season need to be determined experimentally to unravel the mechanisms of observed successional patterns.

Overall, descriptive studies of phylloplane fungal community composition and successional and colonization dynamics highlight the need for a process-oriented investigation of community dynamics that emphasizes and quantifies the role of stochastic immigration events in fungal community development.

13.4 Studies of Immigration Dynamics in Nonphylloplane Systems

The role of immigration (recruitment, settlement) in community dynamics has recently been the subject of investigation and modeling in a number of systems, such as marine invertebrates, tropical reef fish, and lizards (Victor, 1983; Underwood and Denley, 1984; Roughgarden, 1986; Roughgarden et al., 1985, 1987; Hughes, 1990). Like phylloplane ecologists, researchers of these systems have historically tended to focus on within-community interactions and mortality from disturbances or physical stress in studies of population and community dynamics (having assumed that immigrants are either not limiting or are limiting) (Hughes, 1990). We may benefit from a consideration of the concepts and models currently being developed in these open systems. One idea is that of recruitment or immigration limitation. Populations that respond directly to enhancements in recruitment or immigration are by definition immigration limited (Hughes, 1990). Hughes (1990) notes that the role of immigration limitation as a mechanism of regulating population or community size remains controversial. However, phylloplane ecologists need to consider the extent to which immigration may limit population or community size on the phylloplane. We also need to establish criteria for determining that an organisms's population size is truly *immigration* limited (rather than limited by growth, death, or emigration).

Aureobasidium pullulans may provide one example of a phylloplane microorganism that is immigration limited, at least under certain circumstances (Kinkel et al., 1989b). *Aureobasidium pullulans* population sizes on apple leaves in the field onto which large numbers of individuals had been inoculated (immigrants composed of the inoculated population plus the continual introduction of natural inoculum), were significantly larger than population sizes on leaves that had not been inoculated (immigrants composed of natural inoculum only), even after substantial apparent growth of *Aureobasidium pullulans* on the leaves (Kinkel et al., 1989b). Growth of the inoculated populations was suggested by the occurrence of much larger increases in *Aureobasidium* populations on inoculated leaves as compared with uninoculated leaves from 1–10 days following inoculation. This direct and prolonged response to an increase in numbers of immigrants indicates that *Aureobasidium pullulans* is an "immigration-limited" population, as defined by Hughes (1990). Populations may presumably be immigration limited during some periods and not during others. Distinctions between populations that are limited by immigration versus populations that are limited by growth or some other process will be very difficult to determine in a field setting unless immigration (or growth, emigration, or death, individually) is experimentally altered. Ultimately, it is the *balance* among the population processes that is critical to understanding the factors that may limit population sizes on the phylloplane.

Roughgarden (1986), Hughes (1990), and others have developed models to study the role of immigration in population and community dynamics. Roughgarden's model, developed for sessile marine organisms, is particularly interesting because it emphasizes the limiting role of space, rather than nutrients, in population dynamics. This recognition is coupled with an acknowledgment of the difficulties of assessing free space from the perspective of the organism—a problem shared by phylloplane ecologists. The model evaluates the relative roles of immigration (termed recruitment or settlement), growth, and death in community dynamics. Density-dependent and density-independent mortality (death) rates are considered separately. Immigration is considered to be directly proportional to free space in this system, an assumption which probably does not hold for fungi or other microbes on the phylloplane. The model also fails to incorporate emigration, which may be important in phylloplane community dynamics.

Roughgarden's model suggests four basic community types, depending upon the relative predominance of immigration, growth, density-dependent mortality, and density-independent mortality within the community. "High-free-space communities" result when immigration rates are low relative to density-independent mortality rates. These are immigration-limited communities, and as defined by Hughes (1990), fluctuation in immigration rates will affect community size. "Low-free-space communities" are communities in which immigration is not limiting and growth rates are low in relation to density-independent mortality. Basically, immigration rates > density-independent mortality > growth. "Patch mosaic communities" have high immigration rates coupled with low density-independent mortality and high density-dependent mortality rates. Finally, "lattice communities" result from high immigration rates, low density-independent mortality, low density-dependent mortality, and growth that stops when space is exhausted.

The important point of the model for phylloplane ecologists is that communities resulting from different balances of population processes will have distinct characteristics, specifically with respect to the role of resource limitation and density-dependent versus density-independent processes in the community. Thus, understanding the relative quantitative influence of immigration, emigration, growth, and death in the community may provide insight into the mechanisms of community dynamics.

The Hughes (1990) model considers population dynamics as a function of immigration (independent of within-population dynamics), density-independent growth, density-independent death, and density-independent "shrinkage." (Hughes worked with a colonial invertebrate species whose colonies can shrink.) Using growth, shrinkage, and death rates measured from field observations, Hughes showed that population sizes stabilized as a function of immigration rates and size-specific mortality. The model predicts that relatively small increases in immigration will have a large effect on the population size of the invertebrate (for an immigration-limited population). Differential predictions for the effect of a change in immigration rate versus a change in death rate were made by the model for short-lived versus long-lived species, though predicted effects may be a function of the temporal scale at which the populations are assessed. Hughes' model is applicable to phylloplane communities in which growth and death are density-independent (unknown in most cases). The model highlights the point that population sizes may stabilize as a function of immigration and death alone, with little effect of growth on dynamics overall (see also Simberloff, 1969). This result suggests caution in interpreting stable population sizes as being at the "carrying capacity" of the leaf or as evidence for density dependence in death or emigration rates. Unless mortality, emigration, and growth are directly measured as a function of population or community size, conclusions about the density dependence or independence of these processes are inappropriate.

Roughgarden's and Hughes' models illustrate the manner in which a process-oriented approach can enhance our understanding of population and community dynamics. Both models make specific predictions about the relationships between population processes and community or population characteristics, and they suggest fundamental questions on the relationship of immigration to other population processes and to community size. For example, what is the relationship between immigration rates and death rates for fungi on leaves? Under what conditions, if any, are population processes on the phylloplane density dependent? Is free space (leaf area or perhaps specific microsites) limiting for fungi on leaves? How may changes in immigration rates versus changes in death rates, or mortality, affect phylloplane community size? These are basic issues for understanding the functioning and development of phylloplane communities.

13.5 Fungal Immigration and Phylloplane Community Dynamics

Studies on the dynamics of fungal immigration to apple leaves were initiated to provide basic information on the magnitude and variability of immigration to leaves in the field over time (Kinkel et al., 1989a, 1989b). Specific questions considered in this work included: (1) How does immigration to leaves in the field vary over time?; (2) how does immigration vary among leaves during a single time period, and are numbers of immigrants related to leaf area?; and (3) how do developing communities compare quantitatively or qualitatively with the cumulative immigrants to which they have been exposed? Finally, communities of very different "ages" (differing lengths of time of exposure to immigrants) were compared to determine whether the communities had different sizes or compositions.

Direct estimates of immigration in the field have been determined for a number of systems (Bott and Brock, 1970; Schoener et al., 1978; Sutherland and Karlson, 1977). The approach used in our study was to surface disinfest (Kinkel and Andrews, 1988) a large number of apple leaves on a single tree in the field and follow the development of the fungal community on these uncolonized leaf habitats (Kinkel et al., 1989a). Immigration was assessed by enumerating fungal individuals present on leaves 12 hr after surface disinfestation. The 12-hr-time period was based upon growth chamber experiments that showed that significant growth of phylloplane fungi (*Cladosporium cladosporioides* and *Aureobasidium pullulans*) inoculated onto apple leaves did not occur within this time. Immigration values based on 12-hr assessments are likely underestimates of actual immigration because emigration and death of some immigrants may have occurred during this period.

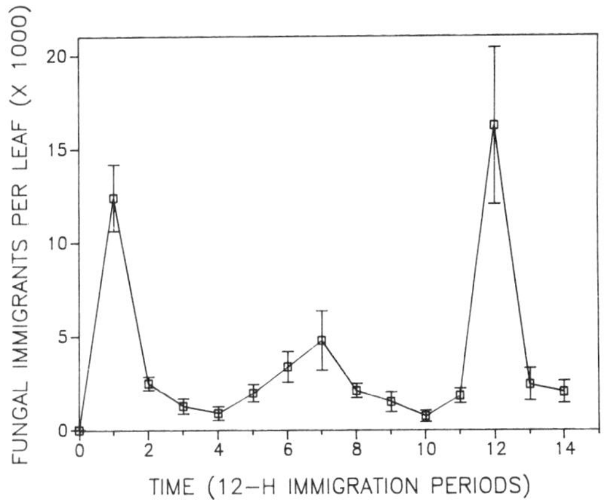

Figure 13.1 Mean numbers of fungal immigrants per leaf for 14 consecutive immigration periods (16–23 June 1986). From Kinkel et al. (1989a).

Immigration was studied during three separate experiments (May, June, July) in a series of consecutive 12-hr "immigration periods." Communities developing on surface-disinfested leaves over the entire series of immigration periods were investigated to determine the contrasts between developing communities and the cumulative immigrants to which they have been exposed. Leaf-surface communities (or immigrants) were assessed using an indirect leaf wash-dilution plating procedure. Total numbers (colony-forming units) of filamentous fungi and yeasts as well as numbers of individuals of predominant fungal genera were determined for each of five individual leaves in every sample. Further details of the experimental design are described elsewhere (Kinkel et al., 1989a, 1989b).

We found that fungal immigration to apple leaves varied substantially through time (Figure 13.1), ranging from an average of 300 to 29,200 individuals per leaf within separate 12-hr immigration periods. Variation among immigration periods in numbers of immigrants per leaf was sometimes correlated with specific environmental events (rainfall, mowing of orchard grass) (Kinkel et al., 1989a). Previous work demonstrated that fungal communities on untreated apple leaves in the field ranged in size from less than 10,000 on small leaves early in the growing season to more than 1,000,000 individuals (Andrews et al., 1987; Kinkel et al., 1987). Thus, quantitatively, numbers of new arrivals per leaf per 12-hr immigration period may range from 0.03% to more than 100% of the established community size (number of fungi already present on the leaf). Numbers of immigrants varied significantly among May, June, and July experiments, with May immigration being on average the lowest (560 immigrants per leaf per 12-hr period) and July immigration the highest (an average of 12,470 immigrants per leaf per 12-hr period). Total numbers of fungal individuals per leaf also tend to increase throughout the growing season (Pennycook and Newhook, 1981; Breeze and Dix, 1981). Whether numbers of immigrants per leaf per 12-hr period contribute on average some relatively consistent proportion of the number of individuals already present on the leaves (some equilibrium between immigration and community size) is unclear. Quantitative investigations of immigration in relation to community size on untreated leaves through time are needed. Given that microbes on the leaf surface may be an important source of the air spora through emigration (Lindemann et al., 1982), phylloplane community size may be one predictor of the magnitude of immigration to the phylloplane.

Numbers of filamentous fungal and yeast immigrants per leaf varied among leaves within single 12-hr immigration periods and were positively correlated with leaf area (Kinkel et al., 1989a). One would predict that larger targets would collect larger numbers of individuals. Previous work does not provide a clear picture of the relationship between fungal community size and leaf area, possibly because data from communities on leaves across entire growing seasons were pooled in area analyses (Kinkel et al., 1987; Andrews et al., 1987). If community size is changing as a function of both leaf area and time (seasonal and leaf age influences) simultaneously, investigation of leaf area influences on community size independent of time may be equivocal.

Given the observation that leaf area influences the immigration process (numbers of immigrant per leaf), an understanding of the relationship between community size (number of individuals) and leaf area may be useful in investigating leaf area influences on emigration, growth, and death. For instance, if community size is not related to leaf area, the implication is that one or more of the other nonimmigration processes must act independently of leaf area. A consistent relationship between community size and leaf area may indicate that communities are immigration driven or that one or more of the leaf surface population processes are also area limited or influenced. The fundamental issue is whether free space is limiting for fungal population processes on the phylloplane. This may not be easy to answer directly. Ruscoe (1971) and Diem (1974) found less than 10% of the physical area of *Nothofagus* and *Hordeum* leaves, respectively, were colonized by fungi, but it is not clear whether other sites are uncolonized for lack of propagules (immigration limited) or are simply not suitable for colonization. Determining the extent to which nonimmigration processes are limited by space or sites (or nutrients) on the phylloplane will be experimentally difficult, but an understanding of the relative relationships among population processes and leaf area may provide some insight into the amount of free space (colonizable sites) existing on leaves (e.g., Roughgarden, 1986).

The variability in immigration among leaves within a single time period may also be useful for investigating whether similar community processes are occurring on separate leaves. First, we should compare the predicted variability among communities based on the cumulative variability in immigration with the variability among developed communities. Communities that are significantly more variable than predicted suggest that different levels of emigration, growth, and death are occurring on different leaves. Communities that become significantly less variable than predicted suggest that similar postimmigration processes occur on separate leaves.

Comparisons of total numbers of fungi, numbers of filamentous fungal propagules, and numbers of yeast individuals in developing phylloplane communities with the estimated cumulative numbers of fungal, filamentous fungal, and yeast immigrants to those communities showed that communities were consistently smaller than predicted based on immigration estimates alone (Figure 13.2). Thus, given that

$$C = I + G - D - E$$

where C = community size (total number of individuals), I = numbers of immigrants, G = inputs due to growth, D = losses due to death, and E = losses due to emigration. Given the observation that

$$C < I$$

then

$$(G - D - E) < 0$$

and therefore

$$(D + E) > G.$$

In other words, community losses due to emigration and death played a quantitatively larger role than growth following immigration for fungi on the phylloplane. Community size continued to increase throughout all experiments, reflecting the continued inputs due to immigration, and probably some levels of fungal growth on the phylloplane as well. Overall, however, the results suggest that, quantitatively, over the time scale of these experiments, immigration > (emigration + death) > growth in phylloplane fungal community development. This does not indicate that growth is absent under these conditions, but only that it is quantitatively less important than the other processes.

Comparison of apple phylloplane community size and composition (filamentous fungi versus yeasts) among communities of very different ages showed that, despite variations in the early phases of community development, the communities become progressively more similar in species composition with time (Kinkel et al., 1989b). Notably, single large immigration events sometimes had large short-term effects on community size and composition, though effects varied among immigrating organisms (e.g., *Aureobasidium*, Kinkel et al., 1989b; and *Microsphaeropsis*, Kinkel et al., 1989a). Apple phylloplane communities with distinct immigration or environmental histories may become similar with time because recent immigrants, rather than early colonists, dominate the community and are similar for all leaves. Alternatively, the communities may become similar with

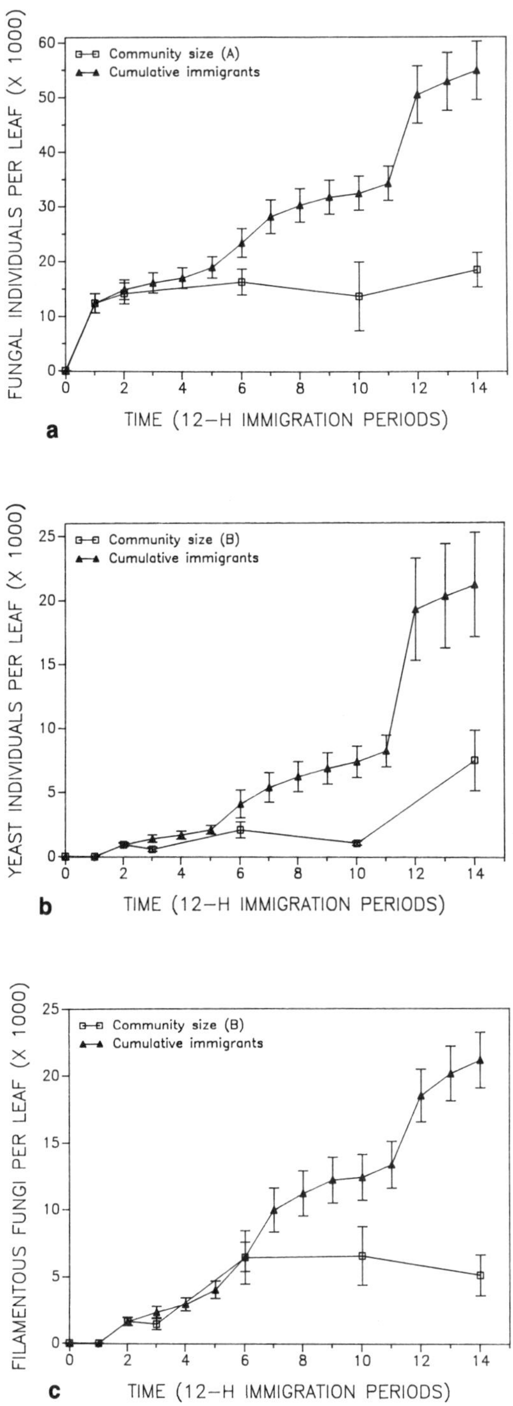

Figure 13.2 Mean numbers of fungal, filamentous fungal, and yeast individuals per leaf in developing communities contrasted with cumulative numbers of fungal, filamentous fungal, or yeast immigrants per leaf over the corresponding times. Time 0 is 7 a.m., 16 June 1986. (a) Community development on one set of leaves (set A); (b and c) community development on a different leaf set (set B) over approximately the same time frame. Leaf sets were distinguished by the time at which they were initially disinfested. From Kinkel et al. (1989a).

time because the selective roles of biotic interactions and physical forces on phylloplane growth and death are similar on all leaves, thereby leading to a characteristic phylloplane fungal community. Further data on the relationship of established communities to recent immigrants may provide insight into which of these two (or some other) alternatives describe the phylloplane system.

13.6 Overview: Immigration and Phylloplane Community Dynamics

Immigration plays a quantitatively significant role in phylloplane fungal community dynamics and appears to be important in maintaining community size on apple leaves in temperate climates. During the early phases of development, apple phylloplane communities are immigration limited, as evidenced by the significant influences of large immigration events on community size (Kinkel et al., 1989a). Continued immigration to the leaves may be crucial to balancing continuous losses due to death and/or emigration. The data indicate that, under the conditions of the experiment, immigration is quantitatively greater than emigration and death. This is certainly not true for all communities under all conditions; community size does not infinitely increase over the growing season. The communities must go through phases during which losses due to emigration and death together are greater than increases due to immigration and growth. This suggests that communities may be in transition between Roughgarden's "low-free-space" and "high-free-space" community types, though differences in the mechanism of immigration (density-dependent versus density-independent) in the systems make conclusions on the amount of free space (or the density dependence of emigration or death processes) in the phylloplane community questionable. The extent to which emigration and death are density dependent might be investigated by considering the relative impacts of immigration on community size as community size increases. Specifically, community size is some function of a series of discrete immigration events:

$$N = x_1(I_1) + x_2(I_2) + x_3(I_3) + \ldots$$

where N = community size, I_1 = number of immigrants during period 1, and x_1 = proportion of immigrants remaining in the community at the end of time 1 (proportion of immigrants that do not emigrate or die). Thus, for example, $N_3 - N_2 = (x_3)(I_3)$ = increase in N (community size) observed during period 3. Assuming no growth (not always justifiable) and increasing N, and assessing community size and immigration at every interval, consistent declines in x over time (such that $x_1 > x_2 > x_3 > \ldots$) illustrate density-dependent death and/or emigration. Increasing x values ($x_1 < x_2 < x_3 \ldots$) indicate either growth or density-dependent declines in death or emigration. Distinct results may be obtained when considering entire communities versus subcomponents of the community; density-dependent interactions may be more important for yeasts than for fungi, for instance. Data indicating the presence or absence of density dependence in phylloplane fungal communities are important in understanding the role of microbial interactions in community development.

If immigration is of fundamental importance in phylloplane community dynamics, we need to consider the factors that influence immigration rates. Simplistically, under dry conditions, immigration is some function of leaf area and the air spora (immigrant source pool):

$$I_a = \text{fn}\ \{ (S_a)\ (A)\ (T_a) \}$$

where I_a = numbers of immigrants during period *a*, S_a = air spora concentration for period *a*, A = effect of leaf area (trap size) on immigration, and T_a = trapping efficiency, or the rate of removal of propagules from the air spora by leaves during period *a*. Using such a simplistic approach, the interesting question may be to evaluate changes in T_a over time. However, at a larger scale, we need to consider the ways in which S itself may be a function of N^*, or the mean phylloplane community size. As noted previously, immigration may itself be related to community size through emigration's effects on the air spora [$S = \text{fn}\ (N^*)$]. Since $N^* = \text{fn}\ (I)$, then, by substitution (above),

$$I = \text{fn}\{[\text{fn}\ (N^*)](A)(T)\}.$$

Ultimately, determining the ways in which the air spora concentration, immigration, and community size are related is fundamental to understanding phylloplane community dynamics. This emphasizes the importance of scale in studying and describing phylloplane community dynamics. Roughgarden (1986) suggests the concept of a "metapopulation," a population consisting of multiple open (space-limited) local populations. We might think of the phylloplane system as a collection of open local communities within a specified location. In fact, we have always been forced by design to study such "populations of communities;" perhaps a process (immigration)-oriented approach can encourage the consideration of dynamics among communities at the metapopulation level as well as the dynamics of individual communities.

This chapter has posed more questions about phylloplane community dynamics than there are data to answer. However, the objective has been to illustrate the ways in which a process-oriented approach to community study that specifically emphasizes the open nature of phylloplane communities can be used to ask questions and gain insight into the mechanisms of phylloplane microbial community dynamics. Traditional models of community dynamics that focus on within-community interactions are not sufficient for understanding phylloplane communities. Future work will gain from a consideration of approaches taken to studying and modeling dynamics in other open systems.

The author thanks David Andersen, Jim Groth, and Elwin Stewart for helpful comments on the manuscript.

References

Abele, L.G., Simberloff, D.S., Strong, D.R., and Thistle, A.B. 1984. Preface. pp. vii–x in Strong, D.R., Simberloff, D.S., Abele, L.G., and Thistle, A.B. (editors), *Ecological Communities: Conceptual Issues and the Evidence*. Princeton University Press, Princeton, New Jersey.

Andrews, J.H. and Harris, R.F. 1986. r- and K-selection and microbial ecology. pp. 99–147 in Marshall, K.C. (editor), *Advances in Microbial Ecology*, vol. 9. Plenum Publishing Corporation, New York.

Andrews, J.H., Kinkel, L.L., Berbee, F.M., and Nordheim, E.V. 1987. Fungi, leaves, and the theory of island biogeography. *Microbial Ecology* 14:277–290.

Bashi, E. and Fokkema, N.J. 1977. Environmental factors limiting growth of *Sporobolomyces roseus*, an antagonist of *Cochliobolus sativus*, on wheat leaves. *Transactions of the British Mycological Society* 68:17–25.

Bott, T.L. and Brock, T.D. 1970. Growth and metabolism of periphytic bacteria: Methodology. *Limnology and Oceanography* 25:333–342.

Breeze, E.M. and Dix, N.J. 1981. Seasonal analysis of the fungal community on *Acer platanoides* leaves. *Transactions of the British Mycological Society* 77:321–328.

Carroll, G.C. 1979. Needle microepiphytes in a Douglas fir canopy: biomass and distribution patterns. *Canadian Journal of Botany* 57:1000–1007.

Collins, M.A. 1976. Colonisation of leaves by phylloplane saprophytes and their interactions in this environment. pp. 401–418 in Dickinson, C.H. and Preece, T.F. (editors), *Microbiology of Aerial Plant Surfaces*. Academic Press, New York.

Collins, M.A. and Hayes, A.J. 1976. Seasonal incidence of microbes on the surface of first year needles of Norway spruce. *Transactions of the British Mycological Society* 66:457–461.

Connell, J.H. and Slatyer, R.O. 1977. Mechanisms of succession in natural communities and their role in community stability and organization. *American Naturalist* 111:1119–1144.

Dickinson, C.H. 1967. Fungal colonisation of *Pisum* leaves. *Canadian Journal of Botany* 45:915–927.

Dickinson, C.H. 1971. Cultural studies of leaf saprophytes. pp. 129–137 in Preece, T.F. and Dickinson, C.H. (editors), *Ecology of Leaf Surface Microorganisms*. Academic Press, New York.

Dickinson, C.H. 1976. Fungi on the aerial surfaces of higher plants. pp. 293–324 in Dickinson, C.H., and Preece, T.F. (editors), *Microbiology of Aerial Plant Surfaces*. Academic Press, New York.

Dickinson, C.H. and O'Donnell, J. 1977. Behaviour of phylloplane fungi on *Phaseolus* leaves. *Transactions of the British Mycological Society* 68:193–199.

Diem, H.G. 1974. Micro-organisms of the leaf surface: estimation of the mycoflora of the barley phyllosphere. *Journal of General Microbiology* 80:77–83.

Fenn, M.E., Dunn, P.H., and Durall, D.M. 1989. Effects of ozone and sulfur

dioxide on phyllosphere fungi from 3 tree species. *Applied and Environmental Microbiology* 55:412–418.
Fokkema, N.J., den Houter, F.G., Kosterman, Y.J.C., and Nelis, A.L. 1979. Manipulation of yeasts on field-grown wheat leaves and their antagonistic effect on *Cochliobolus sativus* and *Septoria nodorum*. *Transactions of the British Mycological Society* 72:19–29.
Frankland, J.C. 1981. Mechanisms in fungal successions. pp. 403–426 in Wicklow, D.T. and Carroll, G.C. (editors), *The Fungal Community—Its Organization and Role in the Ecosystem*. Marcel Dekker, Inc., New York.
Giller, P.S. and Gee, J.H.R. 1987. The analysis of community organization: the influence of equilibrium, scale, and terminology. pp. 519–542 in Gee, J.H.R. and Giller, P.S. (editors), *Organization of Communities Past and Present*. Blackwell Scientific Publications, Palo Alto, California.
Grace, J. and Collins, M.A. 1976. Spore liberation from leaves by wind. pp. 185–198 in Dickinson, C.H. and Preece, T.F. (editors), *Microbiology of Aerial Plant Surfaces*. Academic Press, New York.
Gregory, P.H. 1971. The leaf as a spore trap. pp. 239–243 in Preece, T.F. and Dickinson, C.H. (editors), *Ecology of Leaf Surface Micro-Organisms*. Academic Press, New York.
Gregory, P.H. 1973. *Microbiology of the Atmosphere*, 2nd ed. Leonard Hill, London.
Hirano, S.S. and Upper, C.D. 1989. Diel variation in population size and ice nucleation activity of *Pseudomonas syringae* on snap bean leaflets. *Applied and Environmental Microbiology* 55:623–630.
Hirst, J.M. 1953. Changes in atmospheric spore content: Diurnal periodicity and the effects of weather. *Transactions of the British Mycological Society* 36:375–393.
Hogg, B.M. and Hudson, H.J. 1966. Micro-fungi on leaves of *Fagus sylvatica* I. The micro-fungal succession. *Transactions of the British Mycological Society* 49:185–192.
Hughes, T.P. 1990. Recruitment limitation, mortality, and population regulation in open systems: a case study. *Ecology* 71:12–20.
Hurlbert, S.H. 1984. Pseudoreplication and the design of ecological field experiments. *Ecological Monographs* 54:187–211.
Hurst, J.L., Pugh, G.J.F., and Walton, D.W.H. 1984. The effect of temperature on the growth of *Candida sake* isolated from the leaves of a subantarctic grass. *Microbial Ecology* 10:89–93.
Kinkel, L.L. 1988. *Fungal Immigration and Community Dynamics on Apple Leaves*. Ph.D. dissertation, University of Wisconsin, Madison.
Kinkel, L.L. and Andrews, J.H. 1988. Disinfestation of living leaves by hydrogen peroxide. *Transactions of the British Mycological Society* 91:523–528.
Kinkel, L.L., Andrews, J.H., Berbee, F.M., and Nordheim, E.V. 1987. Leaves as islands for microbes. *Oecologia* 71:405–408.
Kinkel, L.L., Andrews, J.H., and Nordheim, E.V. 1989a. Fungal immigration dynamics and community development on apple leaves. *Microbial Ecology* 18:45–58.
Kinkel, L.L., Andrews, J.H., and Nordheim, E.V. 1989b. Microbial introductions to apple leaves: influences of altered immigration on fungal community dynamics. *Microbial Ecology* 18:161–173.
Last, F.T. 1955a. The seasonal incidence of *Sporobolomyces* on cereal leaves. *Transactions of the British Mycological Society* 38:221–240.
Last, F.T. 1955b. The spore content within and above mildew-infected crops. *Transactions of the British Mycological Society* 38:453–465.
Lindemann, J., Constantinidou, H.A., Barchet, W.R., and Upper, C.D. 1982.

Plants as sources of airborne bacteria, including ice nucleation-active bacteria. *Applied and Environmental Microbiology* 44:1059–1063.
Lindsey, B.I. and Pugh, G.J.F. 1976. Succession of microfungi on attached leaves of *Hippophae rhamnoides*. *Transactions of the British Mycological Society* 67:61–67.
McBride, R.P. 1971. Micro-organism interactions in the phyllosphere of larch. pp. 545–555 in Preece, T.F. and Dickinson, C.H. (editors), *Ecology of Leaf Surface Micro-organisms*. Academic Press, New York.
McBride, R.P. and Hayes, A.J. 1977. Phylloplane of European larch. *Transactions of the British Mycological Society* 69:39–46.
McCune, B. and Allen, T.F.H. 1985. Will similar forests develop on similar sites? *Canadian Journal of Botany* 63:367–376.
Mishra, R.R. and Dickinson, C.H. 1981. Phylloplane and litter fungi of *Ilex aquifolium*. *Transactions of the British Mycological Society* 77:329–337.
O'Neill, R.V. 1989. Perspectives in hierarcy and scale. pp. 140–156 in Roughgarden, J., May, R.M., and Levin, S.A. (editors), *Perspectives in Ecological Theory*. Princeton University Press, Princeton, New Jersey.
Parbery, I.H., Brown, J.F., and Bofinger, V.J. 1981. Statistical methods in the analysis of phylloplane populations. pp. 47–65 in Blakeman, J.P. (editor), *Microbial Ecology of the Phylloplane*. Academic Press, New York.
Pennycook, S.R. and Newhook, F.J. 1981. Seasonal changes in the apple phylloplane microflora. *New Zealand Journal of Botany* 19:273–283.
Pugh, G.J.F. and Buckley, N.G. 1971. The leaf surface as a substrate for colonization by fungi. pp. 431–445 in Preece, T.F. and Dickinson, C.H. (editors), *Ecology of Leaf Surface Micro-organisms*. Academic Press, New York.
Pugh, G.J.F. and Williams, G.M. 1968. Fungi associated with *Salsola kali*. *Transactions of the British Mycological Society* 51:389–396.
Roughgarden, J. 1986. A comparison of food-limited and space-limited animal competition communities. pp. 492–516 in Diamond, J. and Case, T.J. (editors), *Community Ecology*. Harper and Row, New York.
Roughgarden, J., Iwasa, Y., and Baxter, C. 1985. Demographic theory for an open marine population with space-limited recruitment. *Ecology* 66:54–67.
Roughgarden, J., Gaines, S.D., and Pacala, S.W. 1987. Supply side ecology: the role of physical transport processes. pp. 491–518 in Gee, J.H.R. and Giller, P.S. (editors), *Organization of Communities Past and Present*. Blackwell Scientific Publications, Palo Alto, California.
Ruscoe, Q.W. 1971. Mycoflora of living and dead leaves of *Nothofagus truncata*. *Transactions of the British Mycological Society* 56:463–474.
Schoener, A., Long, E.R., and DePalma, J.R. 1978. Geographic variation in artificial island colonization curves. *Ecology* 59:367–382.
Sherwood, M. and Carroll, G. 1974. Fungal succession on needles and young twigs of old-growth Douglas fir. *Mycologia* 66:499–506.
Simberloff, D.S. 1969. Experimental zoogeography of islands: a model for insular colonization. *Ecology* 50:296–314.
Southwood, T.R.E. 1987. The concept and nature of the community. pp. 3–27 in Gee, J.H.R. and Giller, P.S. (editors), *Organization of Communities Past and Present*. Blackwell Scientific Publications, Palo Alto, California.
Sutherland, J.P. and Karlson, R.H. 1977. Development and stability of the fouling community at Beaufort, North Carolina. *Ecological Monographs* 47:425–446.
Thomas, M.R. and Shattock, R.C. 1986. Filamentous fungal associations in the phylloplane of *Lolium perenne*. *Transactions of the British Mycological Society* 87:255–268.
Underwood, A.J. and Denley, E.J. 1984. Paradigms, explanations, and gener-

alizations in models for the structure of intertidal communities on rocky shores. pp. 151–180 in Strong, D.R., Jr., Simberloff, D., Abele, L., and Thistle, A.B. (editors), *Ecological Communities: Conceptual Issues and the Evidence*. Princeton University Press, Princeton, New Jersey.

van den Heuvel, J. 1971. Antagonism between pathogenic and saprophytic *Alternaria* species on bean leaves. pp. 537–544 in Preece, T.F. and Dickinson, C.H. (editors), *Ecology of Leaf Surface Micro-Organisms*. Academic Press, New York.

Victor, B.C. 1983. Recruitment and population dynamics of coral reef fish. *Science* 219:419–420.

Wiens, J.A. 1984. On understanding a non-equilibrium world: myth and reality in community patterns and processes. pp. 439–457 in Strong, D.R., Jr., Simberloff, D., Abele, L., and Thistle, A.B. (editors), *Ecological Communities: Conceptual Issues and the Evidence*. Princeton University Press, Princeton, New Jersey.

Wildman, H.G., and Parkinson, D. 1979. Microfungal succession on living leaves of *Populus tremuloides*. *Canadian Journal of Botany* 57:2800–2811.

14

Bacterial Community Dynamics

Susan S. Hirano and Christen D. Upper

14.1 Introduction

The term *community* has been used in a number of different ways by ecologists (cf. Krebs, 1985; Diamond and Case, 1986). Indeed, Roughgarden and Diamond (1986) state that ". . . a natural unambiguous definition of communities does not exist." They further state that the most inclusive definition of a community is "all the organisms in a prescribed area," a definition that is usually restricted by the criteria used to describe the organisms and the boundaries delimiting the area (Roughgarden and Diamond, 1986). If we follow this definition, the set of organisms that we address is restricted to bacteria; the area to that of a leaf or leaflet. Because a leaf is discrete in space, the boundary of the community is explicitly delineated. Unlike many terrestrial habitats in which the "edge" of the area of interest may be more arbitrarily defined, the leaf margin clearly describes the edge of the phyllosphere habitat. However, each of the many plants in a canopy produces many leaves; each is, in turn, separated in space and inhabited by a distinct assemblage of bacteria (i.e., a community). In a canopy, there exists not a single habitat, but rather populations of habitats. Thus, bacterial communities in the phyllosphere are best described in terms of populations of communities on populations of leaves.

To understand the dynamics of the populations of phyllosphere bacterial communities, we must know what bacteria are present and in what numbers, and how these numbers change as a function of time on populations of leaves. To fully understand these dynamics, we must determine the relative quantitative contribution of immigration, emigration, growth, and death, and the underlying environmental and biological factors that regulate the magnitudes and rates with which these processes occur as they affect the populations of each component within bacterial communities on populations of leaves.

We develop our presentation in two stages: First, bacterial community structure is discussed. Questions that are considered include: What types of bacteria are found in the phyllosphere and in what numbers? What are the current limitations in studies of community structure? Next, the contribution of the population processes to temporal changes in community structure is examined. How important are immigration, emigration, growth, and death to community dynamics? Both topics are discussed in relation to our perspective that phyllosphere bacterial communities exist as *populations* of communities present on populations of discrete habitats.

14.2 Bacterial Community Structure

Identification of community members and enumeration of their relative abundances are requisite steps toward understanding bacterial community dynamics. Both steps are fraught with potential difficulties. Biases may be introduced by the choice of sampling unit (i.e., single leaves versus bulk leaf samples containing several leaves), sampling design (i.e., considerations of temporal and spatial scales), and methods used to isolate, characterize, identify, and enumerate the community members. Many of these issues have been discussed previously (cf. Hirano and Upper, 1983, 1986; Morris and Rouse, 1986) but not necessarily addressed or resolved.

An issue that warrants particular emphasis is the total reliance on culture methods for isolation and enumeration of phyllosphere bacteria from terrestrial plants. The types of bacteria that have been isolated from the phyllosphere may have been restricted by the conditions used to grow them in culture. In view of the increasing numbers of reports on the presence of viable but nonculturable bacteria encountered in other (e.g., aquatic) ecosystems (Colwell et al., 1985; Ward et al., 1990), and the general consensus among microbial ecologists (few of whom work in the phyllosphere) that culture methods are inadequate for studying bacterial communities (Brock, 1987), the question of whether the bacterial assemblages that have been recovered by culture methods are, indeed, representative of the communities in their natural habitat needs to be addressed.

Most of the studies on the population dynamics of phyllosphere bacteria have measured numbers of bacterial components that can be enumerated on a specific medium. For components that can be enumerated by culturing, if the plating efficiency for a particular component that can be enumerated by culturing is reasonably reproducible over time, then dilution plating is a suitable method to use in these experiments. These studies should not be clouded by the fact that culture methods were used. On the other hand, to study communities, it is important to have an accurate assessment of all of the components, or at least, all of the abundant components. Thus, studies of communities that rely solely on culture methods may be biased by the elimination of populations of nonculturable bacteria in the habitat.

Much of the information on the types of bacteria that are found in the phyllosphere comes from studies that have focused on specific species or groups having some physiological trait or function of known interest to the investigator(s). Most notable among these are bacteria with phytopathogenic capabilities. The pioneering works of Crosse (1959) and Leben (1965) have led to the general acceptance of the concept that phytopathogenic bacteria (e.g., pathovars of *Pseudomonas syringae* and *Xanthomonas campestris*) are common inhabitants of the phyllosphere of a wide range of plant species (cf. Hirano and Upper, 1983). Many pathogens are able to grow on healthy leaves, in the absence of disease, in what Leben (1965) termed the epiphytic phase. With the discovery of the ice nucleation activity of certain bacteria, the ubiquity of *P. syringae* and other ice nucleation-active bacterial species such as *Erwinia herbicola* in phyllosphere bacterial communities was further documented (Lindow et al., 1978; Gross et al., 1983).

Two other physiologically interesting groups of bacteria that have been isolated from many plant species are the nitrogen-fixing bacteria and the methylotrophs (i.e., bacteria that are able to grow on reduced one-carbon compounds such as methanol; Green and Bousfield, 1982). Ruinen (1965, 1974) reported that nitrogen-fixing bacteria of the genera *Beijerinkia* and *Azotobacter* were present in large numbers on leaves of many tropical plants in Indonesia and Surinam. Nitrogen-fixing bacteria (e.g., *Enterobacter cloacae, Klebsiella planticola,* and a *Corynebacterium* sp.) also have been isolated from leaf surfaces of Douglas fir (Jones, 1970), rice (Ladha et al., 1983), and *Phaseolus mungo* (Sebastian et al., 1987). The methylotrophs from leaf surfaces are facultative methylotrophs and belong to the genus *Methylobacterium* (Green and Bousfield, 1982). They are able to grow on some organic acids and sugars in addition to one-carbon compounds (Austin and Goodfellow, 1979; Corpe and Basile, 1982). Because of their characteristic pink-pigmented colonies in culture, they are referred to as pink-pigmented facultative methylotrophs (PPFMs) (Green and Bousfield, 1982). The ubiquity of PPFMs on plants is supported by the studies of Corpe et al. (Corpe and Basile, 1982; Corpe, 1985; Corpe and Rheem, 1989). PPFMs (presumably *Methylobacterium organophilum* and/or *M. extorquens*) were isolated from over 50 field- and glasshouse-grown plant species by imprinting leaf samples onto a mineral salts medium containing methanol. What role(s), if any, might these bacteria have in the phyllosphere? It is tempting to speculate that, given the vast land areas covered by plants and the potentially abundant distribution of methylotrophs and nitrogen fixers on leaf surfaces, these bacteria may be of significance to global biogeochemical cycles (i.e., N and C cycles). This possibility awaits further investigations.

While numerous studies have been conducted on specific groups of bacteria across plant species, relatively few attempts have been made to characterize and identify all members of bacterial communities on a particular plant species. Austin et al. (1978) and Ercolani (1978) applied numerical taxonomic methods to phenetic data to identify large numbers of bacteria isolated from leaves of perennial rye (*Lolium perenne*) and olive (*Olea europeae*), respectively (Table 14.1). In both studies, the bacteria were isolated from leaf samples collected over time and cultured on relatively rich media (e.g., glucose yeast extract agar). Each sample was comprised of several leaves (i.e., bulk samples). The bacteria from perennial rye and olive leaves were predominantly pigmented, nonsporeforming heterotrophs. These are characteristics common to bacteria isolated from the phyllosphere of many other plant species (Stout, 1960; Last and Deighton, 1965; Dickinson et al., 1975; Billing, 1976; Goodfellow et al., 1976). On perennial rye, *P. fluorescens* and *X. campestris* each comprised about 20% of the isolates. On olive, almost 68% of the isolates were identified as *P. syringae* pv. *savastanoi*. In neither study did any other species comprise more than 10% of the isolates. This tendency for a few taxa with relatively large numbers of individuals and many taxa with few individuals appears to be generally the case for bacterial communities on other plants (Morris, 1985; Van Outryve et al ., 1989).

Interestingly, of the 621 isolates from perennial rye, approximately 63% were identified at least to the level of genus, whereas over 97% of the 1789 isolates from olive were assigned to the level of species (Table 14.1). The latter situation is quite exceptional. Frequently, identification to the species level has been difficult due to the dearth of information on those phyllosphere bacteria that are neither phytopathogenic nor related to bacteria of medical interest. Further, when some of the much-needed taxonomic studies on phyllosphere bacteria are published, they do not always clarify earlier studies on phylloplane communities. For example, are the "coryneform bacteria" isolated from perennial rye in England (Austin et al., 1978) the same as those isolated from maize (*Zea mays*) and soybean (*Glycine max*) in Iowa, which have been assigned to a new species, *Curtobacterium plantarum* (Dunleavy, 1989)?

As extensive as the studies by Austin et al. (1978) and Ercolani (1978) were, they ignored differences in communities among individual leaves. However, each leaf may be viewed as a discrete habitat which supports a distinct bacterial community. Is this, indeed, the case? Few studies have considered leaf-to-leaf variability in bacterial community structure. A notable exception is the work of Morris and Rouse (Morris, 1985; Morris and Rouse, 1985). In their study, 30 to 50 bacterial isolates were selected at random from dilution plates of washings of *individual* bean leaflets and pods. The isolates were tested for their ability to utilize a range of substrates as sole carbon and nitrogen sources. Some bacteria, such as *P. syringae* and *Enterobacter* spp., were able to utilize as many as 16 to 17 of the compounds tested as sole carbon source; others were less versatile. For example, *Methy-*

Table 14.1 Bacterial community structure on leaves of perennial rye (*Lolium perenne*)[a] and olive (*Olea europea*)[b]

Perennial rye		Olive	
Component or species	Relative abundance (%)[c]	Component or species	Relative abundance (%)[c]
Pseudomonas fluorescens	20.12	*Pseudomonas syringae* pv. *savastanoi*	67.86
Xanthomonas campestris	19.64	*Erwinia herbicola*	8.50
"Coryneform bacteria"	8.37	*Bacillus megaterium*	4.02
Yellow chromogens	4.83	*Micrococcus luteus*	3.63
Flexibacter spp.	4.66	*Xanthomonas campestris* group	3.35
Listeria spp.	4.02	*Arthrobacter globiformis*	2.07
Pink chromogens	3.86	*Lactobacillus plantarum*	1.45
Staphylococcus saprophyticus	1.77	*Klebsiella pneumoniae*	1.40
Other Gram-negative rods	1.61	*Serratia marcescens*	1.34
Klebsiella spp.	0.96	*Acetobacter aceti*	1.23
Acinetobacter spp.	0.96	*Leuconostoc dextranicum*	1.12
Erwinia herbicola	0.80	*Pseudomonas fluorescens*	1.06
Pseudomonas spp.	0.64	*Bacillus subtilis*	0.34
Staphylococcus spp.	0.64	*Pseudomonas delafieldii*	0.11
Bacillus spp.	0.32		
Micrococcus luteus	0.32		
Orange chromogens	0.32		
Unidentified isolates	26.57		

[a] Adapted from Austin et al. (1978).
[b] Adapted from Ercolani (1978).
[c] Total number of isolates tested were 621 for perennial rye and 1789 for olive.

lobacterium mesophilicum used only two to seven of the substrates tested. Different types of nutrient utilization patterns and their frequencies of occurrence were detected among collections of bacteria from individual bean leaflets (Morris, 1985; Morris and Rouse, 1985). For example, collections of bacteria were made from two leaflets harvested at 36 days after planting (Morris and Rouse, 1985). On one leaflet, 29 different patterns were detected among the 32 isolates tested. However, only seven patterns were found among the 50 isolates recovered from the other leaflet. Bacteria differ significantly, both qualitatively and quantitatively, *within* and *among* individual bean leaflets. Although the number of leaflets sampled was small, the results indicate that, at any given time, individual leaves within a canopy may, indeed, support distinct bacterial communities of which probably no two are identical.

The numerical taxonomic method used by Morris (1985) did not identify species per se, but defined taxa within these communities on the basis of nutrient utilization patterns, properties that may possibly be related to function within the community. Few species were actually identified in their study. However, on the leaflet where seven patterns were detected among 50 isolates, all but two of the isolates were identified as *P. syringae* (Morris and Rouse, 1985). This illustrates two additional points: First, the way taxa are defined may influence our view of how the community is constituted. Second, because the ways in which bacteria are assigned to species are not necessarily related to the function of the bacteria in the environment, intra- as well as interspecific variability may become an important consideration in understanding bacterial communities in the phyllosphere.

Additional evidence for the concept that bacterial communities among individual leaves are comprised of distinct assemblages of bacteria is presented in Figure 14.1 (Hirano and Upper, unpublished). The relative abundance of five components and total bacteria on each of 29 snap bean leaflets harvested at the same sampling time were enumerated by dilution plating of homogenates of individual leaflets. The components included phytopathogens (*Xanthomonas campestris* group, Figure 14.1A, and *Pseudomonas syringae,* Figure 14.1B), and saprophytes (*Erwinia herbicola,* Figure 14.1C, Coryneform-like bacteria, Figure 14.1D, and the PPFMs, Figure 14.1E). Although total bacterial population sizes enumerated on King's B medium (King et al., 1954) varied by only 10-fold from one leaflet to the next, the relative proportions of the selected bacterial components varied substantially among the individual leaflets.

The data presented in Figure 14.1 can be used to illustrate the way in which selection of sampling units (i.e., single leaves versus bulk leaf samples) may affect determinations of community structure. The data (Figure 14.1) were recalculated to simulate what would have been found if all of the leaves had been homogenized as a single bulk sample prior to plating (Table 14.2). Thus, the data in Table 14.2 represent the arithmetic average of

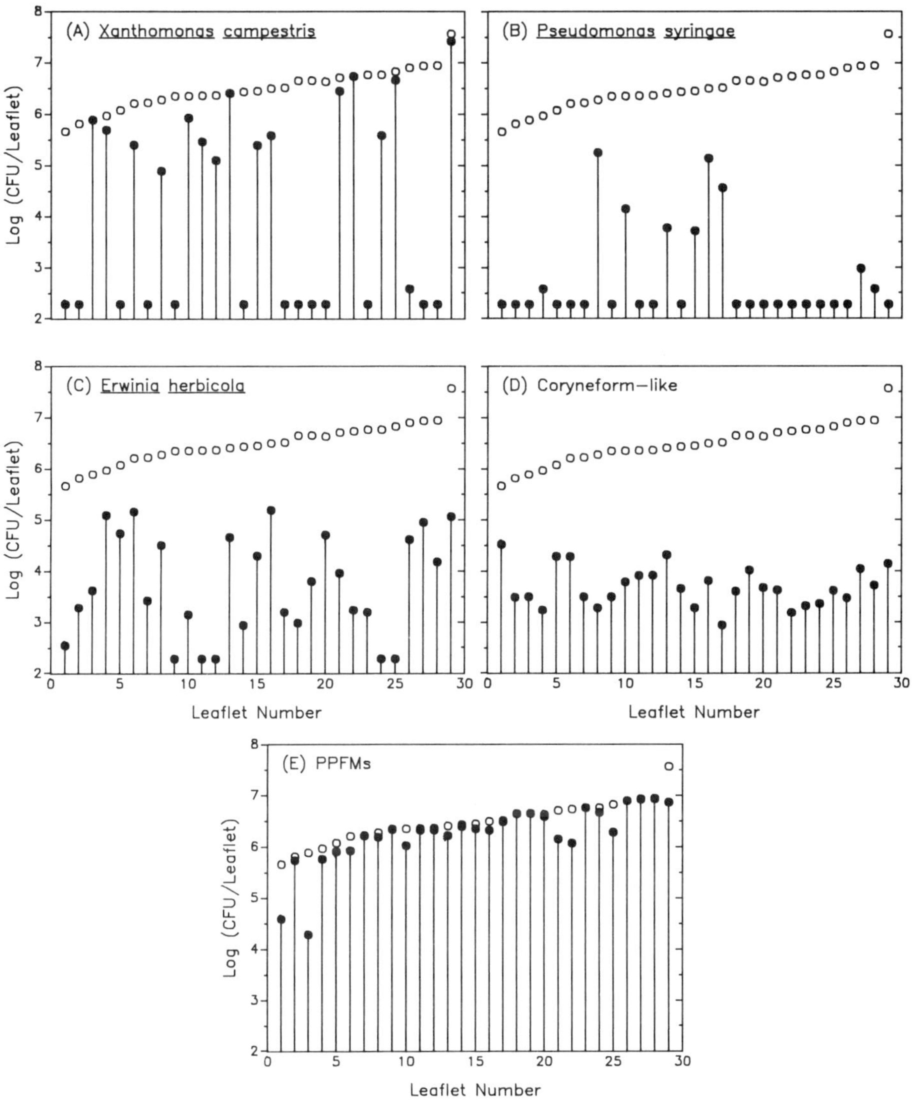

Figure 14.1 Distribution of component and total bacterial populations on individual snap bean leaflets. Leaflets were sampled at 0800, 59 days after planting during the 1988 growing season. Each leaflet was homogenized individually and dilution platings were used to enumerate the components (closed circles) and total bacteria (open circles). The components are: (A) *Xanthomonas campestris* pv. *phaseoli*, (B) *Pseudomonas syringae*, (C) *Erwinia herbicola*, (D) Coryneform-like bacteria, and (E) pink-pigmented facultative methylotrophs (PPFMs).

Table 14.2 Expected means and proportions of component bacterial populations on snap bean leaflets from simulated bulk leaf samples[a]

Component	Mean log (CFU/leaflet)[b]	Relative abundance (%)[c]
Total bacteria	6.66	
Pink-pigmented facultative methylotrophs	6.23	62.17
Xanthomonas campestris	6.19	33.28
Erwinia herbicola	4.49	0.67
Pseudomonas syringae	4.11	0.28
Coryneform-like bacteria	3.85	0.15

[a] Unpublished data of Hirano and Upper.
[b] The mean for each component and total bacteria was calculated by untransforming the data presented in Figure 14.1 and averaging the untransformed data (i.e., CFU per leaflet) over the 29 leaflets. The values presented are the $\log_{10}$ transformations of the arithmetic means. Leaflets with undetectable bacteria were assigned a value of 190 CFU/leaflet, the limit of sensitivity of the assay.
[c] The proportions were calculated using the untransformed means.

the communities on the 29 leaflets, while Figure 14.1 includes the information relative to the communities on the individual leaflets. From Table 14.2, we might conclude that the PPFMs were most abundant, followed by *X. campestris*, *E. herbicola*, *P. syringae*, and the Coryneform-like bacteria. From Figure 14.1E, it is apparent that the PPFMs were, indeed, very abundant on all of the leaflets, but had lower population sizes than *X. campestris* (Figure 14.1A) on six of the leaflets. Thus, the same general conclusion relative to the abundance of the PPFMs might be drawn from both single (Figure 14.1) and bulk (Table 14.2) leaf samples. A similar situation occurs with the Coryneform-like bacteria; the numbers of these bacteria were relatively low on all leaves (Figure 14.1D), and the average population size (Table 14.2) reflects this observation quite reasonably. More important than the average numbers, however, is the relatively small variance in population sizes among the individual leaflets shared by the PPFMs and the Coryneform-like bacteria.

In contrast, the population sizes of *E. herbicola*, *X. campestris*, and *P. syringae* varied greatly among the individual leaflets. The population variances were relatively large for these bacteria. On the average, *E. herbicola* comprised 0.28% (Table 14.2) of the total number of bacteria which corresponds to about 1 cell in 350. However, from Figure 14.1C, it can be estimated that *E. herbicola* ranged from undetectable (less than 1 cell in 10^4) to 14.6% (1 cell in 7; leaflet 4) of the total population on an individual leaflet. The relative contribution of this species in the 29 distinct communities was quite variable, and was not well represented by the mean value. Similar results were obtained for *X. campestris* and *P. syringae*. On average, *X. campestris* was nearly as abundant as the PPFMs (Table 14.2). However, this component was not even detectable on 44.8% of the 29 leaflets (Figure 14.1A). The relative abundance of this component ranged from undetectable (less than 1 cell in 10^5), to greater than 99% of the population (leaflet no. 13). These data illustrate the importance of recognizing that the communities of bacteria may differ substantially in structure among each of the numerous leaves within the canopy, and thus the need to view the canopy as being comprised of populations of bacterial communities on populations of leaves.

Morris and Rouse (1986) discussed the various advantages and shortcomings of different means of estimating diversities of communities. One of the problems that arose in their attempts to measure the diversity of bacterial communities in the phyllosphere relates to the observation that the types of bacteria and their relative abundances vary not only within leaves but among individual leaves. Morris and Rouse (1986), thus, raised the question of whether one should consider the diversity of bacteria within a canopy or the diversities of bacterial communities among populations of individual leaves.

14.3 Population Processes and Bacterial Community Dynamics

The processes of growth, death, immigration, and emigration are usually discussed in relation to attributes of populations (i.e., individuals of the same species) rather than communities (e.g., Krebs, 1985). However, a change in density in any one or more of the component populations relative to other components in a community will necessarily lead to a change in community structure. As relative numbers of the various component bacterial populations on populations of leaves within a canopy change over time, so do the associated communities. Although there is no plant species for which all bacterial community member have been identified or characterized, many studies have been published on patterns of temporal changes of specific species or components and for total numbers of bacteria culturable on a specific medium (Ercolani, 1971; Lindow et al., 1978; Mew and Kennedy, 1982; Smidt and Vidaver, 1986; Smitley and McCarter, 1982; Weller and Saettler, 1980). Examples of such patterns are illustrated in Figure 14.2 for *P. syringae*, the PPFMs, and total bacteria from bean leaflets. In this section, we discuss the extent to which immigration, emigration, growth, and death may contribute to population changes such as those illustrated in Figure 14.2. If the assumption is made that the relative plating efficiency for component and total bacterial populations remains relatively constant throughout the duration of experiments such as that shown in Figure 14.2, then a change in structure in communities from which these bacteria were enumerated can be inferred by comparing numbers of component populations relative to total bacteria.

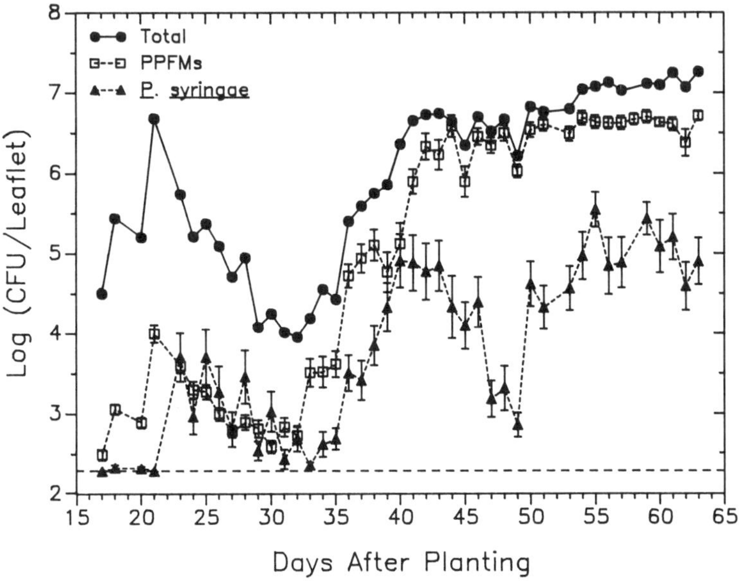

Figure 14.2 Temporal changes in population sizes of total bacteria, pink-pigmented facultative methylotrophs (PPFMs), and *Pseudomonas syringae* on snap bean leaflets. Thirty leaflets were collected at 0800 at each sampling time during the 1986 growing season. Each leaflet was processed individually by dilution plating of leaf homogenates. The data are the means and standard errors for each set of 30 leaflets.

Immigration and Emigration Immigration is critical to the development of phyllosphere bacterial communities. In the absence of immigrants, there can be no community. Once the colonists have arrived, however, can we assess the relative importance of immigration and emigration relative to growth and death in development of bacterial communities in the phyllosphere? There are sufficient data available to ask whether the magnitudes and rates with which phyllosphere bacteria immigrate to and emigrate from leaf surfaces are sufficient to explain the short-term changes in bacterial densities and abundances of individuals within communities (Lindemann et al., 1982; Lindemann and Upper, 1985; Constantinidou et al., 1990).

Lindemann and Upper (1985) measured concentrations of airborne bacteria at canopy height and at ca. 1.5 m above the plant canopy using six-stage Andersen viable samplers. The measurements at two heights allowed estimates of net upward flux of bacteria from the canopy. Bacterial fluxes (i.e., net movement of bacteria away from the canopy source) can be taken as estimates of emigration from the phyllosphere. Deposition of airborne bacteria was measured at canopy height by exposing petri dishes filled to the rim with King's B medium. Although the surface properties of culture medium in petri dishes are not identical to those of leaves, at least some rough estimate of bacterial deposition rates could be obtained (Lindemann and Upper, 1985). These estimates, in turn, can be viewed as rates of bacterial immigration to the phyllosphere. Estimates of immigration and emigration, made during dry weather over three growing seasons, are summarized as a function of time of day in Figures 14.3A and 14.4A, respectively, and the relative frequencies with which these rates were detected are shown in Figures 14.3B and 14.4B (Lindemann and Upper, unpublished). All estimates are for total bacteria.

There was substantial variability in the rates with which bacteria immigrated to bean canopies (Figure 14.3). The highest rates tended to occur during midday (Figure 14.3A; Lindemann and Upper, 1985). Immigration also occurred at night when an inversion layer was often present (Lindemann and Upper, 1985). Most of the time (67% of 130 measurements), immigration rates were between 1 and 10 CFU m^{-2} s^{-1} (Figure 14.3B). This corresponds to approximately 18 to 180 bacterial immigrants arriving per hr on a bean leaflet with an area of 50 cm^2. Less frequently, immigration rates as high as 50 CFU m^{-2} s^{-1} (i.e., 900 CFU per bean leaflet per hr) occurred during dry weather. The estimated numbers of bacterial immigrants that arrive on a bean leaflet per hr or day may appear sufficiently high to contribute to the observed short-term dynamics of phyllosphere bacteria during that period of time when community membership is low (i.e., less than about 10^4 individuals). However, the contribution of immigration needs to be considered relative to that of the other population processes.

During dry weather, emigration of bacteria from bean canopies occurs mainly during the warmest part of sunny days when leaves are dry and wind speeds are greater than 1 m s^{-1} at canopy level (Lindemann and Upper, 1985). Emigration was detected during about one-half of the days during which samples were taken. Similarly to immigration rates, emigration rates were highest during midday (Figure 14.4A; Lindemann and Upper, 1985). Emigration rates most frequently encountered were on the order of about 100 CFU m^{-2} s^{-1} (Figure 14.4B). On one occasion, the rate of emigration was as high as 1100 CFU m^{-2} s^{-1}. These values correspond to the loss of about 1.8×10^3 and 2×10^4 bacteria per hr per bean leaflet, respectively. Thus, substantial numbers of bacteria emigrate from bean canopies during dry weather.

Lindemann and Upper (1985) estimated that a net loss of over 1×10^4 bacterial cells from each bean leaflet occurs daily, atmospheric conditions permitting. The estimate was derived from calculations of the midday average of upward flux of total bacteria (i.e., 195 CFU m^{-2} s^{-1}) for data collected in 1980 and the duration of favorable conditions for upward flux in a given day (i.e., about 6.5 hr, 1000 to 1630). An emigration rate of 195 CFU m^{-2} s^{-1} over a 6.5-hr period would correspond to a loss of about 2.3×10^4 bacteria per leaflet per day. The gain due to immigration was estimated to be only

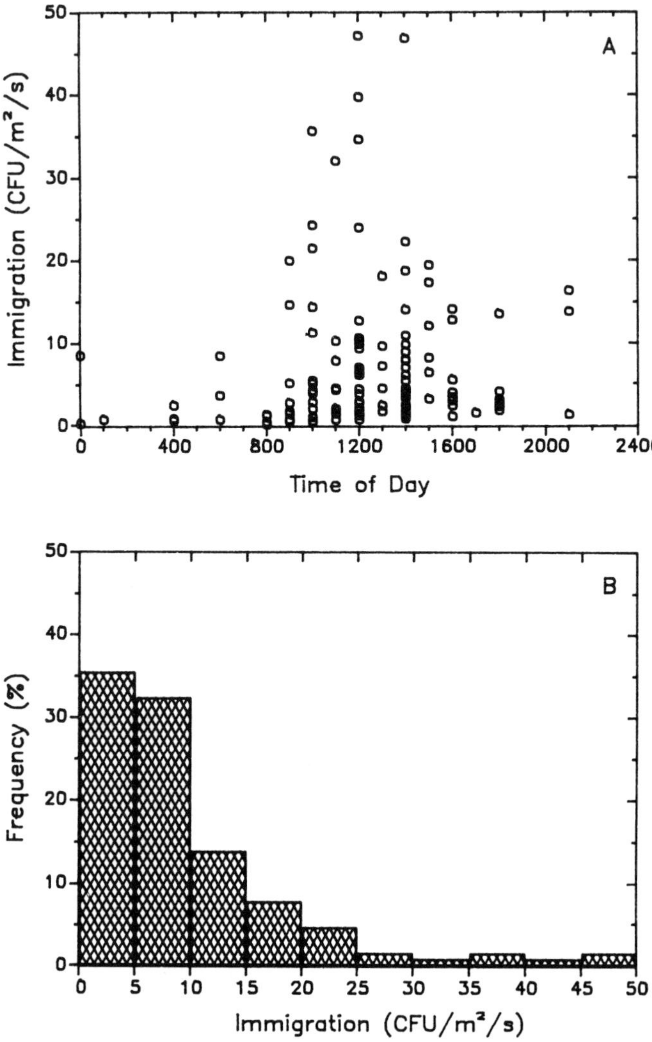

Figure 14.3 Immigration of bacteria to bean canopies. Immigration was measured by the deposition of airborne bacteria onto petri dishes filled to the rim with King's B medium (King et al., 1954). The measurements were made over canopies of snap beans during the 1980, 1981, and 1982 growing seasons. (A) Individual values for immigration measurements as a function of time of day. (B) Frequency of immigration rates. From unpublished data of Lindemann and Upper.

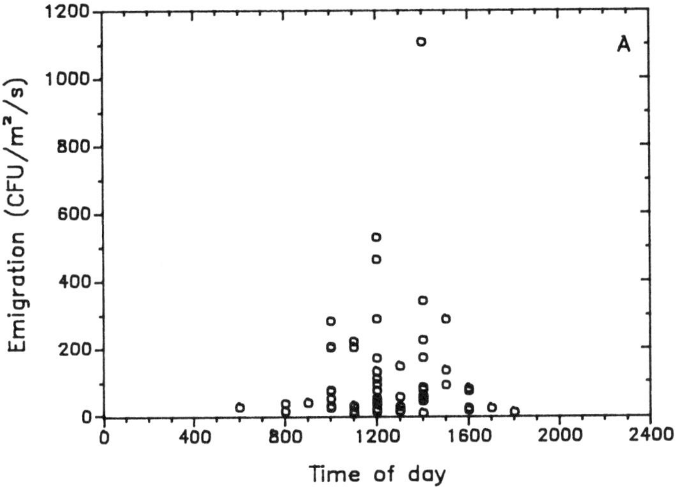

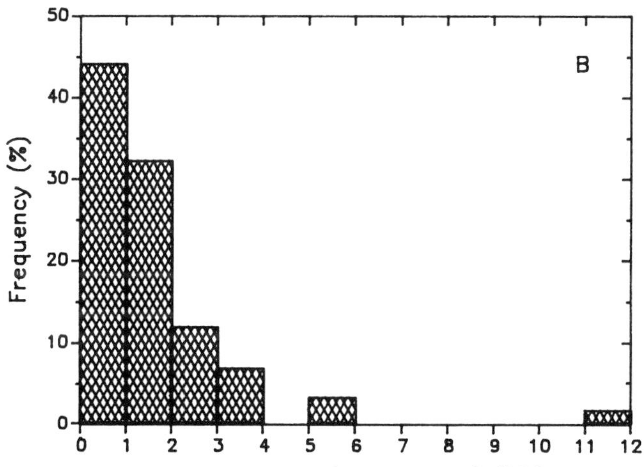

Figure 14.4 Emigration of bacteria from bean canopies. Emigration rates are based on measurements of concentrations of airborne bacteria at canopy height and 1.5 m above the canopy using six-stage Andersen viable samplers. Measurements were made over canopies of snap beans during the 1980, 1981, and 1982 growing seasons. Calculations of net upward flux were done by J. Lindemann as described in Lindemann and Upper (1985). (A) Individual values for emigration measurements as a function of time of day. (B) Frequency of immigration rates. From unpublished data of Lindemann and Upper.

1×10^3 bacteria per leaflet per day. The difference between numbers of immigrants and emigrants, thus, results in a net loss of approximately 1×10^4 bacteria per leaflet per day.

During rain, substantial numbers of bacteria are removed from the phyllosphere (Haas and Rotem, 1976; Lindemann and Upper, 1985; Umekawa et al., 1981; Weller and Saettler, 1980). Some of the bacteria may become immigrants as they are redistributed from one leaf to another by rain-splash. However, because there is a strong net downward flux of bacteria during rain, most of the bacteria that are removed from the phyllosphere are washed off of the leaves and deposited onto the soil (Lindemann and Upper, 1985; Constantinidou et al., 1990). For example, during a rainfall of moderate intensity (0.4 cm hr^{-1}), numbers of bacteria in rainwater collected with funnel-top test tubes increased from about 2 cells ml^{-1} at 2 m above plant canopy, to 7.7×10^2 cells m^{-1} at canopy height, to greater than 10^5 cells ml^{-1} below the canopy (Lindemann and Upper, 1985). Lindemann and Upper (1985) estimated that 10^5 bacteria were lost per bean leaflet in 15 min during this rainfall. Weller and Saettler (1980) estimated that at least 10% of the populations of *Xanthomonas campestris* pvs. *phaseoli* and *fuscans* present on bean leaves were removed during rainfall. Thus, large and rapid decreases in leaf-surface bacterial populations during rain can be attributed to emigration.

Although emigration appears to be the dominant process during rain, immigration, nonetheless, may be of qualitative importance under these conditions. Lindemann and Upper (1985) speculated that a bacterium that arrives on and adheres to a leaf during rain may have a high probability of successfully colonizing the habitat due to the removal of potential resident competitors. Rain may also "cleanse" the habitat of accumulated toxic compounds. These immigrants, then, would play an important role in the subsequent development of the community. Interestingly, during rain, immigrants might arrive, not only from neighboring plants by rain-splash, but also from distant sources by being part of the microflora of the falling raindrops (Constantinidou et al., 1990). Constantinidou et al. (1990) isolated and characterized bacteria from soybean leaves and from rainwater and aerosols present over the soybean canopy during rain. The bacteria in the rain and air over the soybean canopy were different from those on the leaves in terms of the proportion that were fluorescent, ice nucleation active, and pathogenic to soybeans. These differences coupled with the fact that there is a net downward flux of bacteria during rain were taken as evidence in support of the hypothesis that bacteria delivered to a canopy by rain may originate from distant sources (Constantinidou et al., 1990).

The studies on aerial dispersal of phyllosphere bacteria have dealt primarily with fluxes of total bacteria averaged over a particular canopy. As we approach an understanding of the dynamics of populations of bacterial communities on populations of leaves, we must also consider the extent to which these processes affect, not only the average community, but populations of communities. In addition, information is needed on the relative rates of immigration or emigration of specific component populations.

Growth and Death Because immigration rates are insufficient to balance the losses due to emigration, growth must be the driving force in the establishment of bacterial communities in the phyllosphere. The question of when and how rapidly bacteria are likely to multiply in the phyllosphere is particularly intriguing given the fact that these microbes exist in a continuously fluctuating physical environment. Temperature, relative humidity, radiation, wind speed, and other weather parameters change diurnally. The leaf surface alternates through wet and dry periods depending on the availability of free water due to dew, rain, or irrigation. To estimate bacterial growth rates, sampling must be conducted on a time scale of the lifetime of a single bacterial cell (i.e., hours) (Hirano and Upper, 1983, 1986, 1989). In addition, changes in bacterial numbers due to immigration, emigration, and death must be accounted for. Given these criteria, we can only arrive at rough estimates of bacterial growth rates in the phyllosphere since simultaneous measurement of the four population processes has yet to be achieved, and few experiments have been conducted using an appropriate sampling time scale (cf. Hirano and Upper, 1989).

From daily monitoring of population sizes of *Pseudomonas syringae* on snap bean leaflets, we have measured net increases as large as 155-, 500-, and 2300-fold during a 24-hr period (Hirano and Upper, unpublished). If we assume that the increases were due solely to multiplication of the bacterium and that the rate of multiplication was constant during the 24-hr period, then doubling times on the order of 2.1 to 3.3 hr occurred under field conditions. Correction of these values for rates of emigration would result in small increases in the growth rates. The extent to which growth was balanced by death during these periods is unknown, but any correction for the effect of death would increase these growth rate estimates. Sampling of bean leaflets every two hr over a 24-hr period in which population sizes of *P. syringae* increased 28.5-fold indicated that an almost-continuous logarithmic increase in population size can occur during a 24-hr period (Figure 14.5; Hirano and Upper, 1986, 1989). During this 24-hr period, the estimated doubling time was 4.9 hr. Interestingly, population sizes of *P. syringae* increased during the day despite the absence of free water on the leaf surfaces. The possibility that multiplication of the bacterium occurred within the leaf and not on the surface cannot be ruled out since population sizes were determined from leaflet homogenates.

Under field conditions, *P. syringae* may, on occasion, multiply at rates comparable to those measured in broth cultures in the laboratory (e.g., doubling time of 1.27 hr at 28°C; Young et al., 1977). These large, rapid blooms in population sizes of the bacterium occur much less frequently than smaller changes. Similar results were obtained from daily monitoring of populations of *P. syringae* pvs. *syringae* and *morsprunorum* that were applied as a mixture to cherry trees (Sundin et al., 1988). When the total number of bacteria in a community exceeds about 10^4 individuals per leaflet, the small increases are probably dominated by growth. Under these conditions, apparent growth rates of *P. syringae* in the phyllosphere are considerably less than maximal. This appears to be the case for other phyllosphere bacteria such as the PPFMs (see Figure 14.2) and pathovars of *Xanthomonas phaseoli* (Weller and Saettler, 1980). Doubling times of 19.4 and 18.8 hr were estimated for rifampicin-marked strains of *X. phaseoli* pvs. *phaseoli* and *fuscans*, respectively, that were applied to field-grown navy bean (*Phaseolus vulgaris*) plants (Weller and Saettler, 1980). The doubling times were estimated from periods of constant increase in bacterial population sizes based on samples taken every three days. The cumulative effect of small daily increases due to growth can lead to establishment of population sizes approaching and, in some cases, exceeding about 10^5 to 10^6 cells per leaflet. For example, note the daily increase in population sizes of *P. syringae* and the PPFMs from 33 to 40 and from 32 to 42 days after planting, respectively, in Figure 14.2.

From the population trends illustrated in Figure 14.2, it is apparent that population sizes of total and component bacteria do decrease. However, in the absence of direct measurements of changes in numbers of dead cells on leaf surfaces or simultaneous measurements of short-term changes in emigration and population sizes of bacteria on leaves, only rough estimates of death rates can be made. There are times when death appears to be the dominant process. For example, total bacterial populations decreased almost 10-fold from 21 to 23 days after planting corresponding to a loss of about 4×10^6 cells per bean leaflet (Figure 14.2) during a 48-hr period in which no rain occurred. If growth and immigration were minimal and if we accept the assumption of Lindemann and Upper (1985) that, on the average, 10^4 cells may emigrate from a bean leaflet per day, then death contributed substantially to the decrease in total bacterial population sizes during this period. We estimated that the community decreased in abundance by 50% every 2.4 hr.

It is likely that death balanced by growth occurs daily even when population sizes of phyllosphere bacteria appear to have attained a steady state. For example, no net change in population sizes of the PPFMs was observed from daily sampling of bean leaflets from 53 to 61 days after planting. However, when samples were taken every 2 hr during a 26-hr period (Figure 14.5), population sizes of these bacteria decreased during the day and increased at night resulting in no net change during this period. Because the decrease in population sizes of the PPFMs was orders of magnitude higher than could be accounted for solely from estimates of expected rates of emigration (Lindemann and Upper, 1985), death must have occurred. It is interesting to note that two frequently dominant components in bean leaf communities respond differently to changes in the environment. During the 26-hr period shown in Figure 14.5, population sizes

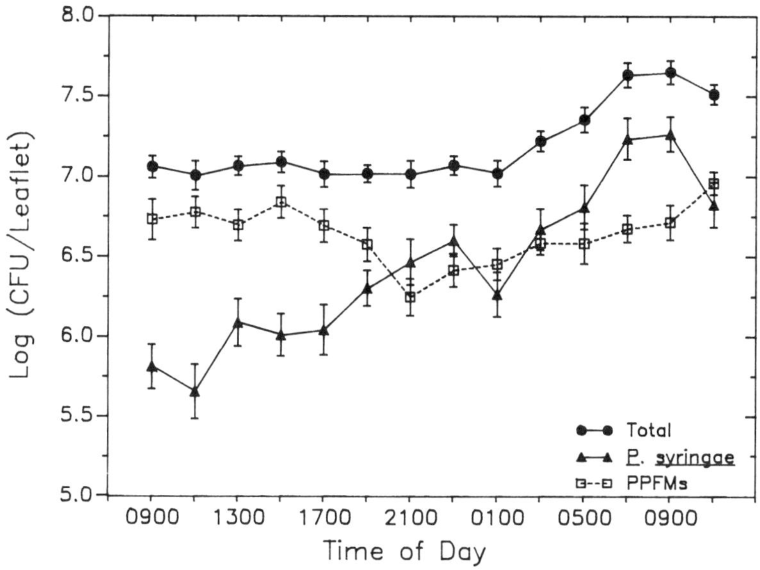

Figure 14.5 Diel changes in population sizes of *Pseudomonas syringae*, pink-pigmented facultative methylotrophs (PPFMs), and total bacteria on snap bean leaflets. Thirty leaflets were collected at each sampling time and processed individually by dilution plating of leaf homogenates. The leaflets were collected at 44 to 45 days after planting during the 1983 growing season. The data are the means and standard errors for each set of 30 leaflets. From Hirano and Upper (1989), with permission.

of the PPFMs decreased during the day and increased at night, while *P. syringae* numbers increased continuously during this same time. In one case (PPFMs), there was no net change in population size from one day to the next; in the other (*P. syringae*), population sizes increased 28.5-fold. Because these two components changed relative to each other *and* relative to total members of the communities on individual leaflets, it is clear that the communities were changing continuously during this 24-hr period.

14.4 Populations of Communities on Populations of Leaves

The relative importance of the four population processes—immigration, emigration, growth, and death—in the dynamics of phyllosphere bacterial communities was discussed earlier in terms of their effects on the average community. However, there is tremendous variability in population sizes of component species, and hence, in community structure from one leaf to the next within a canopy (Figure 14.1). The observed leaf-to-leaf variability

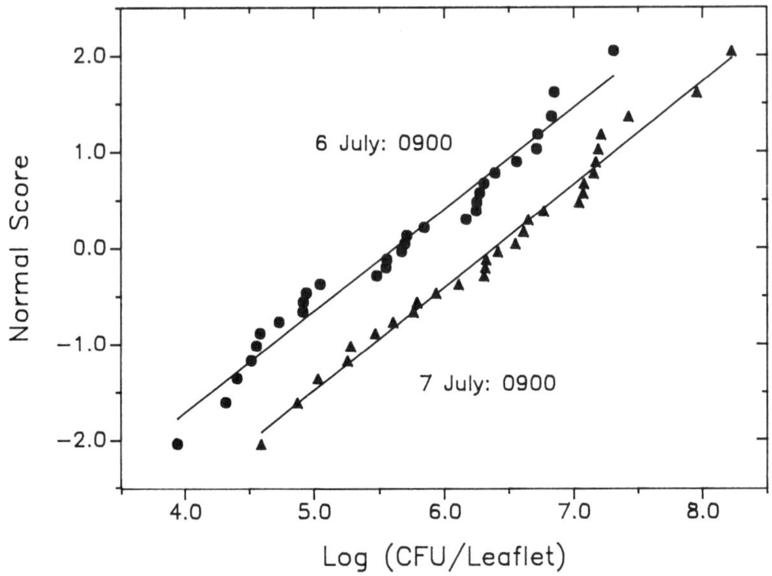

Figure 14.6 Distribution of population sizes of *Pseudomonas syringae* on individual leaflets. The leaflets were collected at 0900 on 6 and 7 July (30 to 31 days after planting) during the 1983 growing season. Each leaflet was processed individually by dilution plating of leaf homogenates. The line for each set of data points (circles or triangles) was obtained by regressing normal scores on bacterial population sizes. From Hirano and Upper (1989), with permission.

in community structure implies that at some time(s) immigration, emigration, growth, and/or death of any component species must occur at different rates on different leaves within a canopy. However, because our sampling method is destructive, we can never know for certain how a particular community changes with time. We can either determine the structure of a community on a particular leaf today or tomorrow, but never both. Given this uncertainty (i.e., the "Phyllosphere Community Uncertainty Principle"), how can we determine whether the population processes occur at different rates on different leaves?

One approach is to view bacterial communities as populations of communities on populations of leaves and frame our questions in terms of probabilities or frequencies. If the frequency distribution of population sizes for a component species among individual leaves at a given time is known, the probability with which a particular density will be likely to be present in the population of leaves can be estimated. For phyllosphere bacterial populations, the mean and variance of the lognormal distribution are useful parameters for this purpose (Hirano et al., 1982). A comparison of the mean and variance of the lognormal for sets of populations sampled at different times can provide some information on the behavior of bacterial popula-

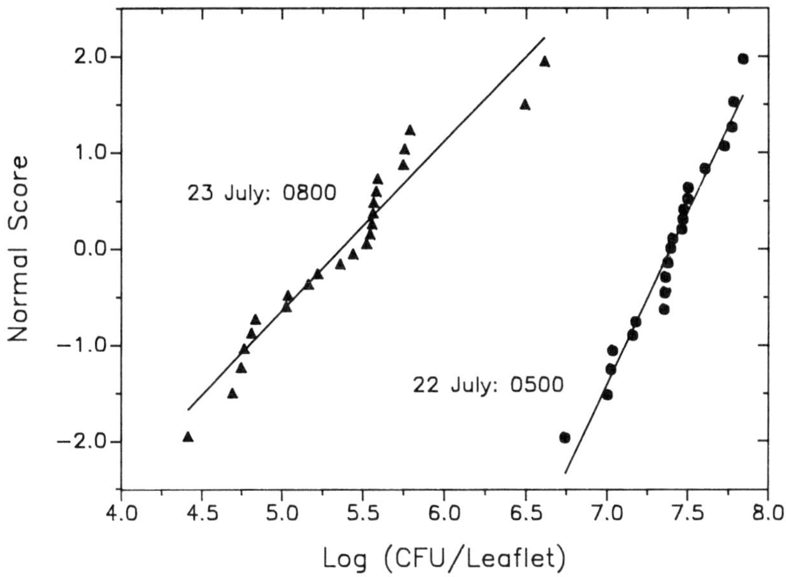

Figure 14.7 Distribution of population sizes of a rifampicin-marked strain of *Pseudomonas syringae* applied to field-grown bean plants. The sampling conducted at 0500 on 22 July was approximately 1.5 hr after the bacteria had been sprayed onto the bean plants. Each leaflet was processed individually by dilution plating of leaf washings. The line for each set of data points (circles or triangles) was obtained by regressing normal scores on bacterial population sizes.

tions among individual leaves. For example, we have examined temporal changes in the distribution of population sizes of *P. syringae* on individual bean leaflets to determine the extent to which changes in densities of the bacterium differ among individual bean leaflets. In Figure 14.6, population sizes of *P. syringae* appeared to increase similarly on all leaflets. In this case, the leaf-to-leaf variability in population sizes remained constant from one day to the next. This can be visualized by the similar slopes of the lines connecting the data points within each set of leaflets. If the increases were due to growth, then the exponential growth rates must have been relatively similar on all leaves and presumably were not affected by initial population size.

In other cases, temporal changes in bacterial densities appeared to differ from one bean leaflet to the next as judged by the different slopes of the lines connecting the data points (Figure 14.7). The distribution of population sizes shown in Figure 14.7 are for a rifampicin-marked strain of *P. syringae* about 2 hr (0500 22 July) and 29 hr (0800 23 July) after it was sprayed onto field-grown bean plants. The decrease in population sizes of this strain of *P. syringae* was probably due to death. The population variance increased from 22 July to 23 July (note the increased slopes of the lines, Figure 14.7) suggesting that the rate of decrease was different among the individual leaves.

In general any number of factors could affect the rates with which the population processes occur on different leaves. They include leaf-to-leaf differences in microclimate, nutrient availability, concentration of toxic compounds, presence of competitors, accessibility to immigrants, and others. The extent to which rates of the population processes are affected by these factors await future experimentation.

14.5 Conclusions and Future Directions

Our understanding of community dynamics is based on the population dynamics of a few components, usually species, for which suitable culture methods are available to quantitate their population sizes. Changes in communities are inferred from changes in relative numbers of these few components. Our ignorance regarding the identity of many of the members of the communities and the lack of suitable methods to specifically enumerate their population sizes has limited expansion of our understanding to most members of bacterial communities in the phyllosphere.

The role of immigration, emigration, growth, and death in phyllosphere bacterial community dynamics has been inferred from the role of these processes in the dynamics of a few component populations. Immigration is critical to initiate community development. Immigration rates, however, are insufficient to account for the relative abundance of component species in most communities. Thus, growth appears to be the driving force underlying community expansion. Most of the time, net rates of growth of component populations under field conditions are much less than optimal growth rates measured under controlled conditions. However, under favorable environmental conditions, estimated growth rates for at least one species (*P. syringae*) were of the same magnitude as those measured in the laboratory. Loss of community members due to emigration and death has been shown to be substantial under some conditions. Emigration and death may be major processes balancing growth during periods when apparent growth rates are less than expected.

While we have gained some insight into the importance of the population processes in the dynamics of an average community, we know very little about how these processes and the factors that affect their rates result in the observed leaf-to-leaf variability in phyllosphere bacterial communities. The variability in community structure must result from the differential rates with which any one or more of the population processes occur or the durations for which they predominate for any one or more of the component populations. In a canopy, each community exists on what may be viewed as a "unique" habitat (i.e., a leaf), no two of which are exactly identical. Individual leaves may differ in their physical, biological, and/or chemical environments. These differences, in turn, may affect rates of immigration, emigration, growth, and/or death. Even slight variations in these rates when accrued over time may result in the observed leaf-to-leaf differences. Similarly, small differences in the duration over which optimal rates occur can produce large differences over time.

A complete understanding of community dynamics in the phyllosphere will require the integration of the effects of individual processes on single components on single leaves into a cohesive view of the sum of the effects of these processes on populations of communities on populations of leaves. Clearly, further progress toward this goal will require substantial research on each process relative to each component. It will also require the generation of models that attempt to integrate the various pieces into the whole. To the extent that these activities are coordinated, both will benefit from exchange of information and ideas among participating scientists.

We thank Julianne Lindemann for allowing us to use her data on aerial dispersal of epiphytic bacteria, Steven Vicen for assistance in preparation of the figures, and the Agricultural Research Service and the Competitive Grants Program of the U.S. Department of Agriculture and the Midwest Food Processors Association for their support of our research.

References

Austin, B. and Goodfellow, M. 1979. *Pseudomonas mesophilica*, a new species of pink bacteria isolated from leaf surfaces. *International Journal of Systematic Bacteriology* 29:373–378.

Austin, B., Goodfellow, M., and Dickinson, C.H. 1978. Numerical taxonomy of phylloplane bacteria isolated from *Lolium perenne*. *Journal of General Microbiology* 104:139–155.

Billing, E. 1976. The taxonomy of bacteria on the aerial parts of plants. pp. 223–273 in Dickinson, C.H. and Preece, T.F. (editors), *Microbiology of Aerial Plant Surfaces*. Academic Press, London.

Brock, T.D. 1987. The study of microorganisms in situ: Progress and problems. pp. 1–17 in Fletcher, M., Gray, T.R.G., and Jones, J.G. (editors), *Ecology of Microbial Communities*. Cambridge University Press, Cambridge, UK.

Colwell, R.R., Brayton, P.R., Grimes, D.J., Roszak, D.B., Huq, S.A., and Palmer, L.M. 1985. Viable but non-culturable *Vibrio cholerae* and related pathogens in the environment: Implications for release of genetically engineered microorganisms. *Bio/Technology* 3:817–823.

Constantinidou, H.A., Hirano, S.S., Baker, L.S., and Upper, C.D. 1990. Atmospheric dispersal of ice nucleation-active bacteria: The role of rain. *Phytopathology* 80:934–937.

Corpe, W.A. 1985. A method for detecting methylotrophic bacteria on solid surfaces. *Journal of Microbiological Methods* 3:215–221.

Corpe, W.A. and Basile, D.V. 1982. Methanol-utilizing bacteria associated with green plants. *Developments in Industrial Microbiology* 23:483–493.

Corpe, W.A. and Rheem, S. 1989. Ecology of the methylotrophic bacteria on living leaf surfaces. *FEMS Microbiology Ecology* 62:243–250.

Crosse, J.E. 1959. Bacterial canker of stone-fruits. IV. Investigation of a method for measuring the inoculum potential of cherry trees. *Annals of Applied Biology* 47:306–317.

Diamond, J. and Case, T. (editors). 1986. *Community Ecology*. Harper and Row, New York. 665 pp.

Dickinson, C.H., Austin, B., and Goodfellow, M. 1975. Quantitative and qualitative studies of phylloplane bacteria from *Lolium perenne*. *Journal of General Microbiology* 91:157–166.

Dunleavy, J.M. 1989. *Curtobacterium plantarum* sp. nov. is ubiquitous in plant leaves and is seed transmitted in soybean and corn. *International Journal of Systematic Bacteriology* 39:240–249.

Ercolani, G.L. 1971. Occurrence of *Pseudomonas savastanoi* (E.F. Smith) Stevens as an epiphyte of olive trees, in Apulia. *Phytopathologie Mediterranean* 10:130–132.

Ercolani, G.L. 1978. *Pseudomonas savastanoi* and other bacteria colonizing the surface of olive leaves in the field. *Journal of General Microbiology* 109:245–257.

Goodfellow, M., Austin, B., and Dickinson, C.H. 1976. Numerical taxonomy of some yellow pigmented bacteria isolated from plants. *Journal of General Microbiology* 97:219–233.

Green, P.N. and Bousfield, I.J. 1982. A taxonomic study of some gram-negative facultatively methylotrophic bacteria. *Journal of General Microbiology* 128:623–638.

Gross, D.C., Cody, Y.S., Proebsting, E.L., Jr., Radamaker, G.K., and Spotts, R.A. 1983. Distribution, population dynamics, and characteristics of ice nucleation-active bacteria in deciduous fruit tree orchards. *Applied and Environmental Microbiology* 46:1370–1379.

Haas, J.H. and Rotem, J. 1976. *Pseudomonas lachrymans* inoculum on infected cucumber leaves subjected to dew- and rain-type wetting. *Phytopathology* 66:1219–1223.

Hirano, S.S. and Upper, C.D. 1983. Ecology and epidemiology of foliar bacterial plant pathogens. *Annual Review of Phytopathology* 21:243–269.

Hirano, S.S. and Upper, C.D. 1986. Temporal, spatial, and genetic variability of leaf-associated bacterial populations. pp. 235–251 in Fokkema, N.J. and Van den Heuvel, J. (editors), *Microbiology of the Phyllosphere*. Cambridge University Press, Cambridge, UK.

Hirano, S.S. and Upper, C.D. 1989. Diel variation in population size and ice nucleation activity of *Pseudomonas syringae* on snap bean leaflets. *Applied and Environmental Microbiology* 55:623–630.

Hirano, S.S., Nordheim, E.V., Arny, D.C., and Upper, C.D. 1982. Lognormal distribution of epiphytic bacterial populations on leaf surfaces. *Applied and Environmental Microbiology* 44:695–700.

Jones, K. 1970. Nitrogen fixation in the phyllosphere of the Douglas fir, *Pseudotsuga douglasii*. *Annals of Botany* 34:230–244.

King, E.O., Ward, M.K., and Raney, D.E. 1954. Two simple media for the demonstration of pyocyanin and fluorescin. *Journal of Laboratory and Clinical Medicine* 44:301–307.

Krebs, C.J. 1985. *Ecology: The Experimental Analysis of Distribution and Abundance*, 3rd ed. Harper and Row, New York. 800 pp.

Ladha, J.K., Barraquio, W.L., and Watanabe, I. 1983. Isolation and identification of nitrogen-fixing *Enterobacter cloacae* and *Klebsiella planticola* associated with rice plants. *Canadian Journal of Microbiology* 29:1301–1308.

Last, F.T. and Deighton, F.C. 1965. The non-parasitic microflora on the surfaces of living leaves. *Transactions of the British Mycological Society* 48:83–99.
Leben, C. 1965. Epiphytic microorganisms in relation to plant disease. *Annual Review of Phytopathology* 3:209–230.
Lindemann, J. and Upper, C.D. 1985. Aerial dispersal of epiphytic bacteria over bean plants. *Applied and Environmental Microbiology* 50:1229–1232.
Lindemann, J., Constantinidou, H.A., Barchet, W.R., and Upper, C.D. 1982. Plants as sources of airborne bacteria, including ice nucleation-active bacteria. *Applied and Environmental Microbiology* 44:1059–1063.
Lindow, S.E., Arny, D.C., and Upper, C.D. 1978. Distribution of ice nucleation-active bacteria on plants in nature. *Applied and Environmental Microbiology* 36:831–838.
Mew, T.W. and Kennedy, B.W. 1982. Seasonal variation in populations of pathogenic peudomonads on soybean leaves. *Phytopathology* 72:103–105.
Morris, C.E. 1985. *Diversity of Epiphytic Bacteria on Snap Bean Leaflets*. Ph.D. thesis, University of Wisconsin, Madison, WI. 210 pp.
Morris, C.E. and Rouse, D.I. 1985. Role of nutrients in regulating epiphytic bacterial populations. pp. 63–82 in Windels, C.E. and Lindow, S.E. (editors), *Biological Control on the Phylloplane*. American Phytopathological Society, St. Paul, MN.
Morris, C.E. and Rouse, D.I. 1986. Microbiological and sampling considerations for quantification of epiphytic microbial community structure. pp. 3–13 in Fokkema, N.J. and Van den Heuvel, J. (editors), *Microbiology of the Phyllosphere*. Cambridge University Press, Cambridge, UK.
Roughgarden, J. and Diamond, J. 1986. Overview: The role of species interactions in community ecology. pp. 333–343 in Diamond, J. and Case, T.J. (editors), *Community Ecology*. Harper and Row, New York.
Ruinen, J. 1965. The phyllosphere. II. Nitrogen fixation in the phyllosphere. *Plant and Soil* 22:375–394.
Ruinen, J. 1974. Nitrogen fixation in the phyllosphere. pp. 121–167 in Quispel, A. (editor), *The Biology of Nitrogen Fixation*. North-Holland Publishing Company, Amsterdam.
Sebastian, J., Chandra, A.K., and Kolattukudy, P.E. 1987. Discovery of a cutinase-producing *Pseudomonas* sp. cohabiting with an apparently nitrogen-fixing *Corynebacterium* sp. in the phyllosphere. *Journal of Bacteriology* 169:131–136.
Smidt, M. and Vidaver, A.K. 1986. Population dynamics of *Clavibacter michiganense* subsp. *nebraskense* in field-grown dent corn and popcorn. *Plant Disease* 70:1031–1036.
Smitley, D.R. and McCarter, S.M. 1982. Spread of *Pseudomonas syringae* pv. *tomato* and role of epiphytic populations and environmental conditions in disease development. *Plant Disease* 66:713–717.
Stout, J.D. 1960. Bacteria of soil and pasture leaves at Claudelands showgrounds. *New Zealand Journal of Agricultural Research* 3:413–430.
Sundin, G.W., Jones, A.L., Olson, B.D. 1988. Overwintering and population dynamics of *Pseudomonas syringae* pv. *syringae* and *P. s.* pv. *morsprunorum* on sweet and sour cherry trees. *Canadian Journal of Plant Pathology* 10:281–288.
Umekawa, M., Watanabe, Y., and Inomata, Y. 1981. Facilitative effect of rainfall on the transmission of the pathogen and the development of angular leaf spot of cucumber. *Annals of the Phytopathological Society of Japan* 47:346–351.
Van Outryve, M.F., Gossele, F., and Swings, J. 1989. The bacterial microflora of witloof chicory (*Cichorium intybus* L. var. *foliosum* Hegi) leaves. *Microbial Ecology* 18:175–186.

Ward, D.M., Weller, R., and Bateson, M.M. 1990. 16S rRNA sequences reveal numerous uncultured microorganisms in a natural community. *Nature* 345:63–65.

Weller, D.M. and Saettler, A.W. 1980. Colonization and distribution of *Xanthomonas phaseoli* and *Xanthomonas phaseoli* var. *fuscans* in field-grown navy beans. *Phytopathology* 70:500–506.

Young, J.M., Luketina, R.C., and Marshall, A.M. 1977. The effects of temperature on growth in vitro of *Pseudomonas syringae* and *Xanthomonas pruni*. *Journal of Applied Bacteriology* 42:345–354.

15

Determinants of Epiphytic Fitness in Bacteria

Steven E. Lindow

15.1 Introduction

The bacteria that inhabit the phylloplane are generally distinctive from those that colonize other habitats and therefore probably have special adaptations which allow them to exploit leaf surfaces. Several studies have noted that a majority of the bacteria isolated from leaf surfaces are pigmented, unlike those recovered from other habitats such as nearby soil (Stout, 1960a, 1960b; Goodfellow and Austin, 1976). It has been speculated (Gibbins and Peterson, 1978) that pigmentation in epiphytic bacteria is an adaptation which allows these microorganisms to tolerate incident ultraviolet (UV) light impinging on the leaf surface (Dickinson, 1986). Presumably, such adaptations would be unnecessary in bacteria inhabiting other habitats in which incident UV radiation was lower.

Surprisingly, few direct tests of the habitat specificity of bacteria have been made. Since the predominant epiphytic bacteria are usually quite different from the predominant soil bacteria nearby (Stout, 1960b; Pieczarka and Lorbeer, 1975), the assumption has commonly been made that bacteria whose normal habitat is the soil do not flourish on leaf surfaces. It is interesting to note that many plants emerge from the soil with very low epiphytic bacterial populations (Lindow et al., 1978a, 1982; Mew and Kennedy, 1982). Epiphytic populations normally increase with time after plant emergence (Lindow et al., 1978b; Lindow, 1982b). Thus, the soil does not appear to serve as an important source of inoculum for epiphytes. Few studies have been done in which bacteria have been transferred from one habitat to another to test whether adaptations permitting exploitation of one habitat would permit the colonization of another. O'Brien and Lindow (1989) found that even bacteria such as *Salmonella, Aeromonas,* and *Rhizobium* species could achieve population sizes as high as "good" epiphytes, such as *Pseudomonas syringae*, on plant surfaces under controlled conditions with continuous free moisture on leaves. However, after the imposition of desiccation stress to the leaf surface, epiphytic bacteria maintained a high population size while nonepiphytic bacteria exhibited greatly reduced viability.

Epiphytic fitness, for the purposes of this review, is defined as the ability of bacteria to grow and/or survive on leaves exposed to a given environmental regime. As defined, fitness is a context-dependent attribute since it will depend on the environment in which it is being assessed. Thus, I will attempt to identify traits which individually determine the behavior of bacteria on leaves under a restricted environmental context rather than to consider fitness in an evolutionary context, in which changes in the environment may require different traits to alternately contribute to bacteria survival. The contribution of single traits to fitness currently can only be easily and unambiguously tested within the context of specific environments. I will not attempt to indicate the relative fitness that different traits may impart to epiphytic bacteria when evolutionary time spans are considered. By focusing on traits that affect short-term behavior of bacteria on leaves, I hope to interest others to undertake future studies that will better address such ecological questions. The epiphytic fitness traits that are discussed in this review are related to the unique features of the leaf surface habitat colonized by bacteria. However, these traits may not specifically confer the ability to survive only on leaf surfaces. For example, while motility may be advantageous to epiphytic bacteria in their acquisition of protected sites and resources, this trait could also be advantageous when the bacterium is in water or soil. However, other traits such as the ability to metabolize specific resources that plants might provide could confer a selective advantage for epiphytic bacteria and are therefore addressed.

15.2 Leaf Ecology and Selective Pressures on Epiphytic Bacteria

The physical, chemical, and biological environment of leaf surfaces, while very poorly understood, has many unique features that probably required adaptation by epiphytic bacteria. A review of microbial adaptations in relation to leaf surface environments was presented by Dickinson (1986). Some of the known or speculated features of the leaf surface that might select for particular traits in bacteria include:

(1) Large and rapid changes in free water on leaf surfaces. Large quantities of free water are present following precipitation or dew formation on leaves. Many such changes in the leaf-surface water content can occur in a 24-hr period. Epiphytic bacteria must therefore be able to cope with rapid changes in water availability. However, the water status of bacteria on plant surfaces is still unknown. For example, if bacteria are located predominate-

ly in crevices formed by topological features on the leaf surfaces, as is commonly observed in scanning electron microscopy (Roos and Hattingh, 1983; Leben, 1965; Leben et al., 1970; Blakeman, 1985; de Cleene, 1989), then free moisture may not occur in their vicinity if the aperture of the crevice in which they are found is sufficiently small. The surface tension of water and the hydrophobicity of the plant surface on which they are located will determine the ability of liquid water to "wet" such a site.

(2) Bacteria do not cover leaf surfaces uniformly. Limited studies have indicated that bacteria occur in small groups or "microcolonies" on a very limited portion of the leaf surface, while much of the leaf surface remains uncolonized (Leben, 1988). This suggests that the resources limiting colonization of leaf surfaces are unequally distributed across the leaf or that bacterial communities develop only after they are able to change the leaf surface habitat. It could be speculated that nonuniform coverage of the leaf surface occurs because colonization proceeds only at the sites of initial successful introductions. Such a situation is analogous to the pattern of human colonization. The majority of humans live in large urban areas, lesser numbers live in small cities, and even smaller numbers are distributed over the rural countryside. Urban centers (which often occur by chance at sites selected by early colonists) develop as commerce, supported by complementary trading and manufacturing activities, permits further growth. Complementary metabolic activities among epiphytic bacteria may similarly support a higher population size than if colonists were found only in isolated sites on the leaf.

(3) The average nitrogen content of organic compounds that are leached from plants onto leaf surfaces is probably much lower than the average nitrogen content of bacterial cells (Tukey, 1970). The population size of bacteria on leaves is probably limited by the required nutrient resources which occur in the most limited quantities. It is likely that nitrogen-containing compounds, which occur in much lower quantities than sugars and other carbohydrates on leaves (Tukey, 1970), limit bacterial growth; e.g., growth cannot occur even if sucrose is abundant if the nitrogen compounds required for growth are exhausted.

(4) The flux of UV light and other ionizing and nonionizing radiant energy is substantially higher on the phylloplane than in other microbial habitats. The amount of short-wavelength UV light at the earth's surface is generally very low (Krinsky, 1976; Dickinson, 1986). Substantially more long-wavelength UV light impinges at the earth's surface. While UV light is damaging to DNA and other cellular constituents (Bridges, 1976; Jagger, 1981), it is unknown whether the flux is sufficient to require a costly adaptation to be made by bacteria inhabiting leaf surfaces.

(5) Bacteria of different species often occur together in mixtures on leaves (Hirano and Upper, 1989); a leaf rarely harbors only a single species or strain of bacteria. If these different species or strains of bacteria coexist at

the same microsites, then they may need to compete for the resources limiting growth of bacteria at that site. The pattern of resource utilization by epiphytic bacteria may determine the bacterial composition of a leaf surface.

(6) The topology of leaf surfaces is irregular (Hallam and Juniper, 1971). Studies by Juniper and Cox (1973) have shown the presence of different types of crystals of epicuticular wax as well as differences in other larger topological features, such as glands, hairs, trichomes, and stomates, on different plants. The growth or survival of epiphytic bacteria, however, is not well correlated with any one topological feature. Bacteria often are found in depressions on the leaf surface, as well as at the bases of trichomes, etc. Such "preferred" sites of colonization may be actively sought out by bacteria.

(7) Many physical and environmental factors, such as temperature and relative humidity, can vary greatly in time spans of less than one min on leaf surfaces (Burrage, 1971, 1976). The laminar boundary layer, which typically has a higher relative humidity than the bulk air, can be disrupted by wind gusts, thus subjecting epiphytic bacteria to the extremes of relative humidity seen in the bulk air. Epiphytic bacteria, therefore, must cope with such frequent changes of their physical environment.

(8) The rate of nutrient release at the sites that bacteria occupy on leaf surfaces is unknown. Bacteria appear to occupy discrete microsites on leaf surfaces. However, are these sites independent from one another? If so, then, only the nutrients released at the sites that bacteria occupy may be available for exploitation. If the rate of release is low and bacterial resources are continually depleted to a growth-limiting level, then competition for the resources may be intense.

(9) Many epiphytes probably are the descendents of bacteria that immigrated to plants as aerosols (Lindemann and Upper, 1985). Thus, bacteria may have evolved special adaptations for survival during their dispersal from one plant to another as dry aerosol particles, on soil particles, or on insects.

15.3 Hypothetical Epiphytic Fitness Traits

The most common approach used to link bacterial traits with the fitness of bacteria on leaves has been guesswork. In this approach, hypotheses are generated regarding traits that might logically affect epiphytic fitness. Bacteria are then tested to determine whether they possess these phenotypes. Alternatively, the phenotypes that bacteria can express in culture are ascertained, and a conjecture is made about the subset of those phenotypes which might logically confer epiphytic fitness. Unfortunately, rigorous tests of these hypothetical phenotypes have seldom been made. The behavior of two strains which differ in their expression of a phenotype could be compared as a test of this conjecture. Such a test is least ambiguous if the phenotype has been eliminated by mutation and the behavior of a mutant and parental strain is compared. Alternatively, genes conferring the phenotype could be cloned and expressed in a recipient strain. The behavior of that strain and its parental strain could then be compared. The preferred method of mutational testing of hypothesized fitness traits is to utilize transposable elements such as the transposon Tn*5*. This transposon inserts randomly into the chromosome of many bacteria and not only physically "tags" the affected gene, but it also insures that a single polar mutation exists in each cell that is analyzed (Berg and Berg, 1983). One should be aware that it is unlikely that any one gene or phenotype will be *sufficient* to confer fitness to an epiphytic bacterium in a given situation but instead it will be only one of several *necessary* contributors to fitness. In such a case, inactivation of such a trait in a strain would greatly reduce fitness, but its transfer to another strain not normally possessing this trait would not necessarily increase fitness.

Extracellular Polysaccharide Production There is considerable evidence that many epiphytic bacteria are surrounded by a layer of extracellular polysaccharide (EPS) while on leaf surfaces. Many scanning electron microscope images reveal strands of amorphous material that emanate from and between bacterial cells (Hattingh et al., 1986; Thomson, 1986; Wilson et al., 1989). These strands probably represent the dehydrated remnants of a more complete matrix which originally surrounded the cell. This matrix, presumably composed of one or more types of EPS molecules (Fett et al., 1989), may not only anchor cells to the leaf surface but might modify the physical and chemical environment around the cell to a more favorable one for bacterial growth or survival. The matrix hypothesized to exist on leaf surfaces may have many analogies to that of biofilms produced by aquatic microorganisms. Biofilms have ion-exchange capabilities which can concentrate nutrients from dilute sources in the vicinity of the cell (Costerton et al., 1987). The biofilm can also provide protection from predators, such as amebas, and shield cells from the action of lytic enzymes, antibiotics, and other inhibitory compounds (Anwar et al., 1989; Caldwell and Lawrence, 1986). If such a matrix were produced by epiphytic bacteria, it could also be beneficial to the cells.

Few critical tests of the involvement of EPS production in epiphytic fitness have been made. Haefele and Lindow (unpublished) produced mutants of a strain of *Pseudomonas syringae* that varied in the amount of EPS produced in culture. These mutants were identified after exposing cells to the chemical mutagen ethyl methane sulfonate (EMS). Selection was made for colony morphology variants. Such variations were assumed to be indicative of changes in the cell surface and/or in extracellular products. Several mutant strains with significantly lower EPS content compared to the parental strain, but with identical growth rates, were identified. No significant differences were observed in either the extent of growth or death of these cells exposed to wet and dry conditions, respectively, on plants in the greenhouse. This indicates that EPS had little effect on epiphytic fitness under these conditions. Three hypotheses as to why this was the case could be made: (1) EPS production itself is not required for fitness, (2) EPS production may contribute to fitness but its effect is so low that the assays used could not detect differences in behavior, or (3) the types of EPS produced on plants may be different than that expressed in culture. It is likely that many different types of EPS can be produced by cells, each made under specific conditions. There is evidence that EPS production by some phytopathogenic bacteria is expressed dissimilarly in culture and on plants (Coplin and Cook, 1990). Thus, it is probable that some EPS species may be produced in culture while others may be produced only on plants. Much more information will be needed on the regulation of EPS production before this phenotype can be unambiguously associated with epiphytic fitness.

UV Tolerance It has long been assumed that pigmented bacteria are more tolerant of UV irradiation on leaves than unpigmented species. The results of critical tests of this logical hypothesis are somewhat contradictory. In one test, Gibbins and Peterson (1978) isolated strains of *Erwinia herbicola* which had lost a plasmid conferring the production of a yellow pigment. The UV sensitivity of the white variants and the pigmented parental strain could therefore be compared without confounding effects of chromosomal genes. Unexpectedly, white variants were slightly more tolerant of long-wavelength UV irradiation. There was no detectable difference in sensitivity of these two genotypes to the more damaging short-wavelength UV light in culture. In a more recent study, genes conferring pigment production from *E. herbicola* were cloned and introduced into the normally nonpigmented species *Escherichia coli* (Tuveson et al., 1988). The UV tolerance of isogenic pigmented and nonpigmented *E. coli* strains were compared. Pigmented *E. coli* strains exhibited increased tolerance of long-wavelength UV irradiation in culture. These results indicate that UV-absorbing pigments can protect cells against UV irradiation. These two studies are not necessarily inconsistent. It is possible that other fitness factors that confer UV tolerance exist in *E. herbicola* strains. Elimination of pigment production in this species may not have a measurable effect on UV tolerance since a secondary defense mechanism may exist, unlike in *E. coli*. The role of pigmentation in epiphytic fitness obviously needs to be further clarified.

Other genes conferring tolerance to UV light have been cloned. The *rec*A gene, involved in the repair of DNA damage such as that conferred by UV irradiation, has been cloned from a number of epiphytic bacteria including *P. syringae* (Hickman et al., 1987). The elimination of *rec*A function in such strains increases their susceptibility to UV light by several orders of magnitude in culture (Willis et al., 1988). The *rec*A gene function may be more efficiently expressed in epiphytes because of the need to repair damage caused by UV irradiation. However, no critical tests of the fitness of *rec*A$^-$ mutants of epiphytes on leaf surfaces have yet been made.

Motility While several studies have demonstrated an important role of motility in the invasion of wounded plants by phytopathogenic bacteria (Panopoulos and Schroth, 1974; Raymundo and Ries, 1981; Bayot and Ries, 1986; Hattermann and Ries, 1989), its role as an epiphytic fitness determinant has been evaluated only recently. If cells multiply or survive at particular protected sites on the surface of the leaf, then the ability to move to such sites would be an important factor in bacterial growth and survival. To test this concept, Haefele and Lindow (1987) produced nonmotile (Mot$^-$) mutants of the normally motile epiphyte *P. syringae* by exposure to EMS. Isogenic Mot$^+$ revertant strains arising from Mot$^-$ mutants were also isolated. Several interesting differences between Mot$^+$ and Mot$^-$ cells were observed when these stains were applied to leaves and exposed to different environments: (1) More Mot$^+$ cells than Mot$^-$ cells were able to move to more distal parts of the leaf or to different leaves of plants exposed to moist conditions. (2) When leaves were kept continuously moist after inoculation with either Mot$^+$ or Mot$^-$ cells, the population size of Mot$^+$ bacteria was only slightly higher than that of Mot$^-$ strains. However, if large population sizes of either Mot$^+$ or Mot$^-$ cells were allowed to develop under wet conditions, a higher death rate and a lower population size of the Mot$^-$ strains than Mot$^+$ that survived during dry conditions was observed. (3) After multiplication of Mot$^+$ and Mot$^-$ strains had occurred on leaf surfaces under moist conditions, a higher fraction of Mot$^-$ than Mot$^+$ cells were killed by subsequent irradiation of leaves by UV light. Mot$^+$ and Mot$^-$ strains, however, exhibited identical sensitivities to UV light in vitro. (4) When coinoculated on leaf surfaces, Mot$^+$ bacteria grew at the expense of Mot$^-$ bacteria, achieving a higher population size after several days growth under moist conditions on the plant. Active acquisition of particular sites in which physical environmental stresses can be avoided leading to cell death is the most likely reason that Mot$^+$ cells survived better than Mot$^-$ cells on plant surfaces. It appears that bacterial multiplication can occur on sites on leaf surfaces where survival is not favored since the population sizes of Mot$^+$ and Mot$^-$ bacteria were similar on wet plants, while the fraction of Mot$^+$ bacteria that could survive the exposure to dryness on leaves was much higher than that of Mot$^-$ strains.

O'Brien and Lindow (1989) also observed that motile species such as *Salmonella*, *Rhizobium*, and *Aeromonas* could achieve high population sizes on wet leaves but could not sustain these populations when leaves were dried. These results support the conjecture that a major factor constraining the population size of microorganisms on leaves is the number and quality of protected sites and the ability of microorganisms to identify and acquire such sites. The fact that Mot$^+$ and Mot$^-$ bacteria exhibit substantially different behavior indicates that motility can occur on leaf surfaces. However, the role of chemotaxis rather than motility per se remains unclear. It is unknown whether sites on the leaf surfaces where bacterial survival is facilitated are sensed by bacteria via chemical signals emitted by the plant. Alternatively, the ability to move in a nondirected fashion may enable bacteria to explore a larger fraction of the leaf surface, thereby increasing their chance of encountering sites where survival might be facilitated. Unfortunately, little is known of the differences in site occupation exhibited by different strains of bacteria on leaf surfaces. There are some data that indicate that bacterial species, such as some strains of *Xanthomonas campestris* pv. *oryzae*, prefer different colonization sites (Mew et al., 1984).

Plant Hormone Production Many plant-associated bacteria can produce plant growth hormones (Ernstsen et al., 1987; Fett et al., 1987; Hunter, 1987; Loper and Schroth, 1986). Gall-forming plant pathogens including *Pseudomonas* and *Agrobacterium* can produce both auxins and cytokinins and thus cause hyperplasia (Roberto and Kosuge, 1987; Weiler and Schroder, 1987; Thomashow et al., 1984). While such galls might provide refuges for the survival of these pathogens, it is unclear why other non-gall-forming pathogenic and nonpathogenic bacteria can produce 3-indoleacetic acid (IAA) and other hormones, or whether they produce them in nature. Exogenous auxins can induce plant-cell-wall loosening and membrane leakiness (Leopold and Kriedemann, 1975). Production of IAA by bacteria may improve their habitats by increasing the rate of plant leaching or by causing local anatomical changes conducive for growth or survival.

Very few direct tests of the role of IAA in the fitness of epiphytic bacteria have been reported. Indirect studies that IAA production might have a role in the epiphytic fitness of *Pseudomonas syringae* pv. *savastanoi* have been reported. IAA$^-$ strains of *P. syringae* pv. *savastanoi* did not grow as well as IAA$^+$ strains on olive leaves (Varvaro and Surico, 1984). IAA production by wild-type strains of this bacterium might have conferred changes in the plant that facilitated bacterial growth. Lindow (1987) found that fruit russetting on pear was correlated with large population sizes of IAA-producing bacteria, particularly strains of *Erwinia herbicola*. These strains can produce higher quantities of IAA in culture than those of gall-forming pathogens (Clark and Lindow, 1989). These bacteria achieve high population sizes when inoculated onto pear (Lindow, unpublished), indicating that they are excellent epiphytes. In preliminary studies, chemically induced mutants of *E. herbicola* strain 299R, which have reduced ability to produce IAA in culture, achieved a lower population size on pear than the parental strain under field conditions (Lindow, unpublished). Near-isogenic IAA$^+$ and IAA$^-$ strains of *E. herbicola* will be needed to confirm a role for IAA production in epiphytic fitness.

Siderophore Production Microorganisms require Fe^{3+} for growth (Neilands and Leong, 1986). While iron is abundant in nature, only very low amounts of ionic Fe^{3+} is thought to be available to microorganisms, due to the propensity of iron to form very insoluble oxidation products under normal environmental conditions (Lindsay and Schwab, 1982; Raymond and Carrano, 1979). Siderophores are low-molecular-weight compounds with high affinity for iron that are excreted by microorganisms. When they are coupled with a specific siderophore-uptake system, they allow microbial acquisition of iron in environments where the free element is in low concentrations (Neilands, 1981). Since most bacteria can produce siderophores in culture, it has been assumed that in nature microbes are at least occasionally limited by iron and must have mechanisms to acquire it. The concentration of iron in the microhabitats that bacteria exploit on leaves, however, is unknown. Therefore, the question of whether siderophores are a necessary fitness factor for acquisition of iron is an open one.

Loper and Lindow (1987) conducted a study to determine whether siderophores are required for the colonization of leaf surfaces. The siderophore of *P. syringae* is a green-yellow fluorescent pigment produced under iron-limiting conditions in culture. This pigment strongly absorbs UV light. It was reasoned that the *P. syringae* siderophore should act as a UV protectant on leaves if it is produced there. Nonfluorescent (Flu^-) mutants of *P. syringae* were identified following EMS mutagenesis that would not grow on media containing very low concentrations of Fe^{3+}. These mutants, which lacked the ability to produce siderophore, were more sensitive to UV light in culture than their Flu^+ parental strains. The siderophore produced by Flu^+ strains, when added to cultures of Flu^- strains, abolished their UV sensitivity. However, both Flu^+ and Flu^- strains exhibited similar sensitivities to UV irradiation on leaves. They were both about as sensitive as that of the Flu^- strain in culture. This observation was not consistent with a high level of siderophore production by *P. syringae* on plant surfaces. When inoculated onto bean plants, Flu^- strains also reached the same population size and produced as many bacterial brown spot lesions on plants as Flu^+ strains. It was concluded that, under the conditions of these greenhouse studies, iron was not so limited that siderophores were produced in quantities similar to that in iron-deficient culture media. An alternative approach to studying siderophore production on plant surfaces will be addressed in Section 15.5.

Osmotolerance If leachates are abundant on leaf surfaces (particularly small molecules with high colligative effects), their concentration upon drying could be sufficiently high that epiphytic bacteria would experience osmotic stress. Osmotic stresses may occur infrequently and are likely to depend on the amount of solutes present on a plant. Bacteria have evolved mechanisms of osmotic stress tolerance (Le Rudulier et al., 1984; Larsen et al., 1987; Landfald and Strom, 1986). Osmotolerance often results from the synthesis and accumulation of betaine, proline, or other compounds within the cell. Skowlund and Lindow (unpublished) produced Tn5-induced mutants of *P. syringae* strain B728A which are more osmotically sensitive (Osm$^-$) than the parental strain. Both Osm$^+$ and Osm$^-$ strains grew normally in all culture media tested except those having high osmolarities. Both Osm$^-$ and Osm$^+$ strains exhibited similar growth rates and maximum population sizes on wet bean leaves. However, in about one-half of all the experiments performed, Osm$^-$ mutants exhibited a higher fractional decrease in viable population size than Osm$^+$ parental strains when leaves were allowed to dry (Skowland and Lindow, unpublished). No difference in behavior between these strains was observed in the other experiments. No dramatic or consistent benefit of the ability to tolerate high osmotic stresses was detected in these experiments.

Antibiotic Production If microbes must compete for limiting resources on leaf surfaces, production of a biocide such as an antibiotic could improve their competitive ability. Many bacteria produce antibiotics or bacteriocins in culture. However, little is known about the production of such compounds in nature. To determine the role of antibiotic production in antagonism, Lindow (1988) isolated mutants deficient in antibiotic production (Ant$^-$) of epiphytic bacteria known to be highly antagonistic towards ice-nucleation-positive (Ice$^+$) *P. syringae* strains on leaves. No requirement for antibiotic production was found for antagonism of *P. syringae* under greenhouse conditions. Both Ant$^+$ parental strains and Ant$^-$ mutants of antagonists inhibited the growth of *P. syringae* strains on plants when they were inoculated prior to *P. syringae*. Ant$^-$ strains also reached as high a population size in the presence of other competing bacteria as did Ant$^+$ strains. Thus, while it is likely that antibiotic production could have a selective advantage for a producing strain, its effect may be small or conditional. Such effects may be difficult to assess in simple laboratory experiments.

Ice Nucleation Some pathogenic and nonpathogenic bacteria can cause ice formation (Lindow, 1983, 1986; Lindow et al., 1978b; Maki et al., 1974; Maki and Willoughby, 1978). Ice formed in sensitive plants can cause injury (Lindow, 1983; Lindow et al., 1982). The ability to cause injury to plants under subfreezing temperatures may facilitate the entry of phytopathogenic bacteria into plants. Ice nucleation has therefore been suggested to be a temperature-conditional virulence factor (Lindow, 1982a; Sule and Seemuller, 1987). Since there must be some metabolic and genetic cost to maintain the ability to produce ice nuclei, there must be some benefit to the cells. However, there is no evidence that this phenotype enhances the epiphytic phase of ice-nucleation-active bacteria. Lindemann and Suslow (1987) showed that Ice$^-$ mutants of *P. syringae* and *Pseudomonas fluorescens* were equally competitive with their Ice$^+$ parental strains on plant surfaces. Similarly, Ice$^-$ mutants of *P. syringae* produced by Lindow (1985) did not differ in growth or survival on any plant species tested under greenhouse or field conditions (Lindow and Panopoulos, 1988). Therefore, no measurable short-term effect of ice nucleation activity on the epiphytic behavior of bacteria has been observed.

***Avr* Genes** *Avr* genes confer a race-specific incompatible interaction of bacterial plant pathogens with plants (Tamaki et al., 1988). This incompatible interaction is dominant over the compatible reaction with virulent pathogens. Host plants may, therefore, sense the presence of pathogens by recognizing such a necessary gene (*avr*). It would be advantageous to the plant to select a necessary pathogen gene; an unnecessary gene could be lost or altered by mutation in the pathogen to circumvent the defense response that the host would mount. Loss of *avr*Bs2 from *Xanthomonas campestris* pv. *vesicatoria* reduced its ability to grow in compatible plants (those lacking a resistance gene corresponding to *avr*Bs2), but not in culture (Kearney and Staskawicz, 1990). It is possible that such *avr* genes might also condition epiphytic survival as well. However, no direct tests of these genes on epiphytic fitness have been made.

Hydrophobicity Many researchers consider the ability of epiphytic bacteria to adhere to plants to be important in survival. Because the leaf surface contains waxy substances (Jeffree et al., 1976), the hydrophobicity of epiphytes might affect their ability to adhere to this surface. While differences in hydrophobicity among bacterial species which occur on leaf surfaces was found (Fett, 1985), there was no clear correlation between hydrophobicity and adherence potential. Hydrophobicity may therefore be one factor, although small relative to other factors, which is necessary for bacteria to adhere to plants. Differences in hydrophobicity could conceivably affect the microhabitats that bacteria might select since there may be differences in the hydrophobicity of the plant surface due to varying patterns of wax structures and composition. The most hydrophobic sites on leaf surfaces might be preferentially colonized by bacteria having the most hydrophobic cell surface.

Catabolic Properties While most epiphytic bacteria are nutritionally diverse, there is little information on the preferential utilization of the nutrients on plant surfaces. It is known, however, that many species of *Pseudomonas* prefer organic acids such as succinate over simple sugars such as glucose as carbon and energy sources (Clarke and Ornston, 1975). However, there have been no definitive tests of the ability of catabolic-deficient mutants of epiphytic bacteria to exploit plants. Morris and Rouse (1985) observed that a wider range of nutrients were used by epiphytic strains of *P. syringae* recovered from bean plants late in the growing season compared to those collected earlier. This observation is consistent with the scenario in which bacterial community structure is simpler on newly colonized leaf surfaces. Competition among subsequent immigrants with more diverse catabolic properties, which allow them to compete for resources not depleted by initial residents results in the development of more complex communities.

Surface-Active-Compound Production Many bacteria can produce compounds with surfactant properties. Production of these surface-active compounds on leaf surfaces has only recently been described (Bunster et al., 1989). These workers found that plants inoculated with such bacteria were more easily wetted by water than were control plants. The hydrophobic leaf surface, and particularly small irregular depressions in it, may be difficult to wet. Because water has a high surface tension, it may seldom enter such depressions unless the surface tension of the water is reduced. Thus, the ability to produce surface-active compounds may be one way (analogous to that of EPS production) in which bacteria could alter the habitat of the leaf surface to allow its more efficient exploitation.

15.4 Identification of Novel Fitness Traits

The previous sections identified methods by which traits hypothesized to be involved in epiphytic fitness could be evaluated. However, novel or unanticipated traits may condition survival or growth of bacteria on leaves. Such phenotypes would not be found by these methods. How are novel epiphytic fitness traits identified? Instead of searching for a gene with a known product (such as a specific catabolic trait), it is also possible to search for genes contributing to a known function (such as epiphytic fitness) but without a known product. Such an approach has been used to identify previously undetected genes that confer virulence in phytopathogenic bacteria (Lindgren et al., 1986). In this approach, all of the bacterial genes are individually inactivated by mutagenesis and those mutants affected in virulence or pathogenicity are identified. Subsequent examination of the inactivated gene often will reveal the traits conferring pathogenicity or virulence. An analogous approach to identifying genes and phenotypes conferring epiphytic fitness would require the estimation of bacterial population size of many thousands of mutants on plant surfaces in contrast to the qualitative presence or absence of disease used to detect virulence genes. For example, it would be necessary to measure the population size of about 5,000 mutants of a typical epiphyte such as *P. syringae* to ensure that there is an 80% chance that a given gene (out of the 3,000 genes contained in such a strain) is inactivated in such a collection. The logistical constraints of ascertaining the population size of such a large number of mutants on plants have prevented this approach from being used until recently.

A novel approach has been developed which permits the rapid estimation of the population size of ice-nucleation-active bacteria on leaf surfaces. Hirano et al. (1985) used an analysis of the distribution of freezing temperatures of a collection of leaves to estimate the frequency of leaves which had a large population of Ice$^+$ bacteria. Since ice nucleation activity at warm freezing temperatures (> -3°C) occurs vary rarely in a population of cells of an Ice$^+$ bacterial strain, only a leaf with a high population size of Ice$^+$ bacteria has a high probability of freezing at such a warm temperature. Lindow (1986) showed that there was a direct relationship between the logarithm of the population size of Ice$^+$ bacteria and the average nucleation temperature of a collection of leaves. By using the freezing temperature of leaves colonized by different bacterial mutants as a rapid estimator of the population size of those mutants, Lindow (unpublished) has been able to ascertain the epiphytic fitness of over 5,000 mutants of an Ice$^+$ *P. syringae* strain after they had been inoculated individually onto bean plants and subjected to alternating wet and dry conditions. Out of 5,228 mutants, 126 exhibited a lower freezing temperature than parental strains when examined by this method. More than 60% of these 126 mutants exhibited a lower population size than the parental strain when examined using classical bacterial enumeration techniques. Only about 15% of the mutants with altered epiphytic fitness were deficient in any of 128 biochemical traits tested in vitro (Lindow, unpublished). This study indicates that many interesting traits which have large individual effects on the epiphytic fitness of bacteria remain to be discovered. It also indicates that it may be very difficult to guess which phenotypes are important in epiphytic fitness.

15.5 Measuring Bacterial Activities on Leaves

To determine the effect of a given phenotype on the ability of bacteria to colonize leaves, it will be necessary to ascertain whether the gene functions while cells are on the leaf. If the inactivation of genes determining a phenotype of interest does not yield a large and definitive effect on epiphytic fitness, it will be necessary to ascertain when and the extent to which the gene is functioning on leaves. Knowledge of the pattern of gene expression on leaves should reveal its potential role in epiphytic fitness. If, for example, a gene has been hypothesized to be involved in epiphytic fitness and yet can be demonstrated not to function under most situations on leaves, it can be eliminated from further consideration as an epiphytic fitness factor. Unfortunately, however, it is difficult to ascertain whether most phenotypes are actually expressed on leaf surfaces. For example, it has not been possible to measure the presence of most antibiotics, siderophores, or any of the other traits discussed above while cells are on leaf surfaces.

Molecular biological techniques have made it possible to measure the transcriptional activity of genes whose products are difficult to measure in natural settings. Genes conferring the phenotype of interest can be fused upstream of promoterless derivatives of structural genes whose products are more easily measured in natural environments. A promoterless ice nucleation gene is a simple, rapid, and sensitive assay for the transcriptional activity of bacterial genes while on plants (Lindgren et al., 1989). Since plants do not contain appreciable numbers of ice nuclei active at temperatures above −5°C (Lindow et al., 1982), plants colonized by bacteria containing such gene fusions will contain no ice nuclei unless the gene of interest is transcriptionally active. If it is expressed, it will direct the transcription of the gene responsible for the expression of ice nucleation activity. Ice nucleation activity of bacteria on plants can be rapidly quantified by a droplet-freezing assay (Lindow, 1990).

An ice-nucleation reporter gene was used to demonstrate the level of siderophore biosynthesis by *P. syringae* strains on plant surfaces (Lindow and Loper, 1990). No evidence was obtained for high levels of siderophore biosynthesis by *P. syringae* strains in the studies described earlier (Loper and Lindow, 1987). However, low levels of siderophore biosynthesis could not have been detected by those methods. It remained unclear what level of Fe^{3+} limitation was sensed by bacteria on leaves. Since the genes conferring the production of the fluorescent pyoverdine siderophore of *P. syringae* had been cloned and partially characterized (Loper et al., 1984), DNA regions which exhibited Fe^{3+}-regulated transcription could be identified. These iron-regulated DNA regions involved in pyoverdine biosynthesis were cloned in front of a promoterless ice nucleation gene (Lindow and Loper, 1990). By comparing the number of ice nuclei produced by cells containing this fusion gene with those of cells containing an ice nucleation gene transcribed from its own iron-insensitive promoter, it was concluded that cells exhibited only about 10% the level of transcriptional activity of genes conferring pyoverdine biosynthesis as under iron-limiting conditions in culture. Thus, while siderophores are apparently made on leaf surfaces, the average amount of siderophore produced by the population of cells was apparently low; on average most cells sensed sufficient amounts of iron that transcription of these genes was repressed. In the future, molecular genetic techniques such as these should shed much light on the biology of epiphytic bacteria on the phylloplane.

15.6 Conclusions and Future Directions

It is clear from this review that the bacterial traits that influence bacterial behavior on plant surfaces are not well understood. Only recently has experimentation departed from activities that might be regarded as "natural history investigations." Much of the previous work was largely descriptive and lacked the power to associate bacterial attributes with functions. More recent studies, however, have attempted to relate bacterial phenotypes with growth and survival under restricted environmental conditions. Even these studies, which have benefitted greatly from the use of molecular genetic methods, have often given inconclusive results. Better methods for accurately measuring the growth and survival of bacteria will need to be devised in order to detect the small effects that individual bacterial traits may have on leaf surface behavior. An *absolute* measure of bacterial behavior frequently may not be needed to determine the effects of a phenotype on bacterial behavior. For example, if isogenic bacterial strains which differ only in the trait of interest are available, a behavioral *difference* of these stains in mixtures on leaves is indicative of the effects of that phenotype on epiphytic fitness. Unfortunately such a comparative approach, while powerful, has rarely been used (Lindemann and Suslow, 1987). Coinoculation experiments using isogenic strains are also amenable to long-term greenhouse or field experiments in which the environment is changed to ascertain fitness differences that are not obvious in more restricted environmental settings.

While improvements in ecological tests of fitness will help *describe* the effects of individual phenotypes on fitness, molecular genetic techniques will be needed to better *understand* how these traits contribute to fitness. The use of reporter genes is a good example of such a new technique. With this tool, it is now possible to ascertain the activity of individual genes in situ. This technique can also be used to develop biological sensors for chemical resources present in the microhabitats that bacteria occupy on leaves. Great improvements in our understanding of bacterial behavior on the phylloplane can be expected when modern molecular genetic techniques are applied to the study of epiphytic bacteria.

Some of the work reported here was supported by grant BSR-85-16511 from the U.S. National Science Foundation, and grant DE-FG03-86ER13518/A004 from the U.S. Department of Energy.

References

Anwar, H., Dasgupta, M., and Costerton, J.W. 1989. Tobramycin resistance of mucoid *Pseudomonas aeruginosa* biofilm grown under iron limitation. *Journal of Antimicrobial Chemotherapy* 24:647–656.

Bayot, R.G. and Ries, S.M. 1986. Role of motility in apple blossom infection by *Erwinia amylovora* and studies of fire blight control with attractant and repellent compounds. *Phytopathology* 76:441–445.

Berg, D.E. and Berg, C.M. 1983. The prokaryotic transposable element Tn5. *Biotechnology* 1:417–435.

Blakeman, J.P. 1985. Ecological succession of leaf surface microorganisms in relation to biological control. pp. 6–30 in Windels, C.E. and Lindow, S.E. (editors), *Biological Control on the Phylloplane*. American Phytopathological Society Press, St. Paul, Minnesota.

Bridges, B.A. 1976. Survival of bacteria following exposure to ultraviolet and ionizing radiations. pp. 183–208 in Gray, T.R.G. and Postgate, J.R. (editors), *The Survival of Vegetative Microbes*. Cambridge University Press, Cambridge, UK.

Bunster, L.L., Fokkema, N.J., and Schippers, B. 1989. Effect of surface-active *Pseudomonas* spp. on leaf wettability. *Applied and Environmental Microbiology.* 55:1340–1345.

Burrage, S.W. 1971. The micro-climate at the leaf surface. pp. 89–101 in Preece, T.F. and Dickinson, C.H. (editors), *Ecology of Leaf Surface Micro-organisms.* Academic Press, London.

Burrage, S.W. 1976. Aerial microclimate around plant surfaces. pp. 173–184 in Dickinson C.H. and Preece, T.F. (editors), *Microbiology of Aerial Plant Surfaces.* Academic Press, New York.

Caldwell, D.E. and Lawrence, J.R. 1986. Growth kinetics of *Pseudomonas fluorescens* microcolonies within the hydrodynamic boundary layer of surface microenvironments. *Microbial Ecology* 12:299–312.

Clark, E. and Lindow, S.E. 1989. Indoleacetic acid production by epiphytic bacteria associated with pear fruit russetting. *Phytopathology* 79:1191.

Clarke, P.H. and Ornston, N. 1975. Metabolic pathways and regulation: II. pp. 263–340 in Clarke, P.H. and Richmond, H. (editors), *Genetics and Biochemistry of Pseudomonads.* John Wiley and Sons, New York.

Coplin, D.L. and Cook, D. 1990. Molecular genetics of extracellular polysaccharide biosynthesis in vascular phytopathogenic bacteria. *Molecular Plant-Microbe Interactions* 3:271–279.

Costerton, J.W., Cheng, K.J., Geesey, G.G., Ladd, T.I., Nickel, J.C., Dasgupta, M., and Marrie, T.J. 1987. Bacterial biofilms in nature and disease. *Annual Review of Microbiology* 41:435–464.

de Cleene, M. 1989. Scanning electron microscopy of the establishment of compatible and incompatible *Xanthomonas campestris* pathovars on the leaf surface of Italian ryegrass and maize. *EPPO Bulletin* 19:81–88.

Dickinson, C.H. 1986. Adaptations of micro-organisms to climatic conditions affecting aerial plant surfaces. pp. 77–100 in Fokkema, N.J. and van den Heuvel, J. (editors), *Microbiology of the Phyllosphere.* Cambridge University Press, Cambridge, UK.

Ernstsen, A., Sandberg, G., Crozier, A., and Wheeler, C.T. 1987. Endogenous indoles and the biosynthesis and metabolism of indole-3-acetic acid in cultures of *Rhizobium phaseoli. Planta* 171:422–428.

Fett, W.F. 1985. Relationship of bacterial cell surface hydrophobicity and charge to pathogenicity, physiological race, and immobilization in attached soybean leaves. *Phytopathology* 75:1414–1418.

Fett, W.F., Osman, S.F., and Dunn, M.F. 1987. Auxin production by plant pathogenic Pseudomonads and Xanthomonads. *Applied and Environmental Microbiology* 53:1839–1845.

Fett, W.F., Osman, S.F., and Dunn, M.F. 1989. Characterization of exopolysaccharides produced by plant-associated fluorescent pseudomonads. *Applied and Environmental Microbiology* 55:579–583.

Gibbins, L.N. and Peterson, D.L. 1978. Responses of *Erwinia herbicola* Y46 and a non-pigmented mutant to ultraviolet radiation and visible light: survival curves and photoreactivation. pp. 443–450 in *Proceedings of the Fourth International Conference on Plant Pathogenic Bacteria.* Institut National de la Recherche Agronomique, Angers, France.

Goodfellow, M. and Austin, B. 1976. Numerical taxonomy of some yellow-pigmented bacteria isolated from plants. *Journal of General Microbiology* 97:219–233.

Haefele, D. M. and Lindow S.E. 1987. Flagellar motility confers epiphytic fitness advantages to *Pseudomonas syringae. Applied and Environmental Microbiology* 53:2528–2533.

Hallam, N.D. and Juniper, B.E. 1971. The anatomy of the leaf surface. pp. 3–39 in Preece, T.F. and Dickinson, C.H. (editors), *Ecology of Leaf Surface Microorganisms*. Academic Press, New York.

Hattermann, D.R. and Ries, S.M. 1989. Motility of *Pseudomonas syringae* pv. *glycinea* and its role in infection. *Phytopathology* 79:284–289.

Hattingh, M.J., Beer, S.V., and Lawson, E.W. 1986. Scanning electron microscopy of apple blossoms colonized by *Erwinia amylovora* and *Erwinia herbicola*. *Phytopathology* 76:900–904.

Hickman, M.J., Orser, C.S., Willis, D.K., Lindow, S.E., and Panopoulos, N.J. 1987. Molecular cloning and biological characterization of the *rec*A gene from *Pseudomonas syringae* pv. *syringae*. *Journal of Bacteriology* 169:2906–2910.

Hirano, S.S. and Upper, C.D. 1989. Diel variation in population size and ice nucleation activity of *Pseudomonas syringae* on snap bean leaflets. *Applied and Environmental Microbiology* 55:623–630.

Hirano, S.S., Baker, L.S., and Upper, C.D. 1985. Ice nucleation temperature of individual leaves in relation to population sizes of ice nucleation active bacteria and frost injury. *Plant Physiology* 77:259–265.

Hunter, W.J. 1987. Influence of 5-methyltryptophan-resistant *Bradyrhizobium japonicum* on soybean root nodule indole-3-acetic acid content. *Applied and Environmental Microbiology* 53:1051–1055.

Jagger, J. 1981. Near-UV radiation effects on micro-organisms. *Photochemistry and Photobiology* 34:761–768.

Jeffree, C.E., Baker, E.A., and Holloway, P.J. 1976. Origins of the fine structure of plant epicuticular waxes. pp. 119–158 in Dickinson, C.H. and Preece, T.F. (editors), *Microbiology of Aerial Plant Surfaces*. Academic Press, London.

Juniper, B.E., and Cox, G.C. 1973. The anatomy of the leaf surface: the first line of defense. *Pesticide Science* 4:543–561.

Kearney, B. and Staskawicz, B.J. 1990. Widespread distribution and fitness contribution of *Xanthomonas campestris* avirulence gene *avr*Bs2. *Nature* 346:385–386.

Krinsky, N.I. 1976. Cellular damage initiated by visible light. pp. 209–239 in Gray, T.R.G. and Postgate, J.R. (editors), *The Survival of Vegetative Microbes*. Cambridge University Press, Cambridge, UK.

Landfald, B. and Strom, A.R. 1986. Choline-glycine betaine pathway confers a high level of osmotic tolerance in *Escherichia coli*. *Journal of Bacteriology* 165:849–855.

Larsen, P.I., Sydnes, L.K., Landfald, B., and Strom, A.R. 1987. Osmoregulation in *Escherichia coli* by accumulation of organic osmolytes: betaines, glutamic acid, and trehalose. *Archives of Microbiology* 147:1–7.

Le Rudulier, D., Strom, A.R., Dandekar, A.M., Smith, L.T., and Valentine, R.C. 1984. Molecular biology of osmoregulation. *Science* 224:1064–1068.

Leben, C. 1965. Influence of humidity on the migration of bacteria on cucumber seedlings. *Canadian Journal of Microbiology* 11:671–676.

Leben, C. 1988. Relative humidity and the survival of epiphytic bacteria with buds and leaves of cucumber plants. *Phytopathology* 78:179–185.

Leben, C., Schroth, M.N., and Hildebrand, D.C. 1970. Colonization and movement of *Pseudomonas syringae* on healthy bean seedlings. *Phytopathology* 60:677–680.

Leopold, A.C. and Kriedemann, P.E. 1975. *Plant Growth and Development*. McGraw-Hill, New York.

Lindemann, J. and Suslow, T.V. 1987. Competition between ice nucleation active wild-type and ice nucleation deficient mutant strains of *Pseudomonas syringae* and *P. fluorescens* biovar I and biological control of frost injury on strawberry blossoms. *Phytopathology* 77:882–886.

Lindemann, J. and Upper, C.D. 1985. Aerial dispersal of epiphytic bacteria over bean plants. *Applied and Environmental Microbiology* 50:1229–1232.

Lindgren, P.B., Peet, R.C., and Panopoulos, N.J. 1986. Gene cluster of *Pseudomonas syringae* pv. *"phaseolicola"* controls pathogenicity of bean plants and hypersensitivity on nonhost plants. *Journal of Bacteriology* 168:512–522.

Lindgren, P.B., Govindarajan, A.G., Frederick, R., Panopoulos, N.J., Staskawicz, B.J., and Lindow, S.E. 1989. An ice nucleation reporter gene system: identification of inducible pathogenicity genes in *Pseudomonas syringae* pv. *phaseolicola*. *EMBO Journal* 8:1291–1301.

Lindow, S.E. 1982a. Epiphytic ice nucleation active bacteria. pp. 334–362 in Lacy, G. and Mount, M. (editors), *Phytopathogenic Prokaryotes* vol. 1 Academic Press, New York.

Lindow, S.E. 1982b. Population dynamics of epiphytic ice nucleation active bacteria on frost sensitive plants and frost control by means of antagonistic bacteria. pp. 395–416 in Li, P.H. and Sakai, A. (editors), *Plant Cold Hardiness*. Academic Press, New York.

Lindow, S.E. 1983. The role of bacterial ice nucleation in frost injury to plants. *Annual Review of Phytopathology* 21:363–384.

Lindow, S.E. 1985. Ecology of *Pseudomonas syringae* relevant to the field use of Ice^- deletion mutants constructed *in vitro* for plant frost control. pp. 23–35 in Halvorson, H.O., Pramer, D., and Rogul, M. (editors), *Engineering Organisms in the Environment: Scientific Issues*. American Society for Microbiology, Washington, D.C.

Lindow, S.E. 1986. Strategies and practice of biological control of ice nucleation active bacteria on plants. pp. 293–311 in Fokkema, N.J. and van den Heuvel, J. (editors), *Microbiology of the Phyllosphere*. Cambridge University Press, Cambridge, UK.

Lindow, S.E. 1987. Severity of pear fruit russetting associated with epiphytic indoleacetic acid-producing bacteria. *Phytopathology* 77:1724.

Lindow, S.E. 1988. Lack of correlation of antibiosis in antagonism of ice nucleation active bacteria on leaf surfaces by non-ice nucleation active bacteria. *Phytopathology* 78:444–450.

Lindow, S.E. 1990. Bacterial ice nucleation measurements. pp. 428–434 in Klement, Z., Rudolf, K., and Sands, D. (editors), *Methods in Phytobacteriology*. Akademia Kiado, Budapest.

Lindow, S.E. and Loper, J.E. 1990. Transcriptional activity of fluorescent siderophore genes from *Pseudomonas syringae in situ* on leaf and root surfaces. *Phytopathology* 80:982

Lindow, S.E. and Panopoulos, N.J. 1988. Field tests of recombinant Ice^- *Pseudomonas syringae* for biological frost control in potato. pp. 121–138 in Sussman, M., Collins, C.H., and Skinner, F.A. (editors), *Proceedings of the First International Conference on The Release of Genetically Engineered Microorganisms*. Academic Press, London.

Lindow, S.E., Arny, D.C., and Upper, C.D. 1978a. Distribution of ice nucleation active bacteria on plants in nature. *Applied and Environmental Microbiology* 36:831–838.

Lindow, S.E., Arny D.C., and Upper, C.D. 1978b. *Erwinia herbicola*: a bacterial ice nucleus active in increasing frost injury to corn. *Phytopathology* 68:523–527.

Lindow, S.E., Arny, D.C., and Upper, C.D. 1982. Bacterial ice nucleation: a factor in frost injury to plants. *Plant Physiology* 70:1084–1089.

Lindsay, W.L. and Schwab, A.P. 1982. The chemistry of iron in soils and its availability to plants. *Journal of Plant Nutrition* 5:821–840.

Loper, J.E. and Lindow, S.E. 1987. Lack of evidence for *in situ* fluorescent pigment production by *Pseudomonas syringae* pv. *syringae* on bean leaf surfaces. *Phytopathology* 77:1449–1454.

Loper, J.E. and Schroth, M.N. 1986. Influence of bacterial sources of indole-3-acetic acid on root elongation of sugar beet. *Phytopathology* 76:386–389.

Loper, J.E., Orser, C.S., Panopoulos, N.J., and Schroth, M.N. 1984. Genetic analysis of fluorescent pigment production in *Pseudomonas syringae* pv. *syringae*. *Journal of General Microbiology* 130:1507–1515

Maki, L.R. and Willoughby, K.J. 1978. Bacteria as biogenic sources of freezing nuclei. *Journal of Applied Meteorology* 17:1049–1053.

Maki, L.R., Galyan, E.L., Chang-Chien, M., and Caldwell, D.R. 1974. Ice nucleation induced by *Pseudomonas syringae*. *Applied and Environmental Microbiology* 28:456–460.

Mew, T.W. and Kennedy, B.W. 1982. Seasonal variation in populations of pathogenic pseudomonads on soybean leaves. *Phytopathology* 72:103–105.

Mew, T.W., Mew, I.C. and Huang, J.S. 1984. Scanning electron microscopy of virulent and avirulent strains of *Xanthomonas campestris* pv. *oryzae* on rice leaves. *Phytopathology* 74:635–641.

Morris, C.E. and Rouse, D.I. 1985. Role of nutrients in regulating epiphytic bacterial populations. pp. 63–82 in Windels, C.E. and Lindow, S.E. (editors), *Biological Control on the Phyllosphere*. American Phytopathological Society Press, St Paul, MN.

Neilands, J.B. 1981. Iron absorption and transport in microorganisms. *Annual Review of Nutrition* 1:27–46.

Neilands, J.B. and Leong, S.A. 1986. Siderophores in relation to plant growth and disease. *Annual Review of Plant Physiology* 37:187–208.

O'Brien, R.D. and Lindow, S.E. 1989. Effect of plant species and environmental conditions on epiphytic population sizes of *Pseudomonas syringae* and other bacteria. *Phytopathology* 79:619–627.

Panopoulos, N.J. and Schroth, M.N. 1974. Role of flagellar motility in the invasion of bean leaves by *Pseudomonas phaseolicola*. *Phytopathology* 64:1389–1397.

Pieczarka, D.J. and Lorbeer, J.W. 1975. Bacterial populations on basal lettuce leaves and in soil from under lettuce plants. *Phytopathology* 65:509–513.

Raymond, K.N. and Carrano, C.J. 1979. Coordination chemistry and microbial iron transport. *Archives of Chemical Research* 12:183–190.

Raymundo, A.K. and Ries, S.M. 1981. Motility of *Erwinia amylovora*. *Phytopathology* 71:45–49.

Roberto, F.F. and Kosuge, T. 1987. Phytohormone metabolism in *Pseudomonas syringae* subsp. *savastanoi*. pp. 371–380 in Fox, J.E. and Jacobs, M. (editors), *Molecular Biology of Plant Growth Control*. Alan R. Liss, Inc., New York.

Roos, I.M. and Hattingh, M.J. 1983. Scanning electron microscopy of *Pseudomonas syringae* pv. *morsprunorum* on sweet cherry leaves. *Phytopathologische Zeitschrift* 108:18–25.

Stout, J.D. 1960a. Biological studies of some Tussock-Grassland soils: XV bacteria of two cultivated soils. *New Zealand Journal of Agricultural Research* 3: 214–223.

Stout, J.D. 1960b. Bacteria of soil and pasture leaves at Claudlands Showgrounds. *New Zealand Journal of Agricultural Research* 3:413–430.

Sule, S. and Seemuller, E. 1987. The role of ice formation in the infection of sour cherry leaves by *Pseudomonas syringae* pv. *syringae*. *Phytopathology* 77:173–177.

Tamaki, S., Dahlbeck, D., Staskawicz, B., and Keen, N.J. 1988. Characterization and expression of two avirulence genes cloned from *Pseudomonas syringae* pv. *glycinea*. *Journal of Bacteriology* 170:4846–4854.

Thomashow, L.S., Reeves, S., and Thomashow, M.F. 1984. Crown gall oncogenesis: evidence that a T-DNA gene from the *Agrobacterium* Ti plasmid pTiA6 encodes an enzyme that catalyzes synthesis of indoleacetic acid. *Proceedings of the National Academy of Sciences USA* 81:5071–5075.

Thomson, S.V. 1986. The role of the stigma in fire blight infections. *Phytopathology* 76:476–482.
Tukey, H.B. 1970. The leaching of substances from plants. *Annual Review of Plant Physiology*. 21:305–324.
Tuveson, R.W., Larson, R.A., and Kagan, J. 1988. Role of cloned carotenoid genes expressed in *Escherichia coli* in protecting against inactivation by near UV light and specific phototoxic molecules. *Journal of Bacteriology* 170:4675–4680.
Varvaro, L. and Surico, G. 1984. Epiphytic survival of wild types of *Pseudomonas syringae* pv. *savastanoi* and their Iaa$^-$ mutants on olive leaves. pp. 20–22 in Panagopoulos, C.G., Psallidas, P.G., and Alivizatos, A.S. (editors), *Proceedings of the 2nd Working Group on Pseudomonas syringae pathovars*. Hellenic Phytopathological Society, Athens, Greece.
Weiler, E.W. and Schroder, J. 1987. Hormone genes and crown gall disease. *Trends in Biochemical Sciences* 12:271–275.
Willis, D.K., Hrabak, E.H., Lindow, S.E., and Panopoulos, N.J. 1988. Construction and characterization of *Pseudomonas syringae rec*A mutant strains. *Molecular Plant-Microbe Interactions* 1:80–86.
Wilson, M., Epton, H.A.S., and Sigee, D.C. 1989. *Erwinia amylovora* infection of hawthorn blossom. II: The stigma. *Journal of Phytopathology* 127:15–28.

16

Evolutionary Perspective on the Ice Nucleation Gene-Encoded Membrane Protein

Paul K. Wolber and Gareth J. Warren

16.1 Bacterial Ice Nucleation

It comes as a surprise to most people that small volumes of water can be "supercooled," that is, cooled below 0°C without freezing. The basis for this phenomenon is the requirement by most liquids for a nucleus to initiate a liquid-to-solid phase transition; liquids that are nucleus-free can supercool. The reason that supercooled water is not commonly observed is that only a single nucleus is required to initiate a freezing event, and ice nuclei are usually present in the environment. Many of these nuclei are of bacterial origin (Lindow, 1982, 1983; Warren, 1987; Warren and Wolber, 1987).

Ice nucleation may operate by a template mechanism: ice nuclei bind water molecules in a configuration which closely mimics the structure of ice (Fletcher, 1970). The number of molecules bound with high fidelity determines the temperature at which nucleation takes place, with larger templates initiating ice crystallization at temperatures closer to 0°C (Fletcher, 1970). Very few substances (with the exception of ice itself) can initiate ice formation at temperatures as close to 0°C as bacterial ice nuclei.

Observed Occurrence Ice nuclei are expressed by relatively few bacterial genera. Ice-nucleation-positive (Ina^+) strains of *Pseudomonas syringae* (Arny et al., 1976), *P. fluorescens* (Maki and Willoughby, 1978), *P. viridiflava* (Anderson and Ashworth, 1986), *Erwinia herbicola* (Lindow et al., 1978), *E. ananas* (Arai et al., 1989), and *Xanthomonas campestris* (Orser et al., 1985) have been isolated. Most of the isolates come from plant surfaces, although soil (Maki et al., 1974) and marine (Fall and Schnell, 1985) isolates have also been described.

Several surveys of the incidence of Ina^+ bacteria in specific temperate locations have been reported (S. Lindow, personal communication). Generally, the incidence of ice nucleating bacteria is high in locations which experience yearly frosts. However, no definitive study of the frequency of occurrence of Ina^+ bacteria with geographical location has been published.

Agronomic and Meteorological Implications Ice nucleating bacteria initiate frost damage to many economically important crop plants. Agricultural losses in the USA due to frost injury have been estimated at over $1 billion yearly (cf. Lindow, 1983). In some cases, frost injury appears to predispose plants to infection by phytopathogenic bacteria (Süle and Seemüller, 1987); in other cases, the ice nucleating bacteria do not cause systemic plant disease, but opportunistically feed on nutrients released from frost-damaged plant tissue (Buttner and Amy, 1989; Lindow, 1983). The use of nonnucleating (Ina^-) mutants of nonpathogenic, Ina^+ bacteria as biocontrol agents for preventing plant frost damage has been explored (Lindemann and Suslow, 1987; Lindow et al., 1983).

The discovery that plant surfaces are a source of airborne (bacterial) ice nuclei resulted from the general interest among atmospheric scientists in ice nucleation (Schnell and Vali, 1972, 1973). To date, the meteorological importance of bacterial ice nuclei remains unproven although it is known that deposition of airborne bacteria by rain is an important dispersal mechanism for ice nucleating bacteria (Constantinidou et al., 1990). This is likely to remain the case until the relative roles of primary and secondary nucleation (by shattered ice) in storm clouds are fully defined.

Evolutionary Questions The existence of bacterial ice nuclei raises several evolutionary questions. Is the Ina^+ phenotype positively selected, or is it an incidental effect of some other, unrecognized phenotype? If the Ina^+ phenotype is positively selected, then what advantage does it impart? Are the bacterial ice nucleation genes observed today the result of divergent or convergent evolution? What was the nature and evolutionary origin of the first structure possessing a recognizable ability to nucleate ice?

Evolutionary studies are handicapped by the difficulty of experimentally repeating evolutionary events. However, nature has performed a great many experiments whose results may often be compared. Using this approach, we shall address the questions posed above.

16.2 Ice Nucleation Genes and Proteins

Gene and Protein Sequences Four ice nucleation genes from different bacterial species have been cloned and sequenced (Arai et al., 1989; Green and Warren, 1985; Warren and Corotto, 1989; Warren et al., 1986). All four are capable of conferring the Ina^+ phenotype on *E. coli*, and all contain a

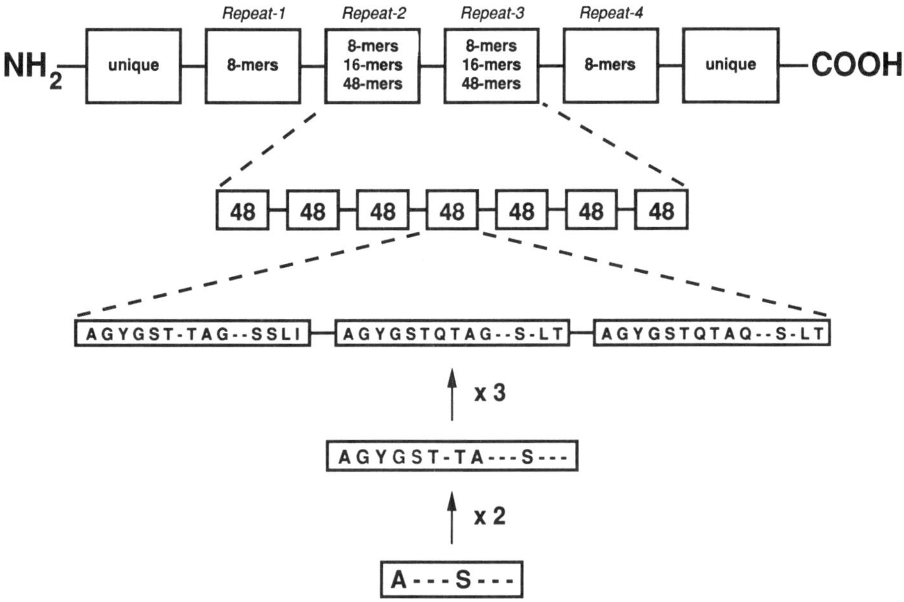

Figure 16.1 The general structure of ice nucleation proteins. Ice nucleation proteins consist of three overall domains: the N-terminal unique, repeating, and C-terminal unique domains. The repeating domain can be further divided into a short linker (Repeat-1), followed by three repeating subdomains with nested orders of repetition. The manner in which the 48 amino acid repeats have been built by duplication and elaboration of a basis 8-mer is shown below the central repeating domain. Sequences are listed using single-letter amino acid abbreviations (Lehninger, 1982); dashes depict positions which may be filled by any of several amino acid residues.

single open reading frame with a length in excess of 3500 base pairs. The open reading frame encodes a single protein with a conserved domain structure and primary sequence.

The consensus domain structure is outlined in Figure 16.1. An ice nucleation protein contains three domains: an N-terminal unique region, a central region rich in alanine, glycine, serine, and threonine that contains nested sets of repeated sequences, and a C-terminal unique region. The first 20 residues of the N-terminal unique region possess some of the properties of a membrane insertion signal sequence (von Heijne, 1984, 1985): the extreme N-terminus is positively charged (net +2 to +3 for residues 1–10), and the predicted secondary structure is an α helix, followed by a turn (Warren et al., 1986). However, the helix is not as hydrophobic as a typical signal sequence. The remainder of the N-terminal unique domain is unremarkable. The nucleotide and predicted protein sequences of the first 60% of this domain are much better conserved among genes than the last 40%.

This last point is not evident in the comparison of two closely related genes (Figure 16.2a), but is clearly demonstrated by the consensus amino acid sequence of three ice nucleation proteins, shown in Figure 16.2b.

The central, repeating domain can be divided into four subdomains, which are, in turn, composed of nested repeated sequences of 8,16, and 48 amino acid residues, in increasing order of fidelity, in subdomains Repeat-2 and Repeat-3. The repetition in subdomains Repeat-1 and 4 lacks the 48-amino acid periodicity. These subdomains are outlined in Figure 16.1, and shown in greater detail in the consensus sequence, Figure 16.2b. The fidelity of repetition increases from the C-terminal subdomain Repeat-4, to subdomain Repeat-2 (where the 48 amino acid repeat is nearly perfect). Subdomain Repeat-1 is an incomplete (32-residue) repeat, which may serve as a transition region between the N-terminal and repeating domains. The transition region between subdomains 2 and 3 is marked by a loss of conservation of primary sequence between different ice nucleation proteins, and is sometimes marked by the presence of a break in the phase of the 48-amino acid repeat. There is also a consistent substitution of basic residues in subdomain 2 for corresponding acidic residues in subdomain 3, or vice versa. This may indicate that these two domains form salt bridges with each other. The transition between subdomains 3 and 4 is marked by loss of the 48-residue repeat. Subdomain 4 also consistently substitutes cysteine for serine in position 5 of the third-from-last 16-amino acid repeat.

The repeating domain has been predicted to form a structure dominated by β-sheets and turns (Warren et al., 1986). On the basis of repetition, symmetry, size, and chemical nature, the repeating domain is expected to be the portion of the ice nucleation protein which binds water into an ice template (Green and Warren, 1985; Warren et al., 1986, 1987). The C-terminal unique domain is strongly positively charged at its N-terminal end (consensus net +3), and strongly negatively charged at its C-terminal end (consensus net −5 to −7). The secondary structure of this domain is predicted to form a turn between these highly charged regions; the domain may therefore fold back on itself, to form internal salt bridges.

Homology Within and Between Ice Nucleation Genes The protein-encoding nucleotide sequences of ice nucleation genes are homologous; the predicted protein sequences are even more strongly homologous. The homologies between the N- and C-terminal unique domains of the four sequenced ice nucleation proteins are shown by nucleotide position within a codon, in Table 16.1. The repeating domain is not included in Table 16.1. because previous analyses of ice nucleation genes have demonstrated that the central domain has probably evolved via repeated duplication and may self-correct silent, third-base mutations (Green and Warren, 1985; Warren et al., 1986); inclusion of this domain would therefore artificially skew the results towards homology. In the unique coding regions, substitutions at the third "silent" codon position occur disproportionately to substitutions

```
                     (a)                                                                 (b)
MKEDKVLILRTCANNMADHGGIIWPLSGIVECKYWKPVKGFENGLTGLIW   N-Terminal  Mk-DKvLvLRTCANNMaDHcGliWP-sG-VEcKyW-ptkk-ENGLtGllW
GKGSDSPLSLHADARwVVAEVdADECIAIETHGWIKFPRAEVLHVGTktS     Unique    G-Gssa-LS-hADARWvVcEV---E-I-lE--g-vKFPRAEVvHVGtK-S
AMQFILHHRADYVACTEMQAGPGsPDVTSEvKvGNRSLPVTDDIDATIES               A--fIs---ad--acs-----p-------------s--lpv-d-------
GSTQPTQTIEI                                                      -------------t

                 ATYGSTLSGTHQSQLI AGYGSTETAGDSSTLI    Repeat-1                    AtYGSTlsg---S-LI AGYGSteTaG--S-lI

AGYGSTGTAGaDSTLV AGYGSTQTAGEESSQM AGYGSTQTGMKGSDLT       |       AGYGSTGTAGsDS-Lv AGYGSTQTAGgESslt AGYGSTQTa--GSnLT
AGYGSTGTAGdDSSLI AGYGSTQTAGEDsSLT^AGYGSTQTAQKGSDLT       |       AGYGSTGTAG-DSSLI AGYGSTQTaGgDSSLT AGYGSTQTAQ-GSdLT
AGYGSTGTAGADSSLI AGYGSTQTAGEESTQT AGYGSTQTAQKGSDLT       |       AGYGSTGTAGsDSSLI AGYGSTQTaGgESslT AGYGSTQTAQ-GSnLT
AGYGSTGTAGDDSSLI AGYGSTQTAGEDSSLT AGYGSTQTAQKGSDLT    Repeat-2   AGYGSTGTAG-DSSLI AGYGSTQTaG-DSsLT AGYGSTQTAQ-GSdLT
AGYGSTGTaGADSSLI AGYGSTQTAGEESTQT AGYGSTQTAQKGSDLT       |       AGYGSTGTAGsDSSLI AGYGSTQTaG-ESslT AGYGSTQTAQ-GSdLT
AGYGSTGTAGDDSSLI{agygstqtagedsslt}AGYGSTQTAQKGSDLT       |       AGYGSTGTAG-DSSLI AGYGSTQTaG-DSsLT AGYGSTQTAQ-GSdLT
AGYGSTSTAGYESSLI AGYGSTQTAGYGSTLT AGYGSTQTAQNESDLI       |       AGYGSTgTAG-ESSLI AGYGSTQTaG-dSsLT AGYGSTQTAq-gSdLt
TGYGSTSTAGANSSLI AGYGSTQTASYNSVLT AGYGSTQTAREGSDLT§      |       aGYGSTgTAGadSSLI AGYGSTQTag-dSsLT AGYGSTQTAq-GS-LT
AGYGSTgTAGsDSSiI AGYGSTQTAsYhSsLT AGYGSTQTArEqSvLT               agygstgtagsdss-i agygstqta---sslt
TGYGSTSTAGADSSLI AGYGSTQTAGYNSILT AGYGSTQTAQEGSDLT                                                 -gygstqta---s-lt
AGYGSTSTAGADSSLI AGYGSTQTAgYnSiLT AGYGSTQTAQEgSdLT       |       -gygst-tag-ds--i AGYGSTQTAGh-SILT AGYGSTQTAqEGSdLt
sGYGSTSTAGADSSLI AGYGSTQTAsYhSsLT AGYGSTQTArEqSvLT       |       aGYGSTsTAG-dSSLI AGYGSTQTAsy-S-LT AGYGSTQTArE-S-Lt
TGYGSTSTAGADSSLI AGYGSTQTAGYHSILT AGYGSTQTAQERSDLT       |       tGYGSTSTAG-nSSLI AGYGSTQTaGy-SILT AGYGSTQTAQErssLt
TGYGSTSTAGADSSLI AGYGSTQTAGYNSILT AGYGSTQTAQENSDLT•   Repeat-3   tGYGSTSTAGy-SSLI AGYGSTQTAGYeStLT AGYGStQTAQEnS-LT
TGYGSTSTAGyDSSLI AGYGSTQTAGYNSILT AGYGSTQTaQENSDLT       |       TGYGSTSTAGYdSsLI AGYGSTQTAGY-SiLT AGYGSTQTAQE-SdLt
TGYGSTSTAGYESSLI AGYGSTQTASFKSTLM AGYGSSQTAREQSSLT       |       tGYGSTSTAGYaSSLI AGYGSTQTAgydStLt AGYGStqTA-E-SsLT
AGYGSTSMAGYDSSLI AGYGSTQTAGYQSTLT AGYGSTQTAEHSSTLT       |       aGYGSTstAGf-SSLi aGYGStQtAGY-STLT AGYGSTQtAE--StLT
AGYGSTATAGADSSLI                                                 AGYGSTaTAGqDSSLI

AGYGSSLTSGIRSFLT AGYGSTLISGLRSVLT AGYGSSLISGRRSSLT       |       AGYGSSLTSGiRSfLT AGYGSTLIsGLrSVLt AGYGSSLtSG-RSsLT
AGYGSNQIASHRSSLI AGPESTQITGNRSMLI AGKGSSQTAGYRSTLI    Repeat-4   AGYGSNQIAShrSSLI AGhEStQiaGnRSMLI AGKGSSQTAG-RSTLI
SGADSVQMAGERGKLI AGADSTQTAGDRSKLL AGNNSYLTAGDRSKLT       |       aGA-SVQmAGERsKLi AGAdS-QTAGDRSKLL AGnNSYLTAGDRSKLT
AGNDCILMAGDRSKLT AGINSILTAGCRSKLI GSNGSTLTAGENSVLI       |       aG-DC-LMAGDrSKLT AGkNsvLTAGarSKLI GS-GSTLsaGEnS-LI

FRCWDGKRYTNVVAKTGKGGIEADMPYQMDEDNNIVNKKEE*           C-terminal  FRcWDGKRYtnvVaKTG--gvEAD-PYq-dED-niv-KpEE--d--------
                                                       Unique
```

Figure 16.2 (a) Comparison of the amino acid sequences of two highly homologous ice nucleation proteins, IceE and InaA. Identical amino acids are capitalized; differences are shown with the residue from IceE in lower case. (b) The consensus sequence for three ice nucleation proteins: InaZ, InaW, and IceE. Upper case letters denote amino acid positions of complete identity, lowercase letters denote positions of identity between any pair of proteins, and dashes denote positions of complete nonidentity. Amino acid abbreviations are as in Figure 16.1.

Table 16.1 Homologies among the ice nucleation genes *iceE* (*E. herbicola*), *inaW* (*P. fluorescens*), *inaZ* (*P. syringae*), and *inaA* (*E. ananas*) by nucleotide position within the codon

Nucleotide position within codon	Homlogous nucleotides (percent homology)[a]			
	iceE/inaW	*iceE/inaZ*	*inaW/inaZ*	*inaA/iceE*
First	123 (78%)	118 (75%)	115 (73%)	154 (98%)
Second	126 (80%)	119 (76%)	117 (75%)	155 (99%)
Third	53 (34%)	56 (36%)	67 (43%)	153 (97%)

[a] 157 codons each from *iceE, inaW, inaZ,* and *inaA* are compared. These 157 codons are drawn from the most homologous portions of the nonrepeating regions of each gene; they correspond to codons 1–116 and 1218–1258 of *iceE*.

at the other two positions. Substitutions which change the encoded amino acid are also clustered, rather than randomly distributed throughout the sequences. An exception to this pattern is the gene pair formed by *iceE* from *Erwinia herbicola* (Warren and Corotto, 1989) and *inaA* from *E. ananas* (Arai et al., 1989). In this case, the nucleotide conservation is very good at all codon positions and in most regions (including the 5′ upstream region of the genes; this region is poorly conserved among all other ice nucleation genes). The predicted protein products of these two genes are compared in Figure 16.2a. In many regions of this protein pair, the sequence conservation is perfect.

These data imply that selection for conservation of the primary protein sequence (and presumably the protein function) has occurred. This protein sequence conservation is also demonstrated by the consensus sequence shown in Figure 16.2b. The striking sequence conservation between *iceE* and *inaA* suggests that either the divergence of the source species is extremely recent (despite their isolation at nearly opposite points of the Northern Hemisphere), or that both species acquired the ice nucleation trait rather recently, from the same third species via conjugation.

The fact that the repeating domain of ice nucleation proteins is rich in serine has been used to argue that all of the ice nucleation genes sequenced so far probably evolved from a common ancestor (Warren et al., 1986). The argument hinges on the fact that serine is encoded by two families of codons: AG(Py) and TCN where Py denotes a pyrimidine, N denotes any base, and A, G, T and C have their usual meanings. It is impossible to move from one family to another via a single nucleotide change. In all ice nucleation proteins, the serines at positions 8N-3 of the 48-amino acid repeat (for $N = 1, 2, \ldots, 6$) are of the AG(Py) family, while serines at positions 11, 14, and 30 are of the TCN family. If ice nucleation genes and proteins had evolved by convergent evolution of dissimilar ancestors, it is highly unlikely that the same patterns of codon bias would have emerged in all genes.

Homology with Other Genes Searching for genes homologous to ice nucleation genes presents special problems: the presence of repeated sequences stymies all current rapid homology search algorithms. In order to circumvent this problem, four searches were performed, using the coding sequence of the *inaZ* gene, the coding sequence for its N-terminal unique region, the coding sequence for one 48-amino acid repeat, and the coding sequence for the C-terminal unique region. The merged GenBank/EMBL gene library was searched, using the algorithm of Wilbur and Lipman (1983) with a masked 8-nucleotide word, MM*MM*MM, where M indicates a perfect match, and * indicates an allowed mismatch. This search pattern tolerates third-base redundancy in coding sequences.

As expected, the repeat and whole sequences confused the homology search program. However, in all cases, the only sequence found with a homology score exceeding the mean score plus one standard deviation was the *inaW* gene sequence (the only other ice nucleation gene sequence currently in the database). Thus, it would appear that, if bacteria possess genes closely related to ice nucleation genes, they have not yet been sequenced and entered into the standard databases. The results of the homology search are consistent with the hypothesis that ice nucleation proteins impart a unique, selected function, which may be the ability to form ice nuclei.

Serology Antisera have been raised against the product of the *inaW* gene (Deininger et al., 1988) and against a synthetic peptide which represents the consensus 8-amino acid repeat from the central domain of ice nucleation proteins (Mueller et al., 1990). Both antisera cross-react with all of the characterized ice nucleation proteins. The antiserum against the InaW protein also cross-reacts with large proteins which appear in concert with the Ina^+ phenotype in other ice nucleating organisms (Deininger et al., 1988).

The only other proteins which cross-react with such antisera are proteolytic degradation products of ice nucleation proteins (Deininger et al., 1988) or ice nucleation proteins from insect larvae (Duman et al., submitted for publication). This latter observation is rather startling, since at least 2 billion years of evolution separate insects and bacteria. The fact that cross-reaction occurs between the insect nucleator and an antiserum raised against a synthetic consensus repeat octapeptide (Mueller et al., 1990) suggests that the homology is with the repeating portion of bacterial ice nucleation proteins. Currently, there are no known bacterial proteins not associated with ice nucleation which cross react with anti-inaW antisera.

Studies of Nucleus Structure and Assembly Fractionation of ice nucleating bacteria has demonstrated that bacterial ice nuclei are localized on the bacterial outer membrane (Lindow et al., 1989). This location is ideal for the nucleation of extracellular ice, which can harm the plant host, thereby releasing nutrients without harming the bacterial cell.

Studies of overexpression in *Escherichia coli* have also demonstrated that active ice nuclei can be assembled on the bacterial inner membrane (Wolber et al., 1986). Delipidation and reconstitution experiments have shown that membrane lipids are required for ice nucleation activity (Govindarajan and Lindow, 1988a). However, lipopolysaccharide, a lipid found only in the outer membranes of Gram-negative bacteria, is not required. Thus, the requirement for lipid is generic, rather than specific. Proteolysis studies have shown that bacterial ice nuclei contain an essential protein component (Phelps et al., 1986).

Studies of γ-ray inactivation of ice nuclei have demonstrated that the target size for radiation damage increases log-linearly with the temperature of nucleation, and it is always greater than or equal to the average predicted size of ice nucleation proteins (about 120 kDa) (Govindarajan and Lindow, 1988b). Studies of the dependence of ice nucleation frequency on the concentration of ice nucleation protein have shown that nucleation frequency at all temperatures increases nonlinearly with ice nucleation protein concentration, indicating that about three monomers must assemble cooperatively to initiate formation of an active nucleus (Lindgren et al., 1989; Southworth et al., 1988). Measurements of the rates of assembly of ice nuclei have confirmed cooperative initiation of nucleus assembly, and have shown that nuclei active at warmer temperatures are assembled more slowly than nuclei active at lower temperatures (Watanabe et al., 1990).

Taken together, the available data on nucleus structure indicate that ice nucleation proteins must associate with a lipid membrane, and then self-associate, in order to become active. The degree of self-association is directly related to how close to 0°C the nucleus operates. The selective pressure that has preserved the homology among ice nucleation proteins has also evidently preserved the ability of ice nucleation proteins to cooperatively form membrane-bound arrays. If this selection is for the ability to nucleate ice, then the need for aggregation is dictated by the physics of ice nucleation (Burke and Lindow, 1990). Any alternative phenotype proposed for ice nucleation proteins should also rationalize the need for outer membrane localization and aggregation into well-ordered arrays.

Mutagenesis Studies Ice nucleation genes have been mutagenized by several methods in order to study the roles of the various protein domains. Early experiments with transposon mutagenesis yielded an unexpected result: none of the mutants were complete nulls. This result was eventually understood to derive from the combination of the extreme sensitivity of ice nucleation assays and the ability of rare recombinational events to reconstruct functional ice nucleation genes from transpositionally inactivated genes (Corotto et al., 1986). These experiments also showed that recombinational repair of the repeating portion of ice nucleation genes is particularly effective; this may explain why there is improved conservation of third, "silent" bases in the repeating portion of ice nucleation genes, and why there is sometimes insertion or deletion of repeats between otherwise highly conservative pairs of nucleation proteins (see Figure 16.2a).

Experiments in which in-frame sequences have been inserted into or deleted from ice nucleation genes have more precisely defined the functions of various protein domains (Green et al., 1988). The data are consistent with a model in which the C-terminal unique region is a protein-folding "nucleus," the repeated domain binds water into an ice template, and the N-terminal unique region promotes aggregation of ice nucleation protein monomers into large arrays. Insertions and deletions in the repeated domain which preserve the 16-amino acid repetition are much less deleterious than similar mutations which break this periodicity. This finding indicates that the folding pattern of an ice nucleation protein can accommodate, to some extent, the addition or deletion of repeated modules.

Mutagenesis data provide support for two hypotheses about the evolution of ice nucleation genes. First, the three protein domains appear to have distinct functions, which link together to yield the observed function of ice nucleation. The autonomous mutability of the N-terminal and repeating domains suggests that the folding of these domains is independent, and that they may evolve with relative independence. The dependence of ice nucleation on the presence of the acidic portion of the C-terminus suggests that the folding of one or both of the other domains is dependent on the C-terminal domain, or that the C-terminus provides some other function essential for ice nucleation. Second, recombination is enhanced in the repeating portion of an ice nucleation gene. This lends support to the notions that this region evolved by repeated duplication of sequences, and that correction of mutations in this region is enhanced by intragene recombination.

16.3 Incidence and Regulation of the Ina$^+$ Phenotype

Incidence and Regulation in Nature Populations of ice nucleating bacteria on plant surfaces are highly variable in relative abundance and activity. Variations in ice nucleation activity are seen as a function of plant species (Hirano et al., 1978) and ambient temperature (Hirano et al., 1989). Generally, natural Ina$^+$ isolates are capable of nucleation at temperatures above −5°C, at which nuclei from other sources are rare (Lindow, 1983). Increased activity of Ina$^+$ bacteria is favored by lower ambient temperatures (Hirano et al., 1989).

Bacterial populations on leaf surfaces often produce nuclei active at the highest temperatures known for bacterial ice nuclei (about −2°C), indicating that the phenotype is well expressed under "natural" conditions (Lindow, 1982). Measurement of how the strength of expression of ice nucleation genes is modulated by field conditions is complicated by the spectrum of ice nucleation temperatures exhibited by a given population of cells, and by leaf-to-leaf variations in populations of ice nucleating bacteria (Hirano et al., 1985; Lindow, 1982). It is known that the warmest temperature of nucleation is roughly predictive of the population size of ice nucleating bacteria on a leaf (Hirano et al., 1985); however, it is also known that the ice nucleation frequency of *Pseudomonas syringae* on snap beans varies with the diurnal temperature cycle, independently of total bacterial populations (Hirano et al., 1989). The variation mimics that seen in vitro (see below): cooler temperatures favor formation of nuclei active at temperatures closer to 0°C.

Regulation In Vitro The frequencies and temperatures of nucleation by Ina^+ bacteria can be profoundly modulated by growth temperature, culture medium composition, and bacterial growth phase (Deininger et al., 1988; Lawless, 1988; Lindow, 1982; Lindow et al., 1978). Studies of the frequencies and temperatures of ice nucleation in organisms expressing ice nucleation genes under the control of heterologous promoters indicate that the average frequency of nucleation at any temperature follows the average concentration of ice nucleation protein according to a simple power relationship (Lindgren et al., 1989; Southworth et al., 1988; Watanabe et al., 1991). The most currently plausible explanation for the variations in ice nucleation seen with culture conditions is that ice nucleation is transcriptionally regulated; this interpretation is supported by the observation that, in *P. syringae* harboring a *lac*-promoted ice nucleation gene, ice nucleation becomes independent of growth phase and medium composition (Southworth et al., 1988).

The regulation of expression of ice nuclei has been particularly well studied in *P. syringae,* where the Ina^+ phenotype is repressed by high concentrations of nitrogen, exponential growth, or higher growth temperatures, and favored by nitrogen starvation, polyol carbon sources, and lower growth temperatures (Deininger et al., 1988; Lawless, 1988; Lindow, 1982). A similar pattern is seen in *Erwinia herbicola,* where the Ina^+ phenotype is expressed less on nutrient agar than on the same medium supplemented with glycerol or glucose (Lindow et al., 1978).

The pattern of regulation of ice nucleation bears some resemblance to that for the *avrB* gene of *P. syringae* pv. *glycinea* (Huynh et al., 1989); this gene determines host cultivar specificity of this pathogenic bacterial species. Transcription of the *avrB* gene is repressed by TCA cycle intermediates and certain amino acids which support rapid growth rates of *P. syringae,* and is inversely correlated with the rate of bacterial growth (e.g., it is maximal in stationary phase). The common theme linking *avrB* transcription and ice nucleus expression is stress: *P. syringae* cells presented with ample nutrients and space promptly repress both functions.

The similarity of regulation of *avrB* transcription and ice nucleus expression suggests that ice nucleation has evolved as a form of opportunistic pathogenesis: when nutrients and temperatures are low, bacteria can cause the release of additional nutrients by inducing frost damage. This hypothesis is supported by the observation that Ina^+ *P. syringae* appear to possess a competitive advantage over otherwise isogenic Ina^- mutants when host plants are subjected to mild supercooling, so that only the plants carrying Ina^+ bacteria suffer frost damage (Buttner and Amy, 1989).

16.4 A Model of the Evolution of Ice Nucleation Genes

Many proteins share one of two key aspects of ice nucleation proteins: the ability to form membrane-bound aggregates, and duplication of a repeated sequence. Examples of proteins which form bacterial membrane-associated aggregates include outer membrane porin from *E. coli* (Rosenbusch, 1974) and the light-driven proton pump, bacteriorhodopsin (Stoeckenius and Bogomolni, 1982). Proteins built up by repetition of a basis peptide include malarial circumsporozoite protein (Godson et al., 1983), the E2 silkmoth chorion protein (Regier, 1986), late-embryogenesis-abundant (LEA) proteins (Dure et al., 1989), antifreeze peptides (De Vries, 1983), and a sea urchin spicule matrix protein (Sucov et al., 1987). All of these proteins share the common property that at least part of their function is structural. In addition, three of the proteins have functions related in part to the function of ice nucleation proteins. The LEA proteins bind water in plant seeds, antifreeze proteins bind to ice, and sea urchin spicule matrix protein may nucleate biomineralization. Thus, it is not unreasonable to imagine that the precursor of ice nucleation proteins was an outer membrane protein which formed periodic arrays and possessed at least a small number of 8-, 16-, or 48-amino acid repeats. This precursor may even have possessed ice nucleation activity at low temperatures (< -8°C).

Suppose, hypothetically, that the selected function of ice nucleation proteins is the opportunistic damage of host plants by ice nucleation. Given that evidence of a competitive advantage for Ina^+ bacteria has only been seen at temperatures above -5°C (Buttner and Amy, 1989), it seems likely that selection for the Ina^+ phenotype did not occur until a protein arose (by chance?) which was capable of nucleating ice formation at temperatures above some threshold for imparting a selective advantage. At that point, the precursor "changed selections," and genetic events which improved ice nucleation activity were preserved. Eventually, a gene optimized to impart ice nucleation evolved by duplication and fine-tuning (via point mutation?) of the repeating domains. This gene may have been directly inherited by all of the Ina^+ bacteria observed today. However, given the evolutionary distance between genera with Ina^+ members, the large number of Ina^- members of all of these genera, and the observed degree of ice nucleation gene and protein homology, it seems more plausible that the Ina^+ phenotype has been horizontally dispersed via conjugation between members of

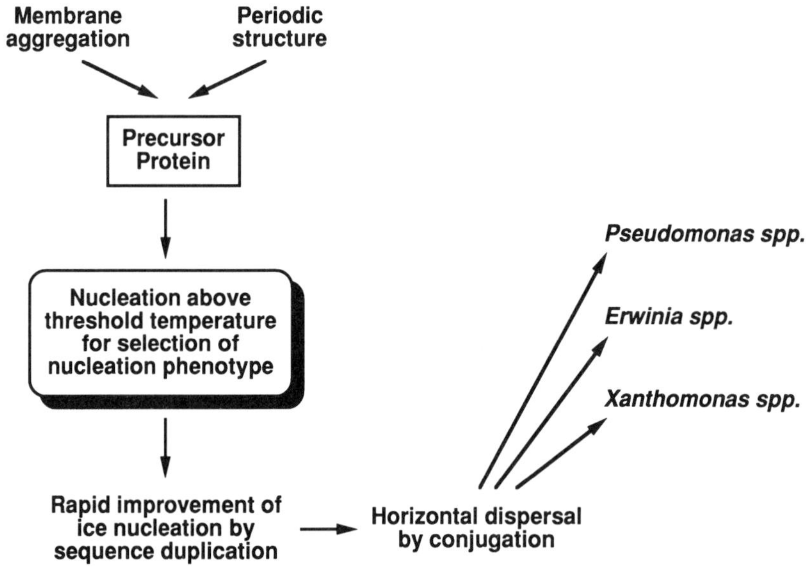

Figure 16.3 A model for the evolution of ice nucleation genes and proteins. In this model, a precursor gene encoded a protein which formed outer membrane aggregates and contained some repeating structure. The function of this protein is unknown. When, by chance, the protein nucleated ice above some threshold temperature, the selection pressure on the gene/protein switched to preservation of the ability to opportunistically damage plant hosts during mild frosts. This function was rapidly optimized by natural selection, and the gene was then dispersed throughout the phyllosphere, most probably by bacterial conjugation in mixed epiphytic communities.

epiphytic communities. This model of the evolution of ice nucleation proteins is shown diagramatically in Figure 16.3.

16.5 Conclusions

Current evidence indicates that the most plausible explanation for the evolution of ice nucleation proteins is that the selective advantage imparted by such proteins is the Ina^+ phenotype, which serves as the basis of a form of opportunistic pathogenesis, by initiating plant damage during mild frosts. The Ina^+ phenotype probably arose only once in bacteria, and then was dispersed through the phyllosphere, possibly by bacterial conjugation. It seems plausible that the first ice nucleation proteins arose by accident from an outer membrane precursor protein which formed periodic arrays and may have had a structural role.

Two facts which do not fit this scenario are the existence of oceanic ice nucleating microorganisms (Fall and Schnell, 1985) which cannot opportunistically damage plant hosts, and the potential homology between prokaryotic and insect larval ice nucleation proteins (although this may simply represent convergent evolution). The determination of the molecular details of ice nucleation in these systems will either strengthen the arguments presented here, or create the need for a reexamination of the evolution of bacterial ice nucleation.

The authors gratefully acknowledge the assistance of J. Lindemann and P. Baldwin. P. Wolber was supported in part by U.S. National Science Foundation SBIR grant ISI-8800969.

Note Added In Proof J. Zhao and C.S. Orser have recently reported the sequence of *inaX*, an ice nucleation gene from *Xanthomonas* (Zhao, J., and Orser, C.S. 1990. Conserved repetition in the ice nucleation gene *inaX* from *Xanthomonas campestris* pv. *translucens. Molecular and General Genetics* 223:163–166). The *inaX* gene shares the properties held in common by the four ice nucleation genes discussed here.

References

Anderson, S.A. and Ashworth, E.N. 1986. The effects of streptomycin, desiccation, and ultraviolet radiation on ice nucleation by *Pseudomonas viridiflava*. *Plant Physiology* 80:956–960.

Arai, S., Abe, K., Watabe, S., Emori, Y., and Watanabe, M. 1989. Molecular cloning of an ice nucleation gene from *Erwinia ananas* and its expression in *Escherichia coli*. *FEMS Microbiological Letters* 61:53–56.

Arny, D.C., Lindow, S.E., and Upper, C.D. 1976. Frost sensitivity of *Zea mays* increased by application of *Pseudomonas syringae*. *Nature (London)*, 262:282–284.

Burke, M.J. and Lindow, S.E. 1990. Surface properties and size of the ice nucleation site in ice nucleation active bacteria: theoretical considerations. *Cryobiology* 27:80–84.

Buttner, M.P. and Amy, P.S. 1989. Survival of ice nucleation active and genetically engineered non-ice nucleating strains of *Pseudomonas syringae* after freezing. *Applied and Environmental Microbiology* 55:1690–1694.

Constantinidou, H.A., Hirano, S.S., Baker, L.S., and Upper, C.D. 1990. Atmospheric dispersal of ice nucleation-active bacteria: the role of rain. *Phytopathology* 80:934–937.

Corotto, L.V., Wolber, P.K., and Warren, G.J. 1986. Ice nucleation activity of *Pseudomonas fluorescens*: mutagenesis, complementation analysis and identification of a gene product. *EMBO Journal* 5:231–236.

Deininger, C.A., Mueller, G.M., and Wolber, P.K. 1988. Immunological characterization of ice nucleating proteins from *Pseudomonas syringae*, *Pseudomonas fluorescens*, and *Erwinia herbicola*. *Journal of Bacteriology* 170:669–675.

De Vries, A.L. 1983. Antifreeze peptides and glycopeptides in cold-water fish. *Annual Review of Physiology* 45:245–260.

Dure, L. III, Crouch, M., Harada, J., Ho, T.H.D., Mundy, J., Quatrano, R.,

Thomes, T., and Sung, E.R. 1989. Common amino acid domains among the LEA proteins of higher plants. *Plant Molecular Biology* 12:475–486.

Fall, R. and Schnell, R. C. 1985. Association of an ice-nucleating pseudomonad with cultures of the marine dinoflagellate, *Heterocapsa niei. Journal of Marine Research* 43:257–265.

Fletcher, N.H. 1970. pp. 73–103 in *The Chemical Physics of Ice*. Cambridge University Press, Cambridge.

Godson, G.N., Ellis, J., Svec, P., Schlesinger, D.H., and Nussenzweig, V. 1983. Identification and chemical synthesis of a tandemly repeated immunogenic region of *Plasmodium knowlesi* circumsporozoite protein. *Nature (London)* 305: 29–33.

Govindarajan, A.G. and Lindow, S.E. 1988a. Phospholipid requirement for expression of ice nuclei in *Pseudomonas syringae* and in vitro. *Journal of Biological Chemistry* 263:9333–9338.

Govindarajan, A.G. and Lindow, S.E. 1988b. Size of bacterial ice-nucleation sites measured in situ by radiation inactivation analysis. *Proceedings of the National Academy of Sciences USA* 85:1334–1338.

Green, R.L. and Warren, G.J. 1985. Physical and functional repetition in a bacterial ice nucleation gene. *Nature (London)* 317:645–648.

Green, R.L., Corotto, L.V., and Warren, G.J. 1988. Deletion mutagenesis of the ice nucleation gene from *Pseudomonas syringae* S203. *Molecular and General Genetics* 215:165–172.

Hirano, S.S., Maher, E.A., Kelman, A., and Upper, C.D. 1978. Ice nucleation activity of fluorescent plant pathogenic pseudomonads. pp. 717–724 in Station de Pathologie Végétale et Phytobactériologie (editor), *Proceedings of the 4th International Conference on Plant Pathogenic Bacteria*. Institute National Recherche Agronomique, Beaucouzé, France.

Hirano, S.S., Baker, L.S., and Upper, C.D. 1985. Ice nucleation temperature of individual leaves in relation to population sizes of ice nucleating bacteria and frost injury. *Plant Physiology* 77:259–265.

Hirano, S.S. and Upper, C.D. 1989. Diel variation in population size and ice nucleation activity of *Pseudomonas syringae* on snap bean leaflets. *Applied and Environmental Microbiology* 55:623–630.

Huynh, T.V., Dahlbeck, D., and Staskawicz, B.J. 1989. Bacterial blight of soybean: regulation of a pathogen gene determining host cultivar specificity. *Science* 245:1374–1377.

Lawless, R.J., Jr. 1988. Fermentation of microorganisms having ice nucleation activity using a temperature change. European Patent Application, no. 0-281-145.

Lehninger, A.L. 1982. p. 96 in *Principles of Biochemistry*, Worth Publishers, Inc. New York, NY.

Lindemann, J. and Suslow, T.V. 1987. Competition between ice nucleation-active wild type and ice nucleation-deficient deletion mutant strains of *Pseudomonas syringae* and *P. fluorescens* Biovar I and biological control of frost injury on strawberry blossoms. *Phytopathology* 77:882–886.

Lindgren, P.B., Frederick, R., Govindarajan, A.G., Panopoulos, N.J., Staskawice, B.J., Lindow, S.E. 1989. An ice nucleation reporter gene system: identification of inducible pathogenicity genes in *Pseudomonas syringae* pv. *phaseolicola*. *EMBO Journal* 8:1291–1301.

Lindow, S.E. 1982. Epiphytic ice nucleation-active bacteria. pp. 335–362 in Mount, M. S. and Lacy, G.H. (editors), *Phytopathogenic Prokaryotes*, vol. 1. Academic Press, New York.

Lindow, S.E. 1983. The role of bacterial ice nucleation in frost injury to plants. *Annual Review of Phytopathology* 21:363–384.

Lindow, S.E., Arny, D.C., and Upper, C.D. 1978. *Erwinia herbicola*: a bacterial ice nucleus active in increasing frost injury to corn. *Phytopathology* 68:523–527.

Lindow, S.E., Arny, D.C., and Upper, C.D. 1983. Biological control of frost injury I: an isolate of *Erwinia herbicola* antagonistic to ice nucleation active bacteria. *Phytopathology* 73:1097–1102.

Lindow, S.E., Lahue, E., Govindarajan, A.G., Panopoulos, N.J., and Gies, D. 1989. Localization of ice nucleation activity and the *iceC* gene product in *Pseudomonas syringae* and *Escherichia coli*. *Molecular Plant-Microbe Interactions* 2:262–272.

Maki, L.R. and Willoughby, K.J. 1978. Bacteria as biogenic sources of freezing nuclei. *Journal of Applied Meteorology* 17:1049–1053.

Maki, L.R., Galyan, E.L., Chang-Chien, M., and Caldwell, D.R. 1974. Ice nucleation induced by *Pseudomonas syringae*. *Applied Microbiology* 28:456– 459.

Mueller, G.M., Wolber, P.K., and Warren, G.J. 1990. Clustering of ice nucleation protein correlates with ice nucleation activity. *Cryobiology* 27:416–422.

Orser, C., Staskawicz, B.J., Panopoulos, N.J., Dahlbeck, D., and Lindow, S.E. 1985. Cloning and expression of ice nucleation genes in *Escherichia coli*. *Journal of Bacteriology* 164:359–366.

Phelps, P., Giddings, T.H., Prochoda, M., and Fall, R. 1986. Release of cell-free ice nuclei by *Erwinia herbicola*. 167:496–502.

Regier, J.C. 1986. Evolution and higher order structure of architectural proteins in silkmoth chorion. *EMBO Journal* 5:1981–1989.

Rosenbusch, J.P. 1974. Characterization of the major envelope protein from *Escherichia coli*. *Journal of Biological Chemistry* 249:8019–8029.

Schnell, R.C. and Vali, G. 1972. Atmospheric ice nuclei from decomposing vegetation. *Nature (London)* 236:163–165.

Schnell, R.C. and Vali, G. 1973. World-wide source of leaf-derived freezing nuclei. *Nature (London)* 246:212–213.

Southworth, M.W., Wolber, P.K., and Warren, G.J. 1988. Nonlinear relationship between concentration and activity of a bacterial ice nucleation protein. *Journal of Biological Chemistry* 263:15211–15216.

Stoeckenius, W. and Bogomoini, R.A. 1982. Bacteriorhodopsin and related pigments of halobacteria. *Annual Review of Biochemistry* 52:587–616.

Sucov, H.M., Benson, S., Robinson, J.R., Britten, R.J., Wilt, F., and Davidson, E.H. 1987. A lineage-specific gene encoding a major matrix protein of the sea urchin embryo spicule. *Developmental Biology* 120:507–519.

Süle, S. and Seemüller, E. 1987. The role of ice formation in the infection of sour cherry leaves by *Pseudomonas syringae* pv. *syringae*. *Phytopathology* 77:173–177.

von Heijne, G. 1984. Analysis of the distribution of charged residues in the N-terminal region of signal sequences: implications for protein export in prokaryotic and eukaryotic cells. *EMBO Journal* 3:2315–2318.

von Heijne, G. 1985. Signal sequences: the limits of variation. *Journal of Molecular Biology* 184:99–105.

Warren, G.J. 1987. Bacterial ice nucleation: molecular biology and applications. pp. 107–135 in Russell, G.E. (editor), *Biotechnology and Genetic Engineering Reviews*. Intercept, Ltd., Wimborne, Dorset, UK.

Warren, G. and Corotto, L. 1989. The consensus sequence of ice nucleation proteins from *Erwinia herbicola*, *Pseudomonas fluorescens*, and *Pseudomonas syringae*. *Gene* 85:239–242.

Warren, G.J. and Wolber, P.K. 1987. Heterogeneous ice nucleation by bacteria. *Cryo-Letters* 8:204–215.

Warren, G., Corotto, L, and Wolber, P. 1986. Conserved repeats in diverged ice nucleation structural genes from two species of *Pseudomonas*. *Nucleic Acids Research* 14:8047–8060.

Warren, G., Wolber, P., and Green, R. 1987. Functional significance of oligonucleotide repeats in a bacterial ice nucleation gene. pp. 1013–1017 in Civerolo, E.L., Collmer, A., Davis, R.E., and Gillaspie, A.G. (editors), *Plant Pathogenic Bacteria*. Martinus Nijhoff Publishers, Boston.

Watanabe, N.M., Southworth, M.W., Warren, G.J., and Wolber, P.K. 1991. Rates of assembly and degradation of bacterial ice nuclei. *Molecular Microbiology*. 5:239–243.

Wilbur, W.J. and Lipman, D.J. 1983. Rapid similarity searches of nucleic acid and protein data banks. *Proceedings of the National Academy of Sciences USA* 80:726–730.

Wolber, P.K., Deininger, C.A., Southworth, M.W., Vandekerckhove, J., Van Montagu, M., and Warren, G.J. 1986. Identification and purification of a bacterial ice-nucleation protein. *Proceedings of the National Academy of Sciences. USA* 83:7256–7260.

17

Endophytes as Antagonists of Plant Pests

Keith Clay

17.1 Introduction

Endophytic bacteria and fungi are common inhabitants of aerial plant tissues. Considerable information has come to light over the past ten years regarding the effects of endophytic microorganisms on their host plants and on other organisms interacting with host plants. Unlike typical plant pathogens, endophytes cause host plants little harm and often provide measurable benefits. Two broad classes of benefit have been documented. Endophyte-induced changes in plant physiology leading to enhanced growth and abiotic stress resistance are known from infected graminoids (Clay, 1987b; West et al., 1988; Arachevaleta et al., 1989). A second major benefit accruing to endophyte-infected plants is enhanced resistance to plant pests such as insects, seed predators, plant pathogens, and nematodes (Clay et al., 1985b; Latch et al., 1985; Bacon et al., 1986a; Cheplick and Clay, 1988; West et al., 1988; Ford and Kirkpatrick, 1989). If the cost of supporting endophytes is less than the damage sustained by endophyte-free plants, natural selection should favor plants with characteristics that enhance infection by and persistence of endophytes. The purpose of this paper is to consider how endophyte infection influences interactions between plants and other species.

Endophyte refers to any organism that exists within plants. In the strictest sense, the term includes only those organisms that complete their entire life cycles without any stages or structures external to the plant. The endophyte *Acremonium coenophialum* of tall fescue grass is an example (Siegel et al., 1984). However, the term endophyte has more usually been applied to organisms that spend most of their life cycle within plant tissues but leave living or dead host tissues to sporulate.

Endophytes are a heterogeneous group of primarily ascomycetous fungi that share certain features: they spend all or most of their life cycle within aerial plant tissues, they cause relatively little or no damage to their hosts and may have beneficial effects, and they occur at high frequencies within plant populations. The distinction between some endophytes and benign or weak pathogens can be vague; for example, *Balansia henningsiana* (endophyte) and many smuts (pathogens) of grasses are systemic, inhibit host reproduction, and have little effect on vegetative growth of the host. Surveys have revealed fungal endophytes in all plants examined including algae, conifers, and flowering plants (see Chapters 9 and 10, this volume). The largest body of data on endophytes comes from grasses infected by clavicipitaceous endophytes, especially the endophyte *A. coenophialum*, which infects tall fescue (*Festuca arundinacea*). This has become a model system for endophyte studies (Bacon and Siegel, 1988; Clay, 1988b, 1990).

There is great diversity among endophytes with respect to their host range, distribution within the plant, life span relative to that of the host, expression of disease symptoms, and effects on host plant fitness. Some endophytes like *Rhabdocline parkeri*, which infects Douglas fir, have scattered infection sites within the plant, low metabolic demand on the plant's resources, and cause little destruction of host cells and tissues (Carroll, 1986). Many different endophyte genotypes and species can coexist in the same host plant (Carroll and Carroll, 1978; Petrini and Carroll, 1981; Carroll and Petrini, 1983; Sieber et al., 1988; Johnson and Whitney, 1989). Contagious spread can be very effective, and wind and water are important agents of dissemination (Carroll and Carroll, 1978; Carroll, 1986; Johnson and Whitney, 1989). At the other extreme are perennial systemic endophytes such as *Epichloë typhina* or *B. henningsiana*, where a single endophyte genotype is present throughout the plant and reduces seed production (Sampson, 1933; Clay, 1988a; Clay et al., 1989). Contagious spread is infrequent (Rykard et al., 1985; Clay, 1990), and the endophytes are often seedborne (Sampson, 1933, 1935; Boursnell, 1950; Neergaard, 1977; White, 1987).

17.2 Ecological Interactions Among Species

Ecologists identify three primary outcomes of the interaction of two species: competition (−/−), antagonism (i.e., parasitism or predation) (+/−), and mutualism (+/+). Many host-plant/endophyte associations are mutualistic—the endophyte gains nutrition and a habitat; the host gains increased survival, growth, and/or reproduction. Plant/pest associations are by definition antagonistic—the pest gains nutrition and a habitat while the host suffers decreased survival, growth, or reproduction.

Mutualistic interactions between two species, such as endophytes and host plants, often depend on the presence of a third species interacting antagonistically with the host (Thompson, 1982). An example was described from the seasonally dry forests of Costa Rica (Janzen, 1966). Acacias bearing swollen thorns and glandular organs called Beltian bodies that secrete substances sought after by ants were colonized, and the ants protected the trees against herbivores. In the absence of herbivores, the energetic investment made by the plants to support the ants may not be recouped and the association is detrimental to the hosts. Supporting this idea is evidence that Acacias do not produce swollen thorns and Beltian bodies where herbivores are lacking (Janzen, 1973). The interaction between host trees and resident ants can be facultatively mutualistic or antagonistic depending on the presence or absence of herbivores. The interaction between endophytes and their host plants may also be facultatively mutualistic, depending on the balance between the cost and/or damage resulting from infection and the benefit resulting from the deterrence of plant pests.

We should expect mutualistic endophyte associations to be most prevalent where pest pressure is chronic. Natural selection will favor hosts that facilitate endophyte infection, if the amount of damage prevented is large compared to the cost of supporting the endophyte, and if avoidance of pests and endophytes is not possible. Consistent with this scenario is the absence of genetically based resistance to endophyte infection in some plants (Latch and Christensen, 1985; Leuchtmann and Clay, 1989). However, if the amount of damage prevented is small compared to the cost of supporting the endophyte, selection should favor endophyte-resistant plants. There are very few empirical data available that bear on the question of whether hosts suffer from endophyte infection relative to uninfected plants in the absence of plant pests (but see Cheplick et al., 1989).

Evolution of Endophytes One selective pressure that would favor commensalistic or mutualistic endophytes over more pathogenic endophytes results from dissemination. There is general correlation between greater virulence and higher rates of contagious transmission in many disease organisms (Ewald, 1983). Parasites with poor powers of contagious spread should not be expected to be extremely debilitating to their hosts because the host could be killed before the parasite successfully reproduces or infects new host plants. Endophytes capable only of seed transmission, such as many *Acremonium* endophytes of grasses, could not persist if they had significant detrimental effects on the fitness of their hosts. When pathogenic microorganisms evolve reduced virulence, for whatever reason, selection for contagious spread may be relaxed. On the other hand, pathogens that have reduced powers of contagious spread, for example, by loss of a vector, may be selected against if they are too virulent.

Another selective pressure favoring mutualistic endophytes is predation of host plants. Classic competition theory suggests that two species can not coexist indefinitely if they utilize the same limited resources. Endophytes, herbivores, and pathogens are heterotrophs dependent upon the same carbon compounds produced by photosynthetic plants. Herbivores and pathogens may also consume endophytes as leaves are eaten or digested. Endophytes that best monopolize plant nutrients would have higher growth and reproductive rates than less efficient endophytes. The growth and reproduction of their hosts may also be increased relative to host plants infected by endophytes without deterrent properties. Endophytes that defend their hosts are not altruistic, but rather they are successfully competing with other fungal pathogens and insect herbivores for the same resources (Janzen, 1977; Wicklow, 1981).

17.3 Effects of Endophyte Infection on Plant Pests

Early researchers mistakenly concluded that endophyte infection had no effects on herbivores (Neill, 1940; Diehl, 1950; Cunningham, 1958). However, since the first studies experimentally documenting resistance of endophyte-infected perennial ryegrass to stem-boring insects (Prestidge et al., 1982), there has been a great deal of research on how endophytes affect plant pests in grasses and other plants. While the data base has grown substantially since 1985, only a tiny fraction of endophyte-infected plants and their pests have been critically examined. Sweeping generalizations must be suspect pending a much larger number of experimental studies. Nevertheless, the available data suggest that endophyte-induced resistance to insect herbivores and other plant pests may be widespread (Table 17.1) (Siegel et al., 1987; Clay, 1988b; Dahlman et al., 1991).

Insect Herbivores A total of 21 grasses and three sedges have been reported in the literature to have increased resistance to insect pests when infected by clavicipitaceous endophytes (Table 17.1). Research on the role of endophyte infection in plant resistance has focused on insect perfor-

Table 17.1 List of grasses and sedges reported to have increased resistance to insect pests when they are endophyte-infected

Bromus anomalus[a]	*Festuca gigantea*[a]	*Lolium perenne*[b]
Cenchrus echinatus[b]	*Festuca glauca*[a]	*Panicum agrostoides*[f]
Cyperus pseudovegetus[c]	*Festuca longifolia*[g]	*Paspalum dilatatum*[b]
Cyperus rotundus[d]	*Festuca obtusa*[f]	*Paspalum notatum*[f]
Cyperus virens[c]	*Festuca paradoxa*[a]	*Poa ampla*[a]
Dactylis glomerata[e]	*Festuca rubra*[f]	*Poa autumnalis*[a]
Danthonia spicata[f]	*Festuca versuta*[f]	*Stipa leucotricha*[b]
Festuca arundinacea[b]	*Glyceria striata*[f]	*Tridens flavus*[f]

[a] Data from Siegel et al. (1990).
[b] Data from Clay et al. (1985a).
[c] Data from Clay et al. (1985b).
[d] Data from Stovall (1987).
[e] Data from Schmidt (1986).
[f] Data from Cheplick and Clay (1988).
[g] Data from Saha et al. (1987).

Table 17.2 Effect of endophyte infection of grasses on aphids

Aphid species	Survival rate on uninfected ÷ rate on infected plants	
	Tall fescue	Perennial ryegrass
Diuraphis noxia[a]	2.0	1.5
Metopholophium dirhodum[b]	1.0	1.0
Rhopalosiphum maidis[c]	1.3	2.4
Rhopalosiphum padi[c]	5.3	1.0
Rhopalosiphum padi[b]	12.0	1.0
Rhopalosiphum padi[d]	20.0	1.0
Schizaphis graminum[c]	4.8	1.1
Schizaphis graminum[d]	15.0	20.0
Sitobion avenae[c]	0.8	0.9

[a]Data from Clement et al. (1991).
[b]Data from Latch et al. (1985).
[c]Data from Johnson et al. (1985).
[d]Data from Siegel et al. (1990).

mance on infected and uninfected plants of the same species (Funk et al., 1983; Clay et al., 1985b; Latch et al., 1985; Carroll, 1986; Cheplick and Clay, 1988). The underlying basis of endophyte-induced resistance to insect pests appears most frequently to be mycotoxins.

Most of the data come from grasses infected by clavicipitaceous fungal endophytes, especially the pasture and turf grasses tall fescue and perennial ryegrass. Several insects have been especially well studied, including aphids, lepidopterans, and the Argentine stem weevil (*Listronotus bonariensis*). The responses of at least six aphid species to endophyte-infected vs. uninfected plants have been examined in several studies. The results are summarized in Table 17.2 (Johnson et al., 1985; Latch et al., 1985; Clement et al., 1991; Siegel et al., 1990). Ratios below one indicate that the aphids performed better on infected plants while ratios above one mean that aphid performance was superior on uninfected plants. Most aphids strongly preferred endophyte-free tall fescue although two or three species were relatively unaffected by the presence of the endophyte. In contrast, fewer aphids discriminated between infected and uninfected perennial ryegrass. The performance of *Schizaphis graminum* on infected vs. uninfected plants varied considerably between two separate studies (Johnson et al., 1985; Siegel et al., 1990). Siegel et al. (1990) recently reported on a large study where two aphid species were screened against a large number of grasses infected with a variety of naturally occurring and artificially inoculated endophytes. There was considerable heterogeneity; *S. graminum* tended to be more sensitive to endophyte infections than did *Rhopalosiphum padi*. These data indicate that endophyte-infected grasses differ in their toxicity and that different insects may perform differently on the same endophyte-infected grass, perhaps resulting from different tolerances to alkaloids.

Lepidopteran responses to endophyte-infected tall fescue and perennial ryegrass have also been examined in some detail. Johnson et al. (1985) reported that the southern armyworm *Spodoptera eridania* was not affected by the endophyte in tall fescue. However, Ahmad et al. (1987) demonstrated that the armyworm survived poorly on infected perennial ryegrass plants. The fall armyworm *S. frugiperda* has also been used in several feeding studies. Larvae preferred uninfected ryegrass to infected ryegrass and developed more slowly on the latter (Hardy et al., 1985). Feeding trials with greenhouse and field-grown plants indicated that the conditions under which plants were raised could influence larval growth (Clay et al., 1985b). Fall armyworm also developed faster on endophyte-free tall fescue plants (Clay et al., 1985b) but the detrimental affect of infection strongly depended on the age of the tissue consumed (Hardy et al., 1986). Additional experiments have shown that fall armyworm larvae are negatively affected by clavicipitaceous endophytes infecting many grasses (Clay et al., 1985b; Cheplick and Clay, 1988). Sod webworms (*Crambus* sp.), another lepidopteran, caused significantly less damage to ryegrass plots with high levels of endophyte infection compared to plots with low levels of infection (Funk et al., 1983).

The Argentine stem weevil is a major pest of perennial ryegrass in New Zealand, and, accordingly, the impact of *Acremonium lolii* on the weevil has been subject to considerable research. Insect populations were found to be inversely correlated with the frequency of endophyte infection in ryegrass pastures (Mortimer and di Menna, 1983). Laboratory studies have shown that Argentine stem weevils strongly prefer endophyte-free plants for feeding and oviposition (Barker et al., 1983, 1984b). When confined to infected plans, they exhibit lower survival, feeding, and oviposition rates (Prestidge et al., 1982; Barker et al., 1983, 1984a; Stewart, 1985; Prestidge and Gallagher, 1988). Chemical studies suggest that the endophyte-produced alkaloid peramine is the primary determinant of stem weevil resistance (Rowan and Gaynor, 1986; Rowan et al., 1986). Endophyte-infection in tall fescue is also effective against stem weevils (Barker et al., 1983).

Other insects that are deterred by endophyte infection include pests of ryegrass turfs such as billbugs and chinch bugs (Funk et al., 1984; Ahmad et al., 1986; Saha et al., 1987). The common domestic cricket suffered 100% mortality within four days when confined to endophyte-infected ryegrass, although survival was quite high on endophyte-free ryegrass (Ahmad et al., 1985). Several leafhoppers and the corn flea beetle (*Chaetocnema pulicaria*) had lower populations where endophyte infection in tall fescue was high (Kirfman et al., 1986).

Fall armyworm survival and development were reduced on a range of grasses besides fescue and ryegrass infected by many different endophytes including species of *Acremonium*, *Atkinsonella*, *Balansia*, *Epichloë*, and *Myriogenospora* (Clay et al., 1985a, 1985b; Cheplick and Clay, 1988). Aphids have also been screened against a variety of grasses showing endophyte-induced resistance (Siegel et al., 1990). In Europe, *Epichloë typhina*-infected *Dactylis glomerata* resulted in reduced growth of black cutworm larvae (*Agrotis segetum*) (Schmidt, 1986). Sedges infected by the clavicipitaceous fungus *Balansia cyperi* also exhibit increased resistance to insects (Clay et al., 1985a; Stovall, 1987).

The database on induced resistance to insects caused by other, nonclavicipitaceous endophytes is relatively small. The endophyte *Rhabdocline parkeri* of Douglas fir was significantly correlated with increased mortality of gall midges (*Contarinia* sp.) (Carroll, 1986). The fungus colonizes the galls but does not actually invade the resident insects. Some endophytes isolated from conifers were toxic to spruce budworm (*Choristoneursa fumiferana*) in laboratory tests (Clark and Whitney, 1987). Sieber and Hugentobler (1987) found that beech trees often were infected by the ascomycetous endophyte *Apiognomonia errabunda* and suggested that it may cause mortality of gall-forming insects. Johnson and Whitney (1989) found no differences in the endophyte floras of red spruce needles from trees infested by spruce budworm vs. uninfested trees. However, when endophyte hyphae from cultures were incorporated into spruce budworm diet, some caused declines in survival and growth of the insect. Carroll (1988) described several examples where endophytes (or latent pathogens) reduce the severity of attack on their hosts compared to uninfected plants (see Diamandis, 1981; Webber, 1981). Infection by plant pathogens and mycorrhizal fungi can also deter insect herbivore attack of host plants (Clay, 1987a).

Several studies have clearly indicated that endophyte infection has little or no effect on some insect herbivores. For example, frit fly larvae (*Oscinella frit*), a major insect pest of ryegrass seedlings in sown pastures or lawns in Britain, were apparently resistant to the endophyte (Lewis and Clements, 1986). In surveys of insect presence and abundance in endophyte-infected and endophyte-free tall fescue pastures in Missouri, some insects were negatively correlated with infection levels but other insect species were uncorrelated or positively correlated with endophyte infection (Kirfman et al., 1986). The aphid *Rhopalosiphum padi* strongly discriminated between infected and uninfected tall fescue while artificial inoculations of tall fescue with the endophyte of ryegrass, *A. lolii*, had no effect on the aphid. Infections by *Phialophora*-like and *Gliocladium*-like endophytes also were ineffectual (Latch et al., 1985). The latter two endophytes are unrelated to the clavicipitaceous *Acremonium* endophytes. Other aphid species are not affected by endophyte infection of tall fescue or perennial ryegrass (see Table 17.2). The large number of endophytic fungi isolated from many plants suggest that it is unlikely that they all have significant pest-deterrent properties.

In the absence of experimental evidence, I would suggest that there are two criteria pointing to an important role for endophytes. One is that quantitative or qualitative variation in resistance of host plants to herbivory is significantly correlated with endophyte infection. The second is that secondary compounds produced in vivo or in vitro by endophytes have toxic effects on herbivores. Mycotoxin-producing endophytes should be considered candidates for deterring plant pests.

Grazing Mammals Clavicipitaceous endophytes are responsible for toxic syndromes of domestic mammals grazing on a variety of grass species (see also Siegel and Schardl, Chapter 10, this volume). Many grasses long known for being poisonous to livestock have been shown in recent years to be endophyte infected (White, 1987; Clay, 1988b). Cattle in the southeastern United States and sheep in New Zealand grazing endophyte-infected tall fescue and perennial ryegrass, respectively, have been especially well studied. The endophytes and the primary or secondary compounds involved differ but the consequences are similarly detrimental for both animals (see Latch, 1985; Bacon and Siegel, 1988). Other grazing animals reported to be affected by endophyte-infected grasses include deer, goats, horses, and oxen (Shaw, 1873; Marsh and Clawson, 1929; Nobindro, 1934; Garrett et al., 1980; Mackintosh et al., 1982). This topic has been reviewed extensively, given its obvious economic significance for meat and dairy production (Bacon et al., 1986; Bacon and Siegel, 1988).

Symptoms of fescue toxicosis include reductions in rate of weight gain, milk production, and forage intake, increased body temperature, respiration, and salivation. Animals often develop rough hair coats and have difficulty conceiving (Siegel et al., 1987). Symptoms are most frequently observed in the hot summer months. In the winter, some animals may become lame and develop gangrene of hooves, ears, and the tail, sometimes losing these parts, as a result (Hoveland et al., 1983; Bacon et al., 1986; Siegel et al., 1987; Bacon and Siegel, 1988). Afflicted animals graze less such that consumption of plant material is reduced, and greater standing biomass is maintained in endophyte-infected pastures (Bond et al., 1984; Read and Camp, 1986). The ergot alkaloid ergovaline appears to be the primary toxin in fescue toxicosis (Lyons et al., 1986). The major symptom of ryegrass staggers, which occurs in sheep and other livestock grazing endophyte-infected perennial ryegrass, are spasms that can prevent animals from walking or standing normally (Latch, 1985). Livestock deaths result from falling, drowning, or becoming tangled in fences. As with infected tall fescue, animal weight gains are reduced and standing plant biomass is greater in endophyte-infected ryegrass pastures (Fletcher, 1983; Mortimer and di Menna, 1983). The primary toxin associated with ryegrass staggers is a class of alkaloids called lolitrems which are found only in endophyte-infected grasses (Gallagher et al., 1981). Other animal toxicoses associated with grazing endophyte-infected grasses also are primarily neurological in nature. The effect of sleepy grass (*Stipa robusta*) on horses and drunk grass (*Melica decumbens*) on cattle are aptly described by their common names (Shaw, 1873; Marsh and Clawson, 1929). Both grasses are frequently infected by an *Acremonium*-type endophyte, although the significant chemical components have not been identified (White, 1987).

Other livestock poisonings of unknown etiologies may result from endophyte infection. For example, annual ryegrass toxicity to sheep in Australia is common (Trotman, 1978), and many annual ryegrasses are endophyte infected (Latch et al., 1987, 1988). Species of canary grass (*Phalaris*) can contain high concentrations of indole alkaloids, which have potent physiological effects on animals (canary grass staggers) (Woods and Clark, 1971; Ball and Hoveland, 1978; Corcuera, 1989). *Phalaris tuberosa* in Australia is often infected by a systemic, clavicipitaceous, fungal parasite, indicating that endophytes could be involved in livestock poisonings (Walker, 1970). A syndrome called bermuda grass tremors has sporadically affected cattle that graze bermuda grass (*Cynodon dactylon*) in the southeastern United States (Diener et al., 1975). Symptoms include twitching, trembling, and loss of motor coordination, similar to ryegrass staggers, but as yet the cause is unknown. Endophyte-produced mycotoxins could be responsible. Dallisgrass (*Paspalum dilatatum*) is a common pasture grass in the southeastern United States that often results in ergot poisoning of cattle, sheep, and horses (Goodwin, 1967). The species is frequently infected by *Claviceps paspali*. However, symptoms of ergot poisoning have been observed where ergot is absent (Luttrell and Bacon, 1977; Bacon et al., 1986). Dallisgrass is also commonly parasitized by *Myriogenospora atramentosa*, a clavicipitaceous endophyte similar to species of *Balansia* which may contribute to ergot poisoning of livestock in the absence of *Claviceps*. In a recent study, an endophyte (*Pseudocercosporella trichachnicola*) was isolated from the grass *Trichachne insularis*, which is commonly called sour grass and is reported to be avoided by herbivores (White et al., 1990). Interestingly, this endophyte is not related to clavicipitaceous endophytes of grasses, indicating that parallel evolution of unrelated, herbivore-deterring endophytes has occurred in the grasses. The presence of trichothecene mycotoxins in shrubs of the composite genus *Baccharis* in Brazil, responsible for cattle poisonings, raises the possibility that phyllosphere or rhizosphere endophytes are involved (Jarvis et al., 1988).

In most agricultural situations, livestock are restricted in their range and in the forage grasses available. Animals may have no choice but to consume endophyte-infected grasses even though under other circumstances they might choose to avoid them. Thus, data on animal toxicities resulting from grazing endophyte infected grasses largely reflect enforced consumption. The toxic effects of endophyte-infected grasses are typically cumulative and can be mitigated or reversed by diluting the forage or substituting an endophyte-free forage (Hoveland et al., 1983; Bush and Burrus, 1988; Stuedemann and Hoveland, 1988).

Previous research on mammalian grazers has largely been motivated by economic concerns. Little is known about the response of wild mammalian grazers to endophyte-infected grasses and other plants, despite their potential significance on wild plant communities. Simple experiments where known endophyte-infected and uninfected plants are marked or transplanted into wild plant communities, and levels of herbivory by wild grazers monitored, could provide valuable data on the response of wild grazers to infected plants and on the relative damage of infected and uninfected plants.

While some people have suggested that grasses (McNaughton, 1986) and other herbaceous plants (Paige and Whitham, 1987) might benefit from grazing compared to ungrazed plants, there is little doubt that mammalian grazers are capable of consuming far more plant tissue over a given time than could insect herbivores in the same area. In mixed populations and communities, ungrazed plants gain a competitive advantage over more heavily grazed individuals (Bentley and Whittaker, 1979; Crawley, 1983; Clay, 1989). It may not be coincidental that endophyte symbiosis is highly developed in the grasses, which have been subject to continuous grazing pressure throughout their evolutionary history (Stebbins, 1981). The major selective force favoring endophyte infection in grasses could be resistance to mammalian grazers, rather than insect herbivores.

Seed Predators Many *Acremonium* endophytes of grasses are maternally inherited through the seeds of their hosts. Fungal alkaloid concentrations can be substantially higher in seeds than in photosynthetic tissues (Bush et al., 1982; Jones et al., 1983), although endophyte-infected seeds are completely viable and have vigorous germination and seedling growth (Clay, 1987b). In laboratory experiments utilizing flour beetles (*Tribolium castaneum*), population growth rates were significantly lower on a diet consisting of ground, endophyte-infected perennial ryegrass seed than on uninfected seed (Cheplick and Clay, 1988). On infected tall fescue seed, beetle populations rapidly perished although they increased rapidly on uninfected seed. The differential performance of the beetles on the two infected grasses suggested that the seeds differed in the identity or amount of toxins present. In a series of experiments, granivorous birds were offered equal amounts of paired seed samples consisting of either millet and uninfected tall fescue, or uninfected tall fescue and infected tall fescue (Madej-Wolock, unpublished). Four of five species tested (juncos, tree sparrows, song sparrows, and house sparrows) exhibited a significant preference for uninfected tall fescue seed when infected seed was the alternative (Table 17.3). Further experiments with juncos showed that birds lost more weight on a diet of infected tall fescue seeds compared to uninfected seeds (Table 17.3). Laboratory feeding trials with rats also have indicated that animal performance is significantly reduced on a diet of endophyte-infected tall fescue seed compared to a diet of uninfected seed (Neal and Schmidt, 1985; Zavos et al., 1986). Feeding trials with cattle where seeds were incorporated into the

Table 17.3 Effect of endophyte infection on bird preference and weight loss on tall fescue seeds[a]

	Preference for seeds[b]	
Bird species	Millet > NI fescue	NI fescue > I fescue
Junco	***	**
Tree sparrow	*	*
Song sparrow	***	*
House sparrow	*	*
Chipping sparrow	*	NS

	Mean weight loss[c] on				
Species	100% millet	50% NI	50% I	100% NI	100% I
Junco	.45 g[d]	.60 g	.75 g	1.25 g	1.60 g

[a] NI = uninfected, I = infected.
[b] One-hr trials. Symbols: *** = $P < 0.0005$, ** = $P < 0.005$, * = $P < 0.05$; NS = $P > 0.05$.
[c] 24-hr trials.
[d] There was a significant difference among food surces ($P < 0.001$, Kruskal-Walls test).
Data from Madej-Wollock, unpublished.

diet were consistent with results from the bird and rat feeding experiments (Jackson et al., 1984). Herbivory of perennial ryegrass seedlings from seeds containing inviable endophyte was significantly lower than herbivory of uninfected seedlings, indicating that there is residual protection of seedlings despite the absence of a living endophyte (Stewart, 1985).

Unfortunately there are no data available on the effect of endophyte infection on seed predation in field studies. Granivorous insects, birds, and rodents may provide a significant selective advantage to endophyte-infected grasses in mixed populations and communities if seeds lying on the ground are avoided in favor of endophyte-free seeds. The proportion of infected versus uninfected seedlings becoming established could be shifted rapidly by intense, discriminatory seed predation. Research is also needed to determine the effect of endophyte infection on the long term viability of dormant seeds.

Many nonclavicipitaceous endophytes are also seedborne (Bose, 1947; Neergaard, 1977), and protection from seed predators could result. Seeds often contain much higher concentrations of secondary compounds than maternal vegetative tissues (Rosenthal and Janzen, 1979). The seed is a vulnerable life cycle stage that can tolerate relatively little damage compared to adult plants (Janzen, 1971). The possibility that defensive compounds in seeds may be of endophyte origin has not been thoroughly investigated. In one possible example, large quantities of a bacterial antibiotic became concentrated in seeds of the plant *Maytenus rothiana*, suggesting that a bacterial endophyte was responsible (Nettleton et al., 1981).

Nematodes Infection by the endophyte *Acremonium coenophialum* is associated with reduced nematode populations in the roots and surrounding soil of tall fescue compared to endophyte-free grasses. In one study, populations of the nematodes *Helicotylenchus dihystera* and *Paratrichodorus minor* in roots of pot-grown tall fescue were 40% and 25% lower, respectively, in endophyte-infected plants than in endophyte-free plants (Pedersen et al., 1988). However, in several cultivars of annual ryegrass (*Lolium multiflorum*, an *Acremonium*-type endophyte had no effect on nematode populations. In another study, endophyte infection was correlated with significantly lower populations of the nematodes *Pratylenchus scribneri* and *Tylenchorhynchus acutus* in field soil from tall fescue plots (West et al., 1988). The greater growth rates and drought tolerance of tall fescue in field and greenhouse studies could reflect differences in nematode damage to root systems (Read and Camp, 1986; Clay, 1987b; West et al., 1988; Arachevaleta et al., 1989).

Molluscs There is one example of endophyte infection that deters molluscan herbivory: The marine green alga *Enteromorpha vexata* is sometimes infected by an endophytic fungus in the genus *Turgidosculum*. Grazing by the marine snail *Littorina scutulata* was greatly reduced on infected algal thalli compared to grazing on uninfected thalli (means of 9.7% of thallus consumed vs. 89.0%) (Cubit, 1974). Many marine fungi form symbiotic associations with ascomycetous endophytic fungi (Kohlmeyer, 1968; Kohlmeyer and Demoulin, 1981; Kohlmeyer and Hawkes, 1983). Kohlmeyer has termed these associations mycophycobioses. At least one of the fungi has been found to produce antiherbivory compounds in culture (Pederson and Fries, 1977). Mycophycobioses may represent defensive mutualisms but as yet only few systems have been examined in any detail.

Plant Pathogens In addition to having antiherbivory properties, some fungal endophytes inhibit fungal plant pathogens in vitro and in vivo. Most data are from grasses and clavicipitaceous fungal endophytes, although the available data base is much smaller than that for herbivore deterrence. Evidence of antimicrobial effects is also found in endophytes from plants other than grasses.

Cultural filtrates of *Acremonium* endophytes from grasses and *Balansia cyperi* from purple nutsedge inhibited in vitro growth of several plant-pathogenic fungi including *Rhizoctonia cerealis*, *R. solani*, and *Fusarium oxysporum* (White and Cole, 1986; Baya et al., 1987; Stovall, 1987). Cultures of endophytes also inhibit various plant-pathogenic and soil-dwelling fungi when grown simultaneously on a single petri plate (White and Cole, 1985; Baya et al., 1987; Stovall, 1987). Several fungitoxic compounds have been isolated from the stromata of *Epichloë typhina* infecting *Phleum pratense* (Yoshihara et al., 1985; Koshino et al., 1987).

Inhibition of fungal pathogens in vivo by grass endophytes has been documented in two studies. Infection of the grass *Panicum agrostoides* by *Balansia henningsiana* was associated with significantly fewer lesions of the leaf spot fungus *Alternaria triticina* than found in endophyte-free plants (Clay et al., 1989). Crown rust (*Puccinia coronata*) disease severity of tall fescue was significantly lower on endophyte-infected compared to noninfected plants (Ford and Kirkpatrick, 1989). Stovall (1987) found that the fungus *Rhizoctonia solani* was frequently isolated as an endophyte from leaves of the sedge *Cyperus rotundus*, but never when plants were infected by *B. cyperi*.

With respect to nonclavicipitaceous endophytes of grasses, various plant-pathogenic and endophytic fungi were isolated from wheat, often many from the same plant (Sieber et al., 1988). When correlations between species pairs were calculated, there was no consistent correlation among endophyte taxa and between endophytes and plant pathogens. In another study, an endophyte isolated from *Vaccinium* exhibited antibiotic activity in culture although in vivo activity was not investigated (Fisher et al., 1984). Isolations from several plant species have revealed an abundance of endophyte taxa, often from the same leaves (Carroll and Carroll, 1978; Carroll and Petrini, 1983; Dreyfuss and Petrini, 1984; Riesen and Sieber, 1985; Petrini, 1986). This diversity of endophytes suggests that competitive exclusion within a single host is uncommon. The localized, as opposed to systemic, infections by most nonclavicipitaceous endophytes might promote coexistence of many endophyte taxa. A comparison of the endophyte flora of grasses either infected or uninfected by systemic, clavicipitaceous endophytes would be worthwhile.

17.4 Mechanisms of Pest Resistance

Endophyte-produced alkaloids and other mycotoxins appear to be responsible in most cases for the resistance of endophyte-infected plants to pests (Bacon et al., 1986; Miller, 1986; see also Siegel and Schardl, Chapter 10, this volume). Clavicipitaceous endophytes of grasses are known to produce at least four classes of alkaloids, although not all four are found in the same grasses. Ergot, loline, lolitrem, and peramine alkaloids have been isolated and identified from pure fungal cultures and/or from endophyte infected plants (Bacon et al., 1981, 1986; Bush et al., 1982; Gallagher et al., 1982; Rowan et al., 1986). Where the compounds are not produced in pure culture, precursors often are, although a synergistic interaction between plant and fungus may be required for the synthesis of the final compound (Weedon and Mantle, 1987). Limited evidence suggests that ergot and lolitrem alkaloids exhibit the greatest activity against mammalian herbivores, while loline and peramine alkaloids are most active against insects (Siegel et al., 1987), although all may have some effects on a broad spectrum of herbivores (Jones et al., 1983; Clay and Cheplick, 1989). The role of alkaloids from clavicipitaceous endophyte-infected grasses has been thoroughly reviewed (Dahlman et al., 1990). Grass endophytes also produce several other biologically active compounds that may be active against plant pests including the steroid compounds ergosterol, ergosterol peroxide, and tetraenone (Dahlman et al., 1990)) and the plant growth regulator auxin (Porter et al., 1985). *Epichloë typhina* also has been reported to produce several fungitoxic sesquiterpenes (Yoshihara et al., 1985).

Toxic metabolites with activity against spruce budworm larvae have been isolated from endophytes from conifer needles (Miller, 1986). Similarly, trichothecene toxins are found in poisonous *Baccharis* species where *Myrothecium* fungi are associated with plant roots, perhaps in a mycorrhizal association (Jarvis et al., 1987, 1988). The endophytic fungus *Mycosphaerella ascophylii* of the brown alga *Ascophyllum nodosum* produces halogenated phenols, which are known to deter herbivores of marine algae (Pederson and Fries, 1977; Hay and Fenical, 1988). The well-known capacity of many filamentous fungi to produce secondary metabolites with activity against animals would suggest that many endophytes are capable of producing toxins. However, the diversity of endophytes within a single plant, the diversity of potential toxins, and many unknown environmental variables affecting toxin production make it difficult to assign pest deterrence to a specific compound in most cases.

Other mechanisms by which endophytes could alter the susceptibility of host plants to pest attack include physical changes in host plant tissues (e.g., moisture content, fiber content) (Lewis, 1984), induced resistance to subsequent insects and fungi (Karban et al., 1987), and the depletion of nutrients from host tissues (Smith et al., 1985). However, these factors have received little attention.

17.5 Environmental Variation and Pest Deterrence

Research on tall fescue and its fungal endophyte has revealed that several factors can affect the relative toxicity of infected plants to herbivores, including genetic variation among endophyte strains (Bacon, 1988). Soil fertility, tissue age, and grazing history are important environmental factors affecting toxicity of tall fescue and by extension may be important in other plant/endophyte associations. Infected tall fescue plants subjected to high rates of nitrogen fertilization in the greenhouse produced significantly higher concentrations of ergot alkaloids than plants receiving lower fertilization levels (Lyons et al., 1986). However, loline alkaloid concentrations in infected plants were not significantly affected by fertilization (Belesky et al., 1987) although weather did appear to influence loline concentrations. Soil moisture levels may be of prime importance; they were found to correlate with both ergot and loline alkaloid concentration in tall fescue (Belesky et al., 1987, 1989).

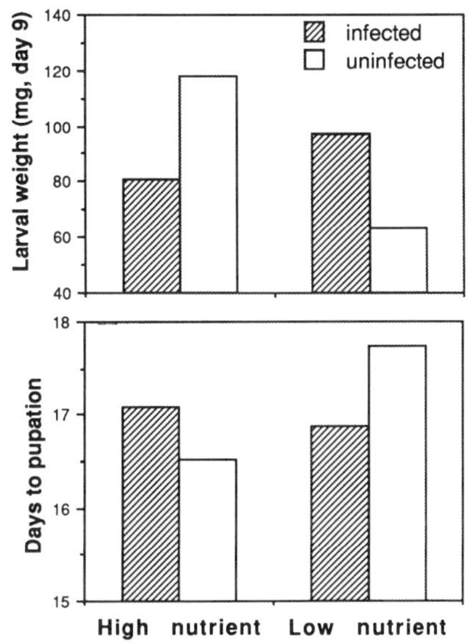

Figure 17.1 Fall armyworm performance on endophyte-infected and uninfected tall fescue grown under high and low nutrient conditions. Bars represent means of 37 to 45 insects. ANOVA indicated a significant infection-time-nutrient interaction for both developmental variables.

Given the observed association between fertilization and ergot alkaloid content and the presumed effect of alkaloids on insect herbivores (Clay and Cheplick, 1989), we conducted an experiment to ascertain the effect of fertilization and endophyte infection on an insect herbivore. Approximately 15 plants each of endophyte-infected and uninfected tall fescue were grown in the greenhouse and watered weekly with 250 ml of full-strength fertilizer. An additional 25 infected and 25 uninfected plants were grown under similar conditions except that they received 1/10 strength fertilizer. Fall armyworm (*Spodoptera frugiperda*) larvae were reared on each of the four classes of grass following well-established protocols (Clay et al., 1985b; Hardy et al., 1985). Survival, growth, and development of larvae were followed through pupation. Weights of nine-day-old larvae and the length of the larval period both indicated that endophyte-infected plants were much more toxic than uninfected plants in the high fertilizer treatment, but, in the low ferilizer treatment, larvae performed better on the infected plants (Figure 17.1). Statistical analyses of these data indicated that there was a highly significant treatment × infection interaction. These results suggest that the relative attractiveness and palatibility of endophyte-infected grasses can vary with soil fertility.

Tissue age is also an important determinant of toxicity in tall fescue. In further experiments with the fall armyworm, larvae were reared on leaves of different ages from endophyte-infected or uninfected plants (Hardy et al., 1986). Leaf ages 1, 2, and 3 referred to the first, second, and third leaves on an individual tiller starting with the youngest fully emerged new leaf (Figure 17.2). In choice tests, armyworm larvae significantly preferred

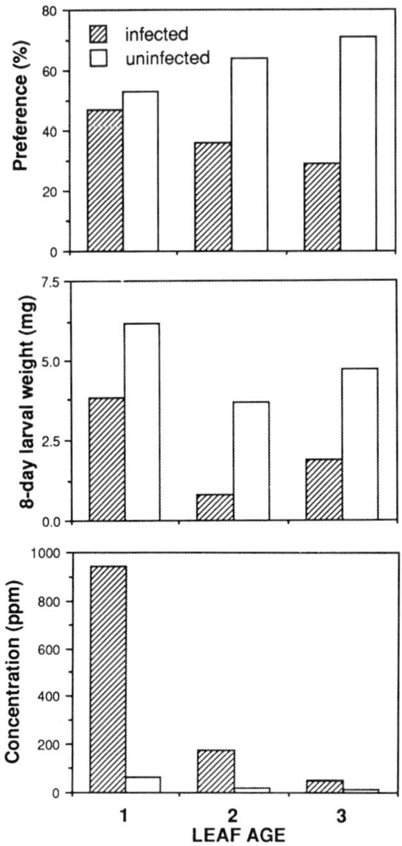

Figure 17.2 Fall armyworm performance on endophyte-infected and uninfected tall fescue leaves of three different ages, and the concentration of loline alkaloids in leaves.

leaves from uninfected plants over leaves from infected plants for leaf ages 2 and 3, but there was no significant preference at leaf age, 1 (Hardy et al., 1986). Rate of larval growth was generally concordant with their preference with the youngest leaves giving the best growth but for both infected and uninfected leaves, leaf age 3 resulted in faster growth than leaf age 2 (Figure 17.2). Both infection and leaf age were significant determinants of larval weight. Concentrations of loline alkaloids in infected leaves was highest in leaf age 1 and lowest in leaf age 3, with significant effects of age, infection, and their interaction. This result seems somewhat contradictory since the youngest leaves were most preferred and gave the greatest growth. It may be that the greater nutritional value of young leaves compensates for their greater alkaloid content. Alternatively, ergot alkaloids, which were not quantified, could be most concentrated in older leaves.

A further factor influencing toxicity of tall fescue and perennial ryegrass is their grazing history. Endophyte hyphae and alkaloids are more concentrated in the leaf sheaths and stem bases than in the leaf blades

(Lyons et al., 1986). Previously grazed pastures are likely to be shorter and have a higher proportion of short leaves and tillers. Grazing on short grasses is more toxic to sheep than grazing on longer grasses (Mortimer and di Menna, 1985). Whether endophytes respond to grazing by facultatively increasing alkaloid production is not known, but it would further increase the relative toxicity of previously grazed grasses.

17.6 Endophyte Effects on Competitive and Mutualistic Interactions

Mutualism In symbioses between plants and mycorrhizal fungi, the fungus obtains carbon compounds while the plant benefits from enhanced resource uptake (Harley and Smith, 1982). There is some evidence that clavicipitaceous fungal endophytes of grasses may interact with mycorrhizal associations involving the same host plant and with neighboring, unrelated plants. Mycorrhizal colonization of pine seedlings was slightly higher with endophyte-infected tall fescue compared to endophyte-free tall fescue (Ford and Kirkpatrick, 1989). All plants were grown in the same pot. In another study, soil from infected tall fescue plots was found to have lower levels of mycorrhizal propagules than soil from uninfected fescue plots (Chu-Chou et al., 1990). Isolates of *Glomus macrocarpum* and *G. mosseae* had lower reproductive rates on endophyte-infected tall fescue. In a similar study, endophyte-infected tall fescue had lower levels of mycorrhizal colonization than uninfected fescue sampled from a field site (Blaney and Clay, unpublished). In a subsequent experiment, individual tillers from infected or uninfected plants were rooted in sterile water and then transplanted into pots containing field soils. After 20 weeks of growth, endophyte-infected plants had approximately 40% fewer vescicles and arbuscles in random samples of fine roots than endophyte-free plants ($P = 0.06$) (Blaney and Clay, unpublished).

Endophyte infection can interact with mycorrhizal infection in complex ways to affect the insecticidal activity of the endophyte. Endophyte-infected, mycorrhizae-free, perennial ryegrass was a poor food plant for Argentine stem weevils compared to endophyte-free and mycorrhizae-free plants (Table 17.4; Barker, 1987). However, mycorrhizal, endophyte-infected plants exhibited no more resistance to Argentine stem weevil than endophyte-free plants. It may be that mycorrhizal infection alters the mineral balance of the plants, indirectly affecting toxin production by the endophyte. Phosphorus did not appear to be involved since differences in P uptake between mycorrhizal and nonmycorrhizal plants was controlled for.

The effect of endophyte infection on other plant mutualisms should be considered in future studies. Are endophyte-infected seeds as successful in animal-mediated dispersal as uninfected seeds? Is animal pollination

Table 17.4 Effect of mycorrhizal infection, endophyte infection, and fertilization on perennial ryegrass resistance to Argentine stem weevil

	Ln feeding scars[a]/5 weevils after 96 hr	
	Endophyte-free	Endophyte-infected
Nonmycorrhizal		
High P	2.89	0.73
Low P	2.60	0.28
Mycorrhizal		
High P	2.84	2.50
Low P	2.41	2.01

Data from Barker (1987).
[a]Natural log-transformed feeding scars.

affected by endophyte toxins? For example, in *Baccharis*, the trichothecene mycotoxins become most concentrated in the female inflorescences (Jarvis et al., 1987). Other rhizosphere symbioses with nonmycorrhizal fungi and bacteria could also be influenced by endophyte infection. Infection by one endophyte could alter the subsequent probability of infection by another endophyte (see Stovall, 1987).

Competition Direct effects of endophyte-infection on host plant physiology, such as enhanced drought tolerance (Arachevaleta et al., 1989), and indirect effects on other species, such as resistance to insect herbivores (Clay, 1988b), can affect the competitive ability of host plants. In an experimental field study, clones of the grass *Danthonia spicata* either infected or uninfected by the endophytic fungus *Atkinsonella hypoxylon* were planted in sunken pots with the common cooccurring grass *Anthoxanthum odoratum*. After two years of growth, infected *Danthonia* plants had produced significantly more tillers and biomass than uninfected plants (Kelley and Clay, 1987). Demographic studies showing that endophyte-infected grasses have higher survival, growth, and/or reproductive rates than endophyte-free plants in the same population are consistent with enhanced competitive ability (Bradshaw, 1959; Clay, 1984, 1990; Read and Camp, 1986). Particularly relevant are studies where the percentage of endophyte-infected grasses in dense turfs increases over time in the absence of seedling establishment (Saha et al., 1987).

Differential herbivory can interact with competition such that the more heavily damaged plants are weakened in competition with undamaged plants. Experiments where combinations of endophyte-infected and endophyte-free plants are grown together at a range of densities and frequencies and where levels of herbivory are manipulated are needed. Herbivory can be increased over normal levels by introducing additional insect or mammalian herbivores. Similarly, herbivory can be reduced by a combination of fencing and insecticide application. Such experiments are capable of separating the direct effects of infection on hosts from indirect effects on plant pests. The increased fitness of endophyte-infected plants resulting from reduced demage by herbivores and pathogens must often be mediated through changes in plant competitive ability.

17.7 Conclusions and Future Directions

Endophyte-infected plants are proving to be widespread examples of a defensive mutualism where secondary fungal metabolites provide the basis for pest defense. This association is paralleled by the association of many plants that produce extrafloral nectaries and domatia with ants that physically defend the plant against herbivore attack (Beattie, 1985). Endophytes may be better mutualists than ants because they are active against a wider range of pests including plant pathogens, and they are less likely to abandon a given host. Endophyte defense against herbivores and other plant pests has certain advantages over inherent plant defenses as well. Many endophytes are capable of rapid population growth and thus have high evolutionary potential, especially compared to long-lived hosts (Carroll, 1988). A given host may be infected by multiple endophytes that provide a greater diversity of chemical defenses than uninfected hosts. Endophytes are capable of producing compounds not found in uninfected plant, which reduces the probability that an herbivore or pathogen will be preadapted to attack the host. The small size and biomass of endophytes mean there is little energetic cost to the host. The potential for contagious spread by some endophytes means that hosts can acquire defenses by horizontal transmission rather than vertical inheritance.

Additional research is required to answer some very basic questions: Are all plants infected by endophytes? What are the patterns of endophyte diversity within an individual plant, within species, and within plant communities? How do phyllosphere communities differ between endophyte infected and uninfected plants? Do all endophytes play a defensive role against plant pests? What are the biochemical bases of pest deterrence? Answers to these questions are fairly well understood in some of the grasses but, even there, most research has focused on just two agriculturally important species. Available data on endophytes clearly indicate that the indirect effects of endophyte infection on other organisms interacting with the host are as important or more important than their direct effects on host plants. The diversity of plants and endophytes and their interactions with other species suggest that there is still a great deal to be learned about the role of endophytes in natural plant communities and how we can exploit endophytes for our own benefit in managed plant communities.

References

Ahmad, S., Govindarajan, S., Funk, C.R., and Johnson-Cicalese, J.M. 1985. Fatality of house crickets on perennial ryegrasses infected with a fungal endophyte. *Entomologia Experimentalis et Applicata* 39:183–190.

Ahmad, S., Johnson-Cicalese, J.M., Dickson, W.K., and Funk, C.R. 1986. Endophyte-enhanced resistance in perennial ryegrass to the bluegrass billbug. *Entomologia Experimentalis et Applicata* 41:3–10.

Ahmad, S., Govindarajan, S., Johnson-Cicalese, J.M., and Funk, C.R. 1987. Association of a fungal endophyte in perennial ryegrass with antibiosis to larvae of the southern armyworm, *Spodoptera eridania*. *Entomologia Experimentalis et Applicata* 43:287–294.

Arachevaleta, M., Bacon, C.W., Hoveland, C.S., and Radcliffe, D.E. 1989. Effect of tall fescue endophyte on plant response to environmental stress. *Agronomy Journal* 81:83–90.

Bacon, C.W. 1988. Procedure for isolating the endophyte from tall fescue and screening isolates for ergot alkaloids. *Applied and Environmental Microbiology* 54:2615–2618.

Bacon, C.W. and Siegel, M.R. 1988. Endophyte parasitism of tall fescue. *Journal of Production Agriculture* 1:45–55.

Bacon, C.W., Porter, J.K., and Robbins, J.D. 1981. Ergot alkaloid biosynthesis by isolates of *Balansia epichloë* and *B. henningsiana*. *Canadian Journal of Botany* 59:2534–2538.

Bacon, C.W., Lyons, P.C., Porter, J.K., and Robbins, J.D. 1986. Ergot toxicity from endophyte-infected grasses: a review. *Agronomy Journal* 78:106–116.

Ball, D.M. and Hoveland, C.S. 1978. Alkaloid levels in *Phalaris aquatica* L. as affected by environment. *Agronomy Journal* 70:977–981.

Barker, G.M. 1987. Mycorrhizal infection influences *Acremonium* induced resistance to Argentine stem weevil in ryegrass. *Proceedings of the New Zealand Weed Pest Control Conference* 40:199–203.

Barker, G.M., Pottinger, R.P., and Addison, P.J. 1983. Effect of tall fescue and ryegrass endophytes on Argentine stem weevil. *Proceedings of the New Zealand Weed Pest Control Conference* 36:216–219.

Barker, G.M., Pottinger, R.P., and Addison, P.J. 1984a. Effect of *Lolium* endophyte fungus infections on survival of larval Argentine stem weevil. *New Zealand Journal of Agricultural Research* 27:279–281.

Barker, G.M., Pottinger, R.P., Addison, P.J., and Prestidge, R.A. 1984b. Effect of *Lolium* endophyte fungus infections on behavior of adult Argentine stem weevil. *New Zealand Journal of Agricultural Research* 27:271–277.

Baya, B.O., Halisky, P.M., and White, J.F. 1987. Inhibitory interactions between *Acremonium* spp. and the mycoflora from seeds of *Festuca* and *Lolium*. *Phytopathology* 77:115 (abstract).

Beattie, A.J. 1985. *The Evolutionary Ecology of Ant-Plant Mutualisms*. Cambridge University Press, London. 182 pp.

Belesky, D.P., Robbins, J.D., Stuedemann, J.A., Wilkinson, S.R., and Devine, O.J. 1987. Fungal endophyte infection-loline derivative alkaloid concentration of grazed tall fescue. *Agronomy Journal* 79:217–220.

Belesky, D.P., Stringer, W.C., and Plattner, R.D. 1989. Influence of endophyte and water regime upon tall fescue accessions II. Pyrrolizidine and ergopeptine alkaloids. *Annals of Botany* 64:343–349.

Bentley, S. and Whittaker, J.B. 1979. Effects of grazing by chrysomeid beetle, *Gastrophysa viridual*, on competition between *Rumex obtusifolius* and *Rumex crispus*. *Journal of Ecology* 67:79–90.

Bond, J., Powell, J.B., and Weinland, B.T. 1984. Behavior of steers grazing several varieties of tall fescue during summer conditions. *Agronomy Journal* 76:707–709.

Bose, S.R. 1947. Hereditary (seed-borne) symbiosis in *Casuarina equisetifolia. Nature* 159:512–514.

Boursnell, J.G. 1950. The symbiotic seed-borne fungus in the Cistaceae. I. Distribution and function of the fungus in the seedling and in the tissues of the mature plants. *Annals of Botany* 14:217–243.

Bradshaw, A.D. 1959. Population differentiation in *Agrostis tenuis* Sibth. II. The incidence and significance of infection by *Epichloë typhina. New Phytologist* 58:310–315.

Bush, L.P. and Burrus, P.B. 1988. Tall fescue forage quality and agronomic performance as affected by the endophyte. *Journal of Production Agriculture* 1:55–60.

Bush, L.P., Cornelius, P.L., Buckner, R.C., Varney, D.R., Chapman, R.A., Burrus, P.B., II, Kennedy, C.W., Jones, T.A., and Saunders, M.J. 1982. Association of N-acetyl loline and N-formyl loline with *Epichloë typhina* in tall fescue. *Crop Science* 22:941–943.

Carroll, G.C. 1986. The biology of endophytism in plants with particular reference to woody perennials. pp. 205–222 in Fokkema, N.J. and Van Den Heuvel, J. (editors), *Microbiology of the Phyllosphere.* Cambridge University Press, Cambridge.

Carroll, G.C. 1988. Fungal endophytes in stems and leaves: from latent pathogen to mutualistic symbiont. *Ecology* 69:2–9.

Carroll, G.C. and Carroll, F.E. 1978. Studies on the incidence of coniferous needle endophytes in the Pacific Northwest. *Canadian Journal of Botany* 56:3034–3043.

Carroll, G.C. and Petrini, O. 1983. Patterns of substrate utilization by some fungal endophytes from coniferous foliage. *Mycologia* 75:53–63.

Cheplick, G.P. and Clay, K. 1988. Acquired chemical defenses of grasses: the role of fungal endophytes. *Oikos* 52:309–318.

Cheplick, G.P., Clay, K., and Wray, S. 1989. Interactions between fungal endophyte infection and nutrient limitation in the grasses *Lolium perenne* and *Festuca arundinacea. New Phytologist* 111:89–97.

Chu-Chou, M., An, Z.Q., Hendrix, J.W., Zhai, Q., and Siegel, M.R. 1990. Effect of the *Acremonium* endophyte on endogonaceous mycorrhizal fungi reproducing on tall fescue. *Mycological Society of America Newsletter* 41:10 (abstract).

Clark, C.L. and Whitney, N.J. 1987. Toxicity of endophytic fungi from conifer needles to the spruce budworm, *Choristoneursa fumiferana. Mycological Society of America Newsletter* 38:91 (abstract).

Clay, K. 1984. The effect of the fungus *Atkinsonella hypoxylon* (Clavicipitaceae) on the reproductive system and demography of the grass *Danthonia spicata. New Phytologist* 98:165–175.

Clay, K. 1987a. The effect of fungi on the interaction between host plants and their herbivores. *Canadian Journal of Plant Pathology.* 9:380–388.

Clay, K. 1987b. Effects of fungal endophytes on the seed and seedling biology of *Lolium perenne* and *Festuca arundinacea. Oecologia* 73:358–362.

Clay, K. 1988a. Clavicipitaceous fungal endophytes of grasses: coevolution and the change from parasitism to mutualism. pp. 79–105 in Hawksworth, D.L. and Pirozynski, K. (editors), *Co-evolution of Fungi with Plants and Animals.* Academic Press, London.

Clay, K. 1988b. Fungal endophytes of grasses: a defensive mutualism between plants and fungi. *Ecology* 69:10–16.

Clay, K. 1989. The impact of parasitic and mutualistic fungi on competitive interactions among plants. pp. 391–412 in Grace, J. and Tilman, D. (editors), *Perspectives on Plant Competition*. Academic Press, London.

Clay, K. 1990. Comparative demography of three graminoids infected by systemic, clavicipitaceous fungi. *Ecology* 71:558–570.

Clay, K. and Cheplick, G.P. 1989. Effect of ergot alkaloids from fungal endophyte-infected grasses on the fall armyworm (*Spodoptera frugiperda*). *Journal of Chemical Ecology* 15:169–182.

Clay, K., Hardy, T.N., and Hammond, A.M., Jr. 1985a. Fungal endophytes of *Cyperus* and their effects on an insect herbivore. *American Journal of Botany* 72:1284–1289.

Clay, K., Hardy, T.N., and Hammond, A.M., Jr. 1985b. Fungal endophytes of grasses and their effects on an insect herbivore. *Oecologia* 66:1–6.

Clay, K., Cheplick, G.P., and Wray, S.M. 1989. Impact of the fungus *Balansia henningsiana* on the grass *Panicum agrostoides*: frequency of infection, plant growth and reproduction, and resistance to pests. *Oecologia* 80:374–380.

Clement, S.L., Pike, K.S., Kaiser, W.J., and Wilson, A.D. 1991. Resistance of endophyte-infected plants of tall fescue and perennial ryegrass to the Russian wheat aphid (Homoptera: Aphidideae). *Journal of the Kansas Entomological Society* 64:646–648.

Corcuera, L.J. 1989. Indole alkaloids from *Phalaris* and other Gramineae. pp. 169–177 in Cheeke, P.R. (editor), *Toxicants of Plant Origin*. CRC Press, Inc., Boca Raton, FL.

Crawley, M.J. 1983. *Herbivory: The Dynamics of Animal-Plant Interactions*. University of California Press, Berkeley, CA. 437 pp.

Cubit, J.D. 1974. Interactions of seasonally changing physical factors and grazing affecting high intertidal communities on a rocky shore. Ph.D. Dissertation. University of Oregon, Eugene, OR. 200 pp.

Cunningham, J.J. 1958. Non-toxicity to animals of ryegrass endophyte and other endophytic fungi of New Zealand grasses. *New Zealand Journal of Agricultural Research* 1:489–497.

Dahlman, D.L., Eichenseer, H., and Siegel, M.R. 1991. Chemical perspectives on endophyte-grass interactions and their implications to insect herbivory pp. 227–252 in Jones, C., Krischik, V., and Barbosa, P. (editors), *Microorganisms, Plants and Herbivores*. John Wiley and Sons, New York.

Diamandis, S. 1981. *Elytroderma torres-juanii* Diamandis and Minter. A serious attack on *Pinus brutia* L. in Greece. pp. 9–10 in Millar, C.S. (editor), *Current Research of Conifer Needle Diseases*. Aberdeen University Press, Aberdeen, Scotland.

Diehl, W.W. 1950. *Balansia and the Balansiae in America*. USDA, Washington, DC. 82 pp.

Diener, U.L., Davis, N.D., Morgan-Jones, G., Cunningham, J.P., Langford, J.W., and Johnson, W.C. 1975. Bermudagrass tremors in cattle in Alabama. *Highlights of Agricultural Research* 22:12.

Dreyfuss, M. and Petrini, O. 1984. Further investigations on the occurrence and distribution of endophytic fungi in tropical plants. *Botanica Helvetica* 94:33–40.

Ewald, P.W. 1983. Host-parasite relations, vectors, and the evolution of disease severity. *Annual Review of Ecology and Systematics* 14:465–485.

Fisher, P.J., Anson, A.E., and Petrini, O. 1984. Novel antibiotic activity of an endophytic *Cryptosporiopsis* sp. isolated from *Vaccinium myrtilus*. *Transactions of the British Mycological Society* 83:145–148.

Fletcher, L.R. 1983. Effects of presence of *Lolium* endophyte on growth rates of weaned lambs, growing on to hoggets, on various ryegrasses. *Proceedings of the New Zealand Grassland Association* 44:237–239.

Ford, V.L. and Kirkpatrick, T.L. 1989. Effects of *Acremonium coenophialum* in tall fescue on host disease and insect resistance and allelopathy to *Pinus taeda* seedling. *Proceedings of the Arkansas Fescue Toxicosis Conference* 140:29–34.

Funk, C.R., Halisky, P.M., Johnson, M.C., Siegel, M.R., Stewart, A.V., Ahmad, S., Hurley, R.H., and Harvey, I.C. 1983. An endophytic fungus and resistance to sod webworms: association in *Lolium perenne. Bio/Technology* 1:189–191.

Funk, C.R., Saha, D.C., and Johnson-Cicalese, J.M. 1984. Association of endophytic fungi with increased persistence and improved performance of perennial ryegrasses and tall fescues in low maintenance turfs seeded in 1976. *Rutgers Turfgrass Proceedings* 15:134–146.

Gallagher, R.T., White, E.P., and Mortimer, P.H. 1981. Ryegrass staggers: isolation of potent neurotoxins lolitrem A and lolitrem B from staggers-producing pastures. *New Zealand Veterinary Journal* 29:189–190.

Gallagher, R.T., Campbell, A.G., Hawkes, A.D., Holland, P.T., McGaveston, D.A., Pansier, E.A., and Harvey, I.C. 1982. Ryegrass staggers: the presence of lolitrem neurotoxins in perennial ryegrass seed. *New Zealand Veterinary Journal* 30:183–184.

Garrett, L.W., Heiman, E.D., Pfander, W.H., and Wilson, L.L. 1980. Reproductive problems of pregnant mares grazing fescue pastures. *Journal of Animal Science* 51:237 (abstract).

Goodwin, D.E. 1967. Ergot poisoning of cattle grazing dallisgrass. *Journal of the American Veterinary Medical Association* 151:204–205.

Hardy, T.N., Clay, K., and Hammond, A.M. Jr., 1985. Fall armyworm (*Lepidoptera*: Noctuidae): a laboratory bioassay and larval preference study for the fungal endophyte of perennial ryegrass. *Journal of Economic Entomology* 78:571–575.

Hardy, T.N., Clay, K., and Hammond, A.M., Jr. 1986. The effect of leaf age and related factors on endophyte-mediated resistance to fall armyworm (*Lepidoptera*: Noctuidae) in tall fescue. *Environmental Entomology* 15:1083–1089.

Harley, J.L. and Smith, S.E. 1982. *Mycorrhizal Symbiosis*. Academic Press, New York. 483 pp.

Hay, M.E. and Fenical, W. 1988. Marine plant-herbivore interactions: the ecology of chemical defense. *Annual Review of Ecology and Systematics* 19:111–145.

Hoveland, C.S., Schmidt, S.P., King, C.C., Odum, J.W., Clark, E.M., McGuire, J.A., and Smith, L.A. 1983. Steer performance and association of *Acremonium coenophialum* fungal endophyte of tall fescue. *Agronomy Journal* 75:821–824.

Jackson, J.A., Hemken, R.W., Boling, J.A., Harmon, R.J., Buckner, R.C., and Bush, L.P. 1984. Loline alkaloids in tall fescue hay and seed and their relationship to summer fescue toxicosis in cattle. *Journal of Dairy Science* 67:104–109.

Janzen, D.H. 1966. Coevolution of mutualism between ants and acacias in Central America. *Evolution* 20:249–275.

Janzen, D.H. 1971. Seed predation by animals. *Annual Review of Ecology and Systematics* 2:465–492.

Janzen, D.H. 1973. Dissolution of mutualism between *Cecropia* and *Azteca* ants. *Biotropica* 5:15–28.

Janzen, D.H. 1977. Why fruits rot, seeds mold, and meat spoils. *American Naturalist* 111:365–374.

Jarvis, B.B., Wells, K.M., Lee, Y.W., Bean, G.A., Kommedahl, T., Barros, C.S., and Barros, S.S. 1987. Macrocyclic trichothecene mycotoxins in Brazilian species of *Baccharis. Phytopathology* 12:111–128.

Jarvis, B.B., Wells, K.M., Lee, Y.W., Bean, G.A., Kommedahl, T., Barros, C., and Barros, S.S. 1988. The mystery of trichothecene antibiotics in *Baccharis* species. *Journal of Natural Products* 51:736–744.

Johnson, J.A. and Whitney, N.J. 1989. A study of fungal endophytes of needles of balsam fir (*Abies balsamea*) and red spruce (*Picea rubens*) in New Brunswick, Canada, using culture and electron microscope techniques. *Canadian Journal of Botany* 67:3513–3516.

Johnson, M.C., Dahlman, D.L., Siegel, M.R., Bush, L.P., Latch, G.C.M., Potter, D.A., and Varney, D.R. 1985. Insect feeding deterrents in endophyte-infected tall fescue. *Applied and Environmental Microbiology* 49:568–571.

Jones, T.A., Buckner, R.C., Burrus II, P.B., and Bush, L.P. 1983. Accumulation of pyrrolizidine alkaloids in benomyl-treated tall fescue parents and their untreated progenies. *Crop Science* 23:1135–1140.

Karban, R., Adamchek, R., and Schnathorst, W.C. 1987. Induced resistance and interspecific competition between spider mites and a vascular wilt fungus. *Science* 235:678–680.

Kelley, S.E. and Clay, K. 1987. Interspecific competitive interactions and the maintenance of genotypic variation within the populations of two perennial grasses. *Evolution* 41:92–103.

Kirfman, G.W., Brandenburg, R.L., and Garner, G.B. 1986. Relationship between insect abundance and endophyte infestation level in tall fescue in Missouri. *Journal of the Kansas Entomological Society* 59:552–554.

Kohlmeyer, J. 1968. Revisions and descriptions of algicolous marine fungi. *Phytopathology* 63:341–363.

Kohlmeyer, J. and Demoulin, V. 1981. Parasitic and symbiotic fungi on marine algae. *Botanica Marina* 24:9–18.

Kohlmeyer, J. and Hawkes, M.W. 1983. A suspected case of mycophycobiosis between *Mycosphaerella apophlaeae* (Ascomycetes) and *Apophlaea* spp. (Rhodophyta). *Journal of Phycology* 19:257–260.

Koshino, H., Togiya, S., Yoshihara, T., and Sakamura, S. 1987. Four fungitoxic C-18 hydroxy unsaturated fatty acids from stromata of *Epichloë typhina*. *Tetrahedron Letters* 28:73–76.

Latch, G.C.M. 1985. Endophytes and ryegrass staggers. pp. 135–140 in Lacey, J. (editor), *Trichothecenes and Other Mycotoxins*. John Wiley and Sons, New York.

Latch, G.C.M. and Christensen, M.J. 1985. Artificial infections of grasses with endophytes. *Annals of Applied Biology* 107:17–24.

Latch, G.C.M., Christensen, M.J., and Gaynor, D.L. 1985. Aphid detection of endophytic infection in tall fescue. *New Zealand Journal of Agricultural Research* 28:129–132.

Latch, G.C.M., Potter, L.R., and Tyler, B.F. 1987. Incidence of endophytes in seeds from collections of *Lolium* and *Festuca* species. *Annals of Applied Biology* 111:59–64.

Latch, G.C.M., Christensen, M.J., and Hickson, R.E. 1988. Endophytes of annual and hybrid ryegrasses. *New Zealand Journal of Agricultural Research* 31:57–63.

Leuchtmann, A. and Clay, K. 1989. Experimental evidence for genetic variation in compatibility between the fungus *Atkinsonella hypoxylon* and its three host grasses. *Evolution* 43:825–834.

Lewis, A.C. 1984. Plant quality and grasshopper feeding: effects of sunflower condition on preference and performance in *Melanoplus differentialis*. *Ecology* 65:836–843.

Lewis, G.C. and Clements, R.O. 1986. A survey of ryegrass endophyte (*Acremonium loliae*) in the U.K. and its apparent ineffectuality on a seedling pest. *Journal of Agricultural Science* 107:633–638.

Luttrell, E.S. and Bacon, C.W. 1977. Classification of *Myriogenospora* in the Clavicipitaceae. *Canadian Journal of Botany* 55:2090–2097.

Lyons, P.C., Plattner, R.D., and Bacon, C.W. 1986. Occurrence of peptide and clavine ergot alkaloids in tall fescue grass. *Science* 232:487–489.

Mackintosh, C.G., Orr, M.B., Gallagher, R.T., and Harvey, I.C. 1982. Ryegrass staggers in Canadian wapiti deer. *New Zealand Veterinary Journal* 36:106–107.

Marsh, C.D. and Clawson, A.B. 1929. Sleepy grass (*Stipa vaseyi*) as a stock-poisoning plant. *USDA Technological Report* 114:1–19.

McNaughton, S.J. 1986. Grazing lawns: on domesticated and wild grazers. *American Naturalist* 128:937–939.

Miller, J.D. 1986. Toxic metabolites of epiphytic and endophytic fungi of conifer needles. pp. 223–231 in Fokkema, N.J. and van den Heuvel, J. (editors), *Microbiology of the Phyllosphere*. Cambridge University Press, Cambridge.

Mortimer, P.H. and di Menna, M.E. 1983. Ryegrass staggers: further substantiation of a *Lolium* endophyte aetiology and the discovery of weevil resistance of ryegrass pastures infected with *Lolium* endophyte. *Proceedings of the New Zealand Grassland Association* 44:240–243.

Mortimer, P.H. and di Menna, M.E. 1985. Interactions of *Lolium* endophyte on pasture production and perennial ryegrass staggers disease. pp. 149–158 in Lacey, J. (editor), *Trichothecenes and other Mycotoxins*. Wiley, New York. Neal, W.D. and Schmidt, S.P. 1985. Effects of feeding Kentucky 31 tall fescue seed infected with *Acremonium coenophialum* to laboratory rats. *Journal of Animal Science* 61:603–611.

Neergaard, P. 1977. *Seed Pathology*. John Wiley and Sons, New York. 1187 pp.

Neill, J.C. 1940. The endophyte of rye-grass (*Lolium perenne* L.). *The New Zealand Journal of Science and Technology* 21:280–291.

Nettleton, D.E., Jr., Balitz, D.M., Brown, M., Moseley, J.E., and Myllymaki, R.W. 1981. Large scale isolation of maytansine and other maytansinoids from the seeds of *Maytenus rothiana* using preparative LC. *Journal of Natural Products* 44:340–347.

Nobindro, U. 1934. Grass poisoning among cattle and goats in Assam. *Indian Veterinary Journal* 10:235–236.

Paige, K.N. and Whitham, T.G. 1987. Overcompensation in response to mammalian herbivory: the advantage of being eaten. *American Naturalist* 129:407–416.

Pedersen, J.F., Rodriquez-Kabana, R., and Shelby, R.A. 1988. Ryegrass cultivars and endophyte in tall fescue affect nematodes in grass and succeeding soybean. *Agronomy Journal* 80:811–814.

Pederson, M. and Fries, N. 1977. Bromophenols identified in a marine ascomycete and its culture medium. *Zeitung Pflanzenphysiologia* 82:363–366.

Petrini, O. 1986. Taxonomy of endophytic leaf fungi of aerial plant tissues. pp 175–187. in Fokkema, N.J. and van den Heuvel, J. (editors), *Microbiology of the Phyllosphere*. Cambridge University Press, London.

Petrini, O. and Carroll, G. 1981. Endophytic fungi in foliage of some Cupressaceae in Oregon. *Canadian Journal of Botany* 59:629–636.

Porter, J.K., Bacon, C.W., Cutler, H.G., Arrendale, R.F., and Robbins, J.D. 1985. In vitro auxin production by *Balansia epichloë*. *Phytochemistry* 24:1429–1431.

Prestidge, R.A. and Gallagher, R.T. 1988. Endophyte fungus confers resistance to ryegrass: Argentine stem weevil larval studies. *Ecological Entomology* 13: 429–435.

Prestidge, R.A., Lauren, D.R., van der Zijpp, S.G., and di Menna, M.E. 1982. An association of Lolium endophyte with ryegrass resistance to Argentine stem weevil. *Proceedings of the New Zealand Weed Pest Control Conference* 35:199–202.

Read, J.C. and Camp, B.J. 1986. The effect of fungal endophyte *Acremonium coenophialum in tall fescue on animal performance, toxicity; and stand maintenance. Agronomy Journal* 78:848–850.

Riesen, T. and Sieber, T. 1985. Endophytische Pilze von Winterweizen (*Triticum aestivum* L.). Ph.D. Dissertation. Eidgenössische Technische Hoschschule, Zurich, Switzerland.

Rosenthal, G.A. and Janzen, D.H. 1979. *Herbivores: Their Interaction with Secondary Plant Metabolites.* Academic Press, New York. 718 pp.

Rowan, D.D. and Gaynor, D.L. 1986. Isolation of feeding deterrents against Argentine stem weevil from ryegrass infected with the endophyte *Acremonium loliae. Journal of Chemical Ecology* 12:647–658.

Rowan, D.D., Hunt, M.B., and Gaynor, D.L. 1986. Peramine, a novel insect feeding deterrent from ryegrass infected with the endophyte *Acremonium loliae. Journal of the Chemical Society D Chemical Communications* 142:935–936.

Rykard, D.M., Bacon, C.W., and Luttrell, E.S. 1985. Host relations of *Myriogenospora atramentosa and Balansia epichloë* (Clavicipitaceae). *Phytopathology* 75:950–956.

Saha, D.C., Johnson-Cicalese, J.M., Halisky, P.M., Van Heemstra, M.I., and Funk, C.R. 1987. Occurrence and significance of endophytic fungi in the fine fescues. *Plant Disease* 71:1021–1024.

Sampson, K. 1933. The systematic infection of grasses by *Epichloë typhina* (Pers.) Tul. *Transactions of the British Mycological Society* 18:30–47.

Sampson, K. 1935. The presence and absence of an endophytic-fungus in *Lolium temulentum* and *L. perenne. Transactions of the British Mycological Society* 19:337–343.

Schmidt, D. 1986. La quenouille rend-elle le fourrage toxique. *Revue Suisse Agriculture* 18:329–332.

Shaw, J. 1873. On the changes going on in the vegetation of South Africa. *Botanical Journal of the Linnean Society* 14:202–208.

Sieber, T. and Hugentobler, C. 1987. Endophytische Pilze in Blattern und Asten gesunder ungeschadigter Buchen (*Fagus sylvatica* L). *European Journal of Forest Pathology* 17:411–423.

Sieber, T., Riesen, T.K., Muller, E., and Fried, P.M. 1988. Endophytic fungi in four winter wheat cultivars (*Triticum aestivum* L.) differing in resistance against *Stagonospora nodorum* (Berk.) Cast. and Germ. = *Septoria nodorum* (Berk.) Berk. *Journal of Phytopathology* 122:289–306.

Siegel, M.R., Latch, G.C.M., and Johnson, M.C. 1987. Fungal endophytes of grasses. *Annual Review of Phytopathology* 25:293–315.

Siegel, M.R., Johnson, M., Varney, D.R., Nesmith, W.C., Buckner, R.C., Bush, L.P., and Burrus, P.B. 1984. A fungal endophyte in tall fescue: incidence and dissemination. *Phytopathology* 74:932–937.

Siegel, M.R., Latch, G.C.M., Bush, L.P., Fannin, N.F., Rowan, D.D., Tapper, B.A., Bacon, C.W., and Johnson, M.C. 1990. Alkaloids and aphid response in grasses infected with fungal endophytes. *Journal of Chemical Ecology* 16:3301–3315.

Smith, K.T., Bacon, C.W., and Luttrell, E.S. 1985. Reciprocal translocation of carbohydrates between host and fungus in bahiagrass infected with *Myriogenospora atramentosa. Phytopathology* 75:407–411.

Stebbins, G.L. 1981. Coevolution of grasses and herbivores. *Annals of the Missouri Botanical Garden* 68:75–86.

Stewart, A.V. 1985. Perennial ryegrass seedling resistance to Argentine stem weevil. *New Zealand Journal of Agricultural Research* 28:403–407.

Stovall, M.E. 1987. An investigation of the fungus *Balansia cyperi* Edg., and its effect on purple nutsedge, *Cyperus rotundus* L. Ph.D. Dissertation. Louisiana State University, Baton Rough, Louisiana.

Stuedemann, J.A. and Hoveland, C.S. 1988. Fescue endophyte: history and impact on animal agriculture. *Journal of Production Agriculture* 1:39–44.

Thompson, J.N. 1982. *Interaction and Coevolution*. John Wiley and Sons, New York. 179 pp.

Trotman, C.H. 1978. Controlling annual ryegrass toxicity. *Journal of Agriculture of Western Australia* 19:84–89.

Walker, J. 1970. Systematic fungal parasite of *Phalaris tuberosa* in Australia. *Search* 1:81–83.

Webber, J. 1981. A natural biological control of Dutch elm disease. *Nature* 292:449–450.

Weedon, C.M. and Mantle, P.G. 1987. Paxillime biosynthesis by *Acremonium loliae*: a step forwards defining the origin of lolitrem neurotoxins. *Phytochemistry* 26:969–972.

West, C.P., Izekor, E., Oosterhuis, D.M., and Robbins, R.T. 1988. The effect of *Acremonium coenophialum* on the growth and nematode infestation of tall fescue. *Plant and Soil* 112:3–6.

White, J.F. 1987. Widespread distribution of endophytes in the Poaceae. *Plant Disease* 71:340–342.

White, J.F. and Cole, G.T. 1985. Endophyte-host associations in forage grasses. III. In vitro inhibition of fungi by *Acremonium coenophialum*. *Mycologia* 77:487–489.

White, J.F. and Cole, G.T. 1986. Endophyte-host associations in forage grasses. IV. The endophyte of *Festuca versuta*. *Mycologia* 78:102–107.

White, J.F., Jr., Morrow, A.C., and Morgan-Jones, G. 1990. Endophyte-host associations in forage grasses. XII. A fungal endophyte of *Trichachne insularis* belonging to *Pseudocercosporella*. *Mycologia* 82:218–226.

Wicklow, D.T. 1981. Interference competition and the organization of fungal communities. pp. 351–375 in Wicklow, D.T. and Carroll, G.C. (editors), *The Fungal Community*. Marcel Dekker, New York.

Woods, D.L. and Clark, K.U. 1971. Variation in the content of tryptamines in clones of *Phalaris arundinacea* L. *Crop Science* 11:121–122.

Yoshihara, T., Togiya, S., Koshino, H., and Sakamura, S. 1985. Three fungitoxic cyclopentanoid sesquiterpenes from stromata of *Epichloë typhina*. *Tetrahedron Letters* 26:5551–5554.

Zavos, P.M., Salim, B., Jackson, J.A., Varney, D.R., Siegel, M.R., and Hemken, R.W. 1986. Effect of feeding tall fescue seed infected by endophytic fungus *Acremonium coenophialum* on reproductive performance in male rat. *Theriogenology* 25:281–290.

18

Beyond Pest Deterrence—Alternative Strategies and Hidden Costs of Endophytic Mutualisms in Vascular Plants

George C. Carroll

18.1 Introduction

Over the past 15 years a varied and increasingly voluminous literature has attested to the ubiquitous occurrence of asymptomatic fungal infections within healthy plant tissues. Often, reports have documented high prevalence of infections and bewildering species diversity of endosymbionts, particularly with samples taken from moist climates. Previous reviews have dealt with endophyte floristics in considerable detail (Carroll, 1986, 1988; Clay, 1988), and the topic is really beyond the purview of the present discussion. However, several notable additions to the endophyte universe should be mentioned. These include a seedborne nonclavicipitaceous endophyte in a tropical grass (White et al., 1990), endophytes in leaves of deciduous trees (Sieber and Hugentobler, 1987), in twigs (Griffith and Boddy, 1988; Sieber, 1989), and in wood of branches and trunks (Chapela and Boddy, 1988a, 1988b; Petrini and Fisher, 1988; Chapela, 1989; Fisher and Petrini, 1990).

The authors of early floristic reports on coniferous needle endophytes speculated that endophytes might serve as protective mutualists, deterring invasion by pathogenic fungi and discouraging herbivory by insects and vertebrates (Carroll et al., 1977; Carroll and Carroll, 1978). Although endophytes in grasses have been long recognized, they were dismissed as a cause of toxicoses in farm animals until quite recently (Fletcher and Harvey, 1981; Clay, 1988). Since 1980 several reports have demonstrated that, indeed, endophytes may serve as antagonists to insects, grazing animals, and disease organisms (e.g., see Chapters 10 and 17, this volume). These reports have dealt largely with grasses and, in any case, concern only a small fraction of the total endophyte flora alluded to above. Carroll (1988) and Clay (1988) have argued that antagonism towards pests may prove a general basis for endophytic mutualism. The evidence for such protective mutualism, however, remains nonexistent or inconclusive for an overwhelming majority of commonly isolated endophytes.

In fact, several authors have looked for negative interactions between endophytes and insects and found none. Kirfman et al. (1986) reported the distribution of at least two species of leafhoppers on tall fescue to be uncorrelated with endophyte presence. Lewis and Clements (1986) showed that the ryegrass endophyte (*Acremonium lolii*) was ineffectual in controlling infestations by the fruit fly. Recently I attempted to determine whether the decline of the *Cryptomeria* gall midge in western Honshu (Japan) could be attributed to needle endophytes in Japanese cedar. Although endophytes proved both abundant and diverse, no correlation between larval mortality in the gall midge and endophyte presence could be demonstrated. Given an understandable reluctance by biologists to publish negative results, such experience may be more common than suggested by perusal of the literature. If fungal endophytes do not serve as protective mutualists, what other benefits might accrue to their host plants in return for an asymptomatic free ride for such fungi?

18.2 Harvest Mutualisms

A major category of mutualism involves a currency of nutrient exchange (Janzen, 1985). Such harvest mutualisms occur widely among microbial symbioses and are best exemplified by mycorrhizal fungi. Could endophytes serve as provisioners of growth-limiting nutrients? Typically, harvest mutualists act: 1) by wresting and concentrating scarce substances from a highly competitive environment (e.g., mycorrhizae); or 2) by synthesizing required low-molecular-weight metabolites for nutritionally fastidious partners (insect endosymbionts); or 3) by degrading recalcitrant substrates such as cellulose (rumen fungi). Endophytes in stems and leaves appear to be largely restricted to plant tissues, and thus they could not function in the canopy as mycorrhizae do in the soil, e.g., in recovering nutrients from throughfall. Vascular plants are usually metabolically self-sufficient and are thus unlikely to require biosynthetic assistance from a fungal endophyte. However, while plants show some capability to resorb nutrients from senescent leaves prior to abscission (Chapin and Kedrowski,

1983), elements tied up in wood appear to be a frozen asset. Endophytes might act to decompose wood and liberate scarce elements such as nitrogen and phosphorus, which are chemically bound to lignocellulose.

Recent studies have revealed a guild of fungi that can invade healthy wood of both aspen and beech and can persist there in a latent form for many years (Chapela and Boddy, 1988a, 1988b; Chapela, 1989). In aspen, several species of *Hypoxylon*, *Cryptosphaeria populina*, and *Cytospora chrysosperma* were dominant. In beech, *Hypoxylon* spp. *Nectria coccinea*, and the basidiomycete *Coniophora puteana* were common. While all of these fungi have been regarded as pathogens which invade through wounds, these studies have shown them actually to be wood endophytes which may form latent infections early in the life of a branch. In healthy trees, such fungi are kept in check by the high water content of the sapwood. Under conditions of slow drying, they are specifically released and colonize the wood rapidly (Chapela and Boddy, 1988b). These fungi are capable of effecting significant decomposition; up to 90% weight loss has been observed in dead but attached beech branches in Europe (Chapela and Boddy, 1988b). Whether such decomposition releases nitrogen or phosphorus directly to the host tree is unknown, and studies to look for endophyte-mediated, internal-nutrient-cycling would probably prove very difficult.

Even if internal-nutrient-cycling does not occur, wood endophytes may play a mutualistic role. While abscission of leaves is a regular, physiologically programmed event, plants apparently have developed no comparable mechanism for the abscission of shaded, photosynthetically inefficient branches. Such branches represent a significant fraction of the nutrient capital in a forest. Efficient nutrient cycling requires that dead branches be returned to the litter layer, where they can be further decomposed and where nutrients thus released become available for capture by mycorrhizal fungi. Under conditions of drought, lower, shaded branches are the first to experience a water deficit (Zimmermann and Brown, 1971). These are exactly the conditions that favor release of the latent wood endophytes described by Chapela and Boddy (1988a, 1988b). Well-rotted, structurally weak, attached branches are highly prone to fall when they become saturated with water after rains. Thus, wood endophytes may promote a sloppy but essential abscission of dead branches.

Some heart-rotting fungi (e.g., *Echinodontium tinctorium* on *Abies grandis*) may behave as wood endophytes during the youth and adolescence of the host tree (Aho and Hutchins, 1977; Aho and Filip, 1982). Janzen (1976) suggested in a brief but provocative paper that heart-rotting fungi may promote internal-nutrient-cycling thus benefiting trees which they attack, especially under the conditions of extreme nutrient deprivation prevalent in certain tropical forests. Fisher (1976) showed that, indeed, tropical trees can produce, within their hollow interiors, internal roots which exploit thoroughly rotted heartwood. Roots of other plants were also found, however, suggesting that mutualism between tree and fungus must be diffuse and easily parasitized.

A few endophytes may grow on substrates other than their reported hosts. In such alternative habitats, endophytes may be involved in nutrient cycling activities which indirectly affect their host plants. Fisher et al. (1986) have reported a significant number of isolates of *Sporormiella* spp. from *Ulex* (gorse) spp. *Sporormiella* is also frequently isolated from dung and from recently burned soil (Wicklow, 1975). Carroll (1988) pointed out that *Ulex* species are adapted to periodic fire and speculated that rapid growth of *Sporormiella* in soil of gorse-dominated habitats after fire might serve to immobilize nutrients. Thus, gorse might promote nutrient retention after fire by continually seeding the habitat with a fungus whose spores are stimulated to germinate by heat. While blooms of phoenicoid fungi (i.e., those following disturbance by fire) in the soil are well documented (Wicklow, 1973; Carpenter et al., 1987), data on nutrient immobilization by such fungi remain regrettably scant. Until the appropriate studies have been done, a mutualistic role for endophytes in fire-prone habitats must remain just an attractive hypothesis.

18.3 Combative Mutualisms

Herbivores and pathogens are not the only biological stressors confronted by higher plants. For most plants, both interspecific and intraspecific competition act as major causes of mortality, particularly during the seedling stages (Harper, 1977). Given this fact, one might well expect that endophytic mutualists would evolve to influence the outcomes of contests between host plants and adjacent competitors. If diffuse mutualists can act indirectly as "friends of friends," surely they can also act as enemies of enemies.

The evidence for such combative mutualism among endophytic fungi comes from the literature of plant pathology as well as from a growing corpus of work on mycoherbicides. The fungi involved in such interactions usually manifest a very broad host range; common genera are *Colletotrichum* (particularly *C. gloeosporioides*), *Fusarium*, and *Phomopsis*. Under agricultural situations, crop plants and weeds always compete and thus set the stage for combative mutualists. Several papers indicate that weeds in soybeans and tomatoes can harbor, as latent endophytes, fungi which are pathogenic to the crop plant (Hepperly et al., 1985; Hartman et al., 1986; Raid and Pennypacker, 1987; McLean and Roy, 1988). Further, these fungi can overwinter and sporulate on the weed residues, thus placing the crop plant at a competitive disadvantage in succeeding years unless rigorous treatment with fungicides is carried out. Conversely, crop plants can harbor, in latent form, mycoherbicides which have been introduced to control weeds (Cerkauskas, 1988).

Combative mutualism probably operates in natural as well as agricultural systems. Cook et al. (1989) provide a persuasive example of such a mutualism in their dissection of the role of fire, habitat, and the heart-rotting fungus *Phellinus weirii* as agents of successional change in conifer forests at mid-elevation sites in the western Oregon Cascade Mountains. In these habitats, *Phellinus* occurs as a long-lived perennial mycelium which advances through the forest at a rate of ca 0.3 m/yr. Three coniferous tree species, lodgepole pine, mountain hemlock, and Pacific silver fir, are common in the area. The relative frequencies of the trees at a given site are determined by elapsed time since the last fire, moisture availability, and differential susceptibility of each species to the fungus. Lodgepole pine is a shade-intolerant pioneer which comes in after fire. In the absence of *Phellinus,* mountain hemlock will eventually dominate dry habitats, while silver fir will prevail in moist habitats. In the presence of *Phellinus,* dry sites are characterized by a diverse collection of pines and fir species, while the moist habitats are again dominated by Pacific silver fir. The effect of the fungus in the latter habitat is to advance succession. This happens because mountain hemlock is much more susceptible to *Phellinus* than silver fir. Thus, silver fir may incur the cost of shortened life span by harboring a fungus in its heartwood, but benefits from the early death of its hemlock competitor in the same habitats.

Numerous other examples of combative mutualism could probably be found if plants which share a common pathogen and compete with each other for moisture and light were examined in natural habitats. Heterocyclic rusts might prove a particularly fertile group for such investigations. Alternate stages of rusts on trees frequently occur on understory plant species in the same habitat (Ziller, 1974).

18.4 Metabolic Enhancement

Endophytes in grasses have been reported to bolster growth and confer drought resistance to their host plants (Clay, 1988; Arachevaleta et al., 1989). Clay (1987) has reported that endophytes in *Lolium* and *Festuca* improve seed germination. These responses to endophyte infection represent yet another distinct category of mutualism, a category whose basis is still not understood. Chapters 10 and 17 in this book deal specifically with grass endophytes, and such effects are mentioned here only for the sake of completeness.

18.5 Costs of Endophytism to Plants

If endophytic associations confer benefits on their host plants, they also entail costs. These costs have probably been greatly underestimated and include both continuous and episodic outlays. While few experimental data exist on which to base quantitative estimates of such tradeoffs, a qualitative assessment can be attempted from a close reading of the literature in plant ecology and plant pathology. Several of these costs are discussed in more detail below.

Production and Maintenance Costs Accurate calculation of maintenance and production costs requires knowledge both of the fungal standing crop within the plant tissues and of the annual turnover rate. These data are very scant for any fungus/plant association, although Carroll (1981) has attempted such calculations for leaf epiphytes. Only a very few endophyte species have actually been seen and localized within host plant tissues; in even fewer species have individual infections been counted and measured. Siegel et al. (1984) estimated the endophyte biomass in tall fescue to be less than 1% of the total plant biomass (dry weight). Carroll (1991) presented calculations based on data from Stone (1987) which suggest that the biomass of infections by *Rhabdocline parkeri* in the needles of Douglas fir is about 0.1–1.0 μg/needle, an amount corresponding to about 0.01–0.1% of the needle dry weight. If the production efficiency is taken to be 20% (Calow, 1977), this biomass would correspond to a maximum 5% of the net annual production for the grass and at most 0.5% of just the needle production in Douglas fir. There is no information about fungal turnover times in either situation. Since even perennial grasses die back each year, the fungal biomass in leaves must turn over at least once annually. *Rhabdocline* infections in Douglas fir appear to persist in foliage for several years, and thus turnover times there probably exceed one year.

Carroll (1986, 1988) has defined endophytic mutualisms as constitutive or inducible, depending on mode of transmission and extent of infection. If grass endophytes are considered typical of constitutive associations, we see that the mutualism may entail moderate metabolic cost for the host (ca. 5% net primary productivity). Conversely, inducible endophytism, such as that of *Rhabdocline* on Douglas fir, probably involves inconsequential direct metabolic costs.

Indirect Metabolic Costs While infection with endophytes may incur only modest direct costs to the host, in some cases the fungi may alter biomass allocation within the plant with substantial consequences for the host. Boerner et al. (1990) have studied the effects of endophyte presence and simulated herbivory on growth and biomass allocation in seedlings of two varieties of perennial ryegrass (*Lolium perenne*), one with a high-endophyte-infection frequency (Repell) and another with a low-endophyte-infection frequency (Pennfine). In the absence of simulated herbivory, total growth and patterns of biomass allocation were the same in both varieties. However, when leaves of the high-frequency variety were subjected to experimental clipping, a significant reduction in the root/shoot biomass ratio was noted compared to unclipped plants. Seedlings of perennial grasses often allocate a large proportion of the first-year net carbon gain towards developing an effective root system. The combination of a reduction in total plant mass caused by herbivory, together with an a diminished allocation for root growth caused by endophyte infection, might well decrease the likelihood that an individual will successfully overwinter following the first growing season. In such situations, endophytes might cause a significant decline in plant fitness even though direct metabolic costs are small.

Endophytes may incur other costs in excess of those associated directly with production or maintenance of fungal biomass. Most endophytes show limited growth within host tissues. Such growth limitation is not well understood, but may result from activation of the same localized host defense mechanisms which prevent colonization by virulent pathogens. Smedegaard-Peterson and Tolstrup (1985) have shown that resistance to pathogens in grain crops entails a cost to the host which can be detected as small but highly significant reductions in yield. Such reductions can occur even if plant defenses are mobilized to fend off leaf penetration by saprophytic fungi which do not normally cause plant disease. It seems likely that infection with endophytes will also involve such costs.

Reserves of fixed carbon and nutrients in vascular plants fluctuate seasonally. Leaves, both in perennial herbaceous plants and in trees, represent a significant fraction of the plant's accessible nutrient capital. Typically a portion of these reserves is mobilized and recovered prior to leaf abscission (Chapin and Kedrowski, 1983; Jonasson, 1989). The efficiency of nutrient withdrawal prior to leaf abscission differs greatly among specific elements and among plant species. However, nitrogen, phosphorus, and potassium usually have declined by about 50% of their maximum pools prior to leaf fall, while total leaf biomass has dropped by 20% (Chapin and Kedrowski, 1983). Fungal domains in senescing deciduous leaves usually appear as green spots against a background of red, yellow, or brown. Such areas are known as green islands, and they appear to be produced in response to the elaboration of cytokinins and other metabolites by the fungi. These metabolites locally retard senescence and impede the movement of carbon and other nutrients back into the main axis of the plant prior to leaf abscission (Kiraly et al., 1967; Goodman et al., 1986). The fungi responsible for green islands are often almost completely asymptomatic prior to leaf fall. It seems likely that known endophytes may also block mobilization of leaf nutrients, an activity that could constitute a significant, but hidden cost to their host plants.

Toxins Where endophytes are known to antagonize pests, some form of acquired chemical defense either has been demonstrated or can be inferred from preliminary data (White, 1985; Miller, 1986; Clay, 1988; Clark et al., 1989; Carroll, 1991). It has been widely assumed that the fungal toxins operative in such defenses act selectively towards the target pest and cause little damage to the host plant. That this need not always be the case can be guessed from a close reading of the literature on *Fusarium*. Asymptomatic *Fusarium* infections are reported in the endophyte literature with increasing frequency (Leuchtmann and Clay, 1988; Leslie et al., 1990). Most species of *Fusarium* produce toxic metabolites (Claydon and Grove, 1984; Marasas et al., 1984; Moss and Smith, 1984; Joffe, 1986), and many of these toxins have proved extraordinarily effective both against insect and/or mammalian grazers. For this reason, metabolites of *Fusarium* endophytes in crop plants are now coming under intense scrutiny (Mirocha et al., 1990).

Two important classes of fungal toxins common in *Fusarium* are enniatins and trichothecenes. Enniatins are well known as insect toxins and have been frequently reported from entomopathogenic fungi (Russell, 1966; Grove and Pople, 1980; Claydon and Grove, 1984; Strongman et al., 1988). Enniatins have also turned out to be toxic to higher plants (Burmeister and Plattner, 1987). Trichothecenes are among the most potent inhibitors of protein synthesis known (Jarvis et al., 1987), and this activity has contributed to their notoriety as mycotoxins and to allegations of their use as agents of biological warfare (Seeley et al., 1985). They are produced not only by *Fusarium* species but by other related genera of fungi imperfecti as well. Numerous reports have demonstrated their toxicity to plants as well as animals, and in some cases their involvement in pathogenesis has been proposed (Cutler and Jarvis, 1985; Jarvis et al., 1987; Cutler, 1988; Desjardins et al., 1989; Kuti et al., 1989; Lamprecht et al., 1989; Visconti et al., 1989). The investigation of toxic metabolites of endophytes is still in its infancy. If host sensitivity to metabolites produced by endophytes should prove common, such sensitivity could represent a major cost of endophytism to the host plant in spite of the small quantities of such substances actually produced.

Toxins need not interfere directly with host plant metabolism to exert a negative effect. The roots of most vascular plants routinely establish mutualistic associations with soil microorganisms, either fungi or bacteria. Under conditions of nutrient starvation, these associations appear highly beneficial and may even be required for the survival of the plant. If endophytes produce systemic, broad-spectrum toxins or antibiotics, these compounds may interfere with beneficial root symbioses. Indeed, a recent study by Chu-Chou et al. (1990) has documented just such an effect: Reproduction by two species of *Glomus* colonizing roots of tall fescue was depressed in plots where plants were infected with the endophyte *Acremonium coenophialum*. If levels of mycorrhizal infection and mycorrhizal competence are also depressed, diminished root function may represent yet another tradeoff for endophytism in the shoot.

Pathogenesis Plants usually mobilize various defensive measures, both general and specific, in response to infection by pathogens. Some of these defenses involve the localized production of phytoalexins, which cause necrosis and death of host cells adjacent to infection sites. Thus, disease symptoms may arise as manifestations of plant self-defense as well as from deleterious changes wrought by the invading fungi. Asymptomatic infection is the essence of endophytic mutualism. Clearly, plant defenses operating during the early phases of fungal infection have been relaxed (and/or, alternatively, been evaded) during colonization by endophytes. If defenses are abated for a mutualistic fungus, are they also abated for potential pathogens? If so, this effect must count as another cost of endophytism.

Carroll (1986) has argued that most endophytes are closely related to plant pathogens and that endophytic mutualists have evolved repeatedly from a number of unrelated pathogenic fungi. In fact, endophytes may become pathogenic under conditions of environmental stress for the host plant, notably drought. Repeated examples of stress-induced pathogenesis can be gleaned from the phytopathological literature. One striking case history concerns *Fusarium moniliforme*, a seedborne endophyte of maize and a causative agent of a stalk rot under conditions of drought. Leslie et al. (1990) have shown that virtually 100% of maize kernels sampled from a wide range of sites were "infected" with the fungus; infection frequencies in young and adult plants were similarly high. Maize seedlings with *Fusarium moniliforme* can be protected from infection by pathogenic *Fusarium graminearum* (Van-Wyk et al., 1988). *Fusarium moniliforme* also produces potent mycotoxins which have been detected in corn, other grains, and soybean, and which presumably deter grazing by animals (Gelderblom et al., 1988; Bacon et al., 1989; Sydenham et al., 1990). Reductions in fitness caused by fungal mutualists turned pathogens must also be counted among the costs of endophytism, although the unpredictability of the stresses which trigger such switches greatly complicates cost accounting (see Section 18.6).

18.6 Cost/Benefit Analyses

As the foregoing discussion makes clear, the boundary between endophytic mutualist and pathogen may appear vague. Ultimately the distinction between the two must depend on the net cumulative advantage/disadvantage which accrues to individual host plants over their individual life spans. Detailed cost/benefit analyses are rendered difficult by several factors. First, mutualists usually exchange materials or services which are cheap for one partner but essential and scarce for the other (Janzen, 1985). In effect, apples and oranges are bartered directly, and the assumption of a common currency of carbon or energy may not be justified. Such analyses are further clouded by the presence of "cheaters," individuals who receive all the benefits accorded true mutualists but provide nothing in return (Mainero and del Rio, 1985). Finally, balancing the books for endophytes in plants, especially for long-lived associations, becomes especially tricky when stochastic events (insect outbreaks, drought, flooding) can alter the final outcome. These last two problems deserve further comment.

Cheaters Mutualistic cheaters (e.g., individual plants whose flowers produce little nectar in populations of high nectar producers) have been well documented, particularly among flowering plants and insects and particularly when the two groups interact during pollination. Cheating may be facilitated in interactions where delays between the offering and receipt of benefits happen; this occurs because delays interfere with the recognition of cheaters by the cheated partner (Mainero and del Rio, 1985). Cheating is also favored when the life spans of the associated organisms differ greatly; cheaters of the shorter-lived species can benefit from a high rate of individual turnover, which again makes identification of cheaters by the longer-lived partner difficult (Mainero and del Rio, 1985). Both of these conditions are likely to occur with endophytes in plants, particularly with seedborne endophytes or endophytes in long-lived trees.

How would one recognize a cheater among mutualistic endophytes? Quite simply, cheaters would be fungal strains that benefit from asymptomatic habitation within plant tissues but provide none of the benefits contributed by mutualistic strains. One could identify such strains merely by examining a collection of endophytic isolates for variance in the character of benefit to the host plant. If the endophytes function as protective mutualists, for example, one might look for variance in the production of insect toxins or feeding deterrents. Data already in the literature suggest that cheaters occur commonly among populations of endophytes.

Clark et al. (1989) screened a number of strains of endophytes from conifer needles collected in eastern Canada for metabolites toxic to spruce budworm larvae. Under the fermentation conditions used, only 5 of the 94 strains tested produced extractable compounds that affected spruce budworm larvae. Under other growth conditions other strains may have produced toxins. Nevertheless, it seems tempting to conclude that at least some of the fungal strains tested had become cheaters. Similar results were obtained when extracts from cultures of multiple strains of *Rhabdocline parkeri* were tested for toxicity to spruce budworm larvae. Siegel et al. (1991) have obtained somewhat similar results in a study with completely different objectives. Here the occurrence of several defined alkaloids and the response of aphids to plants containing these compounds were investigated in a number of grass species and cultivars infected with several different endophytes. Several combinations of grasses and endophytes produced no alkaloids, and these may also be examples of cheaters. A detailed analysis of toxin levels in a population of grass individuals all containing the same endophyte species would be a better test of the cheater hypothesis and a worthy experiment.

Stochastic Events Selection pressures presumed to favor the evolution of endophytic mutualisms are intermittent. Herbivory, intraspecific competition, and drought are not constant stresses during the life of any individual plant. For long-lived plants such as coniferous trees, the intervals between stresses mitigated by endophytes can perhaps be measured in years, not days. Since natural selection operates at the level of individuals, how are endophyte populations maintained when, e.g., herbivores are absent and water is plentiful? Part of the answer lies in the fact that mutualistic organisms do not engage in the day-to-day balancing of costs and benefits that might satisfy the biologist-accountant. Rather, these factors are summed over the lifetime of the individual and expressed in the number of viable progeny which survive to the succeeding generation.

Life History Strategies An interesting corollary to the above fact emerges from a consideration of plant life history strategies. The selection pressure for endophytic mutualism should be strongest during those stages of the life history when future reproduction potential is maximally affected. For plants that reproduce by seed these stages will be the seedling and juvenile phases, when individual plants are most likely to die from biotic and abiotic stress. This may explain the prevalence of endophyte transmission through seed in maize and other grasses. The presence of the endophyte during the most vulnerable stages of plant development is guaranteed. In annual plants, the fate of the somatic portion of the plant becomes, in an evolutionary sense, irrelevant subsequent to reproduction. It is during just such phases of somatic senescence that endophytes may become aggressive and may be perceived as pathogens in agricultural situations. The behavior of *Fusarium moniliforme* in maize becomes completely explicable when viewed in this light. During seedling stages, the endophyte provides protection against soil-inhabiting pathogens (VanWyk et al., 1988). It persists as an asymptomatic endophyte in adult plants, becoming pathogenic only when the host is stressed by drought. In fact, this endophyte turned pathogen may have had little effect on the wild predecessors to maize, since effective seed dispersal might well have taken place prior to the onset of stalk rot. For perennial herbaceous (aclonal) hosts, the situation is less clear-cut, although again one might predict that the endophyte should become less useful and more pathogenic as the host ages and reproductive potential decreases. The data from Cook et al. (1989) on *Phellinus* in *Abies amabilis* can also be interpreted in this light.

Mutualist or Pathogen: Fungal Strategies Fungal endophytes are clearly closely related to pathogens. In some cases, different strains of the same species may have taken different paths (e.g., *Acremonium coenophialum* anamorphs of *Epichloe*). The domestication of a latent pathogen would entail patent benefits for any host plant. The question is, rather, what benefits might accrue to a fungus in adopting a mutualistic mode of existence? Specifically, what increase in fitness could any vigorous pathogen derive from restriction to a niche as marginal as that of, e.g., *Rhabdocline parkeri* in Douglas fir? A comparison of the fitnesses of mutualistic and pathogenic sister species would be instructive. This is attempted below for two species of *Rhabdocline* in Douglas fir.

Reproductive output often serves as a measure of fitness. The information for calculating such output is not available either for *R. parkeri* or its pathogenic congeners. However, in most fungi, reproductive output probably relates directly to annual production; the data for gross estimates of production in these fungi can be gleaned from studies carried out in Oregon over the last 15 years. For this comparison, information on the incidence of *R. parkeri* comes from Stone (1987); comparable information for *R. weirii* and *R. pseudotsugae* is derived from Todd (1984), and from my own unpublished observations. Data on the age-specific production and distribution of Douglas fir needle-surface areas are taken from Pike et al. (1977) and Massman (1982). Estimates of needle biomass in young stands of Douglas fir are based on a regression equation from Espinosa Bancalari and Perry (1987).

R. parkeri normally appears in young stands of Douglas fir at about 5–10 years and remains there until the trees die. Thus, it is prevalent in old-growth stands and production estimates are projected from such stands. If there are 45 trees/ha and each tree has a total needle surface area of 2850 m^2 (both sides of the needle), then 1.28×10^5 m^2 of needle surface area would be available for colonization by the endophyte. Stone (1987) has shown that in young stands (12–15 years of age), infection frequency on needles ranges from 0.1 to 17/mm^2 depending on the tree and age of needles. The age of the oldest needles on these trees is 5 years; in old growth stands, needles persist until age 8 to 9 years. Because endophyte infection rate increases greatly with needle age, the overall infection frequency in an old-growth stand is likely to be much higher than that in a young stand, and may be conservatively estimated at 10/mm^2. Multiplying by total needle surface area yields an estimated 1.28×10^{12} infections/ha. The volume of each infection can be approximated by a cylinder measuring 8 μm in diameter and 100 μm long (Stone, 1987); at the stand level this would correspond to a volume of 7.5×10^3 cm^3. If the volume/dry weight conversion factor is 0.2, that volume corresponds to 1.5 kg/ha of standing crop. Each year 20–25% of the needles in a Douglas fir canopy are lost, with loss increasing as the needles age. Because older needles show a higher frequency of endophyte infection, the turnover rate of endophyte biomass due to needle loss can be estimated at one-third, corresponding to an annual production of 0.5 kg/ha. During senescence and after needle loss, endophyte infections expand and the fungus sporulates; then it disappears before an onslaught of vigorous saprophytic competitors. An average increase in fungal biomass by a factor of 10 during this phase does not seem unreasonable; this would yield an annual production estimate of 5 kg/ha.

Rhabdocline needle blights are diseases of young stands before canopy closure, a fact probably related to age-specific mortality of susceptible trees. Under a regime of severe intraspecific competition, mere growth suppression would suffice for ultimate exclusion from stands older than 100 years. Todd (1984) surveyed a subsample of trees in a progeny test site 12–15 years old. He found that about 25% of the trees suffered some damage from *Rhabdocline*, with severe damage evident in 11%. If the value of 25% is applied to the stands studied by Espinosa Bancalari and Perry (1987), annual production of the pathogen can be calculated as follows: These stands are estimated to have a needle biomass (dry weight) of 12,290 kg/ha. Assume the pathogen is *R. weirii*, which attacks and fruits only on first-year needles. Further assume that in these young stands 30% of the foliage is one year old, and only 50% of the needles are colonized by the fungus on a susceptible tree. Multiplying these factors yields an estimate of 461 kg/ha of infected foliage. *Rhabdocline* forms discrete lesions on individual needles, implying that not all of the needle is infected. If the fungus colonizes on the average 50% of each needle, 230 kg/ha of infected tissue might be present in an afflicted stand. If 20% of the plant tissue is converted to pathogen biomass, an annual production of 46 kg/ha emerges from the calculations.

When these two estimates of production are compared, the endophyte comes off second best. On further consideration, however, the situation must be viewed differently. Entire landscapes, not individual stands, constitute the appropriate basis for such comparisons. Pathogenic *Rhabdocline* spp. exist in a patchy environment in which a supply of susceptible hosts must be constantly renewed through fire and natural seeding or cutting followed by planting. *Rhabdocline parkeri*, in contrast, occurs in a highly predictable and stable environment with sparse but evenly distributed production. Beyond this, pathogenic *Rhabdocline* spp. are dispersed by ascospores in May during a relatively short period of uncertain rainfall; a dry month can lead to a drastic decline in successful needle infection. *R. parkeri*, on the other hand, produces conidia gradually over a few months in the fall and winter, a long period of highly predictable rainfall. If production estimates for these species could be calculated for all Douglas fir forests in western Oregon over a period of 50 years, their equivalent fitness would become evident. Therefore, it appears that natural selection can be disruptive, producing closely related species with equivalent fitnesses but different ecological strategies.

18.7 Conclusions

The transition from mutualist to pathogen and back may be easy among endophytic fungi. Many species now regarded as plant pathogens may be derived from mutualistic endophytes widely present in plant progenitors of modern crops. In these cases, pathogenesis should be regarded as an artifact of plant domestication and modern agricultural practice. Important future lessons about endophytic mutualism are to be expected from the literature of plant pathology.

References

Aho, P.E. and Filip, G.M. 1982. Incidence of wounding and *Echinodontium tinctorium* infection in advanced white fir regeneration. *Canadian Journal of Forest Research* 12:705–708.

Aho, P.E. and Hutchins, A. 1977. Microorganisms from the pith region of suppressed grand fir understory trees. *USDA Forest Service Research Paper PNW-229*, Pacific Northwest Forest and Range Experiment Station, Portland, Oregon.

Arachevaleta, M., Bacon, C.W., Hoveland, C.S., and Radcliffe, D.E. 1989. Effect of the tall fescue endophyte on plant response to environmental stress. *Agronomy Journal* 81:83–90.

Bacon, C.W., Norred, W.P., Hinton, D.M., and Marijanociv, D.R. 1989. Production of fusarin-C on cereal and soybean by *Fusarium moniliforme*. *Applied and Environmental Microbiology* 55:2745–2748.

Boerner, R.E.J., Scherzer, A.J., and Sturgis, B.G. 1990. Responses of endophyte-bearing and endophyte-free varieties of *Lolium perenne* L. to fungicide treatment and simulated herbivory. *Journal of Range Management* 43:29–33.

Burmeister, H.R. and Plattner, R.O. 1987. Enniatin production by *Fusarium tricinctum* and its effect on germinating wheat seeds. *Phytopathology* 77:1483–1487.

Calow, P. 1977. Conversion efficiencies in heterotrophic organisms. *Biological Reviews of the Cambridge Philosophical Society* 52:385–409.

Carpenter, S.E., Trappe, J.M., and Ammirati, J. 1987. Observations of fungal succession in the Mount St. Helens devastation zone, 1980–1983. *Canadian Journal of Botany* 65:716–728.

Carroll, F.E., Müller, E., and Sutton, B.C. 1977. Preliminary studies on the incidence of needle endophytes in some European conifers. *Sydowia* 29:87–103.

Carroll, G.C. 1981. Microbial productivity on aerial plant surfaces. pp. 15–46 in J.P. Blakeman (editor), *Microbial Ecology of the Phylloplane*. Academic Press, London.

Carroll, G.C. 1986. The biology of endophytism in plants with particular reference to woody perennials. pp. 205–222 in Fokkema, H.J., and van den Heuvel, J. (editors), *Microbiology of the Phylloplane*. Cambridge University Press, Cambridge, England.

Carroll, G.C. 1988. Fungal endophytes in stems and leaves: from latent pathogen to mutualistic symbiont. *Ecology* 69:2–9.

Carroll, G.C. 1991. Fungal associates of woody plants as insect-antagonists in leaves and stems. pp. 253–271 in Barbosa, P., Krischik, V.A., and Jones, C.G. (editors), *Microbial Mediation of Herbivore-Plant Interactions*. John Wiley and Sons, New York.

Carroll, G.C. and Carroll, F.E. 1978. Studies on the incidence of coniferous needle endophytes in the Pacific Northwest. *Canadian Journal of Botany* 56: 3034–3043.

Cerkauskas, R.F. 1988. Latent colonization by *Colletotrichum* spp.: Epidemiological considerations and implications for mycoherbicides. *Canadian Journal of Plant Pathology* 10:297–310.

Chapela, I.H. 1989. Fungi in healthy stems and branches of American beech and aspen: a comparative study. *The New Phytologist* 113:65–75.

Chapela, I.H. and Boddy, L. 1988a. Fungal colonization of attached beech branches. I. Early stages of development of fungal communities. *The New Phytologist* 110:39–45.

Chapela, I.H. and Boddy, L. 1988b. Fungal colonization of attached beech branches. II. Spatial and temporal organization of communities arising from latent invaders in bark and functional sapwood, under different moisture regimes. *The New Phytologist* 110:47–57.

Chapin, F.S., III and Kedrowski, R.A. 1983. Seasonal changes in nitrogen and phosphorus fractions and autumnal translocation in evergreen and deciduous taiga trees. *Ecology* 64:386–391.

Chu-Chou, M., An, Z.-Q., Hendrix, J.W., Zhai, Q., and Siegel, M.R. 1990. Effect of the *Acremonium* endophyte on endogonaceous mycorrhizal fungi reproducing on tall fescue. *Mycological Society of America Newsletter* 41:10 (Abstract).

Clark, C.L., Miller, J.D., and Whitney, N.J. 1989. Toxicity of conifer needle endophytes to spruce budworm. *Mycological Research* 93:508–512.

Clay, K. 1987. Effects of fungal endophytes on the seed and seedling biology of *Lolium perenne* and *Festuca arundinacea*. *Oecologia* 73:358–362.

Clay, K. 1988. Fungal endophytes of grasses: a defensive mutualism between plants and fungi. *Ecology* 69:10–16.

Clay, K. 1989. Clavicipitaceous endophytes of grasses: their potential as biocontrol agents. *Mycological Research* 92:1–12.

Claydon, N. and Grove, J.F. 1984. *Fusarium* as an insect pathogen. pp. 117–128 in Moss, M.O. and Smith, J.E. (editors), *The Applied Mycology of Fusarium*. Cambridge University Press, Cambridge, England.

Cook, S.A., Copsey, A.D., and Dickman, A.W. 1989. Response of *Abies* to fire and *Phellinus*. pp 363–392 in Bock, J.H. and Linhart, Y.B. (editors), *The Evolutionary Ecology of Plants*. Westview Press, San Francisco.

Cutler, H.G. 1988. Trichothecenes and their role in the expression of plant-disease. *ACS Symposium Series* 379:50–72.

Cutler, H.G. and Jarvis, B.B. 1985. Preliminary observations on the effects of macrocyclic trichothecenes on plant growth. *Environmental and Experimental Botany* 25:115–128.

Desjardins, A.E., Plattner, R.D., Spencer, G.F., and Beremand, M.N. 1989. Furanocoumarin phytoalexins, trichothecene toxins, and infection of *Pastinaca sativa* by *Fusarium sporotrichioides*. *Phytopathology* 79:170–175.

Espinosa Bancalari, M.A. and Perry, D.A. 1987. Distribution and increment of biomass in adjacent young Douglas fir stands with different early growth rates. *Canadian Journal of Forest Research* 17:722–730.

Fisher, J.B. 1976. Adaptive value of rotten tree cores. *Biotropica* 8:264.

Fisher, P.J. and Petrini, O. 1990. A comparative study of fungal endophytes in xylem and bark of *Alnus* species in England and Switzerland. *Mycological Research* 94:313–319.

Fisher, P.J., Anson, A.E., and Petrini, O. 1986. Fungal endophytes in *Ulex europaeus* and *Ulex galli*. *Transactions of the British Mycological Society* 86:153–156.

Fletcher, L.R. and Harvey, I.C. 1981. An association of a *Lolium* endophyte with ryegrass staggers. *New Zealand Veterinary Journal* 29:185–186.

Gelderblom, W.C.A., Jaskiewicz, K., Marasas, W.F.O., Thiel, P.G., Horak, R.M., Vleggaar, R., and Kriek, N.P.J. 1988. Fumonisins—novel mycotoxins with cancer-promoting activity produced by *Fusarium moniliforme*. *Applied and Environmental Microbiology* 54:1806–1811.

Goodman, R.N., Kiraly, Z., and Wood, K.R. 1986. *The Biochemistry and Physiolo-*

gy of Plant Diseases. University of Missouri Press, Columbia, Missouri.

Griffith, G.S. and Boddy, L. 1988. Fungal communities in attached ash (*Fraxinus excelsior*) twigs. *Transactions of the British Mycological Society* 91:599–606.

Grove, J.F. and Pople, M. 1980. The insecticidal activity of beauvericin and the enniatin complex. *Mycopathologia et Mycologia Applicata* 170:103–105.

Harper, J.L. 1977. *Population Biology of Plants*. Academic Press, London.

Hartman, G.L., Manandhar, J.B., and Sinclair, J.B. 1986. Incidence of *Colletotrichum* spp. on soybeans and weeds in Illinois and pathogenicity of *Colletotrichum truncatum*. *Plant Disease* 70:780–782.

Hepperly, P.R., Kirkpatrick, B.L., and Sinclair, J.B. 1985. *Abutilon theophrasti*: wild host for three fungal parasites of soybean. *Phytopathology* 70:307–310.

Janzen, D.H. 1976. Why tropical trees have rotten cores. *Biotropica* 8:110.

Janzen, D.H. 1985. The natural history of mutualism. pp 40–99 in Boucher, D.H. (editor), *The Natural History of Mutualism*. Croom Helm, London, England.

Jarvis, B.B., Wells, K.M., Lee, Y., Bean, G.A., Kommedahl, T., Barros, C.S., and Barros, S.S. 1987. Macrocyclic trichothecene mycotoxins in Brazilian species of *Baccharis*. *Phytopathology* 77:980–984.

Joffe, A.Z. 1986. *Fusarium Species: Their Biology and Toxicology*. John Wiley and Sons, Inc., New York, N.Y.

Jonasson, S. 1989. Implications of leaf longevity, leaf nutrient re-absorption, and translocation for the resource economy of five evergreen plant species. *Oikos* 56:121–131.

Kiraly, Z., El Hammady, M., and Pozsar, B.I. 1967. Increased cytokinin activity in rust-infected bean and broad bean leaves. *Phytopathology* 57:93–94.

Kirfman, G.W., Brandenburg, R.L., and Garner, G.B. 1986. Relationship between insect abundance and endophyte infection level in tall fescue in Missouri. *Journal of Kansas Entomological Society* 59:552–554.

Kuti, J.O., Ng, T.J., and Bean, G.A. 1989. Possible involvement of a pathogen-produced trichothecene metabolite in *Myrothecium* leaf-spot of muskmelon. *Physiological and Molecular Plant Pathology* 34:41–54.

Lamprecht, S.C., Marasas, W.F.O., Vanwyk, P.S., Thiel, P.G., Sydenham, E.W., and Knox-Davies, P.S. 1989. Toxicity to plants and animals of an undescribed, neosolaniol monoacetate-producing *Fusarium* species from soil. *Plant and Soil* 114:75– 83.

Leslie, J.F., Pearson, C.A.S., Nelson, P.E., and Tousson, T.A. 1990. *Fusarium* spp. from corn, sorghum, and soybean fields in the central and eastern United States. *Phytopathology* 80:343– 349.

Leuchtmann, A. and Clay, K. 1988. Experimental infection of host grasses and sedges with *Atkinsonella hypoxylon* and *Balansia cyperi* (Balansiae, Clavicipitaceae). *Mycologia* 80:291– 297.

Lewis, G.C. and Clements, R.D. 1986. A survey of ryegrass endophyte (*Acremonium loliae*) in the United Kingdom and its apparent ineffectuality on a seedling pest. *Journal of Agricultural Science* 107:633–638.

Mainero, J.S. and del Rio, C.M. 1985. Cheating and taking advantage in mutualistic associations. pp 192–216 in Boucher, D.H. (editor), *The Biology of Mutualism*. Croom Helm, London, England.

Marasas, W.F.O., Nelson, P.E., and Tousson, T.A. 1984. *Toxigenic Fusarium Species*. Pennsylvania State University Press, University Park, Pennsylvania.

Massman, W.J. 1982. Foliage distribution in old-growth coniferous tree canopies. *Canadian Journal of Forest Research* 12:10– 17.

McLean, K.S. and Roy, K.W. 1988. Incidence of *Colletotrichum dematium* on prickly sida, spotted spurge, and smooth pigweed and pathogenicity to soybean. *Plant Disease* 72:390–393.

Miller, J.D. 1986. Toxic metabolites of epiphytic and endophytic fungi of conifer needles. pp 221–231 in Fokkema, N.J. and van den Heuvel, J. (editors), *Microbiology of the Phyllosphere*. Cambridge University Press, Cambridge, England.

Mirocha, C.J., Abbas, H.K., and Vesonder, R.S. 1990. Absence of trichothecenes in toxigenic isolates of *Fusarium moniliforme. Applied and Environmental Microbiology* 56: 520–525.

Moss, M.O. and Smith, J.E. (editors). 1984. *The Applied Mycology of Fusarium*. Cambridge University Press, Cambridge, England.

Petrini, O. and Fisher, P.J. 1988. A comparative study of fungal endophytes in xylem and whole stem of *Pinus sylvestris* and *Fagus sylvaticus. Transactions of the British Mycological Society* 91:233–238.

Pike, L.H., Rydell, R.A., and Denison, W.C. 1977. A 400-year-old Douglas fir tree and its epiphytes: biomass, surface area, and their distributions. *Canadian Journal of Forest Research* 7:680–699.

Raid, R.N. and Pennypacker, S.P. 1987. Weeds as hosts for *Colletotrichum coccodes. Plant Disease* 71:643–646.

Russell, D.W. 1966. Cyclodepsipeptides. *Quarterly Review of the Chemical Society*. 20:559–576.

Seeley, T.D., Nowicke, J.W., Meselson, M., Guillemin, J., and Akratanakul, P. 1985. Yellow rain. *Scientific American* 253(3):128–137.

Sieber, T. and Hugentobler, C. 1987. Endophytische Pilze in Blättern und Ästen gesunder ungeschädigter Buchen (*Fagus sylvatica* L.). *European Journal of Forest Pathology* 17:411–425.

Sieber, T.N. 1989. Endophytic fungi in twigs of healthy and diseased Norway spruce and white fir. *Mycological Research* 92:322–326.

Siegel, M.R., Johnson, M.C., Varney, D.R., Nesmith, W.C., Buckner, R.C., Bush, L.P., Burrus II, P.B., Jones, T.A., and Boling, J.A. 1984. A fungal endophyte in tall fescue: Incidence and dissemination. *Phytopathology* 74:932–937.

Siegel, M.R., Latch, G.C.M., Bush, L.P., Fannin, N.F., Rowan, D.D., Tapper, B.A., Bacon, C.W., and Johnson, M.C. (1991) Alkaloids and aphid response in grasses infected with fungal endophytes. *Journal of Chemical Ecology* 16: 3301–3315.

Smedegaard-Peterson, V. and Tolstrup, K. 1985. The limiting effect of disease resistance on yield. *Annual Review of Phytopathology* 23:475–490.

Stone, J.K. 1987. Initiation and development of latent infections by *Rhabdocline parkeri* in Douglas fir, with observations on uninfected epidermal cells. *Canadian Journal of Botany* 56:2614–2621.

Strongman, D., Strunz, G.M., Giguere, P., Chao-Mei, Y., and Calhoun, L. 1988. Enniatins from *Fusarium avenaceum* isolated from balsam fir foliage and their toxicity to spruce budworm larvae, *Choristoneura fumiferana* (Clem.) (Lepidoptera: Tortricidae). *Journal of Chemical Ecology* 14:753–764.

Sydenham, E.W., Gelderblom, W.C.A., Thiel, P.G., and Marasas, W.F.O. 1990. Evidence for the natural occurrence of fumonisin B1, a mycotoxin produced by *Fusarium moniliforme*, in corn. *Journal of Agricultural and Food Chemistry* 38:285–290.

Todd, D. 1984. *A Field Study of the Endophytic Fungi in Douglas Fir Needles*. Ph.D. thesis, University of Oregon, Eugene, Oregon.

VanWyk, P.S., Schultz, D.J., and Marasas, W.F.O. 1988. Protection of maize seedlings by *Fusarium moniliforme* against infection by *Fusarium graminearum* in the soil. *Plant and Soil* 107:251–257.

Visconti, A., Solfrizzo, M., Logrieco, A., Mirocha, C.J., and Bottalico, A. 1989. Mycotoxins produced by *Fusarium acuminatum*—isolation and characteriza-

tion of acuminatin—a new trichothecene. *Journal of Agricultural and Food Chemistry* 37:1348–1351.
White, J.F., Jr. 1985. Endophyte-host associations in forage grasses. III. In vitro inhibition of fungi by *Acremonium coenophialum*. *Mycologia* 77:487–489.
White, J.F., Jr., Morrow, A.C., and Morgan-Jones, G. 1990. Endophyte-host associations in forage grasses. XII. A fungal endophyte of *Trichachne insularis* belonging to *Pseudocercosporella*. *Mycologia* 82:218–226.
Wicklow, D.T. 1973. Microfungal populations in surface soils of manipulated prairie stands. *Ecology* 54:1302–1310.
Wicklow, D.T. 1975. Fire as an environmental cue initiating ascomycete development in a tallgrass prairie. *Mycologia* 67:852–862.
Ziller, W.G. 1974. *The Tree Rusts of Western Canada*. Canadian Forestry Service Publication No. 1329, Department of the Environment, Victoria, BC, Canada.
Zimmermann, M.H. and Brown, C.L. 1971. *Trees: Structure and Function*. Springer, New York.

Part 5
Modification of the Phyllosphere (Deliberate or Otherwise)

19

Effects of Atmospheric Pollutants on Phyllosphere Microbial Communities

Naresh Magan and Andy R. McLeod

19.1 Introduction

Human activity has resulted in atmospheric pollutants becoming a significant influence directly on plants, their soil environment, and on the microorganisms that colonize aerial plant surfaces (the phyllosphere). Many atmospheric and/or particulate pollutants can impact plant surfaces (see Saunders, 1971; Babich and Stotzky, 1982). However, special attention has been placed on the influence that those pollutants which may be transported over long distances may have on crops and trees. This has occurred because of the increase in acidification of precipitation, particularly at high altitudes, and in incidence of poor growth and premature needle loss (i.e., symptoms of "forest decline"), which have been observed since the 1980s in parts of Europe and North America. The symptoms are not directly attributable to insects or diseases. A large research effort has examined the role of many factors in combination with air pollutants that may be responsible for the observed symptoms (Blank et al., 1989).

The pollutants sulfur dioxide (SO_2) and ozone (O_3), together with acidic precipitation, have received the greatest attention in studies of phyllosphere microorganisms. There are also other phytotoxic air pollutants (including HF, peroxacetylnitrate [PAN], Cl_2, HCl, and H_2S), which have received limited attention in studies on vegetation, but have not been studied in relation to phyllosphere microorganisms. Surprisingly, there have been no studies on the effects of NO_2 on phyllosphere microbial communities. The increase in solar ultraviolet (UV-B) radiation, arising from reduction in stratospheric O_3 due to increased emissions of chlorofluorocarbons (CFCs) may also directly affect the phyllosphere and its microbes.

The effect of atmospheric pollutants on phyllosphere microbial communities has been studied in several ways. Investigations have been conducted at field sites close to emission sources, and in closed and open-top chambers or open-air exposure systems. Closed and open-top chambers are somewhat artificial systems in which plants are either entirely or partially enclosed in some form of chamber and exposed to specific concentrations of pollutants. Often, the environmental conditions in these systems are modified from natural field sites. Open-air systems involve fumigation of field plots without any form of enclosure such that environmental conditions are hardly altered and the plants are exposed to natural variations throughout the experimental treatment. This system makes it possible to expose plants in field sites to specific target concentrations of pollutants for long periods of time.

This chapter covers: (1) the sources of SO_2, NO_x, O_3, and acid precipitation (see Section 19.2); (2) observed effects of pollutants on phyllosphere microbial communities from field surveys close to pollutant sources (see Section 19.3); (3) information available from in vitro experimental fumigation studies of these pollutants on component species (see Section 19.4); (4) effect of pollutants on phyllosphere fungi from experimental pollutant exposure in open-top chambers and open-air fumigation systems (see Section 19.5); and (5) the interaction among atmospheric pollutants, phyllosphere microorganisms, and plant pathogens (see Section 19.6).

19.2 Atmospheric Pollutants and Deposition

The air pollutants that have been of greatest international concern for effects on vegetation are SO_2, NO_x (NO and NO_2), and O_3, directly as gases, and through long-distance transport and transformation into SO_4^{2-} and NO_3^- as contributors to acidic precipitation. SO_2 emissions are typically associated with electricity generation from fossil fuel combustion and smelter operations, NO_x with high temperature combustion and fertilizer production, and O_3 with atmospheric transformation of NO_2 driven by sunlight in the presence of NO_x and hydrocarbons (Ling and Ashmore, 1987). The mean concentrations of SO_2 in central England and in central West Germany was found to be 0.005–0.012 ppm and 0.006–0.017 ppm, respectively. Indeed, global emissions of sulfur from anthropogenic sources have been estimated at 180 m tonnes (Moller, 1984). In Europe, particularly in the Netherlands, NH_3 has also been recognized as an important pollutant produced by intensive animal husbandry (Buijsman et al., 1985). A simplified pathway of the atmospheric changes that occur from emissions of these pollutants to deposition is shown in Figure 19.1.

Pollutant levels can vary from very high gaseous and particulate concentrations, sometimes occurring close to specific point sources, to much lower gaseous concentrations and acidic precipitation. The latter impact

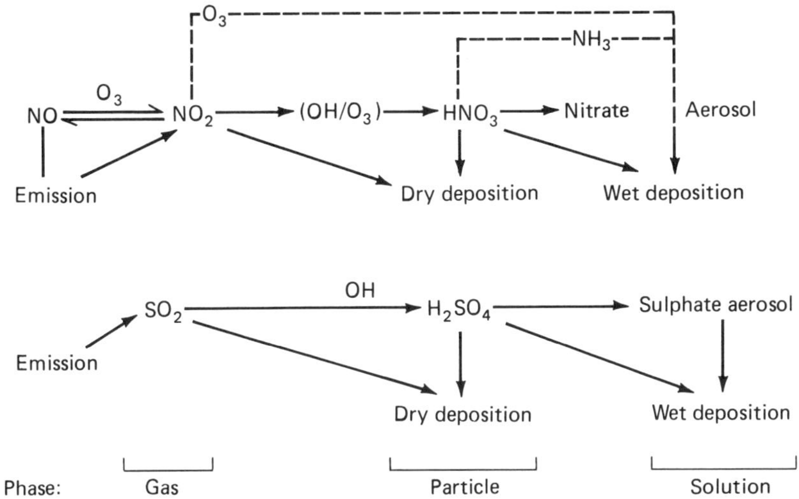

Figure 19.1 Diagram of the pathways of primary pollutants such as SO_2 and NO from emission to formation of secondary pollutants. Nighttime pathways are indicated by broken lines. Adapted from Derwent and Nodop (1986).

large areas and are often associated with longer-range transport, as a result of the introduction of tall chimney stacks and higher capacity industrial plant. Both dry and wet deposition of pollutants are important. Dry deposition is the deposit of the gaseous form of the pollutants directly onto plant and soil surfaces. Wet deposition includes that present in solution in rainfall, the deposition of water droplets from fog, cloud, and mist, termed occult deposition. Evaporation of such deposited mist may cause very high concentrations of pollutants to occur in solution (Unsworth, 1984) on the phyllosphere. The ionic form of the pollutant in solution may also be important for microorganisms, e.g., SO_2 produces two ionic forms, SO_3^- and HSO_3^-, of differing toxicity as well as H_2SO_3 molecules (Vas and Ingram, 1949). The relative amounts may vary with pH, concentration, and time available for equilibration. These factors may all influence pollutant effects on phyllosphere microorganisms.

Rates of gaseous dry deposition very among pollutants in relation to their reactivity. Highly reactive molecules such as HCl and HNO_3 are rapidly absorbed onto leaf and stem surfaces, while SO_2 and O_3 show a limited uptake by the external plant surfaces and deposit readily to internal surfaces through the stomata (Derwent and Nodop, 1986). In the presence of surface moisture and NH_3, however, SO_2 will deposit rapidly. The pollutant species, concentration, plant species, and chemical and physical conditions should all be considered when assessing the impact of atmospheric pollutants on phyllosphere microorganisms. Impacts should also be examined in relation to possible indirect effects via interaction with other abiotic and biotic factors. Wind, precipitation, water availability, and pH of the phyllosphere surface will interact with atmospheric gases. Microbial community structure will also be influenced by the nature and levels of the nutrients in the phyllosphere. These include pollen, organic debris, aphid honeydew, and leaf exudates. For example, atmospheric pollutants have been shown to increase the weathering of conifer needle surfaces and to affect their wax structure, which could influence the level of nutrients available to phyllosphere microorganisms (Cape and Fowler, 1981; Huttunen and Laine, 1983; Rinallo et al., 1986). There are, therefore, many factors which can influence pollutant effects on phyllosphere microorganisms. Numerous experimental studies have examined responses only under specific conditions, while interpretation of results from field studies can be very difficult.

19.3 Field Surveys of Phyllosphere Microbial Communities Near Pollutant Sources

Studies have been conducted at a range of distances (up to 1–2 km) downwind from industrial pollutant sources, in some cases in places where the atmosphere contained an undefined mixture of gaseous and particulate materials (Khanna, 1986), and also where specific point sources produced known pollutants such as SO_2 and NH_3 (Rai and Pathak, 1981; Lettle, 1983; Dowding, 1986). A summary of field observations since 1980 is presented in Table 19.1. Unfortunately, none of these studies included measurements of the actual concentration of pollutants to which sampled plants were exposed, making it difficult to compare these field effects to the results from the controlled fumigation experiments discussed in Section 19.5.

Studies of the phyllosphere of three species of tropical trees at the same site demonstrated that all the tree species at a polluted site had higher populations of actinomycetes and fungi than at an unpolluted site (Khanna, 1986). In contrast, bacterial populations, which were more numerous than actinomycetes or fungi, were markedly reduced. A comparison of the mycoflora at the polluted and nonpolluted site indicated marked differences in the quantity of mycoflora and the relative abundance of different fungi. Genera such as *Acremonium, Arthrinium, Curvularia,* and *Pestalotia* were only isolated from leaves from the nonpolluted site. Sampling of air spora revealed qualitative and quantitative differences in the mycoflora of the air at the two sites. This suggests an effect of pollutants on the ability of phyllosphere microorganisms to grow and sporulate, and thus on concentrations of spores in the air spora. However, no other field survey has considered changes in air-spora patterns in relation to deposition of atmospheric pollutants.

The effect of industrial emissions of SO_2 on the microflora of spruce in Czechoslovakia (Lettle, 1983) contrasts markedly with information from

Table 19.1 Field observations of atmospheric pollutant effects on phyllosphere microorganisms in the 1980s

Pollutant/ source	Vegetation type	Effects on microbes	Reference[a]
Mixed/urban area	*Fraxinus excelsior*	*Sporobolomyces roseus* significantly reduced	1
SO_2/power station	*Solanum tuberosum*	Species diversity similar to control; *Mortierella subtilissima* and *Fusarium semitectum* were resistant; *Streptomyces* absent but bacteria unaffected	2
NH_3/chemical factory	*Solanum tuberosum*	Total fungal populations decreased significantly compared to control; *Fusarium semitectum* resistant	2
Cement dust/ factory	*Solanum tuberosum*	Higher total fungal populations and species diversity compared to control; *Penicillium javanicum* associated with this site only	2
SO_2/industrial site	Spruce	Aerobic and ammonifying bacteria, yeasts, and filamentous fungi significantly increased in polluted site	3
Mixed/factory	*Fraxinus excelsior*	Reduction in *Tilletiopsis* spp. and *S. roseus*	4
SO_2/mixed power station and factories	*Musa sapientum, Tectona grandis, Croton bonplandianum*	Higher populations of fungi and actinomycetes in polluted site, but lower populations of bacteria; *Curvularia lunata, Acremonium,* and *Pestalotia* spp. absent from nonpolluted site	5

[a]References: 1, Dowding and Carvill (1980); 2, Rai and Pathak (1981); 3, Lettle (1983); 4, Dowding (1986); 5, Khanna (1986).

controlled fumigation studies (see Section 19.5) and other field studies. Although the number of different genera of autotrophic oxidizers of sulfur and asymbiotic nitrogen-fixing bacteria decreased on apical shoots of spruce, amylolytic and lipolytic bacteria and yeasts increased markedly. These results seem very unusual considering the detailed evidence of the sensitivity of both pink and white yeasts to SO_2 (Dowding, 1986; Magan and McLeod, 1988, 1991). However, in this study, no information was obtained on changes in species of filamentous fungi in response to pollutants.

Rai and Pathak (1981) carried out an extensive study of the effect of emissions of SO_2, NH_3, and cement dust from industrial sources on the microflora of potato leaves and compared this with a control site some distance from the treatment sites. Unfortunately, no measurements were made of the concentrations of the pollutants deposited from the industrial emission sources. Total numbers of bacteria, actinomycetes, and fungi were assessed for all treatments. All microbial components increased as the leaves matured. Bacterial populations were significantly reduced in the NH_3-polluted site. There was little difference between bacterial populations at the unpolluted control and SO_2 sites. The highest populations of actinomycetes (two *Streptomyces* spp. and a *Nocardia* sp.) were isolated from the NH_3 site with one *Streptomyces* sp. being absent from the SO_2-polluted site. Similarly, while *Cladosporium, Alternaria, Curvularia,* and *Aspergillus* spp. were dominant at all sites (SO_2, NH_3, and cement dust), other fungi such as *Mortierella subtilissima, Penicillium javanicum,* and *Fusarium semitectum* were found only at the SO_2, cement dust, and NH_3 sites, respectively. Both *M. subtilissima* and *F. semitectum* were tolerant of SO_2 and NH_3. Although the concentrations of these pollutants were not determined, the results do suggest that complex interactions between components of the microbial community and air pollutants occur.

19.4 Studies of In Vitro Effects of Atmospheric Pollutants on Microorganisms

Early work with bacteria, fungal saprophytes, and plant pathogens has been extensively reviewed by Babich and Stotzky (1974, 1978), Heagle (1973), and Manning and Keane (1988). Exposure of microorganisms to pollutants has involved gaseous treatments of cultures grown on agar or in liquid media or incorporation of the pollutant into the medium. Results from such methods need to be treated with caution. The phyllosphere environment may experience quite large changes in water availability, and pH, particularly when surface moisture evaporates, resulting in high concentrations of dissolved pollutants (Unsworth, 1984). This may contribute to the "high stress" environment of the phyllosphere (Dickinson, 1986), which does not occur in liquid cultures or on agar media.

Fumigation experiments have been done with concentrations similar to those present near emission sources or, more usually, with pollutant concentrations markedly higher than ambient levels, but for short treatment periods. Numerous studies (see Babich and Stotzky, 1982) have shown a reciprocal relationship between pollutant concentration and length of exposure time. Chronic exposure at low concentrations elicited equivalent biotic responses to short acute exposures at high concentrations. However, both approaches may be relevant because dry deposition and acidic precipitation can result in accumulation of the pollutant on plant surfaces (Garsed, 1985), thus affecting pH and phylloplane microorganisms by both long-term and short-term processes. Therefore, information both on how dry and wet pollutant deposition and pH changes affect spore germination and growth of phyllosphere microorganisms is important.

Table 19.2 The effect of varing concentrations of SO_2 on germination rate and germ tube length of phyllosphere fungi from barley exposed for 24 hr on 1% malt extract agar at 20°C

	Percentage germination[a]				LSD[b]
SO_2 concentration (ppm)	0.002[c]	0.05	0.1	0.2	
Alternaria alternata	100.0	100.0	100.0	99.3	0.6
Cladosporium herbarum	80.7	82.0	71.7	63.0	6.8
Epicoccum nigrum	99.7	99.0	99.0	98.0	2.1
Fusarium culmorum	100.0	100.0	99.7	99.7	0.7
	Germ tube length (μm)				LSD[b]
A. alternata	379.8	297.1	282.5	207.6	52.8
C. herbarum	23.7	15.1	6.6	5.3	2.6
E. nigrum	294.1	218.4	160.2	92.4	34.6
F. culmorum	241.7	199.1	89.9	53.2	27.1

[a] Values given are the mean for 300 spores.
[b] LSD, least significant difference.
[c] Control.
Adapted from Magan and McLeod (1988).

In chamber fumigation studies, phyllosphere yeasts and spores of filamentous fungi on agar media have been exposed to up to 0.2 ppm SO_2. Very different responses were observed between species. Growth of the yeasts *Sporobolomyces roseus* and *Cryptococcus* sp. was significantly inhibited by a 24-hr exposure to between 0.1 and 0.2 ppm SO_2. However, germination of *Alternaria alternata, Epicoccum nigrum,* and *Fusarium culmorum* was unaffected (Table 19.2). Germ tube extension of these spores was markedly reduced by up to 0.2 ppm SO_2 on a 1% malt extract substrate at 10 and 20°C (Magan and McLeod, 1988). Mycelial growth of several fungi from wheat leaves, including *Alternaria alternata, Cladosporium cladosporioides, Epicoccum nigrum, Curvularia lunata,* and *Aspergillus* and *Penicillium* spp., was found to be significantly inhibited by 2.66 ppm SO_2 after exposure for a maximum of 60 min on potato dextrose agar (Singh and Rai, 1990a). However, a 10-min exposure resulted in stimulation of mycelial growth of *Aspergillus flavus, Aspergillus niger,* and *E. nigrum.* Although the inhibitory effect of SO_2 was correlated with exposure times, mycelial growth measurements were made only for an exposure of 24 hr (Singh and Rai, 1990a). Longer periods of exposure and growth measurements may be necessary. In similar chamber fumigation studies on a nutrient-poor water agar substrate, Bos (1986) found stimulation of *S. roseus* at 0.025–0.05 ppm SO_2. In general, the results suggest that high ambient SO_2 concentrations significantly influence the ability of fungi to colonize phyllosphere surfaces.

It has been suggested that undissociated H_2SO_3 is the only effective form of SO_2 against spoilage yeasts and that duration of contact, pH, and binding of SO_2 all influence effects of SO_2 on these microbes (King et al., 1981). Babich and Stotzky (1978) found that bacteria and filamentous fungi were more sensitive to HSO_3^- than to SO_3^{2-}, particularly at pH 4.0. However, SO_2 at levels of 50–100 ppm and pH 5.0 had little effect on the germination or mycelial growth of fungi from conifer needles, suggesting a high tolerance of SO_2 by filamentous fungi (Kirkwood et al., 1989). However, at lower pH, both spore germination and mycelial growth are reduced (Magan and Lacey, 1984). The high sensitivity of phyllosphere yeasts to SO_2 (Dowding, 1986) could also have other implications; for example, yeasts are known to antagonize cereal foliar pathogens (Fokkema, 1978), and reduced yeast populations in polluted environments could allow pathogens to become established more rapidly.

The effect of SO_2 on filamentous fungi may be further modified by other environmental factors. At high SO_2 concentration (> 50 ppm) the sensitivity of an *Alternaria* sp. and of *Botrytis cinerea* increased with rising temperature and humidity, reaching a maximum in the presence of free water (Couey and Uota, 1961; Couey, 1965). For *B. cinerea*, SO_2 fumigation was almost 20 times more effective at 96% than at 75% relative humidity. Unfortunately, no detailed studies on the effect of interaction between water availability and pollutant exposure on germination and mycelial growth have been carried out.

Elevated concentrations of O_3 (0.02–2 ppm) were demonstrated in early work to be toxic to some bacteria, particularly *Escherichia coli* and *Streptococcus* and *Staphylococcus* spp. in culture or in air (see Manning, 1976; Rist and Lorbeer, 1981). However, most of the available information is on yeasts and filamentous fungi. In general, O_3 causes alteration in patterns of conidiation, germ tube extension, or hyphal growth. For example, in a study predominantly concerned with pre/postharvest pathogens, germination of *Botrytis cinerea* conidia was only slightly decreased by an 8-hr exposure to O_3, while germ tube extension was significantly reduced (Hibben and Stotzky, 1969). It was shown that with 0.5–1.0 ppm O_3, fungi with small hyaline spores were most sensitive, while those with pigmented spores, e.g., *Alternaria* spp., were least sensitive. It is now known that electrical discharge methods for producing O_3 also result in the formation of N_2O_5 (McLeod et al., 1989), which could profoundly affect the results obtained. It is not known whether N_2O_5 scrubbers were fitted in these earlier studies. Recently, spores of *Cladosporium herbarum, Epicoccum nigrum, Arthrinium sphaerospermum,* and *Botrytis cinerea* were exposed to 0.05–0.2 ppm O_3, scrubbed of N_2O_5, on 1% malt extract agar in chambers described by Magan and McLeod (1988). After a 24-hr exposure, there was little difference in the germination rate or the germ tube extension of spores exposed to O_3 relative to controls (Kirkwood, Magan and McLeod, unpublished). Furthermore, in many of these studies, rich media such as Sabouraud dextrose, potato dextrose, or malt extract agars have been used (Hibben and Stotzky, 1969; Babich and Stotzky, 1978; Magan and McLeod, 1988). Gener ally, results suggest that up to 1 ppm O_3 has very little effect on germination of phyllosphere fungi, particularly those with pigmented spores and mycelium. The direct effect of O_3 on plant tissue may be of more importance and may indirectly influence fungal activity on plant leaf surfaces. However, to obtain more realistic information, Wainwright (1988) suggested that media with low carbon sources should be used in studies on fungal tolerance/inhibition to pollutants.

Changes in phyllosphere pH due to acid mists or occult deposition may also influence germination and growth of different species of fungi in vitro. Experiments with fungi isolated from the phyllosphere of conifer needles suggest that *Aureobasidium pullulans, Sclerophoma pythiophila* and *Botrytis cinerea* are able to grow in vitro at pH < 3.0 (Kirkwood et al., 1989). Previous work on spore germination and germ tube extension of phyllosphere fungi such as *Alternaria alternata, Cladosporium* spp., *Epicoccum nigrum,* and *Fusarium culmorum* demonstrated that spores took longer to germinate at pH 4.0 than at pH 6.5 (Magan and Lacey, 1984). Germination was also influenced by water availability and temperature. The pH of rain water in low-pollution areas has, in the past, been considered to be close to pH 5.5. However, it may actually vary between pH 4.5 and 5.5 (Charlson and Rodhe, 1982). Acidic deposition, including acid rain, may cause a further decrease in the actual pH of the phyllosphere to between 3.0 and 4.0. However, the influence of atmospheric pollutants such as SO_2 on phyllosphere microbial communities may be affected by deposition history and interactions with the phyllosphere surface. This suggests that modifications of the phyllosphere pH by pollutants could affect phyllosphere fungi differently, influencing interactions among them and altering the species composition of the microbial community and possibly their interactions with foliar pathogens.

In summary, the literature suggests that both dry and wet SO_2 significantly reduce the growth of yeasts and spore germination of some filamentous fungi in vitro. However, some common pigmented phyllosphere fungi appear to be more resistant to elevated concentrations of SO_2. Some fungi are particularly tolerant of very low pH, and this may enable them to actively colonize plant surfaces in high pollution environments. By contrast, yeasts and filamentous fungi have been shown to be tolerant of O_3 in laboratory studies, although less detailed in vitro information is available. Such studies are, however, necessary to help interpret data obtained from plants exposed in open-top and field-exposure systems.

19.5 Controlled Fumigation Studies of Phyllosphere Communities on Plants

Controlled fumigations of crops and trees have been carried out in closed laboratory or glasshouse chambers (Heagle et al., 1973; Rist and Lorbeer,

Table 19.3 Data (since 1980) on effects of phyllosphere microorganisms on plants exposed to atmospheric pollutants

Pollutant	Concentration	Exposure system	Plant species	Effects on microbes	Reference[a]
SO_2	0.125 ppm	Open-top chamber	*Brassica napus*	Phyllosphere yeasts stimulated at <0.05 ppm, but inhibited at higher concentrations	1
SO_2	0.093 ppm	Open-top chamber	*Citrus sinensis*	*Alternaria alternata* and *Cladosporium cladosporioides* populations significantly reduced	2
SO_2	0.03 ppm	Open-air fumigation	*Pinus sylvestris*, *Picea abies*, *P. stichensis*	Total fungal populations reduced on needles; *A. pullulans* and *S. pythiophila* dominant species	3
SO_2	0.05 ppm	Open-air fumigation	*Hordeum vulgare*	White and red yeasts significantly reduced, *A. pullulans* increased, *Cladosporium* spp. unchanged	4
O_3	0.08 ppm	Open-top chamber	*Seguoiadendron giganteum*, *Quercus kelloggii*	No effect on filamentous fungi and yeasts	2
Acid mist	pH 3, 4	Field site	*Betula pubescens*	Significant reduction of populations of *A. pullulans* on leaves at pH 3	5
Acid mist	pH 3, 4	Field site	*Pinus sylvestris*	Species composition not affected by treatment; significant reduction of populations of *A. pullulans* on needles at pH 3	6
Cement dust	10–30 g/m^2	Open-top chamber	*Triticum aestivum*	Significant decrease in filamentous fungi applied to leaves	7

[a] References: 1, Bos (1986); 2, Fenn et al. (1989); 3, Kirkwood et al. (1989); 4, Magan and McLeod (1991); 5, Helander and Rantio-Lehtimaki (1990); 6, Ranta (1991); 7, Singh and Rai (1990b).

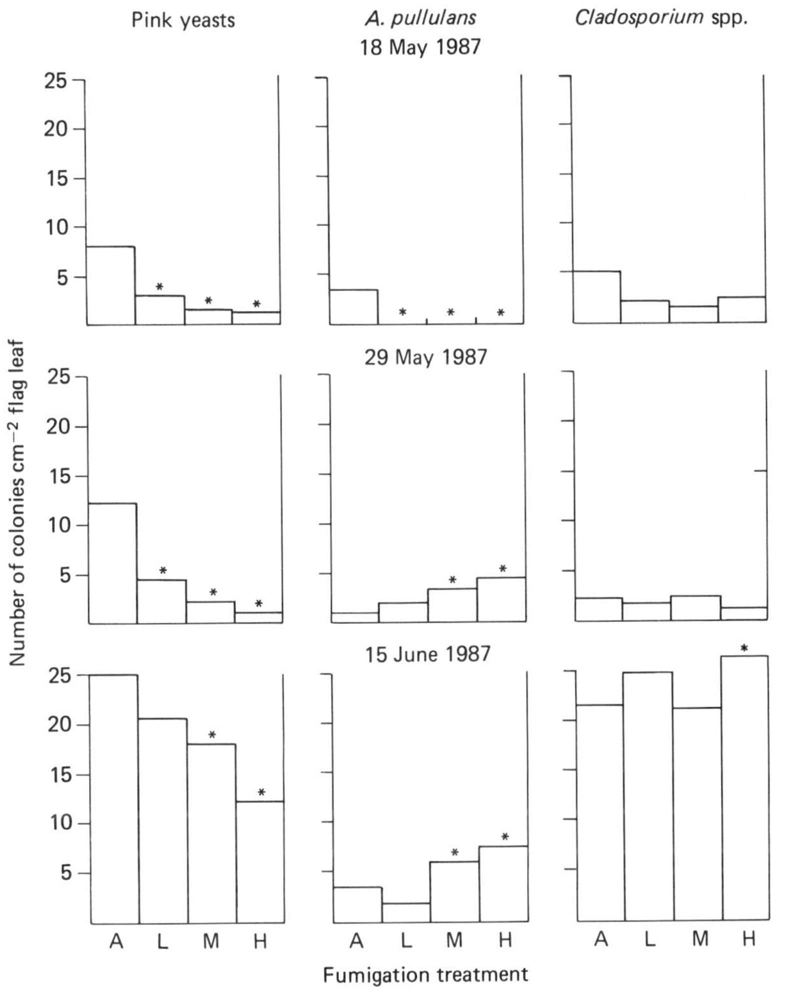

Figure 19.2 Effects of long-term fumigation (throughout growing season) with SO_2 on pink yeasts, *Aureobasidium pullulans*, and *Cladosporium* spp. on flag leaves of winter barley, as assessed by the leaf impression method. SO_2 treatments: A, ambient—0.005 ppm; L, low—0.014 ppm; M, medium—0.028 ppm; H, high—0.038 ppm. The * indicates a significant difference from the ambient plot ($P < 0.05$). Adapted from Magan and McLeod (1991).

1981), open-top field chambers (Heagle, 1973; Manning, 1976; Fenn et al., 1989), and open-air fumigation systems where no enclosures are used (McLeod et al., 1985). Enclosed chambers modify environmental conditions, particularly light, temperature, and humidity. Microbial populations on leaves are lower in enclosed chambers than in open-top or open-air exposure systems. Results from recent studies using open-top and open-air fumigation systems are given in Table 19.3.

Effects of SO_2 and O_3 The effects of SO_2 and O_3 on phyllosphere microorganisms in closed chambers have been reviewed previously by Treshow (1975), Manning (1976), and Rist and Lorbeer (1981).

The effects of three levels of SO_2 (0.01 to 0.05 ppm) on crop growth, pathogens, and mycoflora populations on ripening barley (flag leaves and ears) were determined with an open-air fumigation system (McLeod, 1988; McLeod et al., 1988; Mansfield et al. 1991; Magan and McLeod, 1991). The dominant mycoflora included pink and white yeasts, *Aureobasidium pullulans*, and filamentous fungi such as *Cladosporium* spp. In general, total fungal populations on the ears and particularly on the flag leaves were reduced by the highest concentration of SO_2. Results from three different assessment methods (leaf shadow, leaf impression, and serial dilution) showed that pink yeasts (predominantly *Sporobolomyces roseus*) and, to a lesser extent, white yeasts (predominantly *Cryptococcus* spp.) were sensitive to SO_2 treatments of as little is 0.047 ppm SO_2 (Magan and McLeod, 1991). These results are similar to those obtained previously by Dowding and Carvill (1980) who measured the sensitivity of *S. roseus* to a mixture of atmospheric pollutants, including SO_2 from a factory source, using the spore fall method. Magan and McLeod (1991) also found that populations of *A. pullulans* were significantly higher on flag leaves and ripening ears treated with up to 0.038 ppm SO_2 (Figure 19.2). This suggests that growth of *A. pullulans* may be stimulated by SO_2 or that there may be an interspecific competition between *S. roseus* and *A. pullulans*. *Cladosporium* spp. were relatively unaffected by the SO_2 treatments. *S. roseus* population increased on rape seed plants (*Brassicae napus*) exposed to 0.025 ppm SO_2 but decreased at 0.125 ppm SO_2 (Bos, 1986).

The percentage-green leaf area of flag leaves in the open-air, SO_2-fumigated plots were often significantly higher than in the untreated plot. The control of weakly pathogenic fungi such as *Alternaria alternata* and *Cladosporium* spp. have previously been implicated in increases in the duration of green leaf area and yield after late-season fungicide treatments (Cook, 1980; Magan and Lacey, 1986; Smedgaard-Petersen and Tolstrup, 1986). This suggests that elevated levels of SO_2 might also act as a fungicide, particularly in the phyllosphere of agricultural crops and may also increase yield (McLeod et al., 1988).

Exposure of Valencia orange trees to either ambient air, filtered air (charcoal-filtered), or filtered air plus 0.093 ppm SO_2 for up to four years in open-top chambers resulted in up to 75% reduction in phyllosphere microbial populations (Fenn et al., 1989). The frequency of isolation of *Alternaria alternata, Cladosporium cladosporioides, Aureobasidium pullulans*, and *Epicoccum nigrum* from prewashed leaf discs was significantly lower in the SO_2-treatment chambers than from leaves exposed in filtered air (Figure 19.3). These results contrast with those obtained with barley exposed to elevated SO_2 (0.05 ppm) in the open-air fumigation system described previously. In

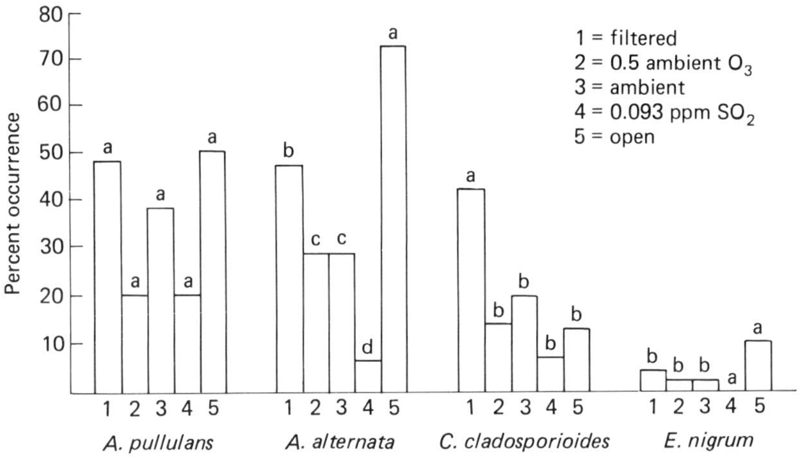

Figure 19.3 Percent leaf discs of Valencia orange trees from which dominant fungi were isolated following a 4-year fumigation period with SO_2 in open-top chambers. A different letter indicates a significant difference ($P < 0.01$). From Fenn et al. (1989), with permission.

the latter case, populations of *A. pullulans* were increased significantly, while little effect was observed on filamentous fungi. However, the methods used in the two studies were not comparable. While Magan and McLeod (1991) used a range of methods based on fungal spore production, Fenn et al. (1989) determined the outgrowth of fungi from washed leaf segments. The different results may also have been due partially to the very different types of plant species used. The limitations of open-top chambers were highlighted in the study of Fenn et al. (1989). Levels of fungi were lower in chambers compared to similar trees grown outdoors. The modified microenvironment of the chambers included differences in solar radiation, air flow, temperature, and humidity and, perhaps, the presence of insects and deposition patterns of fungal air spora from surrounding vegetation. All of these factors could influence the activity of phyllosphere microorganisms as well as modify some effects of pollutants.

A large, open-air fumigation experiment is currently underway in the United Kingdom. Circular plots, each containing 200 trees of three conifer species—Scots pine, Norway spruce, and Sitka spruce—are being treated with SO_2 (0.014 and 0.024 ppm), O_3 (1.5 × ambient), or both, and compared to ambient control plots (Shaw, 1986). Effects of the treatments on total and component microorganisms are being studied with a dilution plating technique and a fluorescein-diacetate (FDA) hydrolysis technique developed by Swisher and Carroll (1980) to estimate changes in microbial biomass. The total mycoflora populations on current-year needles of Scots pine were decreased by the high SO_2 treatment (0.024 ppm). Fumigation with high SO_2 and with O_3 did not result in additional decrease in total fungal populations. The SO_2 effect on total mycoflora of Scots pine needles was confirmed by the FDA technique; with total fungal biomass significantly lower in the high SO_2 treatment. However, the dominant fungi, *Aureobasidium pullulans* and *Sclerophoma pythiophila,* were relatively unaffected by the SO_2 treatment, confirming their tolerance to SO_2 and ability to grow at very low pH levels demonstrated in cultural studies (Kirkwood et al., 1989). Although there were some differences in species composition between tree species, there was no change in dominant mycoflora due to fumigation treatments.

The major effect of O_3 evidently is direct biochemical damage of the plant tissue resulting in plant stress and nutrient leakage. Exposure of pinto bean plants to low concentrations of O_3 (e.g., 0.06 ppm) resulted in leaf injury (see Rist and Lorbeer, 1981). In addition, the frequency of occurrence of individual species increased, especially that of *B. cinerea,* although species diversity was not affected. However, a reevaluation of some of this work may be necessary if the O_3 generation method produced N_2O_5.

Short-term fumigation (12 hr day for three months) of giant Sequoia and California black oak seedlings in open-top chambers with 0.04–0.08 ppm O_3 was found to have very little effect on the total fungal populations or dominant component species such as *Cladosporium* spp., *A. alternata,* and *A. pullulans* when compared to control trees in the chambers (Fenn et al., 1989). However, trees from the area surrounding the experimental chambers were found to have significantly higher populations of yeasts and filamentous fungi than in all the open-top chambers. Because of this marked difference in phyllosphere populations, data from short-term, open-top chamber exposure systems need to be interpreted with some care.

Studies at the open-air fumigation system described earlier also includes an O_3 treatment (1.5 × ambient). The mycoflora populations and the dominant component species, *A. pullulans* and *S. pythiophila,* on O_3-treated conifer needles were found to be more variable than those exposed to SO_2. Populations were similar to or slightly greater than on needles in the control plots. It is suspected that the elevated microbial populations may be due to a stimulating effect of N_2O_5 produced by the O_3 generator (McLeod et al., 1988). However, populations remained variable even after NO_x scrubbers were used. Changes in microbial populations on leaf surfaces may, therefore, be predominantly by direct effects of O_3 injury enhancing nutrient availability for growth.

Effects of Acidic Precipitation Very few controlled studies have been conducted on the effect of acidic precipitation on phyllosphere microorganisms of cultivated crops or trees. In northern Finland, simulated acid rain,

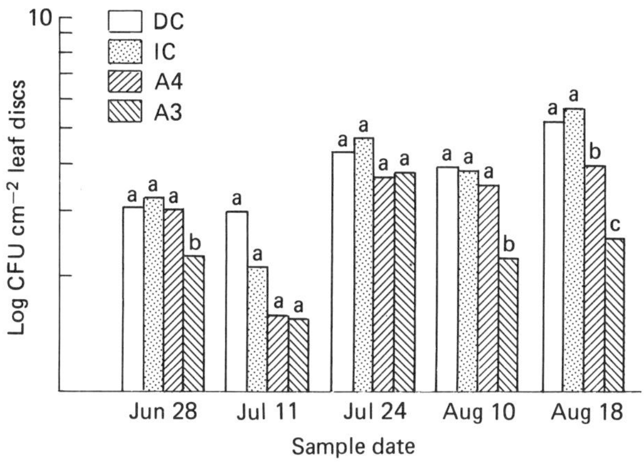

Figure 19.4 Total phyllosphere fungi (colony-forming units; log +1) isolated from birch leaves given four different treatments, three times per week for three years. DC, dry control; IC, irrigated control; A4, pH 4-acid rain treatment; A3, pH 3-acid rain treatment. Means (n = 5) with the same letter are not significantly different ($P < 0.05$). From Helander and Rantio-Lehtimaki (1990) with permission.

consisting of a mixture of H_2SO_4 and HNO_3 at pH 3 or 4, has been applied three times per week for 3 years prior to sampling of beech leaves and Scots pine in field stands (Helander and Rantio-Lehtimaki, 1990; Ranta, 1991). The acid-treated trees were compared with untreated controls and trees treated with spring water at pH 6. The total fungal populations on beech leaves were significantly reduced—on one occasion in the pH 4.0 treatment and on three occasions in the pH 3.0 treatments, out of a total of five sampling times (Figure 19.4). Irrigation with spring water had little effect on the fungal populations of the beech leaves (Helander and Rantio-Lehtimaki, 1990). There was no difference in fungal populations on leaves treated at two different heights, although, as one would expect, older leaves had higher fungal populations. *A. pullulans* represented about 80% of the total fungal population and exhibited the predominant effect of the acid rain treatments.

Similar results were obtained from studies on Scots pine (Ranta, 1991). The dominant fungi were *Aureobasidium pullulans* and *Cladosporium* spp. with other species such as *Acremonium* and *Botrytis* spp. only isolated occasionally. The acid rain treatment significantly reduced fungal population sizes, and the effect was similar at different heights. The treatments had little effect on the range of component species isolated from the needles. Very few bacteria were isolated from the surfaces of conifer needles. This confirms observations by Miller (1981) that while bacteria such as *Bacillus*, *Achromobacter*, and *Pseudomonas* may be present, results from scanning electron microscopic studies suggest that they are very scarce on these plant parts.

Acidic precipitation may modify the quantity and species composition of the phyllosphere of both broad-leaved trees and conifers and might allow tolerant species or pathogens to achieve greater dominance. Barklund and Unestam (1988) reported that reduction in phyllosphere microflora by acid mist treatments favored the development of pathogens such as *Gremmeniella abietina* in conifer shoots. Acidic deposition is also known to promote the leaching of nutrients from foliage as well as to cause tissue damage, particularly at pH 2.0–3.0. An increase in concentrations of cations such as K and Mg relative to sulfate in the rain, and an enhancement of Na, Ca, and nitrate in pine needle leachates have been observed after acidic deposition episodes (Waldman and Hoffman, 1988).

To summarize, conflicting results have been obtained in investigations on the effect of SO_2, O_3, and acid precipitation on microorganisms on different plant species. With flag leaves and ears of barley, yeasts were particularly sensitive to dry deposition of SO_2 (concentrations of up to 0.05 ppm), while *Aureobasidium pullulans* populations were found to increase. Filamentous fungi were significantly reduced on leaf surfaces of orange trees exposed to SO_2 for long periods. The direct effects of elevated O_3 concentrations on the phyllosphere surface may be of more significance to microbial activity than its direct effects on component species. During O_3 generation, N_2O_5 scrubbers need to be used to avoid misinterpretation of results. Acid precipitation treatments (pH 3–4) suggest that *A. pullulans* populations may be significantly reduced even though *A. pullulans* is very tolerant of low acidity. The toxicity of acid precipitation may be due to sulfurous acid and its dissociation products. Undissociated sulfurous acid (H_2SO_3) has been shown to be highly toxic to yeast cells and other microorganisms (Saunders, 1973; King et al., 1981; Dowding, 1986). The proportion of sulfurous acid in a solution increases with decreasing pH. Therefore, both the pH and the form of the acid may be critical in determining the effect of a pollutant on phyllosphere microorganisms.

19.6 Phyllosphere Saprophytes, Pathogens, and Atmospheric Pollutants

Phyllosphere saprophytes can have an antagonistic effect on foliar pathogens and thus function as natural biological control organisms (see Blakeman and Fokkema, 1982). Any disruption in the balance between phyllosphere saprophytes and pathogens brought about by air pollutants could result in a marked change in pathogen activity and disease development (Heagle, 1973; Manning, 1976).

The sensitivity/tolerance of component phyllosphere fungi to pollutants has been shown to influence pathogen activity in the phyllosphere (Hibben and Taylor, 1975). The hyperparasite *Hyalodendron album*, which can inhibit germination and appresorium formation of lilac powdery mildew (*Microsphaera alni*), was found to be more sensitive than the pathogen to O_3 and SO_2. This suggests that disease incidence may be greater in the presence of air pollutants because of their effect on saprophytic antagonists. Rai and Pathak (1981) suggested an inverse relationship between the populations of phyllosphere microorganisms and the severity of early blight of potato (*Alternaria solani*) at sites polluted by NH_3, SO_2, and cement dust. They suggested that the reduction in phyllosphere populations due to NH_3 pollution enabled severe early blight symptoms to occur, resulting in significant yield losses. However, the increase in disease may also have been caused by the direct action of NH_3-causing plant stress and increasing susceptibility to the disease.

Some interesting experiments were conducted in the Netherlands with rapeseed phyllosphere fungi exposed to SO_2 in relation to infection by *Alternaria brassicae* (Bos, 1986). Rapeseed plants were either exposed to SO_2 and then inoculated with *Alternaria brassicae* or inoculated with a mixture of *Sporobolomyces roseus* and *Cryptococcus laurentii*, exposed to up to 0.125 ppm SO_2, and then inoculated with the pathogen. Plants exposed to SO_2 only showed increasing disease levels as the SO_2 concentration was increased from 0.025 ppm to 0.125 ppm. However, less infection occurred in the presence of yeasts at low SO_2 concentrations (<0.05 ppm), while at high SO_2 concentrations (>0.1 ppm), the infection was markedly increased. This work suggested that an interaction between SO_2 concentration, phyllosphere microbial population, and pathogen infection may occur. Low SO_2 thresholds may enable yeast populations to prevent infection of *A. brassicae* while high concentrations may inhibit the saprophytes and directly stress plants sufficiently to enable a rapid increase in infection by the pathogen.

The exposure of cereal flag leaves and ears to up to 0.05 ppm SO_2 in an open-air fumigation system (see Section 19.5) resulted in a decrease in yeast populations while that of *Aureobasidium pullulans* was significantly increased (Magan and McLeod, 1991). This was accompanied by a decrease in *Rhynchosporium secalis* on barley and an increase in infection of wheat leaves by *Erysiphe graminis*, particularly in the high SO_2 plot (0.05 ppm; McLeod, 1988; McLeod et al., 1988; Mansfield et al., 1990). While no correlations were made between the effects on saprophytic populations and on foliar diseases of cereals, the results suggested that some complex interactions may be occurring. The sensitivity or tolerance of such fungi and that of pathogens to dry and wet deposition of pollutants like SO_2 could therefore profoundly influence both the level of infection, disease expression and ultimately yields.

19.7 Conclusions and Future Directions

Studies in the 1980s showed that components of phyllosphere communities are affected by the dominant primary and secondary pollutants from anthropogenic sources. Some microorganisms, particularly yeasts, are sensitive to pollutants such as SO_2; others, including filamentous fungi, may be tolerant of them. The effect of O_3 may be mediated predominantly via a direct effect on the plant which results in a change in nutrient availability on the surface. Acid mist experiments suggest that colonization of the phyllosphere by *A. pullulans*, known to be antagonistic to pathogens, may be significantly decreased. However, very little attention has been given to the interaction between these air pollutants and phyllosphere bacteria. It is important that such data be collected, particularly as bacteria are themselves often considered as candidates for biological control of foliar diseases caused by both bacterial and fungal pathogens.

The effect of pollutants on phyllosphere microorganisms of cultivated crops and trees has been studied, particularly in open-top and open-air fumigation systems. Some fumigation systems involved the exposure of seedlings or young plants, which are not always directly comparable to mature crops or trees; for example, different microbial populations may be present on young compared to mature plants. Most of the information gathered on the effect of atmospheric pollutants from natural emission sites has excluded the range of concentrations impacting on plants and any details of the micrometerology of these areas which would significantly affect microbial community structure. It is surprising that while acidic deposition has been monitored at high altitudes, no attempt has been made to study short-term and long-term effects of such depositions on bacterial, yeast, and filamentous fungal populations on plant surfaces.

The possible interactions between environmental factors, such as pH and water availability, and pollutants such as SO_2, O_3, and acidic precipitation have also not been previously considered. In vitro and whole plant studies are necessary to determine the effect that such factors have on interactions between saprophytic and pathogenic microorganisms. This may be crucial, particularly where biological control agents are being used. Changes in the species composition and abundance of phyllosphere microorganisms due to the impact of atmospheric pollutants could modify and alter the effectiveness of such agents against pathogens.

There is now increasing interest in climate changes, including increases in temperature, changes in patterns of rainfall, and elevated levels of carbon dioxide. The interaction of these factors could have a profound effect on the development of phyllosphere microbial communities on agricultural crops and trees because stomatal formation and functioning as well as patterns of nutrient leaching may be changed (Woodword, 1987). Both short- and long-term studies are necessary in the next decade to understand the mechanisms by which altered plant growth may affect the colonization of the phyllosphere by microorganisms.

References

Babich, H. and Stotzky, G. 1974. Air pollution and microbial ecology. *CRC Critical Reviews in Environmental Control* 4:353–421.

Babich, H. and Stotzky, G. 1978. Influence of pH on inhibition of bacteria, fungi and coliphages by bisulphite and sulphite. *Environmental Research* 15:405–417.

Babich, H. and Stotzky, G. 1982. Gaseous and heavy metal pollutants. pp. 631–670 in Burns, R.G. and Slater, J.H. (editors), *Experimental Microbial Ecology*. Blackwell Scientific Publications, Oxford, U.K.

Barklund, P. and Unestam, T. 1988. Infection experiments with *Gremmeniela abietina* on seedlings of Norway spruce and Scots pine. *European Journal of Forest Pathology* 18:409–420.

Blakeman, J.P. and Fokkema, N.J. 1982. Potential for biological control of plant diseases on the phylloplane. *Annual Review of Phytopathology* 20:167–192.

Blank, L.W., Roberts, T.M., and Skeffington, R.A. 1989. New perspectives on forest decline. *Nature* 336:27–30.

Bos, M. 1986. Effect of SO_2 on the antagonism between saprotrophic phyllosphere organisms and the pathogen *Alternaria brassicae* on agar plates and summer rape. M.Sc. thesis, Institute for Plant Pathology Research, Wageningen, Holland.

Buijsman, E., Maas, J.F.M., and Asman, W.A.H. 1985. Ammonia emission in Europe. Report R-85-I, Institute of Meteorology and Oceanography, State University of Utrecht, Holland.

Cape, J.N. and Fowler, D. 1981. Changes in epicuticular wax of Pinus sylvestris exposed to polluted air. *Silva Fennica* 15:457–458.

Charlson, R.J. and Rodhe, H. 1982. Factors controlling the acidity of natural rainwater. *Nature* 295:683–685.

Cook, R.J. 1980. Effects of late season fungicide sprays on yield of winter wheat. *Plant Pathology* 29:21–27.

Couey, H.M. 1965. Inhibition of germination of *Alternaria* spores by SO_2 under various moisture conditions. *Phytopathology* 55:525–527.

Couey, H.M. and Uota, M. 1961. Effect of concentration, exposure time, temperature and relative humidity on the toxicity of sulphur dioxide to spores of *Botrytis cinerea*. *Phytopathology* 51:815–819.

Derwent, R.G. and Nodop, K. 1986. Long-range transport and deposition of acidic nitrogen species in north-west Europe. *Nature* 324:356–358.

Dickinson, C.H. 1986. Adaptations of microorganisms to climatic conditions affecting aerial plant surfaces. pp. 77–100 in Fokkema, N.J. and van den Heuvel, J., (editors), *Microbiology of the Phyllosphere*. Cambridge University Press, Cambridge.

Dowding, P. 1986. Leaf yeasts as indicators of air pollution. pp. 121–136 in Fokkema, N.J. and van den Heuvel, J. (editors), *Microbiology of the Phyllosphere*. Cambridge University Press, Cambridge.

Dowding, P. and Carvill, P.H. 1980. A reduction of counts of *Sporobolomyces roseus* Kluyve on Ash leaves in Dublin city. *Irish Journal of Environmental Science* 1:65–69.

Fenn, M.E., Dunn, P.H. and Durall, D.M. 1989. Effects of ozone and sulphur dioxide on phyllosphere fungi from three tree species. *Applied and Environmental Microbiology* 55:412–418.

Fokkema, N.J. 1978. Fungal antagonism in the phyllosphere. *Annals of Applied Biology* 89:115–119.

Garsed, S.G. 1985. Sulphur dioxide uptake and transport. pp. 75–95 in Winner, W.E., Mooney, H.A., and Goldstein, R.A. (editors), *Sulphur Dioxide and Vegetation*. Stanford University Press, Palo Alto, CA.

Heagle, A.S. 1973. Interactions between air pollutants and plant parasites. *Annual Review of Phytopathology* 11:365–388.

Heagle, A.S., Body, D.E., and Heck, W.W. 1973. An open top field chamber to assess the impact of air pollution on plants. *Journal of Environmental Quality* 2:365–368.

Helander, M.L. and Rantio-Lehtimaki, A. 1990. Effects of watering and simulated acid rain on quantity of phyllosphere fungi on birch leaves. *Microbial Ecology* 19:119–125.

Hibben, C.R. and Stotzky, G. 1969. Effects of ozone on the germination of fungus spores. *Canadian Journal of Microbiology* 15:1187–1196.

Hibben, C.R. and Taylor, M.P. 1975. Ozone and sulphur dioxide effects on the lilac powdery mildew fungus. *Environmental Pollution* 9:107–114.

Huttunen, S. and Laine, K. 1983. Effects of air borne pollutants on the surface wax structure of *Pinus sylvestris* needles. *Annales Botanica Fennici* 20:79–86.

Khanna, K.K. 1986. Phyllosphere microflora of certain plants in relation to air pollution. *Environmental Pollution* 42:191–200.

King, A.D., Jr., Ponting, J.D., Sanshuck, D.W., Jackson, R., and Mihara, K. 1981. Factors affecting death of yeast by sulphur dioxide. *Journal of Food Protection* 44:92–97.

Kirkwood, I.A., Magan, N., and McLeod, A.R. 1989. Effect of atmospheric sulphur dioxide on the phylloplane mycoflora of conifer needles. pp. 457–459 in Bucher, J.B. and Bucher-Wallin, I. (editors), *Air Pollution and Forest Decline*, vol. II. Swiss Federal Institute of Forestry Research, Birmensdorf, Switzerland.

Lettle, A. 1983. Effect of industrial SO_2 emissions on the epiphytic microflora of spruce. *Folia Microbiologia* 28:187–194.

Ling, K.A. and Ashmore, M.R. 1987. Acid rain and trees—an appraisal of the evidence for damage to native tree species by air pollution and acid precipitation in the United Kingdom. *Focus on Nature Conservancy*, vol. 19. Nature Conservancy Council, Peterborough, England.

Magan, N. and Lacey, J. 1984. The effect of temperature and pH on the water relations of field and storage fungi. *Transactions of the British Mycological Society* 82:71–81.

Magan, N. and Lacey, J. 1986. The phylloplane microbial populations of wheat and effect of late fungicide applications. *Annals of Applied Biology* 109:117–128.

Magan, N. and McLeod, A.R. 1988. In vitro growth and germination of phylloplane fungi in atmospheric sulphur dioxide. *Transactions of the British Mycological Society* 90:571–575.

Magan, N. and McLeod, A.R. 1991. Effect of open-air fumigation with sulphur dioxide on the occurrence of phylloplane fungi on winter barley. *Agriculture, Ecosystems and Environment*. 33:245–261.

Manning, W.J. 1976. The influence of ozone on plant surface microfloras. pp. 159–172 in Dickinson, C.H. and Preece, T.F. (editors), *Microbiology of Aerial Plant Surfaces*. Academic Press, London.

Manning, W.J. and Keane, K.D. 1988. Effects of air pollutants on interactions between plants, insects and pathogens. pp. 365–386 in Heck, W.W., Taylor, O.C. and Tingey, D.T. (editors), *Assessment of Crop Loss from Air Pollutants*.

Elsevier Applied Science, London and New York.

Mansfield, P.J., Bell, N.J., McLeod, A.R., and Wheeler, B.E.J. 1991. Effects of sulphur dioxide on the development of fungal diseases of winter barley in an open-air fumigation system. *Agriculture, Ecosystems and Environment* 33:215–232.

McLeod, A.R. 1988. Effects of open-air fumigation with sulphur dioxide on the occurrence of fungal pathogens in winter cereals. *Phytopathology* 78:88–94.

McLeod, A.R., Fackrell, J.E., and Alexander, K. 1985. Open-air fumigation of field crops: Criteria and design for a new experimental system. *Atmospheric Environment* 19:1639–1649.

McLeod, A.R., Roberts, T.M., Alexander, K., and Cribb, D.M. 1988. Effects of open-air fumigation with sulphur dioxide on the growth and yield of winter barley. *New Phytologist* 109:67–78.

McLeod, A.R., Brown, K.A., Skeffington, R.A., and Roberts, T.M. 1989. Studies of ozone effects on conifers and the formation of oxides of nitrogen during ozone generation. pp. 181–187 in Bucher, J.B. and Bucher-Wallin, I. (editors), *Air Pollution and Forest Decline*. Swiss Federal Institute of Forestry Research, Birmensdorf, Switzerland.

Miller, C.S. 1981. Infection processes on conifer needles. pp. 185–209 in Blakeman, J.P. (editor), *Microbial Ecology of the Phylloplane*. Academic Press, London.

Moller, D. 1984. On the global nature of sulphur emission. *Atmospheric Pollution* 18:29–39.

Rai, B. and Pathak, K.K. 1981. Studies on the phylloplane microflora of potato in relation to air pollutants. *Environmental Pollution* 26:153–166.

Ranta, H.M. 1991. Effect of simulated acid rain on quantity of epiphytic microfungi on Scots pine (*Pinus sylvestris*) needles. *Environmental Pollution*. 67:349–360.

Rinallo, C., Raddi, P., Gellinin, R., and Di Lornardo, V. 1986. Effects of simulated acid deposition on the surface structure of Norway spruce and silver fir needles. *European Journal of Forest Pathology* 16:440–446.

Rist, D.L. and Lorbeer, J.W. 1981. Interactions of ozone, foliar fungi, and the surface and tissues of the leaf. pp. 305–323 in Blakeman, J.P. (editor), *Microbial Ecology of the Phylloplane*. Academic Press, London.

Saunders, P.J.W. 1971. Modifications of the leaf surface and its environment by pollution. pp. 81–89 in Preece, T.F. and Dickinson, C.H. (editors), *Ecology of Leaf Surface Microorganisms*. Academic Press, London.

Saunders, P.J.W. 1973. Effects of atmospheric pollution on leaf surface microflora. *Pesticide Science* 4:589–595.

Shaw, P.J.A. 1986. The Liphook Forest fumigation experiment: Description and project plans. *Central Electricity Research Laboratories Report No.* TPRD/L/2985/R86. Leatherhead, U.K.

Singh, A.K. and Rai, B. 1990a. Effect of SO_2 and NH_3 on growth behaviour of some phylloplane fungi from wheat in vitro. *Water, Soil and Air Pollution* 49:343–347.

Singh, A.K. and Rai, B. 1990b. Effect of cement dust treatment on some phylloplane fungi from wheat. *Water, Soil and Air Pollution* 49:349–354.

Smedgaard-Petersen, V. and Tolstrup, K. 1986. Yield reducing effect of saprophytic leaf fungi in barley crops. pp. 160–171 in Fokkema, N.J. and van den Heuvel, J. (editors), *Microbiology of the Phyllosphere*. Cambridge University Press, Cambridge.

Swisher, R. and Carroll, G.C. 1980. Fluorescein diacetate hydrolysis as an estimator of microbial biomass on coniferous needle surfaces. *Microbial Ecology* 6:217–226.

Treshow, M. 1975. Interactions of air pollutants and plant disease. pp. 307–334 in Mudd, J.B. and Kozlowski, T.T. (editors), *Responses of Plants to Air Pollution*. Academic Press, London.

Unsworth, M.H. 1984. Evaporation from forests in cloud enhances the effects of acid deposition. *Nature* 312:262–264.

Vas, K. and Ingram, M. 1949. Preservation of fruit juice with less sulphur dioxide. *Food Manufacturing* 24:414–416.

Waldman, J.M. and Hoffman, M.R. 1988. Nutrient leaching from pine needles impacted by acidic cloudwater. *Water, Air and Soil Pollution* 37:193–201.

Wainwright, M. 1988. Effects of point source atmospheric pollution on fungal communities. *Proceedings of the Royal Society of Edinburgh* 94B:97–104.

Woodword, F.L. 1987. Stomatal numbers are sensitive to increases in CO_2 from pre-industrial levels. *Nature* 327:617–618.

20

Genetically Engineered Endophytes as Biocontrol Agents: A Case Study from Industry

Jed W. Fahey, Michael B. Dimock, Steven F. Tomasino, Jean M. Taylor, and Peter S. Carlson

20.1 Introduction

Economic, social, and political forces determine which new technologies are utilized. Commercial incentives and environmental standards are the driving forces underlying the development and application of biotechnology. Agricultural biotechnology's customer, the farmer, is a price- and risk-sensitive consumer. Inexpensive, proven technologies that demand little or no change in agronomic practices are, under ordinary conditions, preferred by farmers. Growers are reluctant to increase the portion of their costs devoted to planting materials at the beginning of the growing season. Presently, biotechnology is neither an inexpensive nor a proven method for increased crop production, and its application may alter agronomic techniques. Until proven otherwise, the case can be made that the added value created by many biotechnological manipulations may not be adequately reimbursed by the market.

The agricultural technologies currently used in the developed world produce sufficient foodstuffs. The use of biotechnology to increase food output may not be as readily accepted as the use of biotechnology to redress the negative environmental and ecological impacts of current agricultural practices.

20.2 Endophytes for Crop Protection

Although numerous plant-endophyte interactions have been identified over the past century, only a few species of endophytic microorganisms

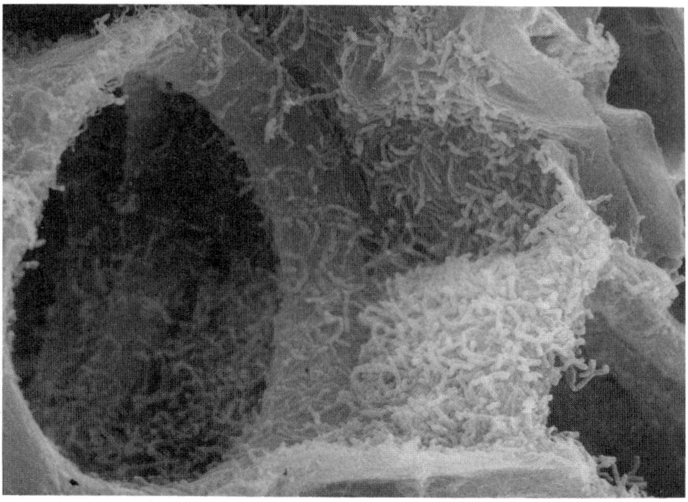

Figure 20.1 Lumen of a xylem vessel from the stem of 4-week-old *Zea mays* var. FR632 colonized with endophytic bacterium *Clavibacter xyli* subsp. *cynodontis*. The freeze-fracture surface reveals annular rings on the left side of the photograph. (Approximate magnification, ×3,500.)

(e.g., *Rhizobium*) have been used commercially for their agricultural potential as symbionts. Genetic engineering of nonpathogenic endophytic bacteria presents an opportunity for the systemic delivery of biopesticides (fungicides and insecticides) within host plant tissues without direct genetic manipulation of the host plant. The advantages of such a strategy include sustained and protected activity of the inhibitory compound.

Such products could be delivered as single-application seed treatments. This approach exploits the biological characteristics of a natural endophytic microbe, which can systemically colonize the xylem (Figure 20.1), and persist in host plants. Because of the containment of the biocontrol agent within the plant's vascular system and its inability to survive when the host dies, effects on nontarget organisms and the environment should be reduced. Thus, the use of endophytes should have significant economic, environmental, and technological advantages over the use of current externally applied agrichemicals. Xylem-inhabiting bacteria are not ubiquitous, but certain species can be recovered from plants in a pattern which suggests that they may be part of the host's normal microflora. For example, bacteria from 13 different genera have been isolated from the xylem of healthy *Citrus* trees at levels of up to 2×10^4 CFU/g of xylem (Gardner et al., 1982; R.M. Zablotowicz, personal communication). In addition, several different bacterial isolates, including species of *Erwinia* and *Bacillus*, have been recovered from symptom-free cotton plants, and they were shown to colonize systemically following inoculation (Misaghi and Donndelinger, 1990). Furthermore, xylem-inhabiting endophytes may be responsible for the fre quent presence of microflora in surface-disinfested explant material. The failure of excised tissues of certain plant species to grow in vitro could in fact be related to a stimulatory effect of an endophytic microbe that cannot be reproduced by culturing isolated, bacteria-free explants.

Methods have been developed by the authors and others at Crop Genetics International (CGI), Hanover, MD, to screen, identify, recover, and characterize endophytic microorganisms. We have conducted an extensive search for endophytes capable of colonizing the major crop species and have developed a collection of putative endophytes. Microorganisms are selected for this collection based on their ability to live inside and inability to survive outside the target crops.

Quantitative isolations from asymptomatic target plant species following rigorous surface disinfestation procedures were used to initially isolate and further characterize bacteria from systemic populations. One such microbe is *Clavibacter xyli* subsp. *cynodontis* (Cxc), a fastidious, Gram-positive, coryneform bacterium. CGI has used Cxc in the first of a planned series of InCide™ biopesticide products. To date, Cxc has only been found to occur naturally in the xylem of bermudagrass (*Cynodon dactylon*) stems, leaves, and roots; it is usually the exclusive occupant of this niche. Species of bacteria that are presently classified as *Clavibacter* were previously assigned to the genus *Corynebacterium* (Davis et al., 1984).

Creating novel crop varieties through biotechnology is generally not a business with an attractive return on investment. No matter how carefully crafted the patent protection may be, the farmer's field is, in effect, a genetic photocopier. The one-time sale of a biotechnology-based new variety cannot support a price structure adequate to cover research costs. A "repeat-sales" mechanism such as hybridization is needed for commercialization. Protecting crops from within via genetically engineered endophytes represents a unique market niche, and InCide biopesticides based on Cxc create their own "repeat-sales" opportunity because they are not seed transmitted.

The first InCide product involves a genetically engineered Cxc capable of producing the delta endotoxin of *Bacillus thuringiensis* subsp. *kurstaki* (Cxc/Bt), a protein that is toxic to the larvae of many species of Lepidoptera (caterpillars). This product can be inoculated into corn to control the European corn borer (ECB) *Ostrinia nubilalis*. There is little published information on Cxc. Thus, CGI has generated extensive data (Experimental Use Permit applications submitted to the U.S. EPA in 1987–1990; see EUP58788-EUP1 1987, -EUP2, 1989, -EUP4, 1990) on the biology, ecology, pathology, and environmental fate of this organism in order to aid in product development and registration.

Cxc has proven to be widespread within the geographic range of its natural host, bermudagrass, based on extensive sampling of bermudagrass populations over the past five years. Cxc has been isolated from bermudagrass in Taiwan, France, Japan, and 26 states within the U.S., including the leading corn-producing states such as Nebraska, Illinois, and Indiana (Chen et al., 1977; EUP58788-EUP1).

Once introduced into corn seedlings by wound inoculation of the stems of young plants, or by seed inoculation, Cxc colonizes the xylem of roots, stems, leaves, and husks. Cxc populations of over 10^9 bacteria per gram of fresh tissue can be achieved within 2–4 weeks of planting inoculated seed. Cxc has been inoculated into over 100 commercial hybrids, all of which were colonized, generally in the range of 10^5–10^9 CFU/g fresh tissue. Since it is a vascular-limited endophyte, Cxc is not present in the seed of colonized plants. Host-range studies have shown that Cxc is primarily capable of colonizing grass species. Recombinant strains of Cxc that are potential InCide candidates show host-colonization patterns very similar to the wild-type organism.

Extensive laboratory, greenhouse, and field studies have been conducted since 1986 to determine the ability of Cxc and/or Cxc/Bt to persist in the environment and disperse beyond the point of introduction (EUP58788-EUP1, 1987; -EUP2, 1989). Survival time of Cxc and a Cxc/Bt strain on the phylloplane of four plant species (corn, radish, marigold, and soybean) was shown to be a maximum of 2 weeks under greenhouse and growth chamber conditions. This characteristic can be attributed to the inability of these microbes to replicate outside a living host plant and the absence of a resting spore stage. Cxc is also short-lived in soil, water, and debris. For example, in 1987 field trials in Maryland, Cxc was undetectable after 2 weeks in soil, 3 weeks in soil-incorporated green residue of colonized corn plants, 5 weeks in buried sections of cornstalks and 7 weeks in cornstalks standing in the field after harvest.

Field studies conducted with a prototype recombinant Cxc (Cxc/Bt) in 1988 again demonstrated poor persistence in soil (1–5 weeks), incorporated plant material (6–8 weeks), and in cornstalks remaining after harvest (6–10 weeks). Cxc/Bt was not detected at all in soil around inoculated plants, even after incorporation of the plant material into the soil.

Results on persistence in 1989 in corn debris were similar to the 1988 data described above. Decline of Cxc and Cxc/Bt over time after harvest was rapid and complete. In 1987 and 1988 field experiments, the endophyte was never detected in runoff water from plots of colonized corn, before or after colonized plant debris was chopped and incorporated. This provides additional evidence that a colonized corn crop is not likely to provide a source of soil inoculum for subsequent crops.

Field trials in 1988 demonstrated that Cxc/Bt was not naturally dispersed from corn to other corn plants or weeds. In an attempt to induce artificial transmission, colonized corn plants were repeatedly cut with shears until the shears were wet with sap, and the shears were then used to trim noncolonized weeds; only in this case was any mechanical transmission observed, and even then the frequency of transmission was quite low. Cxc/Bt did not colonize any of the trap plants (corn and bermudagrass) planted around the perimeter of the test sites used to monitor dispersal of the microbe. Plant-to-plant dispersal of Cxc from inoculum foci was rarely observed, and when it occurred, transmission was by mechanical means (i.e., cultivation designed to exaggerate normal farming practices).

There was no spread of recombinant Cxc/Bt to trap plants in the 1989 or 1990 field studies. These plots were replanted to bermudagrass for monitoring purposes in 1990 and no overwintering of the endophyte was observed.

20.3 Genetic Engineering of Cxc

Bacillus thuringiensis (Bt) is a Gram-positive bacterium that has been used commercially as a biological insecticide for more than three decades. Over 1,000 Bt strains and isolates have been identified, most of them active against larval Lepidoptera but exhibiting different ranges of activity within this group. This extreme specificity is responsible for the attractiveness of Bt as an environmentally safe insecticide which has little or no toxicity toward vertebrates or beneficial insects.

During sporulation, Bt cells produce a crystalline inclusion that is released with the spores upon cell lysis. Most commercial formulations of Bt, such as Dipel™, consist of a mixture of spores and crystals. Upon ingestion by a susceptible caterpillar, the proteinaceous crystal is degraded in the alkaline, reducing environment of the midgut, releasing a protoxin that can be digested by the insect's proteolytic enzymes (Fast, 1981). This process releases one or more delta endotoxins, the number and specificity of which depend on the Bt strain and the insecticidal crystal protein genes expressed (Hofte and Whiteley, 1989). An important step in toxicity appears to be the binding of toxin molecules to receptor sites on the brush border membrane of the midgut epithelium (Hoffman, 1988). Subsequent steps in the process are still under debate, but the end result is loss of integrity of the epithelial cells and consequent disintegration of the gut wall. The insect stops feeding and either dies immediately due to catastrophic increase in pH of the hemolymph (due to leakage of gut contents), starves to death, or succumbs to septicemia when opportunistic microflora (including Bt) invade the hemocoel from the gut (Fast, 1981).

Recombinant DNA techniques were used to modify wild-type Cxc to produce delta-endotoxin proteins of *Bacillus thuringiensis* subsp. *kurstaki* strain HD-73. The Cxc/Bt recombinant strains contain either: (1) the intact cryIA(c) protoxin gene of HD73 (coding for the 130-kDa protoxin that is broken down via proteolysis in the alkaline gut of the corn borer to form the activated toxin); (2) gene fusions combining the toxic domain of HD-73 with various marker genes; or (3) the toxic domain itself (coding for the active endotoxin). Molecular geneticists at CGI have constructed artificial plasmids that include: (1) the toxin coding region; (2) regulatory sequences that control transcription of the genetic code to messenger RNA (promoters); and (3) marker genes that confer selectable traits (such as resistance to antibiotics) for detection of transformants. The common enteric bacterium *Escherichia coli* is used as a host for transformation with the cloning vectors for the initial construction of these expression cassettes. Successful cassettes are then cloned into proprietary integration vectors which contain a segment of DNA homologous to a segment of the chromosomal DNA of Cxc.

When the integration vector is inserted into the Cxc cell, crossing-over occurs between the homologous regions of the vector and the host chromosome, resulting in the stable insertion of the engineered DNA sequence into the Cxc chromosome. The resulting Cxc/Bt recombinant produces HD-73 toxin proteins that can be identified by Western blotting (Burnette, 1981) against crystal protein purified from sporulated HD-73 cultures.

Because the wild-type parental strain of Cxc/Bt has no detectable plasmids or prophage (which could mobilize recombinant genes in nature), and because Cxc isolates have proven unable to transmit or exchange integrated DNA sequences with other bacteria (EUP58788-EUP1, 1987), there is minimal risk of genetic exchange of recombinant toxin genes between Cxc/Bt and other microorganisms. Cxc/Bt recombinant strains have also been shown to segregate spontaneously, losing the engineered gene sequences at a low frequency. Segregants (which can be distinguished from recombinants by use of selective media) are able to out-compete recombinants, so that the recombinant genes are eventually lost from a Cxc population in the host plant (EUP58788-EUP2, 1989; -EUP4, 1990). This reversion phenomenon occurs at a rate slow enough to ensure product performance (i.e., sufficient Cxc/Bt populations) within a growing season, but rapidly enough so that toxin genes could not persist in the environment in the unlikely event that Cxc/Bt were transmitted by mechanical means to a non-crop host plant such as bermudagrass.

20.4 Effects of Cxc/Bt on European Corn Borer

Unlike Bt, Cxc/Bt does not sporulate and release crystal toxins, which can then be ingested by caterpillars. Cxc/Bt does not secrete its toxin, so that the entire bacterial cell must be digested by the insect in order to release its active ingredient. However, once this occurs, the symptoms are similar to those observed in larvae ingesting Bt: feeding slows and eventually stops, and larvae die from starvation or from invasion of the hemocoel by opportunistic microorganisms (Fast, 1981).

Experiments have demonstrated that Cxc/Bt can prevent or reduce European corn borer (ECB) damage to inoculated field corn under conditions of artificial infestation in both greenhouse and field. In greenhouse experiments (described in detail in EUP58788-EUP3, 1989), field corn seedlings were inoculated with Cxc approximately 2 weeks after planting (V3 to

V4 stage) by injection 1 to 2 cm above the soil line with 10^7 to 10^8 CFU in phosphate buffered saline. Plants that were confirmed to be Cxc-colonized by phase contrast microscopy of leaf midrib sap were infested 5 or 6 weeks after inoculation by placing neonate ECB larvae in holes drilled into the stalks (Chiang, 1959). Plants were dissected 3 to 4 weeks later to assess tunneling damage and number of surviving insects. Inoculation with the Cxc/Bt recombinant led to significant ($P \leq 0.05$) reduction in survival and damage. When averaged across three separate trials, Cxc/Bt-colonized plants contained 60% fewer live ECB larvae and tunnels, and total tunnel length per plant (cm) was reduced by 70%, relative to controls inoculated with wild-type Cxc or sterile buffer.

A fourth trial was infested by placing neonate larvae into leaf axils at or near pollen shed to simulate more realistically a natural infestation by second brood ECB (Showers et al., 1989). In this trial, the effect of Cxc/Bt inoculation on ECB parameters was less pronounced (35% to 50% reduction relative to controls), but was nevertheless statistically significant. A field test (described more fully in EUP58788-EUP4, 1990) was conducted in a manner similar to the fourth greenhouse trial (i.e., infested in leaf axils at pollen shed), using the same three inoculation treatments in six different corn hybrids, with similar results. Inoculation with Cxc/Bt caused significant overall reduction in borer survival and tunnel damage, although the magnitude and consistency of this effect varied between hybrids.

Current efforts are aimed at development of more potent strains of Cxc/Bt (producing more of the Bt toxin), elucidation of the relationship between Cxc/Bt population and Bt expression levels in plants and activity against ECB, and selection of corn hybrids which interact most favorably with Cxc/Bt colonization to produce the desired insect suppression and yield benefit.

20.5 Seed Inoculation Technology

Since 1986, CGI has developed proprietary methods for inoculation of Cxc/Bt into corn seeds for delivery to growers via seed company licensees. Early efforts focused on existing seed delivery systems. Various methods for external seed application were examined, including seed pelleting and coating with Cxc contained in a variety of polymers, oils, and powders. Both needle-less injectors and microparticle guns were used for direct injection into the seed. None of these methods appeared commercially useful for producing colonized plants.

The use of a pressure differential to force Cxc-containing suspensions into dry seeds was successful, although this method produced only low percentages of colonized plants and resulted in a precipitous drop in germination and seedling vigor. However, if seeds are subjected to a period of imbibition in water prior to pressure inoculation with Cxc (Figure

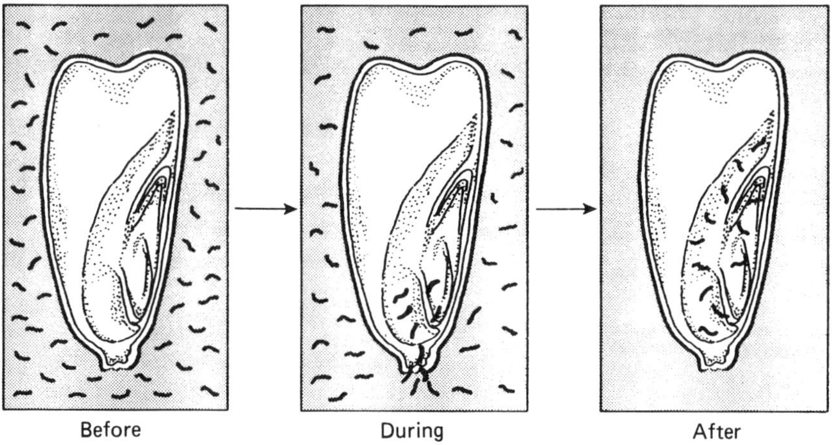

Figure 20.2 Seed inoculation with endophytic bacteria. Seeds are immersed in a buffered, high-density suspension of Cxc. A pressure differential is then applied, forcing the bacteria into the seed via the pedicel. Upon removal from the pressure chamber, bacteria remaining in the embryo lead to colonization of the resultant plant.

20.2), up to 100% of the seeds treated can produce vigorous, endophyte-colonized plants. The current inoculation protocol involves imbibition followed by application of a pressure differential in a vessel containing a buffered suspension of Cxc cells. Bacterial cells can be either harvested from petri plates or from a fermenter using standard procedures and SC or S8 media (Davis et al., 1980). Seeds are then removed from the inoculation suspension and dried on a forced air dryer after which conventional seed treatments can be applied (Figure 20.3). Field studies have demonstrated that a wide range of commercially important field and sweet corn hybrids can be successfully inoculated. Application of commonly used seed treatments (Captan™, NuGro™ Insecticide, and Color Coat Red™) appears to have no adverse impact on bacterial survival or efficacy of inoculation. Storage of inoculated seeds for longer than a year is possible, with only a gradual reduction in bacterial titer and no significant effect on seed germination. Although shelf life for crops is anticipated to vary, a shelf life of one season is expected for the first commercial corn seed product.

Development of an effective seed inoculation technology made larger-scale field trials possible in 1989. These trials were conducted in Maryland, Illinois, Minnesota, and Nebraska with plants from seed inoculated with Cxc and Cxc/Bt. Stand quality and seedling vigor were affected by the inoculation process in some environments, with some hybrids, but this inoculation effect disappeared as the season progressed. There were no apparent effects of the inoculation process at the time of harvest. Although there were yield reductions across test sites, which could be attributed to

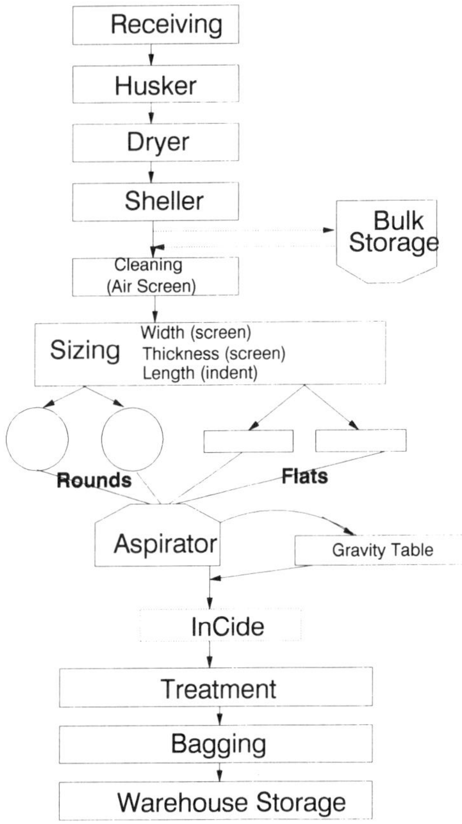

Figure 20.3 Integration of endophytic delivery to seeds into the operation of a seed conditioning plant. Bulk, harvested corn cobs are husked, dried, shelled, cleaned, and sized. InCide™ endophytes produced by classical fermentation will be added at this point in the process, residual added moisture will be removed, and seeds will reenter the conditioning train to be treated with prophylactic chemicals to prevent storage and seedling fungal and insect damage. Seeds are then bagged and stored.

the presence of the endophyte, certain hybrids performed well and will be candidate hybrids for initial product development.

Industrial-scale seed inoculation will be performed with custom-built machinery designed to be incorporated into the lines of current seed conditioning plants with only minimal need for redesign of existing facilities. CGI is currently working with four seed companies on the development and field testing of InCide technology. These companies are DeKalb Plant Genetics (DeKalb, Illinois), NC+ Hybrids (Lincoln, Nebraska), and Hoegemeyer Hybrids (Hooper, Nebraska), and Rogers Brothers Seed Co. (Boise, Idaho).

20.6 Conclusions and Future Directions

CGI expects that the following advantages of InCide biopesticide technology will make InCide products commercially successful compared to externally applied insecticides and fungicides: (1) *Environmental safety*:

Endophyte-based plant vaccines will be safe for farm workers, consumers and wildlife and will not pollute the environment. (2) *Farmer convenience*: Farmers will not need to spray insecticides or fungicides against targeted pests because endophyte-based vaccines will be applied to seeds by seed companies using CGI's proprietary technology. (3) *Minimum disturbance*: The use of InCide products will not require alteration of current agricultural practices. (4) *Consistent dosage*: Endophyte-based delivery systems will provide consistent, continuous levels of protection against the targeted pest. In contrast, externally applied pesticides can be either over-applied or under-applied. (5) *Season-long protection*: Endophyte vaccines will live within the vascular system for the life of the plant, where they are protected from abiotic factors. In contrast, externally applied products can be rendered ineffective by rain, sunlight, wind, and other environmental forces and often require multiple applications. (6) *Low-cost manufacturing*: Only a minute amount of the endophyte is required for each seed because the endophyte multiplies inside the growing plant. Externally applied pesticides require a large-scale manufacturing process.

Endophytes can be found throughout the plant kingdom. We are using the tools of biotechnology to add specific beneficial qualities to carefully chosen endophytes. By selective enhancement, endophytes can be engineered to help solve some of agriculture's most pressing problems.

The authors acknowledge the assistance of their colleagues and cooperators in performing the studies summarized in this chapter, especially: J. Anderson, M. Beach, M.A. Erickson, D. Fay, G. Hurd, G. Johnson, J. Kelly, K. Koenig, S. Kostka, J. Lampel, D. Miller, P. Reeser, A. Ricotta, J. Russin, A. Shenvi, R. Stearman, G. Sundin, J. Turner, and B. Uratani.

References

Burnette, W.N. 1981. "Western blotting": electrophoretic transfer of proteins from sodium dodecyl sulfate-polyacrylamide gels to unmodified nitrocellulose and radiographic detection with antibody and radioiodinated protein A. *Analytical Biochemistry* 112:195–203.

Chen, T.A., Su, H.J., Raju B.C. and Huang, W.C. 1977. A new spiroplasma isolated from Bermuda grass. *Proceedings of the American Phytopathological Society* 4:231 (Abstr.)

Chiang, H.C. 1959. Survival of European corn borer larvae, *Pyrausta nubilalis* (Hubn.) in artificial tunnels in stalks of field corn, *Zea mays* L. *Annals of the Entomological Society of America* 52:631–632.

Davis, M.J., Gillaspie. Jr., A.G., Vidaver, A.K. and Harris, R.W. 1984. *Clavibacter*: a new genus containing some phytopathogenic coryneform bacteria, including *Clavibacter xyli* subsp. *xyli* sp. nov. subsp. nov. and *Clavibacter xyli* subsp. *cynodontis* subsp. nov., pathogens that cause ratoon stunting disease of sugarcane and bermudagrass stunting disease. *International Journal of Systematic Bacteriology* 3:107–117.

Davis, M.J., Gillaspie, A.G., Jr., Harris, R.W. and Lawson, R.H. 1980. Ratoon stunting disease of sugarcane: Isolation of the causal bacterium. *Science* 210:1365–1367.

EUP58788-EUP1. 1987. Application for an Experimental Use Permit to Ship and Use a Pesticide for Experimental Purposes Only. Document Control Office, Office of Pesticide Programs, U.S. Environmental Protection Agency, Washington, D.C.

EUP58788-EUP2. 1989. Application for an Experimental Use Permit to Ship and Use a Pesticide for Experimental Purposes Only. Document Control Office, Office of Pesticide Programs, U.S. Environmental Protection Agency, Washington, D.C.

EUP58788-EUP4. 1990. Application for an Experimental Use Permit to Ship and Use a Pesticide for Experimental Purposes Only. Document Control Office, Office of Pesticide Programs, U.S. Environmental Protection Agency, Washington, D.C.

Fast, P.G. 1981. The crystal toxin of *Bacillus thuringiensis*. pp. 223–248 in H.D. Burges (editor), *Microbial Control of Pests and Plant Diseases, 1970–1980*. Academic Press, New York.

Gardner, J.M., Feldman, A.W. and Zablotowicz, R.M. 1982. Identity and behavior of xylem-residing bacteria in rough lemon roots of Florida citrus trees. *Applied and Environmental Microbiology* 43:1335–1342.

Hofmann, C., Vanderbruggen, H., Hofte, H., van Rie, J., Jansens, S., and Van Mellaert, H. 1988. Specificity of *Bacillus thuringiensis* delta-endotoxins is correlated with the presence of high-affinity binding sites in the brush border membrane of target insect midguts. *Proceedings of the National Academy of Sciences (USA)* 85:7844–7848.

Hofte, H. and Whiteley, H.R. 1989. Insecticidal crystal proteins of *Bacillus thuringiensis*. *Microbiological Reviews* 53:242–255.

Misaghi, I.J. and Donndelinger, C.R. 1990. Endophytic bacteria in symptom-free cotton plants. *Phytopathology* 80:808–811.

Showers, W.B., Witkowski, J.F., Mason, C.E., Calvin, D.D., Higgins, R.A., and Dively, G.P. 1989. European corn borer: development and management. *North Central Regional Extension Publication No. 327*, Iowa State University, Ames, Iowa.

21

Interactions Among Fungicides, Pathogens, Yeasts, and Nutrients in the Phyllosphere

Aleid J. Dik

21.1 Introduction

The phyllosphere is colonized by various saprophytic and pathogenic microorganisms. In temperate climate zones, the saprophytes can be divided into several groups, i.e., pink and white yeasts (mainly *Sporobolomyces* and *Cryptococcus* spp., respectively), yeastlike fungi (e.g., *Aureobasidium* spp.), filamentous fungi (e.g., *Cladosporium* spp.), and bacteria. In general, newly expanded leaves and flowers are first colonized by single-cell organisms like yeasts and bacteria, while populations of fungi such as *Cladosporium* spp., *Alternaria* spp. and *Aureobasidium* spp. increase as the leaves grow older (Blakeman, 1985).

Antagonism between saprophytes and pathogens in the phyllosphere has been reviewed by Reinecke (1981), Blakeman and Fokkema (1982), and Blakeman (1985). The antagonistic effect of saprophytes is usually based on competition for nutrients. Competition for nutrients is effective primarily against unspecialized necrotrophic pathogens, which use nutrients from the phyllosphere for germination or superficial mycelial growth before penetration of the host and not against biotrophic pathogens (Blakeman, 1985). In modern agriculture, however, the situation is more complex, because fungicides, which are used frequently to control diseases, may affect saprophytes as well as pathogens. In this chapter, the effect of fungicide application on the balance between saprophytes, pathogens, and nutrients in relation to plant damage caused by aphids and the efficacy of fungicides against necrotrophic pathogens will be discussed.

21.2 Effect of Fungicides on Nontarget Microorganisms

The effect of fungicides and other agrochemicals on leaf saprophytes has been reviewed by Dickinson (1973), Hislop (1976), and Andrews (1981); therefore, only recent studies on this subject will be discussed here.

Application of fungicides can cause two effects, both of which are alleged to involve saprophytes. First, yields may be increased in crops with low disease intensity. This has been ascribed to delayed senescence due to reduction of *Cladosporium* spp. and other saprophytes (Dickinson and Wallace, 1976; Smedegaard-Petersen and Tolstrup, 1986; Magan and Lacey, 1986). These reports are contradicted by results from Dik and Van Pelt (submitted for publication), who observed no effect of triadimenol, prochloraz, maneb, or anilazine, or of a combination of those fungicides on yields of winter wheat in crops with low disease intensities. In these experiments, the maximum population density of *Cladosporium* spp. was approximately ten thousand colony forming units per cm^2. The possible detrimental effect of *Cladosporium* spp. will only occur when this fungus is abundant. However, the observed yield increases might also be caused by a reduction of latent infections by pathogens or a reduction of endophytic microorganisms. Second, reduction of saprophytes by fungicides may decrease their antagonistic effect against pathogens. In the case of competition for nutrients, this will affect the amount of nutrients in the phyllosphere (see Section 21.3).

The fungicides that have been tested in the field for their effect on saprophytes belong to the groups of the dithiocarbamates (maneb), phthalmid derivatives (captafol and captan), phtalonitrils (chlorothalonil), benzimidazoles and thiophanates (benomyl, carbendazim), phenylamides (metalaxyl), and sterol biosynthesis inhibitors (SBI). The SBI fungicides may be derivatives of imidazole (prochloraz), morpholine (fenpropimorph), or triazole (propiconazole, triadimefon, triadimenol). The dithiocarbamates, the phthalmid derivatives, and chlorothalonil are conventional fungicides with a broad spectrum, while the others are more selective, systemic fungicides.

The effect of several fungicides commonly used in cereals on population densities of yeasts, bacteria, and filamentous fungi on flag leaves of field-grown wheat has been described by Fokkema and De Nooij (1981), Meunier and Meyer (1985), Fokkema (1988), and Magan and Lacey (1986). These studies show that the density of the saprophytic mycoflora was most strongly reduced by captafol, maneb, propiconazole, and chlorothalonil, and by mixtures containing one of these fungicides. The reduction of the saprophytic population density lasted for several weeks. The SBI-fungicides fenpropimorph, triadimefon, and prochloraz had only a small effect on the population density of the saprophytic mycoflora. Meunier and Meyer (1985) reported reduction of the bacterial population by chlorothalonil, maneb, propiconazole, fenpropimorph, and triadimefon, and by mixtures containing one of these fungicides, while captafol and benomyl had no effect. Ducommun and Corbaz (1982) reported strong (ca. 60–85%) reduction in population densities of *Cladosporium* spp., *Sporobolomyces* spp., *Tilletiopsis* spp., and *Torulopsis* spp. on field-grown tobacco leaves by maneb and a moderate reduction (ca. 10–35%) by metalaxyl. The population densities of saprophytic filamentous fungi on *Lolium perenne* were reduced by chlorothalonil and benomyl, while triadimefon had no effect (Thomas and Shattock, 1986). A variable reduction of the population density of saprophytic filamentous fungi on field-grown lettuce following one application of maneb was reported by Mercier and Reeleder (1987). The reduction of the density of the filamentous fungi increased after two or three applications of maneb. Population sizes of bacteria were not affected by maneb.

The reduction of saprophytic population densities by some fungicides can be used to create differences in saprophytic population densities, making it possible to study the effect of naturally occurring saprophytes on disease and yield. Fokkema et al. (1987) applied carbendazim to field-grown winter wheat at 7- to 14-day intervals to create differences in yeast population densities on wheat. Initially, carbendazim reduced the yeast population densities approximately three- to fourfold, but the differences observed early in the season decreased with time due to resistance in the yeasts.

The effect of several pesticides on the saprophytic microorganisms on wheat leaves was assessed in field experiments conducted in The Netherlands in three successive years (1987–1989). Yeasts, filamentous fungi, and bacteria were monitored by plating of leaf washings (Dik et al., 1991b; Dik and Van Pelt, submitted for publication). To reduce the naturally occurring saprophytic population and thus create differences in saprophytic population densities between treatments, maneb was used in 1987 and maneb alternated with the broad-spectrum fungicide anilazine (a triazine compound) was used in 1988 and 1989. When maneb was applied approximately weekly during the growing season (1987), the population density of the yeasts was initially reduced five- to 15-fold, but at the end of the growing season, after six to seven maneb applications, the difference, relative to the control treatment, decreased to less than twofold. Plating of washings of maneb-treated leaves on maneb-containing agar revealed that the percentage of yeast cells that were resistant to this fungicide increased to 40–80% (Dik and Van Pelt, in preparation). Bacteria were not consistently influenced by maneb. In 1988 and 1989, when maneb was alternated with anilazine, yeast population densities were reduced three- to 30-fold, although again, resistance (10–80%) to maneb was found. In 1988, two frequencies of spraying were tested. Yeasts were reduced to the same extent with three applications as with six applications per season, in spite of numerous days with rain. The reduction of the filamentous fungi population density by the fungicides was approximately half the reduction of the yeast population densities. The selective fungicides prochloraz and triadimenol had no effect on the saprophytes (Dik et al., 1991b; Dik and Van Pelt, submitted for publication).

Information on the influence of herbicides and insecticides on saprophytes is scarcely available. Shukla et al. (1988) reported that after application of the herbicides benthiocarb, 2,4-D, and fluchloralin to potato leaves, the densities of bacterial and fungal populations were reduced ca. 45–78% and ca. 10–55%, respectively. The effect was visible for 4 weeks after application. Mercier and Reeleder (1987) found no effect of the insecticide carbaryl on filamentous fungi or bacteria on lettuce. In the field experiments in The Netherlands, mentioned above, the insecticide pirimicarb had no effect on either yeasts, bacteria, or filamentous fungi during the initial growth of the populations, but the maximum population density of the yeasts was 2 to 4 times lower than in the control treatment. This indicates that pirimicarb has no direct effect on the saprophytes, but limits population density by the reduction of honeydew deposition.

The lack of information on the effect of pesticides on nontarget microorganisms observed by Andrews (1981) still exists. The available data indicate that especially broad-spectrum fungicides, such as maneb, captafol, and chlorothalonil, generally reduce population densities of yeasts and, to a lesser extent, filamentous fungi. The effect on epiphytic bacteria is variable. The fact that several herbicides, which are commonly used in agriculture, have a strong effect on phyllosphere saprophytes, indicates that more research on the effect of this group of pesticides on nontarget microorganisms is needed (Shukla et al., 1988). The reduction of saprophytes by agrochemicals interferes with the balance between saprophytes and pathogens and will also affect the amount of nutrients in the phyllosphere.

21.3 Removal of Nutrients from the Phyllosphere by Yeasts

Nutrients in the phyllosphere can originate either from the underlying leaf or from the surrounding environment, like pollen and aphid honeydew. Aphid honeydew consists mainly of carbohydrates (ca. 98%) and a mixture of amino acids (ca. 2%) (Rossing and Van de Wiel, 1990). Since the use of broad-spectrum fungicides will mainly affect yeasts, their role in removal of nutrients from the phyllosphere is of most interest.

Fokkema et al. (1983) applied pink and white yeasts (*Sporobolomyces* spp. and *Cryptococcus* spp., respectively) and a mixture of sucrose and yeast extract to wheat leaves in a controlled environment. The yeasts rapidly removed the nutrients from the phyllosphere. In these experiments, the environmental conditions were optimal for the yeasts and the nutrients were applied only once. In the field, however, nutrients are deposited constantly and environmental conditions change, so the situation is more complicated. To evaluate the role of yeasts and other saprophytes in consumption of nutrients under field conditions, experiments were carried out on winter wheat in several locations in The Netherlands in 1987, 1988, and 1989 (Dik et al., 1991b; Dik and Van Pelt, submitted). One treatment consisted of spraying with maneb (1987), or with maneb alternating with anilazine (1988 and 1989), to reduce yeast population densities. Additional treatments with either insecticide or a combination of maneb/anilazine with insecticide were applied. For all treatments, the amount of nutrients on the leaves was assessed weekly by washing the leaves in an ultrasonic waterbath and measuring the amount of carbohydrates in the leaf washings with anthrone reagent. Population densities of yeasts, filamentous fungi, and bacteria were monitored by plating of leaf washings.

The results of one experiment (1987) are shown in Figure 21.1. As discussed in section 21.2, the yeast population densities were reduced by application of maneb or maneb/anilazine. In periods of dry weather, carbohydrates accumulated on leaves treated with maneb. The main source of these carbohydrates is assumed to be aphid honeydew, since carbohydrate levels were low in the insecticide treatment in which deposition of honeydew is prevented. In the control treatment, the naturally occurring yeasts largely prevented this accumulation. On one sampling date (13 July), the percentage leaf area covered with honeydew was estimated visually before washing the leaves. The ratio between leaf area covered with honeydew and the amount of honeydew in the leaf washings was the same in treatments with natural and reduced yeast population densities. However, the amount of honeydew was more than twice as high in the maneb treatment as in the control. This means that the yeasts in the control treatment actually reduced the leaf area covered with honeydew and not only the thickness of the honeydew layer on the leaves. In one experiment, however, the aphid population grew very rapidly and honeydew also accumulated in the control treatment. This suggests that honeydew consumption by the yeasts was slower than honeydew production by the aphids.

The only other report on nutrient levels on field-grown plants comes from Rodger and Blakeman (1984). They measured the amount of carbohydrates and amino acids on sycamore leaves and monitored the saprophytic population density. In their experiment, carbohydrates accumulated when large numbers of aphids were present and the saprophytic population density was still below ten thousand colony forming units per cm^2. The carbohydrate level decreased when the saprophytic population density increased and the aphid population declined. Apparently, yeasts can prevent accumulation of nutrients in the phyllosphere. The exception would be a situation in which nutrient deposition is very rapid, or where reduction of the naturally occurring saprophytes leads to accumulation of these nutrients.

Rabbinge and Vereijken (1980) ascribed up to 50% of the yield losses caused by aphids in winter wheat to honeydew deposition on the leaves. Honeydew decreases rates of photosynthesis by blocking the stomata and

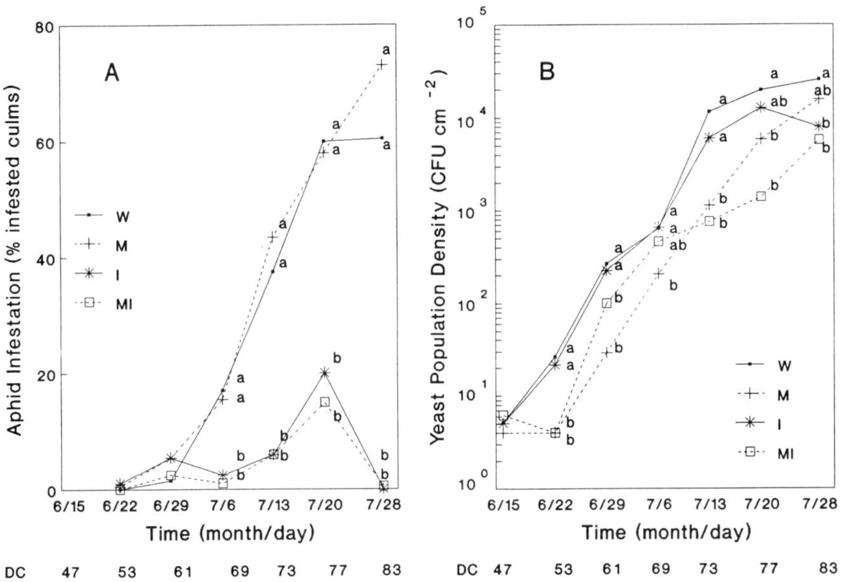

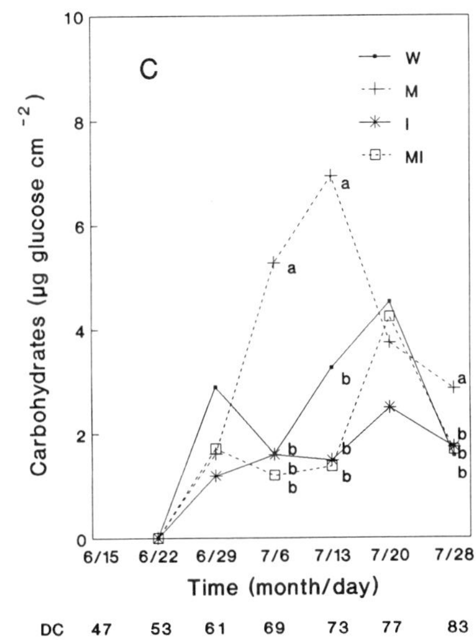

Figure 21.1 Effect of different treatments (W = control treated with water; M = maneb; I = insecticide; MI = maneb + insecticide) on: (A) aphid infestation; (B) yeast population density on the flag leaf; and (C) amount of carbohydrates on the flag leaf of winter wheat in a field experiment in The Netherlands in 1987. On sampling dates with significant differences between treatment means ($P = 0.05$), the means are followed by different letters. DC indicates the Decimal Code for the developmental stage of the wheat (Zadoks et al., 1974). From Dik et al. (1991b) with permission.

also increases leaf senescence (Rabbinge et al., 1981). Accumulation of honeydew will increase its direct detrimental effects on host physiology. Since naturally occurring yeasts reduce the leaf area covered with honeydew, this will decrease the number of stomata blocked by honeydew. For two of the field experiments described above, the yield loss caused by the aphids was calculated by comparing integrated aphid infestation and yields in the presence of the naturally occurring yeasts and with decreased yeast populations. Aphid damage is expressed in yield loss per hectare per 'aphidday,' i.e., the presence of one aphid per culm during one day. In both experiments, the yield loss was considerably higher when yeast population densities were reduced. In the experiment in Figure 21.1, aphids caused a yield loss of 1.34 kg (dry weight) per ha per 'aphidday' when yeasts were present and 1.85 kg (dry weight) per ha per aphidday in treatments with reduced yeast population densities. In most of the experiments, elimination of the aphids with insecticide throughout the growing season caused no increase in yield when the naturally occurring yeast population was preserved, with the exception of one experiment in which honeydew accummulated in the control treatment (mentioned above). These results show that naturally occurring yeasts can reduce the direct detrimental effect of honeydew on host physiology by preventing accumulation of honeydew. Leaf exudates and pollen are not likely to have the same direct effect on the host. Accumulation of nutrients following reduction of saprophytic populations, however, may have other harmful consequences, such as stimulation of pathogens and reduction of fungicide efficacy.

21.4 Effects of Nutrients on Fungicide Activity

The activity of fungicides against pathogens may be influenced by compounds present in the phyllosphere. When the toxicity of the fungicide is enhanced, the effect is called synergism. When the toxicity of the fungicide is decreased, the compound is said to be antagonistic to the fungicide (Tammes, 1964). Mixtures of agrochemicals have been tested in this respect, but in this chapter only the interaction between nutrients and fungicides is discussed.

Several authors have described decreased toxicity of fungicides on leaves compared to glass. According to Hislop (1976), this effect was ascribed mainly to redistribution and solubilization of the fungicides on the leaves. Richardson (1966) demonstrated a reversal of toxicity of thiram against *Glomerella cingulata* by exudates from seeds and seedling roots of pea and soybean. Beynon and Brown (1969) and Dunn et al. (1971) ascribed a decreased effect of three fungicides against *Alternaria brassicola* on leaves compared to glass slides to a stimulating effect of the leaf exudates on the pathogen. This effect was attributed to the presence of glucose in the exudates. Grover et al. (1969) found a synergistic effect of leaf exudates on *Cladosporium fulvum*. The effect of various carbohydrates and amino acids, present in leaf exudates and in aphid honeydew, on fungicide activity against necrotrophic pathogens like *Alternaria, Rhizoctonia, Fusarium, Aspergillus,* and *Pythium* spp. has been described. Most carbohydrates reduce the effect of fungicides, except for fructose and lactose (Chopra and Jhooty, 1974; Grover et al., 1969; Grover and Kumar, 1965; Jhooty and Bains, 1976,

Kataria and Grover, 1978). Amino acids can be either synergistic or antagonistic to fungicide activity (Chopra and Jhooty, 1974; Grover and Sidhu, 1965; Jhooty and Bains, 1976). When a mixture of amino acids is present, the net effect of the mixture depends on the balance between synergistic and antagonistic components in the mixture. Other substances like thiols and lipids are usually antagonistic to fungicide activity (Buchenauer, 1980; Mailman et al., 1971).

Practically all the knowledge on the interaction between nutrients and fungicides is based on observations on spore germination and germ tube growth in experiments under controlled conditions and may not be very relevant to the field situation. The effect on infection is not known, especially not under field conditions. However, results of Rabbinge et al. (1984) show a decreased effect of captafol against yeasts and *Cladosporium* spp. on wheat leaves in the presence of honeydew. Furthermore, Prew et al. (1983) reported increased fungicide efficacy on winter wheat in combination with insecticide. Such observations make it interesting to take a closer look at the effect of honeydew on fungicide efficacy.

Dik and Fokkema (1988) and Dik et al. (1991a) researched the effect of aphid honeydew on the activity of the broad-spectrum fungicide captafol and the more selective fungicides triadimefon, propiconazole, and prochloraz on germination and germ tube growth of *Septoria nodorum* and *Cochliobolus sativus* in vitro and in vivo. All fungicides were antagonized by the honeydew, and this effect was well simulated by a mixture of sucrose and yeast extract. The effect of honeydew on the activity of prochloraz against infection of flag leaves of wheat by *Septoria nodorum* was then studied more extensively. Aphid honeydew, sugars, and a mixture of amino acids, either alone or in combination were included in the study. In in vivo experiments, several concentrations of the nutrients were combined with increasing concentrations of fungicide.

Honeydew stimulated infection by the pathogen. This stimulation increased with an increase in the ratio between honeydew and fungicide. The effect of sucrose alone showed the same tendency, although the effect was smaller than that of sucrose with amino acids and of honeydew in comparable sugar concentrations. Amino acids only stimulated infection in combination with fungicide, not in the fungicide-free treatments.

To quantify the interaction, dose-response curves for prochloraz were calculated for each concentration of sugars, amino acids, and honeydew. With these response curves, the ED_{50}, defined as the prochloraz concentration necessary to obtain reduction of infection to 50% of the infection in the treatment without fungicide and without nutrients, was calculated. The ED_{50} in the presence of nutrients was divided by the ED_{50} in the absence of nutrients, yielding the relative ED_{50} at each nutrient concentration in each experiment, which makes it possible to compare the results of different experiments. This relative ED_{50} is plotted against nutrient concentration, yielding so-called antagonism isoboles. These isoboles were con-

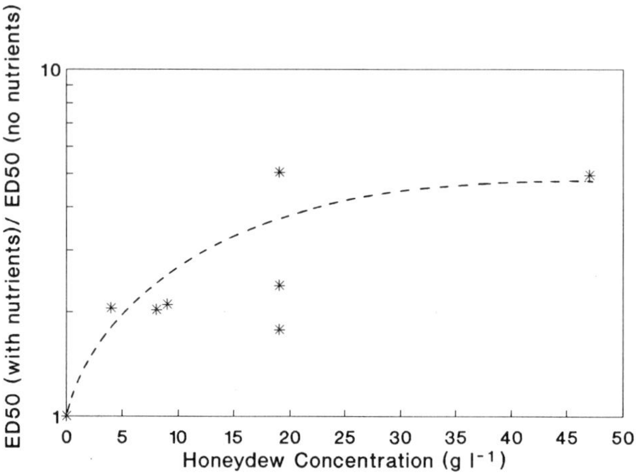

Figure 21.2 Antagonism isobole for the effect of aphid honeydew on the relative ED_{50} for prochloraz against *Septoria nodorum* infection of wheat leaves. The data are ratios of ED_{50} values in the presence and absence of nutrients. From Dik et al. (1991a) with permission.

structed for all three nutrient sources (sucrose, amino acids, honeydew) (Dik et al., 1991a). The concept of isoboles for the quantification of the interaction between chemicals has been discussed by Tammes (1964) and De Waard (1985), but it has not yet been applied to the interaction between a stimulating and a toxic component. Although many combinations of concentrations of the two substances need to be tested, it offers a better method of quantification than others, such as the Colby method (Colby, 1967; Dik et al., 1991a).

Figure 21.2 shows the antagonism isobole for aphid honeydew and prochloraz against *Septoria nodorum* infection. The data were generated in three different experiments. The relative ED_{50} of prochloraz increased up to 5-fold in the presence of honeydew; in other words, fungicide activity was reduced by honeydew. Both the carbohydrates and the amino acids contribute to the antagonistic effect of honeydew on fungicide activity, probably by stimulation of the metabolism of the pathogen. De Waard (1985) mentions several possible mechanisms for antagonism to fungicides: (1) physicochemical interactions outside the fungus; (2) stimulation of the physiology of the fungus resulting in decreased transport across the plasma membrane; (3) stimulation of the energy-dependent efflux of fungicide from the fungus; (4) increased synthesis of detoxifying enzymes; and (5) a change in the permeability of the fungal cell wall. The antagonistic effect of honeydew was observed against several fungi in combination with various fungicides, which indicates that it is most likely caused by stimulation of

Table 21.1 Effects of various treatments on the extent of infection of winter wheat by *Septoria nodorum*[a]

Treatment[b]	18 July 1988		1 August 1988	
	NLA[c]		NLA	
	Flag leaf	Leaf 2	Flag leaf	Spikelets per ear
W	81.5 a	100 a	100 a	2.3 a
MA	17.1 d	56.0 b	76.6 b	2.0 a
S	47.9 b	98.1 a	99.7 a	2.1 a
SI	32.4 c	95.2 a	100 a	1.4 b
SMA	14.7 d	47.3 bc	74.1 b	1.2 b
SMAI	19.5 d	39.2 c	58.5 c	0.7 c

[a] Data are treatment means from a field experiment in The Netherlands. Treatment means followed by the same letter are not significantly different (P = 0.05).
[b] W = water; MA = maneb/anilazine; S = selective fungicide prochloraz; I = insecticide.
[c] NLA = percentage necrotic leaf area.

the metabolism of the fungi. Presumably, the carbohydrates stimulate the metabolism of the fungus, while the amino acids may increase the amount of detoxifying enzymes.

The interaction of honeydew with fungicide efficacy was also assessed in field experiments. Different combinations of fungicides and insecticide were applied to winter wheat. The fungicides were expected to have more effect on necrotrophic pathogens in combination with insecticide than when they were applied separately, because the deposition of honeydew and thus the effect of honeydew on fungicide activity would be prevented. However, this effect will be most apparent when the yeasts are reduced, because the naturally occurring yeast populations will largely prevent the accumulation of honeydew. Table 21.1 shows the amount of necrotic leaf area as a measurement of the extent of *Septoria nodorum* infection on two sampling dates. These values should be compared for the treatments S vs. SI and SMA vs. SMAI (S = selective fungicide prochloraz, I = insecticide, and MA = maneb/anilazine). The largest difference is expected for the last comparison, since in these treatments the yeasts were reduced by maneb and anilazine. Prochloraz had no effect on yeasts. Apparently, necrosis caused by *Septoria nodorum* is indeed reduced less by the fungicides when honeydew is present. In the other experiments, the effect of maneb/anilazine on necrosis caused by *Septoria* spp. was repeatedly enhanced by the addition of insecticide (Dik and Van Pelt, submitted). Together with the results of Prew et al. (1983) and of Daling and Meems (1990), this indicates that in normal wheat cultivation, honeydew can interfere with the efficacy of fungicides.

21.5 Computer Simulation as a Method of Evaluation and Prediction of the Effect of Saprophytes as Biocontrol Agents

The multiple interactions that occur in the phyllosphere among saprophytes, pathogens, nutrients, and agrochemicals create a complex dynamic situation. In wheat, the net effect of honeydew deposition by aphids and honeydew consumption by yeasts depends on the balance between the rates of these two processes. Honeydew deposition by the aphids will be influenced by the nutritional status of the crop, by climatic conditions that influence aphid population growth, and by natural enemies of the aphids. The removal of nutrients from the leaves by yeasts is dependent on yeast population growth, which is influenced by relative humidity and temperature. The interaction between nutrients and fungicide effectiveness depends on the ratio between honeydew concentration and fungicide concentration. The complexity of the interactions makes it difficult to predict the effect of saprophytes in field situations. The same problem arises when microorganisms that are selected as good antagonists in laboratory screening programs are used in field crops. Often, the effect of the antagonists in field studies is disappointing. This is usually due to poor survival of antagonists under field conditions (Tronsmo, 1986). Much of the work that is done during the selection of the antagonists will then be useless. Modern techniques for computer simulation of dynamic processes offer a good method to evaluate and predict the behavior of antagonists under field conditions.

Dynamic simulation models have been used to predict epidemics and resulting yield losses. However, thus far, little modelling has been done on epiphytic microorganisms. Carroll (1981) developed a model for the growth, uptake of nitrogen, and loss of microbial cells in the throughfall during the course of single rainstorms in an old-growth Douglas fir forest. Hyperparasitism of *Puccinia recondita* (brown rust) by *Darluca filum* has been simulated by Kranz (1981). Knudsen and Hudler (1987) developed a dynamic simulation model to describe the population dynamics of *Pseudomonas fluorescens* in the phyllosphere of red pine and for germination of *Gremmeniella abietina* spores. The model predicted the growth of the antagonist in a controlled environment and in the field. The population growth rate was described by the logistic equation and moderated by temperature and relative humidity.

A dynamic simulation model was developed to describe the population dynamics of yeasts on wheat leaves and the resulting consumption of aphid honeydew (Dik, 1990). Basically, simulation models comprise three kinds of variables, i.e., state variables, rate variables, and independent variables. In the yeast population model, the density of the populations of pink and white yeasts, the number of aphids per culm, and the amount of honeydew on the leaves are the state variables. The rate variables are the rates at which the state variables change. They are influenced by the independent or steering variables, like temperature, relative humidity, and developmental stage of the crop.

Several assumptions were made in constructing the model. It was assumed that the yeasts are homogeneously present on the leaves and are not washed off by rain. This assumption is based on observations by Fokkema et al. (1979) and Dik et al. (1991b), who observed no decrease in the yeast population density after days with rain. All the honeydew is washed off when daily rainfall exceeds 5 mm, and 25% of the honeydew is washed off for rain amounts of 1 to 5 mm per day. The additional nutrients from pollen and leaf exudates are neglected in the model. No competition for carbohydrates is assumed between yeasts and bacteria or filamentous fungi. Immigration of yeasts from the air is neglected once the model is initiated. Obviously, this model is a simplification of the field situation.

Data on the influence of humidity and temperature on the growth rates of *Sporobolomyces roseus* and *Cryptococcus* spp. were obtained from the literature (Bashi and Fokkema, 1977) and from additional experiments in growth chambers (Dik, 1990). The influence of temperature was tested at temperatures between 12 and 24°C, with intervals of 3°C. *S. roseus* grew most rapidly at 18°C and *Cryptococcus* spp. at 21°C. Both yeasts grow only when relative humidity is higher than approximately 70%. The rate of conversion of honeydew by the yeasts was calculated from in vivo experiments. It was 3.0×10^{-4} μg glucose cell^{-1} for pink yeasts and 4.2×10^{-4} μg glucose cell^{-1} for white yeasts. The simulation model describes the population growth for pink and white yeasts separately by the Monod equation for substrate-limited growth (Monod, 1949). The half-saturation constant in this equation was estimated from data obtained in laboratory experiments. The growth rate is modified by temperature and humidity. The number of aphids in the field was assumed to increase linearly between sampling dates, and these rates were calculated for each period between two sampling dates. Data from Rabbinge and Coster (1984) and Vereijken (1979) were used to calculate the rate of honeydew production, which was dependent on the developmental stage of the crop.

Initialization of the model occurs with values for all state variables. Temperature and relative humidity are read by the model from a data file and at each time-step (1h), the population growth rates are calculated. The rate of honeydew consumption is dependent mainly on the growth rate of the yeasts. After one time-step, new values are calculated for the state variables, dependent on the calculated rates of change. This process starts again with every time-step.

The model was validated with field experiments at two locations for two years. Figure 21.3 shows the results for one of the field experiments conducted in 1988. The amount of carbohydrates on 14 June 1988 is underestimated in the simulation, perhaps because not only carbohydrates from aphid honeydew, but also carbohydrates from pollen were present at this

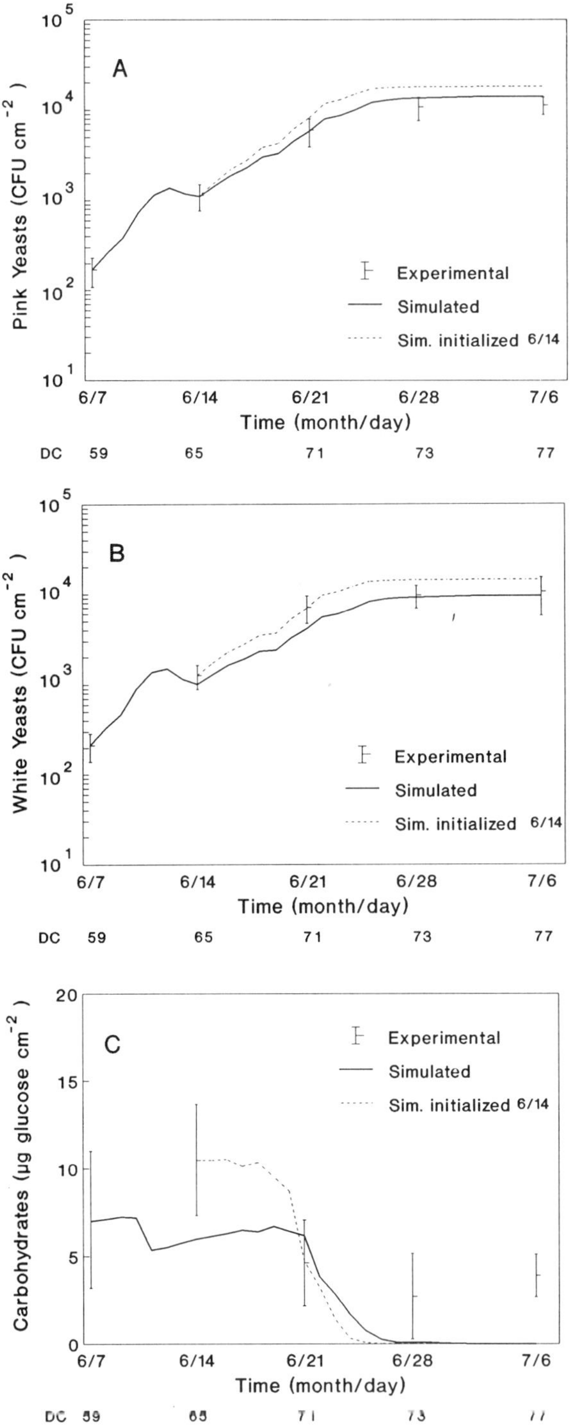

Figure 21.3 Data for simulated and experimental (A) pink yeast population density, (B) white yeast population density, and (C) amount of carbohydrates on flag leaves of winter wheat in a field experiment in The Netherlands in 1988. Experimental data are shown with the 95% confidence interval. DC indicates the Decimal Code for the developmental stage of the wheat (Zadoks et al., 1974).

developmental stage of the crop. Therefore, the simulation was also initialized with data obtained on 14 June 1988. In other field experiments, the simulation generally showed the same trend as the field observations, although not all simulation curves were within the 95% confidence interval of the field data. Generally, the yeasts grew a little too slowly at the start of the experimental period in 1987. This is probably because immigration is neglected. In 1987, when the growth rate of the yeasts was lower than in 1988, the immigration of yeasts would have had relatively more effect on total yeast densities. However, it appears that the model describes the population growth of the naturally occurring yeasts and the subsequent consumption of honeydew reasonably well.

The simulation model for the yeasts is a simple model for microbial population dynamics. Once it simulates the field situation reasonably well, the model can be used for evaluating the effect of the yeasts on nutrient levels. By introducing, for example, different growth rates for the aphid population in the model, the growth rate of the aphids at which the yeasts can still prevent accumulation of nutrients can be calculated. Thus, the explanation for field observations that with rapid increase in the numbers of aphids, honeydew accumulates in the presence of the naturally occurring yeast population, can be verified. Furthermore, the effect of application of saprophytes or nutrients to enhance antagonistic effects can be evaluated by models like this.

The use of simulation models offers a good method of evaluation of the effect of naturally occurring saprophytes in the phyllosphere. It also offers the possibility of using results from laboratory experiments to predict the survival and effect of applied antagonists (Knudsen and Hudler, 1987). Field trials can then be limited to antagonists which are likely to survive well in the field and thus the number of field experiments can be reduced. Although the collection of data for the model may require additional laboratory experiments, the reduction in field studies and the prevention of disappointing results from applied antagonists makes dynamic simulation models a useful tool in research on biological control.

21.5 Conclusions

Multiple interactions among yeasts, necrotrophic pathogens, nutrients, and fungicides have been discussed. The reduction of saprophytes by various agrochemicals decreases the removal of nutrients by saprophytes. The subsequent accumulation of nutrients, observed in wheat, increases a possible direct detrimental effect on host physiology, demonstrated for aphid honeydew. Furthermore, the accumulated nutrients reduce the effect of fungicides that are used against necrotrophic pathogens. Not only aphid honeydew, but also leaf exudates and pollen can interfere with fungicide activity. Therefore, preservation of the naturally occurring saprophytes will not only have beneficial effects by a direct antagonistic effect on pathogens, but also when fungicides are used against the pathogens. To take these multiple interactions into acount, the effect of all agrochemicals on nontarget microorganisms should be known. The use of agrochemicals that eliminate useful saprophytes should be avoided. Dynamic simulation models can be used to evaluate and predict the effect of naturally occurring saprophytes and to predict the effect of applied antagonists. Therefore, this aspect of microbial ecology of the phyllosphere certainly deserves attention in future research on biological and integrated control.

The author gratefully acknowledges that much of the work described in this paper was done with financial support by the Foundation for Biological Research, which is subsidized by the Netherlands Organization for Scientific Research. Thanks also are expressed to Dr. N.J. Fokkema and Professors R. Rabbinge and B. Schippers for critical discussions and to Hans van Pelt and Jos Raaijmakers for practical assistance.

References

Andrews, J.H. 1981. Effects of pesticides on non-target microorganisms on leaves. pp. 283–305 in Blakeman, J.P. (editor), *Microbial Ecology of the Phylloplane*. Academic Press, London.

Bashi, E. and Fokkema, N.J. 1977. Environmental factors limiting growth of *Sporobolomyces roseus*, an antagonist of *Cochliobolus sativus*, on wheat leaves. *Transactions of the British Mycological Society* 68:17–25.

Beynon, K.I. and Brown, K.F. 1969. The identification of a fungicide antagonist in leaf exudate. *Phytopathologische Zeitschrift* 64:213–220.

Blakeman, J.P. 1985. Ecological succession of leaf surface microorganisms in relation to biological control. pp. 6–30 in Windels, C.E. and Lindow, S.E. (editors), *Biological Control on the Phylloplane*. American Phytopathological Society, St. Paul, MN.

Blakeman, J.P. and Fokkema, N.J. 1982. Potential for biological control of plant diseases on the phylloplane. *Annual Review of Phytopathology* 20:167–192.

Buchenauer, H. 1980. Interaction of different lipid components with various fungicides. *Zeitschrift fur Pflanzenkrankheiten und Pflanzenschutz* 87:335–345.

Carroll, G.C. 1981. Microbial productivity on aerial plant surfaces. pp. 15–47 in Blakeman, J.P. (editor), *Microbial Ecology of the Phylloplane*. Academic Press, London.

Chopra, B.L. and Jhooty, J.S. 1974. Antagonism between zineb and water soluble leachates of spores of *Alternaria cucumerina*. *Indian Phytopathology* 27:208–214.

Colby, S.R. 1967. Calculating synergistic and antagonistic responses of herbicide combinations. *Weeds* 15:20–22.

Daling, J. and Meems, J. 1990. Resultaten gewasbeschermingsonderzoek 1988. Internal Report, Ministerie van Verkeer en Waterstaat, no. 1990–3 Liw. Rijkswaterstaat Directie Flevoland, The Netherlands.

De Waard, M.A. 1985. Fungicide synergism and antagonism. pp. 89–95 in *Fungicides for Crop Protection*. BCPC Monograph no. 31. BCPC, Croydon, U.K.

Dickinson, C.H. 1973. Interactions of fungicides and leaf saprophytes. *Pesticide Science* 4:563–574.
Dickinson, C.H. and Wallace, B. 1976. Effects of late applications of foliar fungicides on activity of microorganisms on winter wheat flag leaves. *Transactions of the British Mycological Society* 67:103–112.
Dik, A.J. 1990. *Population Dynamics of Phyllosphere Yeasts: Influence of Yeasts on Aphid Damage, Diseases and Fungicide Activity in Wheat.* Ph.D. thesis, University of Utrecht, The Netherlands.
Dik, A.J. and Fokkema, N.J. 1988. Reduced effect of fungicides on perthotrophic pathogens on wheat in the presence of nutrients. *Acta Botanica Neerlandica* 37:542–543.
Dik, A.J., Fokkema, N.J., and Van Pelt, J.A. 1991a. Interference of nutrients with fungicide activity against *Septoria nodorum* on wheat leaves. *Plant Pathology* 40:25–37.
Dik, A.J., Fokkema, N.J., and Van Pelt, J.A. 1991b. Consumption of aphid honeydew, a wheat yield reducing factor, by phyllosphere yeasts under field conditions. *Netherlands Journal of Plant Pathology*. 97:209–232.
Ducommun, P. and Corbaz, R. 1982. Influence de fongicides anti-mildiou sur la mycoflore du phylloplan tabacole. *Phytopathologische Zeitschrift* 104:19–26.
Dunn, C.L., Beynon, K.I., Brown, K.F., and Montagne, J.Th.W. 1971. The effect of glucose in leaf exudates upon the biological activity of some fungicides. pp. 491–507 in Preece, T.F. and Dickinson, C.H. (editors), *Ecology of Leaf Surface Microorganisms*. Academic Press, London.
Fokkema, N.J. 1988. Agrochemicals and the beneficial role of phyllosphere yeasts in disease control. *Ecological Bulletins* 39:91–93.
Fokkema, N.J. and De Nooij, M.P. 1981. The effect of fungicides on the microbial balance in the phyllosphere. *EPPO Bulletin* 11:303–310.
Fokkema, N.J., Den Houter, J.G., Kosterman, Y.J.C., and Nelis, A.L. 1979. Manipulation of yeasts on field-grown wheat leaves and their antagonistic effect on *Cochliobolus sativus* and *Septoria nodorum*. *Transactions of the British Mycological Society* 72:19–29.
Fokkema, N.J., Riphagen, I., Poot, R.J., and De Jong, C. 1983. Aphid honeydew, a potential stimulant of *Cochliobolus sativus* and *Septoria nodorum* and the competitive role of saprophytic mycoflora. *Transactions of the British Mycological Society* 81:355–363.
Fokkema, N.J., Dik, A.J., and Daamen, R.A. 1987. Use of carbendazim and carbendazim-resistant yeasts to create different yeast densities on wheat leaves for field studies on biological control. *Netherlands Journal of Plant Pathology* 93:273–283.
Grover, R.K. and Kumar, M. 1965. Effect of carbohydrates on the toxicity of cycloheximide against *Pythium aphanidermatum*. *Indian Journal of Microbiology* 5:25–32.
Grover, R.K. and Sidhu, J.S. 1965. Effect of nitrogen sources on the toxicity of cycloheximide against *Pythium aphanidermatum*. *Indian Journal of Microbiology* 5:9–16.
Grover, R.K., Batra, G.K., and Jhooty, J.S. 1969. Modification of toxicity of cycloheximide and dodine to *Cladosporium fulvum* by tomato leaf exudates. *Phytopathologische Zeitschrift* 64:355–361.
Hislop, E.C. 1976. Some effects of fungicides and other agrochemicals on the microbiology of the aerial surfaces of plants. pp. 41–75 in Dickinson, C.H. and Preece, T.F. (editors), *Microbiology of Aerial Plant Surfaces*. Academic Press, London.
Jhooty, J.S. and Bains, S.S. 1976. Modification of toxicity of benomyl by seed exudates of Mung. *Indian Journal of Microbiology* 16:73–75.

Kataria, H.R. and Grover, R.K. 1978. Reversal of toxicity of some systemic and non-systemic fungitoxicants by chemicals. *Zeitschrift fur Pflanzenkrankheiten und Pflanzenschutz* 85:76–83.

Knudsen, G.R. and Hudler, G.W. 1987. Use of a computer simulation model to evaluate a plant disease biocontrol agent. *Ecological Modelling* 35:45–62.

Kranz, J. 1981. Hyperparasitism of biotrophic fungi. pp. 327–352 in Blakeman, J.P. (editor), *Microbial Ecology of the Phylloplane*. Academic Press, London.

Magan, N. and Lacey, J. 1986. The phylloplane microflora of ripening wheat and effect of late fungicide applications. *Annals of Applied Biology* 109:117–128.

Mailman, R., Hodgsen, E., and Huisingh, D. 1971. Effect of thiols in reversing the inhibition of methyl-1-(butylcarbamoyl)-2-benzimidazolecarbamate on *Saccharomyces cerevisiae*. *Pesticide Biochemistry and Physiology* 1:401–408.

Mercier, J. and Reeleder, R.D. 1987. Effects of the pesticides maneb and carbaryl on the phylloplane microflora of lettuce. *Canadian Journal of Microbiology* 33:212–216.

Meunier, S. and Meyer, J.A. 1985. Modification de la flore epiphytique de feuilles de froment d'hiver suite a des traitements fongicides. *Mededelingen van de Faculteit der Landbouwwetenschappen van de Rijksuniversiteit Gent* 50:1039–1043.

Monod, J. 1949. The growth of bacterial cultures. *Annual Review of Microbiology* 3:371–394.

Prew, R.D., Church, B.M., Dewar, A.M., Lacey, J., Penny, A., Plumb, R.T., Thorne, G.N., Todd, A.D., and Williams, T.D. 1983. Effects of eight factors on the growth and nutrient uptake of winter wheat and on the incidence of pests and diseases. *Journal of Agricultural Science, Cambridge* 100:363–382.

Rabbinge, R. and Coster, G. 1984. Some effects of cereal aphids on growth and yield of winter wheat. pp. 163–169 in Barley, P. and Swinger, D. (editors), *Proceedings of the Fourth Australian Applied Entomology Research Conference*. University of Adelaide, Australia.

Rabbinge, R. and Vereijken, P.H. 1980. The effect of diseases or pests upon the host. *Zeitschrift fur Pflanzenkrankheiten und Pflanzenschutz* 87:409–422.

Rabbinge, R., Drees, E.M., Van der Graaf, M., Verberne, F.C.M., and Wesselo, A. 1981. Damage effects of cereal aphids in wheat. *Netherlands Journal of Plant Pathology* 87:217–232.

Rabbinge, R., Brouwer, A., Fokkema, N.J., Sinke, J., and Stomph, T.J. 1984. Effects of the saprophytic leaf microflora on growth and productivity of winter wheat. *Netherlands Journal of Plant Pathology* 90:181–197.

Reinecke, P. 1981. Antagonism and biological control on aerial surfaces of the Gramineae. pp. 383–399 in Blakeman, J.P. (editor), *Microbial Ecology of the Phylloplane*. Academic Press, London.

Richardson, L.T. 1966. Reversal of fungitoxicity of thiram by seed and root exudates. *Canadian Journal of Botany* 44:111–112.

Rodger, G. and Blakeman, J.P. 1984. Microbial colonization and uptake of ^{14}C label on leaves of sycamore. *Transactions of the British Mycological Society* 82: 45–51.

Rossing, W.A.H. and Van de Wiel, L.A.J.M. 1990. Simulation of damage in winter wheat caused by the grain aphid *Sitobion avenae*. 1. Quantification of the effects of honeydew on gas exchange of leaves and aphid populations of different size on crop growth. *Netherlands Journal of Plant Pathology* 96:343–364.

Shukla, A.K., Tiwari, B.K., and Mishra, R.R. 1988. Effect of foliar application of herbicides, Benthiocarb, 2,4-D and Fluchloralin on phyllosphere microflora of potato. *Plant and Soil* 106:277–280.

Smedegaard-Petersen, V. and Tolstrup, K. 1986. Yield-reducing effect of saprophytic leaf fungi in barley crops. pp. 160–171 in Fokkema, N.J. and Van den Heuvel, J. (editors), *Microbiology of the Phyllosphere*. Cambridge University Press, Cambridge.

Tammes, P.M.L. 1964. Isoboles, a graphic presentation of synergism in pesticides. *Netherlands Journal of Plant Pathology* 70:73–80.

Thomas, M.R. and Shattock, R.C. 1986. Effects of fungicides on *Drechslera* spp. and leaf surface filamentous saprophytic fungi on *Lolium perenne*. *Plant Pathology* 35:120–125.

Tronsmo, A. 1986. Use of *Trichoderma* spp. in biological control of necrotrophic pathogens. pp. 348–377 in Fokkema, N.J. and Van den Heuvel, J. (editors), *Microbiology of the Phyllosphere*. Cambridge University Press, Cambridge.

Vereijken, P.H. 1979. *Feeding and Multiplication of Three Cereal Aphid Species and Their Effect on Yield of Winter Wheat*. Ph.D. thesis, Wageningen Agricultural University, The Netherlands.

Zadoks, J.C., Chang, T.T., and Konzak, C.F. 1974. A decimal code for the growth stages of cereals. *Weed Research* 14:415–421.

22

Biocontrol of Foliar Fungal Diseases with Compost Extracts

Heinrich C. Weltzien

22.1 Introduction

Modern plant protection techniques are largely based on application of pesticides. Their frequent and often excessive use has resulted in widespread side-effects. Toxic residues on food and feed, health hazards during the pesticide handling and application processes, elimination of nontarget organisms from the agroecosystem, contamination of ground and drinking water resources, environmental pollution through evaporation, and intercontinental dispersal followed by reconcentration through food chains up to toxic levels, are among the most frequently observed problems associated with this technology.

Though the beneficial effects of pesticide application through increased and stabilized yields, avoidance of postharvest losses and promotion of industrialized agrobusiness operations are clearly recognized, concerns are now frequently raised about the long-term sustainability of pesticide-driven agricultural systems. It is within this context that integrated pest and disease management was promoted, in order to minimize the application of toxic chemicals. "Organic agricultural systems" that exclude synthetic pesticides altogether are also increasing.

These trends have resulted in an increased interest in biological management and control of disease. In many cases, arthropod pests have been successfully controlled by biological and biotechnical methods. However, the lack of biocontrol methods for fungal or bacterial diseases has often limited the practical application of integrated systems. If chemical control schemes are to be used for epidemic diseases, insect pests will often also need to be chemically controlled, as pesticide costs are rather low and tank mixtures allow low-cost application (Zadoks, 1989). Organic growers, having no option of chemical intervention, are even more dependant on the development of biocontrol systems for plant diseases. This situation has prompted the study of potential ways for improved biocontrol of fungal plant diseases of major economic and epidemic significance.

22.2 Methods for Biological Control of Pathogens

Pathogens can be depressed by competitive antagonists or by hyperparasites. Both mechanisms have been studied with soilborne and airborne pathogens. Soil fungistasis, which prevents most fungal spores from germinating in biologically active soil (Lockwood, 1988), is the most generally effective antagonism. Application of suspensions of *Trichoderma* or *Gliocladium* spp. has been tried repeatedly with more or less success (Papavizas, 1985). These fungi can be applied to soil or seeds as root dips or aerial application on above-ground plant parts. Obligate parasites, such as powdery mildews (*Erysiphaceae*) and rusts (*Uredinales*), are frequently subject to hyperparasitic attack by *Ampelomyces quisqualis, Verticillium lecanii, Tilletiopsis* spp., and other fungi (Hijwegen and Buchenauer, 1984; Phillipp and Hellstern, 1986; Adams, 1990). Microbes or microbial metabolites have also been found to be able to induce resistance mechanisms in plants, thus widening the potential tools for biological disease management (Ouchi, 1983; Schönbeck and Dehne, 1986).

22.3 Use of Compost for Disease Suppression

Of the various mechanisms by which nonpathogenic microorganisms can suppress or control plant pathogens, substrates with extraordinary high microbe populations, such as composts, have received some attention. Compost is commonly understood to be a mixture of organic materials, usually plant and animal debris, which is degraded by microbial processes. It is also a highly valued substrate for plant cultivation. Composting processes are attracting new attention in view of the necessity to enhance biologically driven recycling processes. Agricultural production systems are expected to increasingly use compost as part of an international effort to promote sustainable agriculture and minimize environmental impact. Also, reduction of purchased inputs for plant nutrients can be achieved by on-farm recycling of organic materials through composting. A review of the subject was written by Gottschall (1988).

Some authors have suggested that compost application improves plant health. Earlier reports were based on empirical observations and have frequently been met with skepticism (Howard, 1948; Seifert, 1971). However, more recent studies on the effect of organic soil amendments on plant health have yielded encouraging results.

Hoitink et al. (1977) reported the suppression of *Phytophthora cinnamomi* by composted hardwood bark. When lupine seedlings were grown in an inoculated peat-sand medium, severe root rot developed. Seedlings grown in an inoculated bark-compost-sand medium were less affected. It has also been shown that the percentage of 3-week-old apple seedlings killed by *Phytophthora cactorum* was significantly lower in bark-compost-container medium than in a peat medium after inoculation (Spring et al., 1980).

Lumsden et al. (1983) noted the effect of composted sewage sludge on several soilborne pathogens and diseases. *Rhizoctonia* root rot on beans, cotton, and radish; *Sclerotinia* drop of lettuce; *Fusarium* wilt of cucumber; and *Phytophthora* crown rot of pepper were all significantly decreased by the amendment of soil with compost.

An undefined "domestic waste" compost reduced infection of *Phaseolus vulgaris* and *Pisum sativum* by *Rhizoctonia solani* and *Pythium ultimum* (van Assche and Uytterbroeck, 1981). Hadar and Mandelbaum (1986) have shown that composted licorice roots can suppress *Pythium aphanidermatum* damping off in cucumber. Compost samples taken during the thermophilic stage were not suppressive, and sterilized compost lost its ability to suppress *Pythium* damping off. Apparently the suppression of plant pathogens in compost was due to a biological mechanism developed during the later stages of composting.

Hoitink and Fahy (1986) concluded that the microflora of composts plays the major role in suppression of soilborne plant pathogens. Fungal antagonists that are most effective for control of various plant pathogens in bark compost substrates were *Trichoderma* spp. and *Gliocladium virens*. Various bacterial antagonists were also involved, such as *Flavobacterium balustium, Pseudomonas putida,* and *Xanthomonas maltophilia,* all of which rapidly colonized organic matter.

Research on suppression of *Rhizoctonia* damping off has shown that microbial populations in media containing hardwood bark composts are responsible for disease control (Nelson et al., 1983). Heat (60°C) and gamma-radiation (275 krad) treatments, which reduced microbial populations, also eliminated disease suppression. Suppression was restored by adding 10% (v/v) unheated compost or propagules of microorganisms isolated from selective media.

Budde and Weltzien (1988) demonstrated in controlled inoculation studies that the susceptibility of barley and wheat to powdery mildew, *Erysiphe graminis,* was significantly decreased, if plants were grown in compost-enriched substrate. The number of pustules/primary leaf was reduced by 80% as compared to plants grown in ordinary field soil. Increasing the compost content of the substrate correlated directly with increasing plant resistance. The increased resistance remained unchanged when detached leaves from plants grown in compost substrate were transferred to benzimidazol agar and challenged with *E. graminis* conidia. Similar experiments with cucumber powdery mildew, *Sphaerotheca fuliginea*, confirmed these findings (Samerski, 1989). Field soil amended with 75% compost significantly reduced the disease incidence by 41%. To date these are the only reports of reduced leaf disease incidence after applying compost as a planting substrate.

The first experiments that involved the direct application of watery compost preparations on above-ground plant parts were reported by Weltzien and Ketterer (1986). They treated detached grapevine leaves with extracts of horse manure compost. Leaves inoculated with suspensions of sporangia of the downy mildew fungus of grapevines, *Plasmopara viticola*, showed a highly significant reduction in area diseased. The best treatments reached efficiencies of 98.5%. Dip and spray applications were equally effective. Since then, we have studied the potential of watery compost extracts for plant disease suppression and control more systematically (Weltzien and Ketterer, unpublished observations).

22.4 Compost Extraction Process

To produce liquid compost preparations suitable for spray applications on plants, a standard procedure was developed as follows (Figure 22.1): Tap water and untreated, moist compost from the field were mixed in a suitable container. Ratios of 1:5 to 1:10 of compost to water were found to be suitable in most cases. The mixture was usually stirred only once and left at ambient temperatures in open containers for various time spans. Larger quantities were usually kept outdoors during the summer months, while smaller containers were kept indoors, at temperatures usually varying from 15 to 20°C. Various extraction times were tested with several different host-pathogen systems. In many cases, extraction times of 5 to 8 days were most suitable. Immediately before use, the extracts were filtered through cheesecloth or milk filters to remove all larger particles, which otherwise could have clogged spray nozzles. The extracts were then applied to plants or plant parts using ordinary spray equipment for laboratory, greenhouse or field use. If artificial inoculation of the hosts was performed, the time span between extract application and inoculation was determined and is usually referred to as induction time, assuming that some induction of resistance mechanisms may take place during this period. The flow diagram shown in Figure 22.1 summarizes the process.

The composts used were of various raw materials and origins. Horse, pig, and goat manure composts were our own preparations. Manure-straw mixtures were amended with about 15% field soil or old compost and set up in piles of 2 m width, 1.6 m height, and various lengths. They were not turned and were used after about 6-months-decomposition time. Cattle manure compost was a commercial product of the Backhus Company,

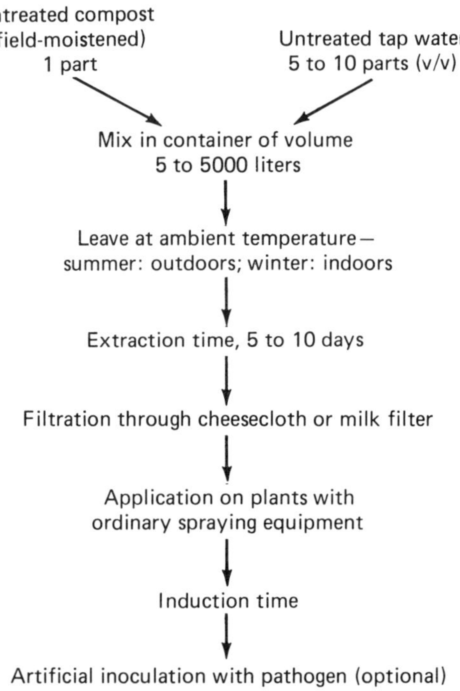

Figure 22.1 A flow diagram for the preparation and use of liquid compost.

Edewecht-Jeddeloh II, Germany, under the trade name Bio-Kuhdung^{R}, while grape marc compost was produced by the Trever Company (Austria) under the trade name Biovin^{R}.

22.5 Microbial Composition of the Extracts

Several studies have dealt with the microbial composition of the extracts. Ketterer (1990) examined the total number of colony-forming units (CFUs) on a nonselective broad-spectrum medium (Caso-Agar, Merck Company). He used extracts based on composted horse manure, cattle manure, and grape marc. The extraction times varied from 1 to 21 days (Table 22.1). In all three extracts, the CFUs increased from extraction days 1 to 14 and decreased afterwards. The highest count was 1.1×10^{11} per ml after 14 days from horse-manure compost extract, an increase of 10^3 from day 1.

Stindt (1990) used selective media and differentiated among groups of microorganisms (Table 22.2). He used the same three composts as indicated in Table 22.1, and extraction times of up to 16 days. Table 22.2 summarizes his data for extraction day 8. The highest count of total mesophilic aerobic bacteria was found in cattle manure extracts, but otherwise the differences

Table 22.1 Total number of bacterial and fungal colonies/ml compost extract on Caso-Agar for three different composts and various extraction times[a]

Extraction time (days)	Compost type		
	Horse manure[b]	Cattle manure[c]	Grape marc[d]
1	1.4×10^8	3.6×10^6	1.1×10^8
3	5.3×10^8	1.8×10^7	6.4×10^8
5	2.2×10^9	8.1×10^7	1.9×10^9
7	1.4×10^{10}	2.8×10^8	3.1×10^{10}
10	3.5×10^{10}	6.8×10^8	3.4×10^{10}
14	1.1×10^{11}	1.4×10^9	5.2×10^{10}
18	7.5×10^9	2.4×10^8	1.1×10^{10}
21	4.4×10^7	4.5×10^6	4.9×10^8

[a] All composts were extracted at a compost : water ratio of 1 : 9 (v/v). All data based on six replicates.
[b] Prepared by author.
[c] Bio-Kuhdung[R], Backhus Co., Germany.
[d] Biovin[R], Trever Co., Austria.
Adapted from Ketterer (1990).

Table 22.2 Counts of microbes/ml for three different composts after 8 days of extraction time

Compost type	Total mesophilic aerobic bacteria	Aerobic bacilli	Entero-bacteriaceae	Pseudo-monads	Filamentous fungi and yeasts
Cattle manure	6×10^8	9×10^5	2×10^6	4×10^6	5×10^5
Grape marc	2×10^8	9×10^5	3×10^5	2×10^6	9×10^5
Horse manure	1×10^7	5×10^6	2×10^5	1×10^6	1×10^5
Medium used	Merck no. 1621	Merck no. 1621	Merck no. 10275	Merck no. 10230	Merck no. 5448

All data are averages of three replicates. Adapted from Stindt (1990).

between the bacterial counts of the three composts studied were not significant. Maximum densities of pseudomonads, enterobacters, and mesophilic aerobes were 4×10^6, 2×10^6 and 6×10^8 CFUs/ml respectively. The highest number of filamentous fungi and yeasts was found in grape marc compost with 9×10^5 CFUs/ml.

It is apparent from these data that compost extracts contain very rich and varied microbial populations. It can be expected that further studies will reveal more details on the quantity and quality of these populations and their effect on disease suppression. Studies are presently being conducted to increase some microbial fractions selectively by the use of nutrients or growth stimulants such as supplines (Schlegel, 1985).

Table 22.3 Percent of empty sporangia with three different compost extracts[a]

Fungus tested	Exposure time (hr)	Empty sporangia (%) Control	Horse manure	Cattle manure	Grape marc
Plasmopara viticola	1	51 a	3.0 c	2.3 c	6.3 b
	4	82 a	19.0 c	15.0 d	28.0 b
Phytophthora infestants	1	66 a	8.7 bc	5.5 c	11.0 b
	4	87 a	31.0 c	33.0 c	41.0 b

[a]Sporangia suspended in 20 ml extract or water (for the control). Sporangia density: 50,000/ml for *P. infestans*; 70,000/ml for *P. viticola*. Four replicates of 100 sporangia were each counted/treatment. Different letters within a row indicate significant differences at 5% level; Duncan's (1955) multiple range test.
Adapted from Ketterer (1990).

22.6 In Vitro Effects of Extracts on Fungi

Several experiments have been conducted to test the direct effect of compost extracts on pathogenic fungi. Ketterer (1990) suspended sporangia of *Phytophthora infestans* and *Plasmopara viticola* in three different compost extracts with a sporangial density of 50,000 and 70,000/ml respectively. Zoospore release was determined by counting the number of empty sporangia. Data for 1- and 4-hr exposure time are presented in Table 22.3. Treatments with extracts of horse manure, cattle manure, and grape marc compost all reduced zoospore release rates significantly as compared with the untreated checks. However, inhibition was never complete and germination rates increased from 1 to 4 hr of exposure time. Grape marc compost extracts were significantly less effective than the other two. As no significant decrease of zoospore release by *Plasmopara viticola* sporangia was found in earlier, less-detailed studies (Weltzien and Ketterer, 1986) one may conclude that direct inhibition depends on still-unknown extract qualities and cannot be held fully responsible for disease inhibition as observed in vivo.

Stindt (1990) used *Botrytis cinerea* as a test organism. Compost extracts were mixed 1:1 (v/v) with water agar and inoculated with spore suspensions at a density of 10^6 and incubated for 48 hr at 20°C with a 16-hr light period. Inhibitory efficiency in percent was determined against an untreated control. Extraction times between 1 hr and 32 days were compared for five different composts. Inhibition at 1 hr varied from 52 to 83%. The most effective extract came from cattle manure compost and reached 100% inhibition after 8 days of extraction time. Horse manure and grape marc compost extracts reached 85% inhibition after 8 days and 100% after 16 days of extraction time.

For mycelium growth tests, the extract-water agar mixtures were covered with 8 ml standard malt agar/petri dish and inoculated with mycelium disks. After 16 days of extraction time, extracts of all three composts reduced mycelial growth to 4.5% of the control, and even a 1-hr extraction time from cattle manure compost inhibited mycelial growth by 50%. It can be concluded that direct inhibition of spore germination and mycelial growth is the main factor in *Botrytis* inhibition by compost extracts. However, very different results were obtained when extracts were tested on powdery mildew fungi. Samerski and Weltzien (1988) studied their effects on the cucumber powdery mildew, *Sphaerotheca fuliginea*. None of the extracts from different composts or extraction times reduced the germination rate of conidia as compared to the untreated checks.

22.7 Studies on Detached Leaves

In vivo effects of extracts on disease development were studied first on detached leaves for various host-pathogen systems. Thus, standardized environmental conditions could be maintained while factors such as extraction and induction times were varied. Weltzien and Ketterer (1986) used greenhouse-grown, mature, grape vine leaves (cv Müller-Thurgau) for tests on *Plasmopara viticola*. Horse manure compost extract was sprayed to full coverage on the lower-leaf surface. Extraction times varied between 1 and 3 days, induction times between 1 and 24 hours. Sporangia were sprayed in watery suspension of 8×10^4/ml. Leaves in petri dishes were then incubated at 20°C with 16 hr of light of 1700 lux. An extraction time of 3 days resulted in reduction of the infected leaf area by 95% compared to the control. More detailed data of Ketterer (1990), including data for *Phytophthora infestans* on tomato leaves, are summarized in Table 22.4. The importance of a suitable extraction time for good disease suppression is apparent. For *Plasmopara viticola*, there is a clear optimum of 3 days, with 1 day and 7 days being significantly less effective. The efficacy for *Phytophthora infestans* was generally lower, the optimum extraction time being 14 days. Once again, the induction time remained largely insignificant.

Winterscheidt et al. (1990) used the cucumber downy mildew *Pseudoperonospora cubense* as a test organism. They achieved 85% disease reduction with cattle manure compost extract after 3 days of extraction time. They showed for the first time that compost age may be an important factor. Horse manure compost aged for 6 months proved to be significantly more effective than 1-year-old material.

Botrytis cinerea is another pathogen for which biocontrol through microorganism-rich substrates seems possible. Wood (1951) showed that soilborne microorganisms can suppress *B. cinerea* on lettuce leaves. More recently Redmond et al. (1987) successfully controlled *B. cinerea* on roses with bacteria and yeasts isolated from the phyllosphere of roses.

Weltzien et al. (1987) reported suppression of *B. fabae* infections on *Vicia faba* leaves when horse manure compost extract was applied after 7

Table 22.4 Suppressive effects of horse manure compost extract on downy mildew of grapes (*Plasmopara viticola*) and tomato leaf blight (*Phytophthora infestans*) as assayed by the detached leaf method[a]

Pathogen/host system	Induction time (hr)	Diseased leaf area (%) Extraction time (days) 1	3	7	14	Control
P. viticola/ grapevine leaves	1	84.0 ab	1.6 d	55.0 c	83.0 b	87.0 a
	3	86.0 b	3.4 c	86.0 b	83.0 b	91.0 a
	24	75.0 ab	7.2 c	69.0 b	82.0 a	75.0 ab
P. infestans/ tomato leaves	3	75.0 a	63.0 b	39.0 c	29.0 d	73.0 a
	24	75.0 a	46.0 b	19.0 c	9.5 d	77.0 a
	48	73.0 a	48.0 b	14.0 c	8.5 c	70.0 a

[a]Inoculum density: 8×10^4/ml; detached leaves incubated for 7 days at 20°C, 16 hr of light at 1700 lux. Different letters within a row indicate significant differences at the 5% level, Duncan's multiple range test.
Adapted from Ketterer (1990).

days of extraction time. Work was then continued with *Botrytis cinerea* on bean (*Phaseolus*) leaves (Stindt and Weltzien, 1988b; Stindt, 1990). Detached leaves of the cultivar Processor were inoculated with 50 μl droplets of 10^6 conidia/ml. Distilled water with 1% malt extract was used to allow good infections of the untreated controls. Incubation was at 20°C, 16 hr of light, and saturated humidity for 3 days. Extraction time turned out to be a major factor. Though even a 1-hr extraction time resulted in a significant decrease in lesion size, only 16-hr extraction was sufficient to suppress all visible infections or lesions. Horse manure and cattle manure compost were superior to other composts with no animal manure components.

The efficacy also depended on the inoculum density (Table 22.5). Grape marc compost again was less effective than cattle or horse manure compost. With the latter, lesions started to occur at inoculum densities above 5×10^7 conidia/ml. This was also the case if the fungicide dichlofluanid (Euparen[R]; Bayer) was applied prophylactically (Table 22.5). Cattle manure compost extract and the fungicide did not differ significantly in their efficacy. *Botrytis* strains resistant to Dichlofluanil (or benomyl or procymidon) were also tested. They were all fully suppressed by cattle and horse manure compost extract. Cattle manure compost extract also equalled the curative effects of procymidon, applied as Sumisclex[R] (Bayer), if used 24 hr after infection and observed for 7 days.

One can conclude that compost extracts are very promising materials for biocontrol of fungal plant diseases. This includes downy mildews, powdery mildews, and *Botrytis* rots. Under controlled conditions they proved to be highly effective and in some cases equalled the effects of fungicide treat-

Table 22.5 Effect of *Botrytis cinerea* inoculum density on lesion size on detached bean leaves after treatment with three different compost extracts and a fungicide[a]

	Lesion size as percentage of the control			
	Inoculum density (conidia/ml)			
Compost type[b]	5×10^5	5×10^6	5×10^7	5×10^8
Cattle manure	0	0	5.7 d	12.9 d
Horse manure	0	0	27.1 c	37.1 c
Grape marc	1.8 b	29.0 b	59.0 b	64.3 b
Fungicide[c]	0	0	2.9 d	10.0 d

[a]Inoculation was with 1% malt extract in water; droplet size, 50 μl. Incubation was for three days, 16 hr of light at 20°C, saturated humidity. Different letters within a column indicate statistical significance at the 5% level; with Duncan's multiple range test, all values are statistically different from the water control.
[b]Extraction time was 16 days.
[c]Dichlofluanid was applied as Euparen[R] (Bayer; 0.25% w/v).
Adapted from Stindt (1990).

ments. Detached leaf tests proved to be well suitable to study various host-pathogen systems and to screen for suitable composts, and extraction and application methods.

22.8 Efficacy of Extracts in Field Experiments

From the outset of our studies we hoped to develop new biological methods for plant disease suppression and thus have tested our material on growing plants, both in the greenhouse and in the field. Thus far, the most intensively studied crop plants have been grape vines, potatoes, and strawberries.

Studies on Grape Vines Potted grape vines were treated under greenhouse conditions with horse manure compost extracted for 3 days. A 3-day induction time was used before artificial inoculation with *Plasmopara viticola*. The extract reduced leaf blight to 11% of the untreated control. Plants exposed to powdery mildew, *Uncinula necator*, showed only 6% diseased leaf area, as compared with the untreated controls (Ketterer, 1990).

These data encouraged us to perform field experiments starting in 1986. As no experimental vineyards were available, all work had to be done in commercial plantations. This meant that the use of larger control plots was difficult, and artificial infection was impossible. A spontaneous, serious epidemic of downy mildew developed in 1987, which allowed us to collect a full set of data. Results from a vineyard near the Ahr River (Germany) planted with the red cultivar Blauer Spätburgunder are shown in

Table 22.6 Effect of compost extracts with and without microorganisms and of fungicides on *Plasmopara viticola* on grapevines

Treatment	Percentage diseased[a] Leaves 25 Aug.	Percentage diseased[a] Leaves 19 Sept.	Percentage diseased[a] Fruit 22 Sept.	Yield (kg/plant)
Control (untreated)	51 a	85 a	84 a	0.2 a
Horse manure compost extract[b]	20 b	32 b	20 b	0.8 b
As above, but amended with microorganisms	5.8 c	15 c	11 c	1.1 c
Fungicides as recommended[c]	5.4 c	14 c	8 c	1.1 c

[a] Different letters within a column indicate statistical significance at the 5% level, Duncan's multiple range test; six replicates/treatment.
[b] Extraction time, 3 days; weekly application, 10 times from 18 May to 27 August 1987.
[c] Five applications from 18 May to 5 September 1987.
Adapted from Ketterer (1990).

Table 22.6. All treatments were replicated six times at an application rate of 1000 to 1500 1/ha. Horse manure compost extract was applied 10 times, starting 18 May and ending 27 August 1987. Because microorganisms had been identified as the most effective components in the extracts (see Section 22.9), we tried to increase efficacy by adding mixtures of isolates of four fungi, two bacteria, and one yeast, which had been isolated from the extract and grown in pure culture. Extracts were inoculated with about 2% (v/v) of the microbial culture immediately before application. Both treatments were compared with the recommended fungicide treatments, applied five times between 18 May, and 5 September 1987 (Table 22.6).

Plants in the untreated control plots were virtually destroyed by the pathogen. Only 15% of the foliage and 16% of the fruit survived, and the yield was negligible (Table 22.6). Regular compost extract treatments significantly reduced the disease intensity and quadrupled the yield, but the level of disease suppression achieved cannot be considered sufficient for commercial growers. Only when the extracts were amended with active microorganisms did the results equal those achieved by regular fungicide application. Though 10 extract applications are compared with only five fungicide applications, it must be noted that this, to our knowledge, is the first report of a successful biological control of grape downy mildew in a commercial orchard under severe infection pressure.

Encouraging results with *Uncinula necator* in field plots were achieved for the first time in 1989 but they have not been fully evaluated and published yet. However, in the years 1986 and 1988, leaf blotch caused by

Pseudopeziza tracheiphila occurred frequently. In all cases, a highly significant biological control of this pathogen was achieved by horse manure compost extract. Though fungicides were superior in their performance, the biocontrol system could be satisfactory for commercial growers.

Studies on Potato Late Blight There were no reports available on the successful biological control of potato late blight, *Phytophthora infestans*, when we began our work in 1985. Based on the encouraging results summarized above, we started our field experiments in 1987. Field plots were available in our experimental farm for organic agriculture and in a conventionally managed commercial field, both near Bonn. In 1987 an extremely serious, country-wide epidemic occurred, due to the wet and cool weather conditions prevailing during most of the summer.

We used rather large plot sizes of 70 m^2 in the organic farm and of 25 m^2 in the conventional farm with six replicates. Treatments included horse manure compost extract, 7-day extraction time, with and without the amendment of microorganisms. The same microorganisms were used for amendment as described above. Three fungicides were used in addition on the conventional farm: mancozeb + metalaxyl as Ridomil MZ^R (Ciba Geigy), 2 kg/ha; fentinacetat + maneb as Brestan 60^R (Hoechst), 0.6 kg/ha; cymoxalil + mancozeb as Ciluan^R (Cela-Merck), 2 kg/ha. Extracts were applied 11 times between June and 25 August 1987 on the organic farm, and seven times between 16 June and 17 August 1987 on the conventional farm. Fungicides were applied five times between 23 June and 7 August 1987 (Ketterer, 1990).

Experiments in 1988 were carried out on two organic farms and one conventional farm. The infection pressure during this year was lower. Horse manure compost extract was used with and without amendment of microorganisms and applied 8–10 times. Disease severity was assessed as the percentage of diseased foliage, using the index from the official guidelines for pesticide evaluation by Martin et al. (1976). The results are summarized in Table 22.7. Results using the three fungicides as described above are also included.

In 1987 the compost extract delayed the epidemic significantly until August 10. After this date, the foliage was destroyed and no yield increases were achieved. However, extracts amended with active microorganisms were very effective in both locations. Infection levels never increased beyond 11–16% and yields increased to 221 and 134% compared with the untreated control. None of the three fungicides tested were more efficient than the amended extract.

The efficacy of amended extracts was confirmed in 1988 under lower infection pressure. When cattle manure extract without amendment was used, the disease was also well controlled, and the yield increased to 168% more than that of the control. This indicates that extracts with high biological activity can be used successfully without amendment of micro-

Table 22.7 Effect of compost extracts with and without microorganisms and of three fungicides on potato late blight, *Phytophthora infestans*

Location; cultivar	Treatment	Leaf area diseased (%)[a]			Yield (dt/ha)
		15 July 1987	10 Aug. 1987	21 Aug. 1987	
Organic farm; Hansa	Control	16 a	72 a	93 a	185 a
	Horse manure compost extract[b]	7 b	63 b	90 a	210 a
	As above, amended with microorganisms[c]	1 c	8 c	17 b	410 b
			6 Aug. 1987	21 Aug. 1987	Yield
Conventional farm; Sieglinde	Control		70 a	96 a	250 a
	Horse manure compost extract amended with microorganisms[b]		4 c	11 c	336 c
	Ridomil MZ[d]		4 c	8 c	341 c
	Brestan 60[e]		9 b	19 b	331 c
	Ciluan[f]		9 b	18 b	286 b

			12 Aug. 1988	30 Aug. 1988	Yield
Organic farm; Hansa	Control		36 a	65 a	230 a
	Horse manure compost extract[g]		29 b	62 a	228 a
	As above, amended with microorganisms[h]		5 c	16 b	413 b
		10 Aug. 1988	22 Aug. 1988	1 Sept. 1988	Yield
Organic farm; Granola	Control	32 a	49 a	88 a	182 a
	Cattle manure compost extract[g]	14 b	21 b	38 b	306 b

[a] Index of leaf area diseased was taken from Martin et al. (1976). Different letters within a column for each year indicate statistical significance at the 5 % level; Duncan's multiple range test.
[b] Seven applications weekly.
[c] Eleven applications weekly.
[d] Mancozeb + metalaxyl (Ciba-Geigy); 2 kg/ha, five applications.
[e] Fentinacetal + maneb (Hoechst); 0.6 kg/ha, five applications.
[f] Cymoxanil + mancozeb (Cela-Merck); 0.6 kg/ha, five applications.
[g] Eight applications weekly.
[h] Ten applications weekly.
Adapted from Ketterer (1990).

organisms. The results clearly demonstrate that biological control of *Phytophthora infestans* in the field is possible, even under very heavy infection pressure.

Studies on Grey Mold of Strawberries Grey mold of strawberries caused by *Botrytis cinerea* presents a major disease problem for strawberry production, both in conventional and organic farms. Biocontrol systems are not available, while resistance to fungicides frequently limits their efficacy. The results reported above encouraged us to set up a series of field experiments in 1987 and 1988. Plots were only available on organic farms, thus fungicides could not be applied for comparison (Stindt and Weltzien, 1988a, 1988b; Stindt, 1990). Plot size was 8 m^2 with four replicates. In 1987 extracts were applied seven times; in 1988, five times, always from the onset of flowering in weekly intervals. In 1987 we used horse manure compost extract, in 1988, cattle manure compost extract. While weather conditions were very favorable for grey mold in 1987, the infection pressure was lower in 1988.

Fruit losses by grey mold in 1987 reached 20% by weight in the untreated controls (Table 22.8). They were reduced to 11% by extract applica-

Table 22.8 Effect of compost extract on grey mold of strawberries (*Botrytis cinerea*)[a]

Treatment	Fruits diseased (% by weight)	Disease suppression (%)	Yield (kg/ha)
	1987; location 1		
Control	20 d	–	31 a
Horse manure compost extract[b]	11 b	45	107 b
	1987; location 2		
Control	13 a	–	87 a
Horse manure compost extract[b]	6 b	54	121 b
	1988		
Control	6 a	–	71 a
Cattle manure compost extract[c]	2 b	67	95 b

[a]Different letters within a column for each year indicate statistical significance at the 5 % level; Duncan's multiple range test.
[b]Extraction time, 12 weeks; seven applications.
[c]Extraction time, 16 days; five applications.

tion, a control efficiency of 45%. In location 2, with slightly lower infection pressure, the control efficiency reached 54%. Yield responses were very high, reaching 345 and 139% that of the control, respectively. In 1988, with only 6% fruit losses by weight, the control efficiency was 67%, but yield responses remained lower with only 2.4 kg/ha. All differences were highly significant. This suggests that the biological control of strawberry grey mold is possible in organic agricultural systems with a high rate of return through increased yield.

In summary, the field experiments showed the high potential of microbiologically active compost extracts for plant disease suppression under field conditions, using various host-pathogen systems, different locations, and several years with different weather conditions. Future research with field experiments will concentrate on practical improvements, such as reduced number of sprays, increased efficacy of the extracts, different environments representing different climatic conditions, and other formulations of the material used.

22.9 Role of Microbes in Compost Extract Efficacy

Earlier observations indicated that the presence of microorganisms in the extract is necessary for their full activity. For example, mycelium growth of *Rhizoctonia solani* was 80% inhibited by extracts of untreated horse manure compost, whereas heat sterilization stimulated growth by 30% (Stübler and Weltzien, unpublished). Ketterer (1990) used successive filtrations of the same extract and tested them against *Phytophthora infestans* on detached tomato leaves. Their inhibitory effect was compared with untreated and heat-sterilized extracts. As shown in Table 22.9, the stepwise elimination of microorganisms through membranes of decreasing pore sizes resulted in a gradual loss of activity. Extracts sterilized by filtration through a pore size of

Table 22.9 Effect of horse manure compost extract treated in various ways on *Phytophthora infestans* infections on detached tomato leaves[a]

	Unfiltered extract	Pore sizes of filter membranes (μm)[b]						Control	Heat-sterilized extract[c]
		50	10	5	1	0.45	0.2		
Diseased leaf area (%)	8e	8e	19de	22d	40c	50b	75a	81a	78a

[a]Different letters indicate statistical significance at the 5% level; Duncan's multiple range test. Extraction time, 14 days; induction time, 3 days.
[b]Cellulose acetate filters (Schleicher & Schüll).
[c]Autoclaved for 40 min at 120°C.
Adapted from Ketterer (1990).

Table 22.10 Effect of horse manure compost extract, of different extraction times after heat treatment, on *Botrytis cinerea* infection of detached *Phaseolus* bean leaves (cv. Processor)[a]

	Healthy leaf tissue (%)[b]					
	Heat treatment, 30 min at:					
Extraction time	20°C	35°C	50°C	65°C	80°C	95°C
1 hr	75 b	75 b	40 d	18 e	10 f	2 g
1 day	78 b	79 b	36 d	21 e	8 f	3 g
4 days	81 b	80 b	64 c	17 e	6 f	4 g
16 days	100 a	100 a	97 a	55 c	3 g	4 g

[a]Different letters indicate statistical significance at the 5% level, Duncan's multiple range test.
[b]Each value is the average of 24 leaflets with three droplets each. Droplet size, 50 μl; 10^6 conidia/ml; incubation for 3 days at 20 °C; 95% R.H.; 16 hr light/day.
Adapted from Stindt (1990).

0.2 μm and through heat sterilization by autoclaving for 40 min at 120°C lost their activity completely.

Heat-treated extracts also did not inhibit *Botrytis cinerea* infections on detached *Phaseolus* bean leaves. Stindt (1990) tested a temperature range from 20 to 95°C, applied in a water bath for 30 min. He compared horse manure compost extract using four different extraction times between 1 hour and 16 days (Table 22.10). Increasing temperatures clearly reduced the extract efficacy with all extraction times. Treatment at 95°C resulted in almost complete removal of the inhibitory effects. However, as the extraction time increased, the heat susceptibility at medium temperatures decreased. Treatment for 30 min at 50°C reduced lesion sizes to 36% for 1-day extraction time, while the inhibitory effect remained intact after 16-day extraction time. It is apparent that longer extraction times result in a more stable biological activity, if *Botrytis cinerea* is used as a test organism.

Stindt (1990) confirmed these results using sterile filtration. However, he also demonstrated that, with increasing extraction times, sterile filtration had less effect on the activity of the extracts. After 16 days, sterile filtrates had the same inhibitory capacity as the nonfiltered controls. The effect was similar on mycelial growth and appressoria formation. Thus it can be concluded that active microorganisms are of high significance for *Botrytis* suppression in extracts from extraction times up to 4 days, but that inhibitory substances, which accumulate when extraction times are extended, may add to the infection-suppressing capacity. Future studies must clarify if these different mechanisms are specific for *Botrytis cinerea* or are of a more general nature.

We have also repeatedly isolated microorganisms from compost extracts and tested their activity in different host-pathogen systems. Ketterer (1990) isolated four filamentous fungi, two bacterial strains, and one yeast with high inhibitory capacity. When they were tested against *Phytophthora infestans* on detached potato leaves, they suppressed the disease development as well as untreated extracts. A mixture of these isolates was later used to amend compost extracts for field application and proved to be highly effective (Tables 22.6 and 22.7).

Stindt (1990) tested 54 bacterial isolates from cattle manure compost extract against *Botrytis cinerea* in spore germination tests. He found 33 which significantly suppressed germination. Of these, there were 11 pseudomonads, 7 enterobacters, 2 bacilli, 2 actinomycetes; the others were unclassified. They were also effective when tested on detached *Phaseolus* bean leaves. Four of the strains also had curative effects; when applied 24 hr after inoculation, they stopped lesion growth for the following 7 days.

We have only recently started to analyze the phyllosphere of treated and untreated plant parts. So far these studies have been restricted to *Phaseolus* bean leaves and strawberry flowers (Stindt, 1990). The total number of colony-forming units per cm^2 leaf area increased from 4×10^3 on the untreated control leaves to 8×10^5 units 1 hr after application of cattle manure compost extract. After 8 days at 60% R.H. and 1 hr leaf wetness daily, the values were reduced to 4×10^2 and 1×10^4, respectively. Thus, although the absolute numbers decreased with time, the difference between the treatments remained large. If plants were kept in a saturated atmosphere, the total number of colony-forming units on control leaves increased from 4×10^3 to 1×10^5 during a period of 8 days. In the extract-sprayed samples, the numbers remained constant at about 8×10^5 throughout the test period. *Pseudomonas* spp. increased from 0.6% to 9.4% of the total count under moist conditions in the controls and from 0.3% to 27% in the extract-treated samples. Correlation tests comparing the effects of microbe population on disease-symptom expression were highly significant for total colony-forming units, pseudomonads, and enterobacteria, showing the negative correlation between increase in microbes on leaves and decrease in disease symptoms.

Studies on the microflora of strawberry flowers were performed outdoors during the 1989 season. Total colony-forming units of 6×10^5 per g of flowers in the controls were increased to 2×10^7 1 hr after application of cattle manure compost extract. Microbial densities varied during a 10-day observation period with changing weather conditions. The difference between two treatments diminished from 2×10^6 to 9×10^6. Filamentous fungi and yeasts were clearly suppressed by extract treatment. Control levels increased during the 10-day period from 3×10^3 to 4×10^5, while in compost treatments both starting and end values were 2×10^3. Pseudomonads and enterobacteria were favored by the extract sprays while the number of bacilli remained relatively constant.

22.10 Conclusions and Outlook

The experimental evidence available to date allows the conclusion that watery fermentation extracts of different composts are suitable substrates to suppress plant-pathogenic fungi in vivo and in vitro. The number and quality of living microorganisms in the extracts are largely, but not exclusively, responsible for disease suppression. They are effective against a wide range of pathogens including downy mildews, powdery mildews, and *Botrytis* spp. When applied regularly, control in field experiments equalled the effects of modern fungicides. This was especially true when the extracts were enriched with pure cultures of active microorganisms isolated from the same extract. It was demonstrated that bacterial populations in the phyllosphere were increased for up to 10 days after extract application, while fungal populations were suppressed. Thus biological control of epidemic fungal plant diseases should be considered for further development.

Future research should concentrate on various subjects. From a practical point of view, there is an urgent need for guided or controlled fermentation processes of organic materials to ensure a high level of effective microorganisms in the extracts. Additives to the extracts such as supplines (Schlegel, 1981) may also be considered as stimulants for the desired microbe populations. More stable formulations of the composts or extracts are also needed. Field experiments under different climatic conditions will allow us to draw more general conclusions. If possible, the number of sprays should be reduced by use of modern forecasting and warning methods. A wider scope of host-pathogen systems can he considered. Test series have to ensure the environmental and toxicological safety of the proposed disease control system. The commercial development of highly effective microorganisms or of their metabolites for disease suppression is another possibility that can be seriously considered.

From a scientific point of view, phyllosphere studies should be expanded to improve our understanding of the population dynamics of useful, pathogenic, and neutral microbes on plant surfaces and their interactions. The mechanisms involved in the disease suppression need a much better and deeper understanding. Direct inhibition through exudates, nutrient competition, induced resistance, and disrupted communication between pathogen and host seem all to be involved, but the significance of each factor seems to be different for each host-pathogen system. There is also very little understanding of the chemical composition of effective extracts and their direct or indirect role in the control processes described. Plant pathology and plant protection may well move a large step beyond chemical pesticide application, once the phenomena described here are better understood.

References

Adams, P.B. 1990. The potential of mycoparasites for biological control of plant diseases. *Annual Review of Phytopathology* 28:59–77.

Budde, K. and Weltzien, H.C. 1988. Untersuchungen zur Wirkung von Kompostextrakten und Kompostsubstraten im Pathosystem Getreide-Echter Mehltau (*Erysiphe graminis*). *Mitteilungen der Biologischen Bundesanstalt* 245:366.

Duncan, P.B. 1955. Multiple range and multiple f-tests. *Biometrics* 11:1–42.

Gottschall, R. 1988. *Kompostierung*. Verlag C.F. Müller, Karlsruhe, 296 pp.

Hadar, Y. and Mandelbaum, R. 1986. Suppression of *Pythium aphanidermatum* damping-off in container media containing composted liquorice roots. *Crop Protection* 5:88–92.

Hijwegen, T. and Buchenauer, H. 1984. Isolation and identification of hyperparasitic fungi associated with Erysiphaceae. *Netherlands Journal of Plant Pathology* 90:79–83.

Hoitink, H.A.J. and Fahy, P.C. 1986. Basis for control of soilborne plant pathogens with composts. *Annual Review of Phytopathology* 24:94–114.

Hoitink, H.A.J., van Doren, D.M., and Schmitthenner, A.F. 1977. Suppression of *Phytophthora cinnamomi* in a composted hardwood bark potting medium. *Phytopathology* 67:561–565.

Howard, A. 1948. *Mein landwirtschaftliches Testament*. Siebeneicher Verlag, Frankfurt. 253 pp.

Ketterer, N. 1990. Untersuchungen zur Wirkung von Kompost-Extrakten auf den Blattbefall der Kartoffel und Tomate durch *Phytophthora infestans* sowie auf den Befall der Weinrebe durch *Plasmopara viticola, Pseudopeziza tracheiphila* und *Uncinula necator*. Dissertation, University of Bonn. 111 pp.

Lockwood, J.L. 1988. Evolution of concepts associated with soil borne plant pathogens. *Annual Review of Phytopathology* 26:93–121.

Lumsden, R.D., Lewis, J.A., and Miller, P.D. 1983. Effect of composted sewage sludge on several soilborne pathogens and diseases. *Phytopathology* 73:1543–1548.

Martin, J., Maykhuss, F., Stingl, H., and Thiede, H. 1976. Vorläufige Richtlinien für die Prüfung von Fungiziden gegen Kraut- und Knollenfäule (*Phytophthora infestans*) an Kartoffeln, Nr. 4-3.1.1.1, Aco Druck Braunschweig. 6 pp.

Nelson, E.B., Kuter, G.A., and Hoitink, H.A.J. 1983. Effects of fungal antagonists and compost age on suppression of *Rhizoctonia* damping off in container media amended with composted hardwood bark. *Phytopathology* 73:1457–1462.

Ouchi, S. 1983. Induction of resistance or susceptibility. *Annual Review of Phytopathology* 21:289–315.

Papavizas, G.C. 1985. *Trichoderma* and *Gliocladium*: Biology, ecology and potential for biocontrol. *Annual Review of Phytopathology* 23:23–54.

Phillipp, W.D., and Hellstern, A. 1986. Biologische Mehltaubekämpfung mit *Ampelomyces quisqualis* bei reduzierter Luftfeuchtigkeit. *Zeitschrift für Pflanzenkrankheiten und Pflanzenschutz* 93:384–391.

Redmond, R.D., Marois, J.J., and MacDonald, J.D. 1987. Biological control of *Botrytis cinerea* on roses with epiphytic microorganisms. *Plant Disease* 71: 799–802.

Samerski, C. 1989. Untersuchungen zu Wirkung und Wirkungsmechanismen von Kompostextrakten im Pathosystem Gurke (*Cucumis sativus* L.)—Echter Gurkenmehltau (*Sphaerotheca fuliginea*) (Schlechtendal ex. Fr.) Pollacci. Dissertation, University of Bonn. 96 pp.

Samerski, C. and Weltzien, H.C. 1988. Untersuchungen zu Wirkung und Wirkungsmechanismen von Kompostextrakten im Pathosystem Gurke—Echter Mehltau. *Mededelingen Faculteit Landbouwwetenshappen Rijksuniversiteit Gent* 53:373–377.

Schlegel, H.G. 1985. *Allgemeine Mikrobiologie*. Thieme Verlag, Stuttgart. 571 pp.

Schönbeck, F. and Dehne, H.W. 1986. Use of microbial metabolites including resistance against plant pathogens. pp. 361–375 in: *Microbiology of the Phyllosphere*. Fokkema N.J. and van den Heuvel, J. (editors), Cambridge University Press, Cambridge, U.K.

Seifert, A. 1971. *Gärtnern—Ackern—ohne Gift*. Biedersteinverlag München. 210 pp.

Spring, D.E., Ellis, M.A., Spotts, R.A., Hoitink, H.A.J., and Schmitthenner, A.F. 1980. Suppression of the apple collar rot pathogen in composted hardwood bark. *Phytopathology* 70:1209–1212.

Stindt, A. 1990. Untersuchungen zur Wirkung und zu den Wirkungsmechnismen von Kompostextrakten auf *Botrytis cinerea* Pers. ex Nocca & Balb an Erdbeeren, Kopfsalat und Buschbohnen. Dissertation, University of Bonn. 169 pp.

Stindt, A. and Weltzien, H.C. 1988a. Der Einsatz von Kompostextrakten zur Bekämpfung von *Botrytis cinerea* an Erdbeeren—Ergebnisse des Versuchsjahres 1987. *Gesunde Pflanzen* 40:451–454.

Stindt, A. and Weltzien, H.C. 1988b. Der Einfluß von wässrigen, mikrobiologisch aktiven Extrakten von kompostiertem organischen Material auf *Botrytis cinerea*. *Mededelingen Faculteit Landbouwwetenschappen, Rijksuniversiteit Gent* 53:379–388.

Van Assche, C. and Uytterbroeck, P. 1981. The influence of domestic waste compost on plant diseases. *Acta Horticultura* 126:169–178.

Weltzien, H.C. and Ketterer, N. 1986. Control of Downy Mildew, *Plasmopara viticola* (de Bary) Berlese et de Toni, on grapevine leaves through water extracts from composted organic wastes. *Journal of Phytopathology* 116:186–188.

Weltzien, H.C., Ketterer, N., Samerski, C., Budde, K., and Medhin, G. 1987. Untersuchungen zur Wirksamkeit von Kompostextrakten auf die Pflanzengesundheit. *Nachrichtenblatt für den deutschen Pflanzenschutzdienst* 39:17–19.

Winterscheidt, H., Minassian, V, and Weltzien, H.C. 1990. Untersuchungen zur biologischen Bekämpfung des Falschen Mehltaus der Gurke—*Pseudoperonospora cubensis* (Berk. et Curt.) Rost.—mit Kompostextrakten. *Gesunde Pflanzen* 42:235–238.

Wood, R.K.S. 1951. The control of diseases on lettuce by the use of antagonistic organisms. 1. The control of *Botrytis cinerea* Pers. *Annals of Applied Biology* 38:203–216.

Zadoks, J.C. 1989. Does IPM give adequate attention to disease control, in particular on the farmer level? *FAO Plant Protection Bulletin* 37:144–150.

23

Manipulation of Microbial Communities in the Phyllosphere

Christen D. Upper

23.1 Introduction

This chapter is intended to be a speculative discussion of paths to the microbial biocontrol of plant diseases. A reader may view this chapter as commentary or as the not-always-well-documented bias of one opinionated individual; either would probably be appropriate. I will have accomplished my purpose in writing this chapter if I can convince you to give careful consideration to the views that follow. Further, this chapter is not intended as a general review of biocontrol. Ample general reviews of biocontrol of plant diseases (cf. Baker and Cook, 1974; Cook and Baker, 1983; Handelsman and Parke, 1989; Hoitink and Fahy, 1986; Papavizas and Lumsden, 1980; Van Alfen, 1982; Weller, 1988; Wilson and Wisniewski, 1989) and of biocontrol in the phyllosphere (Blakeman and Fokkema, 1982; Andrews, 1990; Knudsen and Spurr, 1988; Windels and Lindow, 1985) are currently available.

The concept of the "silver bullet" has been rejected by many workers in biocontrol of plant diseases, yet the search for silver bullets continues—although perhaps by a different name. The extensive natural variability within and among phyllosphere communities in a single field is a real and important part of the system that must be managed. We must view the problem of biocontrol in the phyllosphere as the management of immense numbers of communities on vast numbers of individual leaves. We must expect each of these myriad communities to differ, at least quantitatively, from every other, and we must develop management strategies that are robust to the native dynamics and diversity in these communities. If consideration of the multidimensionality of microbial activities in the phyllosphere leads us back to the silver bullet, then the bullets we choose must not be scattered by the variability they must penetrate. Perhaps more likely, however, will be alternatives designed to manage communities, not individual microorganisms.

23.2 Is Chemical Control of Plant Disease a Good Model for Biocontrol?

Is the "chemical model" for disease control a good model for biocontrol? Regardless of how outlandish this question may appear on first reading, our knowledge of chemical control of plant disease has, either consciously or subconsciously, channelled our thinking about biocontrol throughout the past two decades, and it continues to influence the breadth of our view of the possible. To a degree, this model has proven useful. However, because it has created unnecessary conceptual limitations on the set of alternatives that might be considered for use in biological control strategies, it has probably also been a detriment.

How has the chemical model influenced development of biocontrol? When interest in biocontrol of plant diseases was renewed a few decades ago, chemical control of a number of diseases, particularly foliar diseases, was quite routine. It was quite natural to adapt the basic concepts and methodology of chemical control to the search for biological means to control diseases. Screens for biocontrol activity often were similar to those that were used to search for new chemical control agents (Table 23.1). Indeed, antibiosis was (and often still is) considered a sufficiently important requisite for biocontrol that many searches for biocontrol agents started with a screen for antibiotic production. Conceptually, the biocontrol agent was

Table 23.1 Similarities between screening for chemical control agents and for biocontrol agents

	Screen for	
Level	Biological control	Chemical control
1	Antibiosis	Antibiosis
2	Extensive qualitative screen for biological control of disease in greenhouse	Extensive qualitative screen for chemical control of disease in greenhouse
3	Intensive tests for biological control of disease in greenhouse	Intensive tests for chemical control of disease in greenhouse
4	Small-scale field trials	Small-scale field trials
5	Extensive field trials	Extensive field trials

viewed as an in situ producer of a chemical control agent. The means of disease control was still viewed as chemical control. Some advocates of modern biotechnology have spoken glowingly of the advantages of microorganisms for sophisticated, appropriately placed systems for production and delivery of chemical control agents. According to this scenario, the mechanism of disease control remains chemical control; only the chemical manufacturing plant and the delivery system are changed.

Strains that produce antibiotics against the potential pathogen are often advanced to a growth chamber or greenhouse test for biocontrol. In these tests, the potential biocontrol agent is tested for decreased disease in an assay involving only the plant, the pathogen, and the potential protective agent. Only strains that are judged promising after testing in this way are advanced to tests in small field plots. Strains that succeed in the first field plot experiments are advanced to additional small-scale, and eventually to larger-scale field testing. Thus, the initial steps in many of the screens that have been utilized for identification of biocontrol have involved antibiosis as a putative mechanism; nearly all have required control in the greenhouse or growth chamber before tests are conducted in the field.

To the extent that these screens have lead to successful biocontrol, the model has been useful. Where screens of this kind will work, they should be used. They are conceptually and physically well defined, manageable, and relatively simple. However, in the USA, there are no biocontrol agents registered for use in the phyllosphere for control of plant diseases. Indeed, the total number of biocontrol agents available for commercial application is very small (Andrews, 1990), the number of advanced candidates likely to be commercialized in the near future is small, and the majority of these will probably be for control of postharvest diseases (Wilson and Wisniewski, 1989).

Where potential biocontrol agents have been identified by following procedures similar to those outlined above, some have proven highly successful in selected field trials. Indeed, one thing that we have learned as the result of many different experiments to control a broad range of diseases is that conventional screening procedures often identify microorganisms that can successfully diminish disease development for a number of different diseases on a large number of hosts. This result is substantially reproducible and does not need further repetition (Hirano and Upper, 1985).

Equally reproducible has been the extensive variability of the results of biocontrol in the field. A particular biocontrol agent may be highly effective in controlling the target disease in one field, only to prove ineffective in the same field at a different time, or in a different field at the same time. Because diseases are not reproducibly controlled in all fields, commercialization of the biocontrol agent falters or fails. Nonetheless, the fact that the disease is sometimes dramatically controlled remains to tantalize scientists, so the experiment continues to be repeated.

We should accept these results; diseases can be controlled biologically

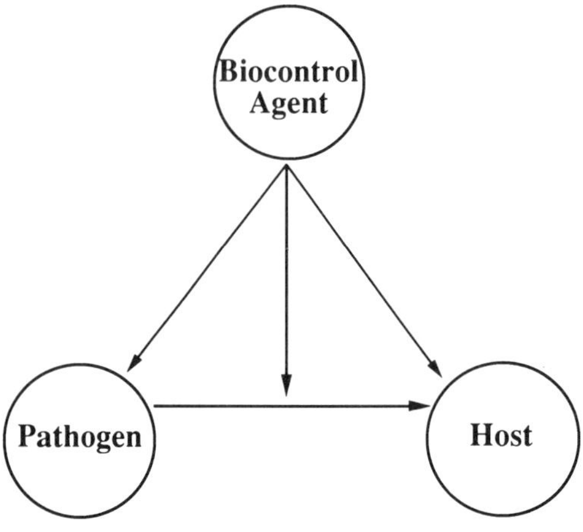

Figure 23.1 A traditional model for biocontrol of diseases in the phyllosphere.

under laboratory conditions. Finding microorganisms that can do so usually requires only a relatively small screen (a few tens to a few hundreds of microorganisms). A reasonable number of the microorganisms that are successful in the laboratory are often also successful in the field in some trials. However, the success of biocontrol in the field is usually highly variable from time to time and place to place. These quite reproducible, quite general results suggest that we may achieve stable, reproducible biocontrol more readily by reexamining the way we approach biocontrol than by repeating the above sequence of experiments with yet another series of microorganisms to control still another plant disease—sometimes.

Explicit use of the chemical model simplifies conceptualization of biocontrol to include only three players: the pathogen, the host, and the putative biocontrol agent (Figure 23.1). According to this model, biocontrol can be visualized as a simple, three-dimensional problem. The object is to adjust the activities of the biocontrol agent in ways that diminish the harmful effects of the pathogen. The three-dimensional model can be restated as an hypothesis: It should be possible to choose a biocontrol agent that can be applied to a given plant species to control a particular disease. This has probably served as the single most commonly tested hypothesis in plant pathology over the past two decades. The hypothesis has been tested in the phyllosphere (Blakeman and Fokkema, 1982; Andrews, 1990; Knudsen and Spurr, 1988; Windels and Lindow, 1985) and in the rhizosphere (Baker and Cook, 1974; Cook and Baker, 1983; Handelsman and Parke, 1989; Papavizas and Lumsden, 1980; Weller, 1988). As described above, the reproducibility of the results has only been surpassed by the consistency of the unwillingness of the community of scientists involved to accept the generality of their results. We should accept the result! The view of biocontrol as a three-dimensional system will often yield highly successful disease control, but the control will be highly variable, and it often will not be consistent enough to be commercially practical.

One interpretation of the variability in success of biocontrol is that success occurs when the three-dimensional model is sufficient to explain the system, and that failure occurs when some other "factor" interferes with the interaction between the pathogen and the biocontrol agent. For example, disease progress is frequently influenced by the environment. One approach to finding improved biocontrol efficacy, then, is to identify the conditions under which the biocontrol agent can be relied upon to control disease effectively and to restrict the use of the particular agent to the appropriate conditions. This approach is being explored with apparent success to the control of "take-all" of wheat by application of a fluorescent pseudomonad to seed (Ownley et al., 1990). It should be considered for use in the phyllosphere as well. At the very least, it will help us to understand why a particular agent fails to control the disease under some conditions, and it may provide sufficient insight to devise an improved control strategy.

Are there alternatives to the chemical model? Let us begin our search for alternatives by reviewing what it is that we wish to accomplish, and what is known about the system that we wish to modify.

Microbial biocontrol of plant diseases is the manipulation of populations or communities of microorganisms associated with plants in ways that diminish the harmful activities of plant pathogens (National Research Council, 1989). Thus, an appropriate strategy for biocontrol should include provision for manipulation of these communities of plant-associated microorganisms. The principal drawback to results to date is the extensive variability and lack of predictability of the results of biocontrol in the field. Let us consider what we know about the microbial communities associated with the phyllosphere, and the natural variability in these communities. Is it possible to identify processes that may explain the kinds of variability that have been associated with biocontrol? Is it possible to suggest alternative strategies for microbial biocontrol of plant diseases?

23.3 The Phyllosphere as a Variable System

A great body of evidence has accumulated that, taken together, describes the phyllosphere as a variable system, and the microbial communities on the phyllosphere as highly variable communities (see Chapters 13 and 14, this volume). Any discussion of microbial communities in the phyllosphere should include consideration of this variability. In the context of developing new strategies for biocontrol, I will attempt to illustrate the kind of information that is contained within this variability, and how it might be utilized for designing biocontrol strategies.

The Physical Environment One of the most obvious aspects of variability in the phyllosphere is the environment in which microbial communities must exist. One of the early attitudinal barriers to the study of microorganisms in the phyllosphere was the misconception that this is a "hostile environment" unsuited for microorganisms (cf. Blakeman, 1982). Unlike aquatic or even soil habitats, the phyllosphere provides little environmental buffering. The temperature, radiation load, and water vapor pressure at which microorganisms on a leaf habitat must survive and grow are those that the instantaneous weather at a given location happens to provide for them. Rapid changes in temperature, radiation, wetness, relative humidity, and ventilation may occur several times each day. Levels of radiation that may be lethal to microorganisms that are adapted to other habitats may occur every day (Burrage, 1971). Periods with near-zero radiative load occur every night. Temperature changes on an individual leaf near the top of the canopy may increase by 20°C or more between sunrise and midday, and decrease by a like amount by the following sunrise. Diel temperature fluctuations for a leaf inside the canopy may be only about one-half as great (cf. Burrage, 1971). Heavy dew may evaporate from leaves in less than one hour. The microbial community on a single leaf may be bathed in a heavy layer of dew at, say, 0800, blown about in a dry wind before 1000, and once again moistened by dew before 2000 the same day.

Why is this so remarkable? Surely the aerial parts of the plants themselves are subjected to the same environmental extremes as are the microorganisms that reside upon them. Other terrestrial organisms, such as insects and other animals also exist in the same fluctuating environment. Yet there is little concern that the temperate terrestrial environment of agricultural fields is too harsh for the survival of well-adapted plants or insects. Perhaps this concern with the harshness of the phyllosphere environment has been invented by microbiologists without adequate consideration (or even understanding) of the physical needs of the microorganisms themselves. Indeed, is relative harshness the only important contrast between the kinds of physical environments in those microbial habitats that have been extensively studied (marine and fresh water aquatic systems, soil, sediment, etc.) and the phyllosphere? I think not. There is another important difference that only emerges when the rates and frequencies of these fluctuations are scaled, temporally, to a microbial generation. Thus, although organisms with both long generation times (plants) and short generation times (bacteria) exist in the same environment, there is probably a substantial difference in the way in which the environment affects evolution of these two classes of organisms. The long generation individual must be able to survive all of the environmental conditions the weather provides during one generation, and must have suitable conditions for reproduction at some time during its life. The short generation organism must be able to complete its life cycle under each of several very different physical conditions, but is less likely to experience a broad range of environmental conditions during a single generation. For example, one day may represent only 1/100 th of the lifetime of an annual plant. The plant must be able to survive the physical fluctuations during each day that it lives, but each day will represent only a brief part of the total process that culminates with successful reproduction of the plant. Indeed, plants often survive days during which little growth occurs. On the other hand, a day may represent five to twelve generations for some species of bacteria (see Chapter 14, this volume). Thus, cells of the same bacterial clone may complete their life cycles between midnight and 0400, or between 1000 to 1400. During each of these two periods the physical environment will probably vary little. (For example, from midnight to 0400 the temperature may vary from 10 to 9°C, and the leaf may be bathed in dew. From 1000 to 1400 the temperature may change from 27 to 29°C, with intense, relatively constant solar irradiation and low relative humidity). A successful bacterial colonist must compete successfully under both of these two, different conditions. Such bacteria must have evolved under conditions of alternation of these very different conditions every two to six generations. The selection pressures for these bacteria then, have required successful reproduction in both environments, not just one or the other, and certainly not in an average of the two. Similarly for organisms with a generation time of 5 to 15 hrs, alternating generations must compete in these very different conditions. Slow-growing bacteria, yeasts, and some relatively fast-growing filamentous fungi may fall in this category. Only the very slow-growing fungi, including many pathogens, require sufficient time for a generation that the average conditions begin to have a major effect on selection each generation. Strategies for manipulation of communities of microorganisms in the phyllosphere should consider, and if possible exploit the effect of the fluctuating environment on the likely life strategies of these microorganisms.

If one strategy for adaptation to the phyllosphere has been relatively slower growth, to wait for favorable conditions, then there must have been a concomitant coselection for survival and persistence during unfavorable periods. Organisms that have adapted such a slow growth strategy must also be able to cope with any advantage gained by those fast-growing organisms that continue to multiply during periods that are unfavorable for the organism that only waits for favorable conditions. We might expect such slow-growing microorganisms to be little influenced by competition from the more abundant, fast-growing phyllosphere residents, or that these slow-growing residents may frequently escape competition from the fast-growing portion of the community. This latter alternative is consistent with the variable nature of biocontrol; at times the biocontrol agent is able to diminish the activities of the pathogen, and at other times it is not.

Further, the environment within the canopy of a field, or even within a single plant may differ substantially. Diel fluctuations in temperature and radiation may be very much less for leaves deep within a plant canopy than for those at the top of the same plant. Leaves near the top of a plant may be wet with dew for more than 10 hr during a night in which dew does not form at all deep within the canopy (Alberga, 1987). Leaves within the canopy may remain wet from rain for hours longer than those near the top of the same plant. Thus, within a single plant, the physical quality of the habitat provided by one leaf may be strikingly different from that provided by another a few centimeters away. Surely, if disease is likely on leaves within the canopy as well as those near the top, a successful biocontrol strategy must be equally effective in both environments.

Microbial Populations and Communities The extent of variability among population sizes of bacteria on leaves was first studied by Crosse (1959). More recently, this variability has been modeled as approximating a lognormal (Hirano et al., 1982), or a Weibull (Ishimaru et al., 1991). The studies of Hirano and Upper (1986, 1989; also see Chapter 14, this volume) have, in general, shown that leaf-to-leaf variability occurs with all bacterial components thus far examined, that the variability of components is usually greater, and sometimes much greater, than that associated with the total populations platable from leaves, and that the variability of a given component may be characteristic of that component (see Chapter 14, this volume). Even a relatively small number of classes (for example, five species) allowed to assume independent lognormal distributions with modest to large variances, provides sufficient variability in the relative numbers of bacteria present on a population of leaves to support the expectation that the ratios of microorganisms among a very large population of leaves (all of the ca. 2×10^8 individual leaflets in a 60-ha bean field) might be different. In the sense that these major components represent the lion's share of the individuals in bacterial communities on these leaflets, each community then can be expected to be unique, because there will be only a very small probability that any other leaflet in the entire field will have the same relative numbers of even the five most abundant bacterial species.

Nonetheless, there are bounds on this variability. Nearly all of the leaves will be dominated by one of these five predominant species. Thus, despite the tremendous variability among communities, there is also substantial similarity. Although each bean leaflet may support a different bacterial community than any other, most will probably be more like each other than like any of the unique bacterial communities on the leaves of a nearby corn field.

The important lesson from this discussion is that sizes of communities and of populations of components vary from leaf to leaf in a manner that can be modeled. Further, the greater variability in population sizes of individual components than in sizes of total communities, and the observation that the extent of leaf to leaf variability is characteristic of a particular component, indicate that aspects of the relationships among these communities probably follow predictable patterns. Thus, generalizable information can be gained from the study of the variability in population sizes within and among these communities that should be broadly applicable to the management of these communities.

Population sizes of yeasts (Fokkema and Schippers, 1986) and probably of fungi (Kinkel et al., 1987) appear to be somewhat less variable than those of bacteria in the phyllosphere. Thus, it is apparent that differences in the variability of the various classes of phyllosphere microorganisms may be important in our consideration of the mechanisms important to development of phyllosphere communities. Nonetheless, there appears to be sufficient variability among the various components to contain substantial information about what is occurring in the phyllosphere. Note, however, that much less data are available comparing the numbers of individuals or of community sizes among populations of leaves for either yeasts or filamentous fungi than for bacteria. In addition, before such studies can be carried out for filamentous fungi, some method other than dilution plating will be needed to quantitate hyphal biomass. Thus, substantial information about the development of communities of fungi and yeasts in the phyllosphere awaits careful examination of the distribution of these organisms among populations of leaves.

Phyllosphere communities, particularly those of bacteria, are often highly dynamic (see Chapter 14, this volume). Indeed, sampling frequencies have been sufficiently frequent to ascertain the true dynamics of few systems (cf. Hirano and Upper, 1986, 1989; Sundin et al., 1988; see also Chapter 13, this volume). As components change, so do the structures of communities (see Chapter 14, this volume). As knowledge of the factors that regulate these dynamics increase, our models need to account for changes in populations of microorganisms on populations of leaves over time. Knowledge of how component populations within communities are regulated in nature will increase our understanding of how to control these populations to minimize the hazardous activities of pathogens.

Microorganisms, as well as disease, are probably not randomly distributed in space. Chapter 6 illustrates that several different patterns of disease can exist in the same field, one superimposed upon another. Effective biocontrol, therefore, needs to take into account the spatial variability of microbial communities, and particularly the spatial variability in likelihood of disease. (A control procedure that is highly effective in the least diseased portions of a field, but relatively ineffective in heavily diseased areas may not provide the desired protection against disease.) Spatial variability in microbial communities is relatively unstudied. Spatial patterns of disease are recognized as existing, but means for modeling them and measuring them are only now being developed (see Chapter 6, this volume; Hudelson et al., 1989; Hudelson, 1990). A number of possible reasons for the existence of spatial patterns of disease have been proposed, most of which assume nonrandom patterns of inoculum. Few mechanisms of formation of spatial patterns have been unequivocally elucidated, however. We need thorough studies of patterns of microbial communities and of patterns of the members of those communities on populations of leaves. When this information is available, we need models to relate these patterns to other aspects of variability in microbial communities in the phyllosphere. When mechanisms of formation of spatial patterns of disease are understood, the nature of these mechanisms will not only improve our understanding of why diseases appear where they do, but also of how to affect their appearance altogether.

23.4 Alternatives to the Chemical Model

In the preceding pages I have discussed the extensive variability in the physical environment of the phyllosphere, in the communities of microorganisms that reside there, in the dynamics of their populations, and in their spatial arrangements. Surely, these several sources of variability are sufficient to explain a substantial portion of the observed variability in the success of single genotypes of single species in achievement of biocontrol (Hirano and Upper, 1985). In the vast majority of such cases, the result should be variable! In search of biocontrol, we should make full use of our knowledge of the range and distribution of the communities that reside on the phyllosphere as we devise biocontrol strategies. If we know how communities of microorganisms are regulated on leaves, we will be well on our way toward modifying that regulation to our own advantage. If we know how the different members of these communities respond to the physical environment, we should be able to choose those that will be likely to be active during periods when a crop is at risk to a particular disease. The answers, in other words, lie in a more thorough understanding of the communities of microorganisms within which pathogens must act. In Figure 23.2, the original three-component system—host, pathogen, and biocontrol agent—has been modified to include the microbial community and all of its variability. In this more complete view of the biocontrol problem, the need to regulate development of the natural community becomes apparent, and the "biocontrol agent" may be any one of a number of devices to achieve this regulation, and not necessarily a single microorganism.

Such a control strategy should take advantage of the commonality among these communities and of their variabilities in ways that would allow manipulation of all of these communities in a generally parallel direction. Hence, by understanding and modeling the variability among communities of microorganisms on populations of leaves, we should gain insight into ways to manipulate these communities to minimize the relative abundance of pathogens.

Stated another way, if we wish to fix a machine, we must know how it functions. We do not know how to achieve stable changes in microbial

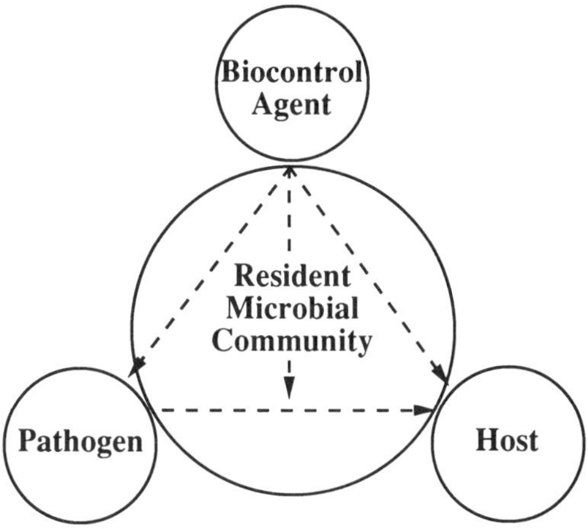

Figure 23.2 An expanded model for biocontrol of diseases in the phyllosphere.

function on leaves because we do not understand microbial communities in the phyllosphere and their regulation.

The search for mechanisms of biocontrol should be broadened. In our search for mechanisms of biocontrol, we need to heed the warning that reductionist methodology only searches for an answer within that universe to which the problem is reduced. If that universe does not contain the answer, the investigator will probably not find it! If all that we search for is antibiosis, all that we will find will be antibiosis! There has been substantial hope that the answer to many of our problems resides in improved methods or increased use of molecular methods. Where these will improve our search for biocontrol, they should be adapted and used. The most important innovation, however, and the one that will lead to satisfying answers most readily, is careful consideration of exactly what the question is that needs to be asked, and what is the minimal universe to which the question can be reduced safely without excluding the answer. Reductionism, the experimental method, is the only way we have to find mechanisms, and thus it is the path we must follow. However, there is every reason to believe that the real secret to understanding biocontrol of plant diseases is to exercise caution during the process of reducing the system, so that the answer will remain within the universe to which the problem is reduced.

The kind of research necessary to achieve what I propose is quite different from that which has previously been directed toward manipulation of microbial activity in the phyllosphere to minimize the harmful effects of the pathogen. The needs are for a larger scale of research, greater emphasis on interdisciplinary approaches, and a greater emphasis on basic research in the early stages of this effort, to provide the basic knowledge necessary to manipulate the microbial communities in the phyllosphere.

References

Alberga, A.H. 1987. *Microclimate Influences on* Pseudomonas syringae *Bacteria on Snap Beans*. Ph.D. thesis, University of Wisconsin, Madison.

Andrews, J.H. 1990. Biological control in the phyllosphere: realistic goal or false hope? *Canadian Journal of Plant Pathology* 12:300–307.

Baker, K.F. and Cook, R.J. 1974. *Biological Control of Plant Pathogens*. W.H. Freeman & Company, San Francisco.

Blakeman, J.P., 1982. Phylloplane interactions. pp. 337–307 in Lacy, M.S., and Mount, G.H. (editors), *Phytopathogenic Prokaryotes*, vol. 1. Academic Press, New York.

Blakeman, J.P. and Fokkema, N.J. 1982. Potential for biological control of plant diseases on the phylloplane. *Annual Review of Phytopathology* 20:167–192.

Burrage, S.W. 1971. The micro-climate at the leaf surface. pp. 91–101 in Preece, T.J. and Dickinson, C.H. (editors), *Ecology of Leaf Surface Micro-organisms*. Academic Press, London.

Cook, R.J. and Baker, K.F. 1983. *Nature and Practice of Biological Control of Plant Pathogens*. American Phytopathological Society, St. Paul, MN.

Crosse, J.E. 1959. Bacterial canker of stone-fruits. IV. Investigation of a method for measuring the inoculum potential of cherry trees. *Annals of Applied Biology* 47:306–317.

Fokkema, N.J. and Schippers, B. 1986. Phyllosphere versus rhizosphere as environments for saprophytic colonization. pp. 137–159 in Fokkema, N.J. and van den Heuvel, J. (editors), *Microbiology of the Phyllosphere*. Cambridge University Press, Cambridge.

Handelsman, J. and Parke, J.L. 1989. Mechanisms in biocontrol of soilborne plant pathogens. pp. 27–61 in Kosuge, T. and Nester, E.W. (editors), *Plant-Microbe Interactions: Molecular and Genetic Perspectives*, vol. 3. McGraw-Hill, New York.

Hirano, S.S. and Upper, C.D. 1985. Ecology and physiology of *Pseudomonas syringae*. *Bio/Technology* 3:1073–1078.

Hirano, S.S. and Upper, C.D. 1986. Temporal, spatial, and genetic variability of leaf-associated bacterial populations. pp. 235–251 in Fokkema, N.J. and Van den Heuvel, J. (editors), *Microbiology of the Phyllosphere*. Cambridge University Press, Cambridge.

Hirano, S.S. and Upper, C.D. 1989. Diel variation in population size and ice nucleation activity of *Pseudomonas syringae* on snap bean leaflets. *Applied and Environmental Microbiology* 55:623–630.

Hirano, S.S., Nordheim, E.V., Arny, D.C., and Upper, C.D. 1982. Lognormal distribution of epiphytic bacterial populations on leaf surfaces. *Applied and Environmental Microbiology* 44:695–700.

Hoitink, H.A.J. and Fahy, P.C. 1986. Basis for the control of soilborne plant pathogens with composts. *Annual Review of Phytopathology* 24:93–114.

Hudelson, B.D. 1990. *Spatial Patterns of Bacterial Brown Spot of Snap Bean*. Ph.D. thesis, University of Wisconsin, Madison.

Hudelson, B.D., Clayton, M.K., Smith, K.P., Rouse, D.I., and Upper, C.D. 1989. Nonrandom patterns of bacterial brown spot in snap bean row segments. *Phytopathology* 79:674–681.

Ishimaru, C., Eskridge, K.M., and Vidaver, A.K. 1991. Distribution analyses of naturally occurring epiphytic populations of *Xanthomonas campestris* pv. *phaseoli* on dry beans. *Phytopathology* 81:262–268.

Kinkel, L.L., Andrews, J.H., Berbee, F.M., and Nordheim, E.V. 1987. Leaves as islands for microbes. *Oecologia* 17:405–408.

Knudsen, G.R. and Spurr, H.W., Jr. 1988. Management of bacterial populations for foliar disease biocontrol. pp. 83–92 in Mukerji, K.G. and Garg, K.L. (editors), *Biocontrol of Plant Diseases*, vol. I. CRC Press, Boca Raton, FL.
National Research Council. 1989. *The Ecology of Plant-Associated Microorganisms*. National Academy Press, Washington, D.C.
Ownley, B.H., Weller, D.M., and Alldredge, J.R. 1990. Influence of soil edaphic factors on suppression of take-all by *Pseudomonas fluorescens* 2–79. *Phytopathology* 80:303.
Papavizas, G.C. and Lumsden, R.D. 1980. Biological control of soilborne fungal propagules. *Annual Review of Phytopathology* 18:289–310.
Sundin, G.W., Jones, A.L., and Olson, B.D. 1988. Overwintering and population dynamics of *Pseudomonas syringae* pv. *syringae* and *P. s.* pv. *morsprunorum* on sweet and sour cherry trees. *Canadian Journal of Plant Pathology* 10:281–288.
Van Alfen, N.K. 1982. Biology and potential for disease control of hypovirulence of *Endothia parasitica*. *Annual Review of Phytopathology* 20:349–362.
Weller, D.M. 1988. Biological control of soilborne plant pathogens in the rhizosphere with bacteria. *Annual Review of Phytopathology* 26:379–407.
Wilson, C.L. and Wisniewski, M.E. 1989. Biological control of postharvest diseases of fruits and vegetables: An emerging technology. *Annual Review of Phytopathology* 27:425–441.
Windels, C.E. and Lindow, S.E., (editors) 1985. *Biological Control on the Phylloplane*. American Phytopathological Society, St. Paul, MN.

Part 6
Conclusion

24

Future Research Directions in Phyllosphere Ecology

John H. Andrews

24.1 Introduction

Phyllosphere ecology can be subdivided into four aspects, each of which presents several research issues (Table 24.1). In this chapter, these four aspects are considered in the order in which they appear in the table. For each one, I provide a very brief overview of the "state of the art" (for details, see Chapter 1, this volume) as a point of departure for considering future research directions. The discussion represents my opinion and is neither intended to summarize the preceding chapters nor to be a consensus report, though each contributor is cited where appropriate.

24.2 The Atmospheric Milieu

The atmosphere is the medium through which most microbes must reach the leaf. Some microorganisms are transported by vectors; in unusual situations this inoculum, though quantitatively small, can initiate a community composed of a few dominant microbial types (see Chapter 8, this volume). For all practical purposes, then, the atmosphere acts as a microbial reservoir (i.e., as both a "source" and "sink;" see Chapter 3, this volume). Microbes originating from leaves and other substrates are withdrawn from or deposited into this reservoir under the influence of various atmospheric processes such as precipitation and wind (see Chapter 11, this volume). Not only is the medium uncolonizable but conditions inhibitory to microbial growth on the leaf surface can be imposed by the direct or indirect effects of atmospheric pollutants (see Chapter 19, this volume). Finally, the scale over which the events occur that influence phyllosphere microbes is enormous. This encompasses phenomena ranging from the effects of the boundary layer within a few mm of the leaf surfaces, to global weather

Table 24.1 The four aspects of phyllosphere ecology and their major components

Aspect	Components
Atmospheric milieu	Transport processes; atmosphere as a microbial reservoir; microscale and microscale meteorological phenomena; pollutants
Leaves as a habitat	Spatial and temporal dynamics in morphology, anatomy, physiology; nutritional environment; external vs. internal domains; role in selecting microflora
Microbial inhabitants	Species composition; adhesion; resident/transient status; organizing forces; biological control/growth promotion; role in leaf senescence and decomposition
Ecosystem processes	Plant-soil continuum; energy flow; trophic levels; nutrient cycling; anthropogenic impacts

patterns (see Chapter 3, this volume; see also Gregory, 1973). Overall, the atmospheric realm is probably the least known and most complex of the phyllosphere aspects identified in Table 24.1.

Research issues pertinent to the atmosphere in phyllosphere ecology seem limitless, but four are of particular interest. First, the role of rainfall and perhaps other factors such as dew in redistributing epiphytic microbes should be examined. The spatial pattern of microflora at any time will be some function of the initial deposition pattern, differential growth, redistribution, and removal of live and dead cells. Much is known about the deposition and redistribution of chemical residues on leaves (Hess and Falk, 1990), and while obviously the expectations for microbial displacement are different from those for a pesticide, this literature is valuable as a point of departure. Research on redistribution poses many challenges. For example, rainfall delivers microorganisms to leaves, moves them about on leaves, and washes them from leaves to the atmosphere. The work will require models developed from simulated rainfall and better techniques for mapping microbes in situ. Accurate maps could perhaps be developed fairly quickly by exploiting advances in image analysis (Greenberg, 1984; Cantonic et al., 1988) and remote sensing methodology (Clark and Steven, 1990).

Second, gusts of wind (turbulence) seem to be particularly effective in releasing microbes, which are swept from the leaf or dislodged by more or less violent leaf flutter (see Chapter 3, this volume). This mechanism should be investigated further. The role of gusts, or any other mechanism for release, may be quite different for pathogens—many of which modify surface features by forming lesions or fruiting structures—than for epiphytes, which typically do not cause such gross disruption of local topography.

Third, the role of vectors in transport needs to be examined. *Drosophila* carry yeasts to decaying cacti (see Chapter 8, this volume), and may be important agents for yeast dispersal in other situations (sap flux from trees). In the fascinating microbial community of cacti, yeasts provide nutritional benefits to the vectors, and evidently this relationship drives the close association that has evolved among microbe, insect, and plant. This appears to be an extreme if not a unique situation. Nevertheless, we should look much more closely at cases where insects and microarthropods are attracted to aerial surfaces (e.g., to flowers or injuries, especially those involving exudates). Bees must contribute microbes along with pollen, and presumably birds release microbes from their body surfaces and in their execreta.

Finally, there is the issue of pollution, a theme addressed in previous volumes and in this one (see Chapter 19, this volume). With increasing international concern about acid rain, pollution should be among the most important topics in phyllosphere research, at least for the coming decade. Advances have been made in determining the sources of pollutants, their study under controlled conditions, understanding their fate in the atmosphere, and assessing their impact over relatively short periods on the phyllosphere community. Further effort is needed to ascertain: (1) the long-term effects on the leaf and leaf litter communities; (2) the direct impact on microflora versus the indirect effect mediated through changes to the leaf surface (e.g., Adams et al., 1990); and (3) the larger implications to ecosystem processes much as nutrient cycling and energy flow (see Section 24.5).

24.3 The Leaf as a Habitat

Leaves are transient, patchy habitats that change through time and space in terms of size, anatomical/morphological features, and physiological (including nutritional) condition. Nutrients are more or less continuously resupplied through the cuticle and stomatal pores to the phyllosphere. This source is supplemented by nutrient pulses that occur at irregular intervals and sites, through the deposition of materials such as pollen or honeydew. However, relative to the rhizosphere (Fokkema and Schippers, 1986) or the surfaces of submerged aquatic plants (see Chapter 4, this volume), the aerial habitat seems nutrient-poor and starvation is probably a common cause of death. The evidence for this is mainly indirect and includes the presence of a prolific epiphytic community on submerged macrophytes and the copious extracellular slime produced by many aquatic plants, especially seaweeds (Lucas, 1947). It is striking that up to 40% of the photosynthate of seaweeds can be excreted (Khailov and Burlakova, 1969). Introduced to this same surface milieu are chemicals, from the plant itself or elsewhere, that inhibit microbial growth (reviewed by Kenerley and Andrews, 1990). In terrestrial systems, microbial biomass is further limited by death from desiccation and erosion by wind and rain. It is little wonder that to an epiphyte the phylloplane must appear as a mountainous, chemically dynamic, "coarse-grained" environment (Andrews, 1991a). The endophyte within the sanctuary of the leaf escapes some of this heterogeneity (see Chapter 2, this volume).

The first challenge in understanding the leaf as a habitat is to get much more information on the nutrient status of the phyllosphere. Unfortunately, pioneering studies (Tukey, 1971) have not been exploited as a basis for more detailed work. It is a lapse that, despite several decades of work, there is still no good, generally accepted medium based on the actual nutrient environment of the leaf. A start has been made (Lindow, 1988), but the existing "leaf" media seem to have been developed empirically and are based on the composition of diffusates from other plant organs (Deverall, 1967). What is the nutrient composition and flux on the phylloplane? How does it vary by hour, by season, by position across a leaf and among leaves on a tree, by plant species? Acquiring answers to these questions will facilitate an understanding of the potential for population growth in situ and for meaningful interpretation of assays of microbial properties or antagonism. The latter issues probably hinge critically on reproducing a realistic nutritional environment (Lindow, 1988).

The role of the leaf in shaping the spatial pattern of its epiphytic flora needs to be understood. Topographic features such as veins and depressions between epidermal cells have long been known from descriptive studies to be "preferred" sites for microbial colonization. But how important are they quantitatively and why? Is the pattern a result of differential deposition, or growth, or dislodgement? If only a small portion of the leaf surface is colonized, as the dogma states, is the remainder uncolonizable? Do surface features send fundamently different signals to pathogens (see Chapter 5, this volume) than to epiphytes? By applying the recent advances in image analysis and remote sensing technology noted earlier, it should now be possible to produce relatively rapidly, extensive, accurate maps of microbes on numerous kinds of leaves under different environmental conditions.

Do leaf surfaces have receptors for certain microbes and not others? In rhizosphere biology, the *Rhizobium*-legume model has established that the host plant contains the entire genetic complement for nodule development. The bacterium merely "switches on" this program. The nodulation signal has recently been shown to be a modified oligosaccharide (Lerouge et al., 1990). There is no convincing evidence to date for strong host species- or strain-specificity on the part of epiphytic microbes. Indeed, specificity to the degree exhibited between host and biotrophic pathogen would not be expected for epiphytes (Harley and Smith, 1983). There is some suggestion of specificity between at least some plants and some endophytes (see Chapter 9, this volume). This avenue of research should be pursued, and, if specificity is confirmed, the molecules in plant and microbe responsible for recognition need to be identified and the underlying genetic basis elucidated.

Finally, research on a model leaf system to facilitate experimentation would be a good investment. Something is needed to bridge the gap between the petri dish and the plant itself. The ideal system would offer a more realistic model than current in vitro methods, yet a more reproducible, tractable, and logistically efficient approach than is possible with studies directly in planta. Ideally such a system would incorporate biologically meaningful variation (e.g., in microsite features) with a high signal-to-noise ratio to reduce the large number of replicates typically required to see a statistical effect when plants are the experimental units. At one extreme, this could take the form of a differentially permeable membrane system, perhaps coated with epidermal cells or cutin to simulate the leaf surface, and superimposed on a solid or liquid nutrient reservoir. Such membrane systems could be removed for examination or for transfer to other incubation environments. The other extreme would be a detached-leaf-culture system. These systems could be used to answer questions about the physiological ecology of epiphytes; about the difference between resident versus transients on the leaf surface; and possibly in the screening of biocontrol candidates (Andrews, 1991b).

24.4 The Microbial Inhabitants

If one were to name the major thread linking reports in phyllosphere microbiology, it would be that extreme variability exists in time and space among microbial populations. Ruinen's (1961) observation in her classic paper on phyllosphere epiflora of plants in the equatorial rain forest is typical: "Not only do different plant species show striking differences in their phyllosphere populations, but different leaves of one and the same plant show variations depending on the position of the leaf in the vegetation and on the age of the leaves" (p. 85).

Variation in population density and in species composition occurs at all levels, e.g., upon a given leaf, among leaves on a plant, and among plants in a field, and presumably reflects mainly the variation in habitat features and microclimate (discussed in Sections 24.2 and 24.3; see also Chapter 6, this volume). Probably several dozen species of bacteria and fungi occur on leaves of a particular plant in a temperature zone over the course of a season, though the great majority of these seem to be relatively rare and communities are usually dominated by a few prominent members (see Chapters 13 and 14, this volume). Both fungi and bacteria are undoubtedly underrepresented, perhaps grossly so, in the prevailing picture, which is based on culture-dependent methods and subject to much criticism in microbial ecology (e.g., Ward et al., 1990). Finally, the view of the phyllosphere community has recently been broadened to include endophytes as key components. These fungi and bacteria have been moved from being considered largely as artifacts a few decades ago to being at the center of attention today (Chapters 9, 10, 17, and 18, this volume). The aerial and other surfaces of plants are also a source of inoculum for allergens (see Chapter 11, this volume) and human pathogens (see Chapter 12, this volume). Unfortunately, this topic has been neglected for too long, a victim of the disciplinary gap between researchers in medical microbiology and plant pathology.

The first research challenge is to redefine the leaf surface community by exploiting culture-independent, molecular methods that have recently become available (see Chapter 7, this volume). Additionally, such techniques offer a powerful new means of monitoring and quantifying selected populations, including those of genetically engineered microbes introduced to nature. The advantage of these new nucleic acid probe methods over the existing approaches such as those utilizing drug resistance is mainly the avoidance or reduction of pleiotropic effects and, frequently, an increased sensitivity without the need to culture the microbe.

Why are some microbes, which might be called residents, common on leaf surfaces while others (transients) occur only sporadically or are perhaps present but quiescent? What are the general and host-dependent phenotypic traits that influence colonization success and are these identifiable and subject to experimental confirmation? Such features may not be what they are supposed to be from laboratory assay of physiological properties (see Chapter 15, this volume), and we may be selecting inadvertently against traits of importance by conventional mass culture methods (Andrews, 1990). One such attribute may be adhesiveness. Much is known about the role of adhesion in the preinfection stage of plant pathogens, but virtually nothing about adherence of epiphytic bacteria or fungi. Residents, or certain growth forms of residents, may adhere better than transients; alternatively, differential adhesion may play little role and residents may simply be able to regenerate biomass quickly, e.g., by having a higher growth rate than transients. To what extent do microbial biofilms (Costerton and Cheng, 1982) occur on leaves?

The resident/transient issue leads into the larger matter of the respective roles of microbial birth, death, immigration, and emigration on the phylloplane. There is evidence that immigration is important for fungi (see Chapter 13, this volume), whereas growth is clearly important for bacteria (see Chapter 14, this volume) and may outweigh the role of immigration. A fascinating analogous system, which is instructive from the standpoint of hypothesis-testing, is the population biology of barnacles on rocks in the marine intertidal zone (Roughgarden et al., 1985; Possingham and Roughgarden, 1990). In brief, the barnacle (*Chthamalus* and *Balanus* spp.) has a two-phase life cycle that alternates between sessile, macroscopic adults that live on rocks, and pelagic, motile, microscopic larvae that live frequently up to about 140 km offshore. Influx of larvae back to the rocks ("settlement rate") may or may not be the limiting factor in determining population size. Some sites (seaward rocks contacted first by inflowing larvae) are exposed to abundant barnacle inoculum. Here immigration is a nonlimiting factor and local conditions of predation, physiological "stress," or physical disruption determine community structure. Other sites (nearshore rocks in the leeward or "settlement shadow") are exposed to low influx rates because many larvae have been deposited on the outer rocks. Such locations are supply-limited and the compositional pattern reflects largely the population dynamics (e.g., as influenced by ocean currents) of the larval phase in coastal water.

Thus, we might imagine that in some leaf situations, either for certain populations (e.g., fungi vs. bacteria; filamentous fungi vs. yeast) or for all populations on certain leaves, the magnitude of immigration would differ and thus so would the resulting community. What then is the relative effect on population numbers of organizing forces such as competition for space, nutrients, or abiotic factors, which operate directly at the leaf surface as opposed to factors that influence the supply of immigrants?

A further research area in the realm of microbial inhabitants is new initiatives in biocontrol. If biocontrol is to succeed, a better understanding is needed of the interaction of antagonists with pesticides and other chemicals on the leaf surface (see Chapter 21, this volume). A strategy based on the chemical model for screening and implementation has not produced many successes and may be misguided (see Chapter 23, this volume). Screens based on field assays would perhaps resolve some of these deficiencies but are logistically intractable, provide no clue as to mechanism, and the results may not be generalizable (Andrews, 1991b). Weltzien (see Chapter 22, this volume) has obtained good control with compost extracts. The mechanism is not yet clear but may involve induced resistance, a phenomeon that should be investigated further (Dixon and Lamb, 1990; Sequeira, 1990). Another possibility is that microbes in the compost establish a novel, antagonistic consortium on the leaves. Whatever the basis, it needs to be identified and the principle applied elsewhere.

Another unexplored possibility is to inactivate pathogens and to use the pacified organism as a biocontrol agent. This strategy should work well, based on predictions from niche overlap theory. Unfortunately, the problems with this approach are: (1) We still know little about determinants of pathogenicity; (2) back-mutation or complementation of mutants is possible; and (3) there would likely be major regulatory obstacles to the release of such agents. Nevertheless, the way is now open to engineer endophytes or epiphytes for growth-promoting or biocontrol activity (see Chapter 20, this volume), although the degree to which this is realistic may be overstated (Andrews, 1990). The ultimate refinement in biocontrol may be utilizing only a few molecules of the agent rather than populations of the entire organism! For example, the coat protein gene of viruses such as TMV, PVX, and PVY has been expressed in transgenic plants where it conveys appreciable resistance to infection (e.g., Powell-Abel et al., 1986; Kaniewski et al., 1990). This strategy involves the molecules of the *pathogen*, but the principle may be relevant in analogous situations, e.g., where the appropriate nucleic acid sequences of an *antagonist* (such as those encoding antibiotics) can be identified, transferred, and expressed in a plant.

24.5 Ecosystem Processes

One of the most challenging and interesting areas for future research is the role of phyllosphere microorganisms in energy flow and nutrient cycling in terrestrial and aquatic ecosystems. Energy flow relationships provide the basis for trophic structure of communities ("who eats whom") and are generally subdivided into grazer (herbivore) chains (i.e., plant → herbivore → first carnivore → . . . . top carnivore) and decomposer (saprovore) chains (i.e., dead organic matter → microscopic and macroscopic scavengers → . . . . top carnivore). The proper placement of microbial epifauna or epiphytes, not to mention plant or animal pathogens, in such schemes has always posed somewhat of a dilemma to ecologists. This and the fact that microbes are invisible means that they tend to be overlooked (Andrews, 1991a). Little effort has been made to quantify microbial productivity (i.e., biomass per unit area per unit time) on or in leaves. Are epiphytic microbes grazed selectively to any degree or do they simply die and enter the saprovore chain? Where mats of epiphytes occur, as in the tropics (Ruinen, 1961), grazing could be significant. What role might epiphytes ingested incidentally by herbivorous insects and vertebrates play in the nutrition of the grazer? Many endophytes appear to be toxic (see Chapter 10, this volume). Apart from nitrogen fixation in the phyllosphere (Ruinen, 1974), little attention has been paid to the role of epiphytic microbes in nutrient cycles (Turner, 1975; Choudhury, 1988).

In the ecosystem context it must be remembered that soil forms a continuum with the plant. Soil conditions influence the health of the plant and, through the composition of leaf exudates, ultimately the phylloplane flora. In turn, nutrients and other chemicals washed from living leaves influence the soil (Ruinen, 1961; Gaber and Hutchinson, 1988), and fallen leaves are a major component of soil organic matter. Phyllosphere microbes are influenced by senescence of the leaf and may actually promote aging (Skidmore and Dickinson, 1973). Decomposition processes are underway before leaf abscission and influence subsequent events (Hudson, 1968; Pugh, 1974). Thus, whatever affects one component in the leaf-soil cycle may well affect others. We still know relatively little about the nontarget impact of pesticides on leaf and litter communities, and even less about other indirect effects and how changes may cascade through the system.

Related to this is the need to determine how microbes released for biocontrol or growth-promoting activities affect the ecosystem. What effects do introduced organisms have *other* than the one for which we are selecting? It is naive to think that there will be none, either beneficial or deleterious. It is appropriate that we be challenged by the scientific community with these broader ecological questions, and it is our responsibility to provide the answers.

24.6 Conclusion

Why Study Phyllosphere Microbiology? Studies of the microbial ecology of leaves are important for many reasons. From the basic science perspective, these microbial communities are an integral part of the trophic structure of herbivore (grazer) and saprovore (decomposer) chains, as noted above. Microbes influence energy flow and nutrient cycles in both terrestrial and aquatic systems. The most obvious example of this is the conversion of plant to microbial biomass by plant pathogens, but the involvement of microbes is certainly not restricted to pathogens (Andrews, 1991a).

Research on epiphytes or endophytes of leaves, and inferences of such work to general ecology, are fostered by the accessible, relatively discrete, and easily replicated nature of leaf "islands" as the experimental unit. The comparatively simple biology and the rapid growth rates of the test organisms (bacteria, fungi), and the degree to which agroecosystems can be manipulated also facilitate ecological studies. As discussed earlier in this chapter, the microbe/plant model can be used to test basic ecological theories (e.g., of succession; island biogeography; r- and K-selection and other life history strategies; invasion and colonization processes). From an applied science perspective, many of the phyllosphere microbes affect plant health negatively (pathogens) or positively (endophytic mutualists; commensals such as antagonists which can be used in biocontrol). Animal health, in turn, can also be influenced negatively (toxic compounds elaborated by endophytes) or positively (food source for grazing microarthropods).

Future Challenges and How to Approach Them If the phyllosphere is not to remain "an ecologically neglected milieu" (Ruinen, 1961), the challenge that lies ahead is fourfold. First, we must acquire considerably more *basic* information about the leaf system. Examples discussed above include the need for a detailed understanding of the nutritional environment on the phyllosphere, and of the species composition of the epiphytic and endophytic inhabitants (which to date has been based on culture-dependent methods and is therefore incomplete, perhaps considerably so). Second, work must move forward into the largely uncharted waters of realms such as ecosystem processes (trophic structure; energy flow; nutrient cycling); community assembly; molecular recognition phenomena, including the genetic basis underlying interactions between plant and epiphyte or endophyte; and biological control. Third, it is imperative that the relationship

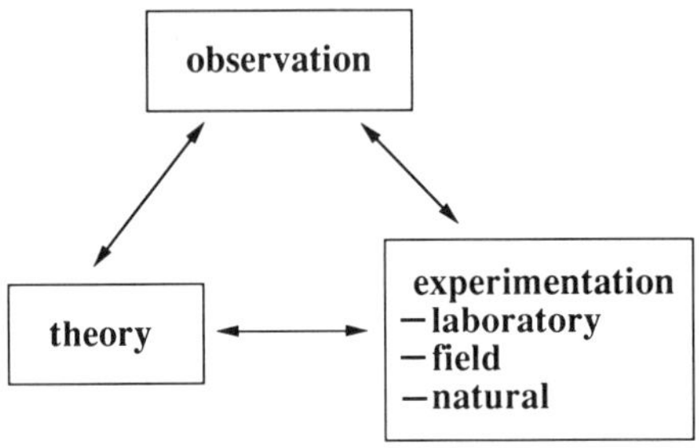

Figure 24.1 The relationships among observation, experimentation, and theory in research.

between general ecology and leaf ecology be understood. Phyllosphere researchers need to become familiar with the literature—not only from other subdisciplines of microbial ecology—but with theories and analogs from plant and animal ecology that can serve as guiding principles for our research. Conversely, the potential role that phyllosphere-microbe systems could play as a paradigm for testing and expanding general ecological concepts needs to be exploited. Finally, advances in technology and in methodological (including statistical and analytical) approaches should be applied to phyllosphere research as soon as they emerge in allied disciplines (for examples, see Chapters 5, 7, and 16, this volume). The most obvious current examples are the amazing breakthroughs in molecular genetics, which present new opportunities to identify, monitor, quantify, and engineer microbial populations.

The approach to this challenge must be, in the most general terms, the approach to scientific investigation, namely a recognition of the interplay among observation, experimentation, and theory (Figure 24.1). While this operational framework has always been implicit in science, it is worthwhile making explicit the contributions and limitations of each component. Traditionally, phyllosphere research has consisted largely of observation, i.e., of descriptive accounts. There is still a role for the natural biologist and for "theory-free" data collection. For example, complete, accurate descriptions of the complex phyllosphere communities of rain forests, building on earlier work (e.g., Ruinen, 1961; Parbery and Brown, 1986), are needed. It would also be interesting to know the differences in microflora among plants of different growth forms (e.g., thallophytes vs. herbs vs. tall trees) and from different ecological formations (e.g., deserts, bogs, moist coastal zones, and littoral zones of lakes or oceans). Description provides a benchmark or basis for experimentation; conversely, experiments may point out deficiencies in understanding that can be remedied by more thorough observation. Hence, there is mutual reinforcement between these two components (Figure 24.1).

We are gradually moving away from observation towards an experimental, hypothesis-testing approach (Figure 24.1) in phyllosphere research, and this is desirable. Diamond (1986) has compared the roles of laboratory (or growth chamber), field, and natural experiments in community ecology. The former two involve deliberate perturbations or manipulations by the researcher. The latter involves disturbances initiated naturally or by persons other than the experimenter. Examples include monitoring changes on an island following a volcano; changes in a plant community following the invasion of a pathogen or herbivore; or changes in species diversity and abundance following eutrophication. In all cases of natural experiments the experimental design should be such that observations at the disrupted site are matched with those at a control site chosen to be as similar as possible in aspects other than the perturbation.

The important message from Diamond's assessment is that all these types of experimentation should be used in phyllosphere studies because each is necessarily a compromise to varying degrees among regulation of experimental variables, realism, and generality. While, again, this is implicit and mainly a matter of common sense, one rarely sees in the literature any explanation of the rationale for a particular approach, *or the basis for the inferences drawn by the author.*

Theory (in this context, *ecological* theory) should play a key role in driving observation and experimentation (Figure 24.1), but unfortunately has been neglected for the most part in phyllosphere work and biological control (Andrews, 1990). Theory also supplies new ideas and the opportunity to see old stories from new vantage points. It offers a conceptual framework for synthesis of disparate facts. And, it can provide a basis for interpreting the outcome of experiments. Theory is of course dynamic and subject to being overturned or modified depending on the results of observation or experimentation; hence the arrows (Figure 24.1) move in both directions.

Renewed efforts at an experimental as opposed to a descriptive approach, integration of relevant advances from other disciplines, and recognition of the interplay among theory, experimentation, and observation, will be the essential ingredients for expanding our knowledge of phyllosphere ecology in the coming decades.

References

Adams, C.M., Caporn, S.J.M., and Hutchinson, T.C. 1990. Crystal occurrence and wax disruption on leaf surfaces of cabbage treated with simulated acid rain. *New Phytologist* 114:147–158.

Andrews, J.H. 1990. Biological control in the phyllosphere—realistic goal or false hope? *Canadian Journal of Plant Pathology* 12: 300–307.

Andrews, J.H. 1991a. *Comparative Ecology of Microorganisms and Macroorganisms.* Springer-Verlag, New York. 302 pp.

Andrews, J.H. 1991b. The screening of microorganisms to phytopathogens. Pesq. Agropec. Brasil. In press.

Cantoni, V., DiGesu, V. and Levialdi, S. (editors). 1988. *Image Analysis and Processing II.* Plenum, New York.

Choudhury, D. 1988. Herbivore induced changes in leaf-litter resource quality: a neglected aspect of herbivory in ecosystem nutrient dynamics. *Oikos* 51:389–393.

Clark, J.A. and Seven, M.D. (editors). 1990. *Applications of Remote Sensing in Agriculture*. Butterworth, Boston.

Costerton, J.W. and Cheng, K.-J. 1982. Microbe-microbe interactions at surfaces. pp. 275–290 in Burns, R.G. and Slater, J.H. (editors), *Experimental Microbial Ecology*. Blackwell, Oxford.

Deverall, B.J. 1967. Biochemical changes in infection drops containing spores of *Botrytis* spp. incubated in the seed cavities of pods of bean (*Vicia faba* L.). *Annals of Applied Biology* 59:375–387.

Diamond, J. 1986. Overview: Laboratory experiments, field experiments, and natural experiments. pp. 3–22 in Diamond, J. and Case, T.J. (editors), *Community Ecology*. Harper and Row, New York.

Dixon, R.A. and Lamb, C.J. 1990. Molecular communication in interactions between plants and microbial pathogens. *Annual Review of Plant Physiology and Plant Molecular Biology* 41:339–367.

Fokkema, N.J. and Schippers, B. 1986. Phyllosphere versus rhizosphere as environments for saprophytic colonization. pp. 137–159 in Fokkema, N.J. and van den Heuvel, J. (editors), *Microbiology of the Phyllosphere*. Cambridge University Press, New York.

Gaber, B.A. and Hutchinson, T.C. 1988. Chemical changes in simulated raindrops following contact with leaves of four boreal forest species. *Canadian Journal of Botany* 66:2445–2451.

Greenberg, S.D. (editor). 1984. *Computer-Assisted Image Analysis Cytology*. Karger, New York.

Gregory, P.H. 1973. *The Microbiology of the Atmosphere*, 2nd edition. Leonard Hill Books, Aylesbury, England. 377 pp.

Harley, J.L. and Smith, S.E. 1983. Specificity and recognition in symbiotic systems. pp. 357–386 in Harley, J.L. and Smith, S.E. (editors), *Mycorrhizal Symbiosis*. Academic Press, New York.

Hess, F.D. and Falk, R.H. 1990. Herbicide deposition on leaf surfaces. *Weed Science* 38:280–288.

Hudson, H.J. 1968. The ecology of fungi on plant remains above the soil. *New Phytologist* 67:837–874.

Kaniewski, W., Lawson, C., Sammons, B., Haley, L., Hart, J., Delannay, X. and Tumer, N.E. 1990. Field resistance of transgenic Russet Burbank potato to effects of infection by potato virus X and potato virus Y. *Bio/Technology* 8:750–754.

Kenerley, C.M. and Andrews, J.H. 1990. Interactions of pathogens on plant leaf surfaces. pp. 192–217 in Hoagland, R.E. (editor), *Microbes and Microbial Products as Herbicides*. Amer. Chem. Soc., Washington, D.C.

Khailov, K.M. and Burlakova, Z.P. 1969. Release of dissolved organic matter by marine seaweeds and distribution of their total organic production to inshore communities. *Limnology and Oceanography* 14:521–527.

Lerouge, P., Roche, P., Faucher, C., Maillet, F., Truchet, G., Prome, J.C. and Denaire, J. 1990. Symbiotic host-specificity of *Rhizobium meliloti* is determined by a sulfated and acetylated glucosamine oligosaccharide signal. *Nature* 344:781–784.

Lindow, S.E. 1988. Lack of correlation of in vitro antibiosis with antagonism of ice nucleation active bacteria on leaf surfaces by non-ice nucleation active bacteria. *Phytopathology* 78:444–450.

Lucas, L.E. 1947. The ecological effects of external metabolites. *Biological Reviews of the Cambridge Philosophical Society* 22:270–295.

Parbery, I.H. and Brown, J.F. 1986. Sooty moulds and black mildews in extra-tropical rainforests. pp. 101–120 in Fokkema, N.J. and van den Heuvel, J. (editors), *Microbiology of the Phyllosphere*. Cambridge University Press, New York.

Possingham, H.P. and Roughgarden, J. 1990. Spatial population dynamics of a marine organism with a complex life cycle. *Ecology* 71:973–985.

Powell-Abel, P., Nelson, R.S., De, B., Hoffman, N., Rogers, S.G., Fraley, R. T., and Beachy, R.N. 1986. Delay of disease development in transgenic plants that express the tobacco mosaic virus coat protein gene. *Science* 232:738–743.

Pugh, G.J.F. 1974. Terrestrial fungi. pp. 303–336 in Dickinson, C.H. and Pugh, G.J.F. (editors), *Biology of Plant Litter Decomposition*, vol. 2. Academic Press, New York.

Roughgarden, J., Iwasa, Y. and Baxter, C. 1985. Demographic theory for an open marine population with space-limited recruitment. *Ecology* 66:54–67.

Ruinen, J. 1961. The phyllosphere. I. An ecologically neglected milieu. *Plant and Soil* 15:81–109.

Ruinen, J. 1974. Nitrogen fixation in the phyllosphere. pp. 121–167 in Quispel, A. (editor), *The Biology of Nitrogen Fixation*. North-Holland Pub. Co., Amsterdam.

Sequeira, L. 1990. Induced resistance: Physiology and biochemistry. pp. 663–678 in Baker, R.R. and Dunn, P.E. (editors), *New Directions in Biological Control: Alternatives for Suppressing Agricultural Pests and Diseases*. Liss, New York.

Skidmore, A.M. and Dickinson, C.H. 1973. Effect of Phylloplane fungi on the senescence of excised barley leaves. *Transactions of the British Mycological Society* 60:107–116.

Tukey, H.B., Jr. 1971. Leaching of substances from plants. pp. 67–80 in Preece, T.F. and Dickinson, C.H. (editors), *Ecology of Leaf Surface Micro-organisms*. Academic Press, New York.

Turner, B.D. 1975. Energy flow in aboreal epiphytic communities. An empirical model of net primary productivity in the alga *Pleurococcus* on larch trees. *Oecologia* 20:179–188.

Ward, D.M., Weller, R. and Bateson, M.M. 1990. 16S rRNA sequences reveal numerous uncultured microorganisms in a natural community. *Nature* 345:63–65.

Index

C

E

Brock/Springer Series in Contemporary Bioscience

(Continued from page ii)

John H. Andrews
COMPARATIVE ECOLOGY OF MICROORGANISMS AND MACROORGANISMS

Ryszard J. Chróst (Editor)
MICROBIAL ENZYMES IN AQUATIC ENVIRONMENTS

Alan G. Williams and Geoffrey S. Coleman
THE RUMEN PROTOZOA

John H. Andrews and Susan S. Hirano (Editors)
MICROBIAL ECOLOGY OF LEAVES